4250

THEORY OF ELEMENTARY ATOMIC
AND MOLECULAR PROCESSES IN GASES

THEORY OF ELEMENTARY ATOMIC AND MOLECULAR PROCESSES IN GASES

BY
E. E. NIKITIN

TRANSLATED BY
M. J. KEARSLEY

CLARENDON PRESS · OXFORD
1974

PHYSICS

Oxford University Press, Ely House, London W.1

GLASGOW   NEW YORK   TORONTO   MELBOURNE   WELLINGTON
CAPE TOWN   IBADAN   NAIROBI   DAR ES SALAAM   LUSAKA
ADDIS ABABA   DELHI   BOMBAY   CALCUTTA   MADRAS   KARACHI
LAHORE   DACCA   KUALA LUMPUR   SINGAPORE   HONG KONG
TOKYO

ISBN 0 19 851928 1

TRANSLATION © OXFORD UNIVERSITY PRESS 1974

All rights reserved. No part of this publication may be reproduced, stored in a retrieval system, or transmitted, in any form or by any means, electronic, mechanical, photocopying, recording or otherwise, without the prior permission of Oxford University Press

FIRST PUBLISHED IN RUSSIAN BY
'KHIMIYA', MOSCOW, 1970

PRINTED IN GREAT BRITAIN BY
J. W. ARROWSMITH LTD., BRISTOL, BS3 2NT

PREFACE

THE book is devoted to calling the attention of readers to the theory of elementary processes determining the kinetics of physical and chemical relaxation in the gaseous phase. By these terms is understood the approach to a thermodynamical equilibrium, disturbed, by some means, at the initial point in time (for example, by a sudden change of temperature, by optical selective excitation of certain states, by the non-equilibrium assignment of the concentrations of the gaseous reactants, and so on). The range of problems, which should be considered, in connection with the microscopic interpretation of non-equilibrium processes, is extremely wide and diverse. It is therefore absolutely necessary to introduce limitations in regard to the possible types of processes under discussion. At the same time, it is desirable to give an account of the theory in a sufficiently general form. The character of the exposition of the material is shaped, to a considerable extent, by the attempt to find a compromise solution to the contradiction between the deductive construction of the theory and the inductive character of its development.

The intensive experimental investigation of non-equilibrium processes, of late, has clearly shown that the calculation of the rate constants of chemical reactions, and their interpretation in terms of molecular parameters, demands a knowledge of the elementary stages of energy transfer in collisions. It has become clear that the construction of a theory of inelastic molecular collisions must precede the creation of a theory of dissociation, recombination, and exchange reactions. Of all the processes taking place in such collisions, usually only the slowest, connected with the excitation and deactivation of the vibrational and electronic states of the partners by collision, is of interest. These processes are considered in Chapters 2 and 3. The reservation should be made that only a very small part of the book is about atomic collisions, the theory of which itself represents a whole field of physics. In general, no mention is made of the many well-studied subtle effects of atomic, and also molecular, collisions, if they have no direct relevance to the theory of chemical reactions, given in Chapters 4–8, on the basis of the transition state method and of the theory of inelastic collisions.

In the classification of elementary chemical reactions, it is expedient to distinguish two important types—processes without redistribution of the atoms amongst the partners (unimolecular reactions, Chapters 4) and with redistribution (bimolecular reactions, Chapter 8). At the same time, certain general approaches to problems of the intramolecular and intermolecular redistribution of the energy are reported, independent of the specific type of reaction.

At present, of all the chemical gas phase reactions, the most fully studied is the thermal decomposition of diatomic molecules. Since these reactions present great practical interest, a separate seventh chapter has been assigned to the theories of thermal decomposition.

Taking into consideration the importance and abundance of exchange reactions, they should be paid even more attention than is done in the book. In particular, the reactions in molecular beams, one of the fields of study in the theory of bimolecular reactions, could have been presented in more detail. However, it would have been expedient to do this only in close connection with a general theory of scattering, which is hardly considered in the book.

In the account of the material, particular attention has been paid to unity of notation. However, this has not been successfully achieved in full, since, in a number of cases, it seemed desirable to preserve the notation generally accepted in the literature. For the reader's convenience, a table of notation for the fundamental physical quantities has been appended.

The theory of elementary processes contained in this monograph mainly concerns reactions of neutral particles and covers an energy range from hundredths of an electron volt to several electron volts. Processes involving free electrons are not touched on in the book. These and also other processes involving charged particles are discussed in the book by B. M. Smirnov.[†]

The phenomenological theory of relaxation and the formal theory of chemical kinetics are not considered in the book. For acquaintance with these problems, the reader is referred to the monographs of Ya. B. Zel'dovich and Yu. P. Raizer, E. V. Stupochenko, S. A. Losev, and A. I. Osipov, and V. N. Kondrat'ev.[‡] Books by E. E. Nikitin and D. Bunker may serve as a brief introduction to the microscopic theory of elementary reactions.[§]

The book includes literature published prior to the autumn of 1966. References to experimental work are cited only with a view to enlisting them as illustrations of the theory. Furthermore, a short survey of literature on topics having a direct bearing on the problems presented is given in the form of a separate appendix; also given is a list of literature published during the

[†] B. M. Smirnov (1968). *Atomic collisions and elementary processes in plasma.* Atomizdat, Moscow.

[‡] Ya. B. Zel'dovich and Yu. P. Raizer (1965). *The physics of shock waves and high-temperature hydrodynamical phenomena.* Fizmatgiz.

E. V. Stupochenko, S. A. Losev, and A. I. Osipov (1965). *Relaxation processes in shock waves.* Fizmatgiz.

V. N. Kondrat'ev (1958). *Chemical kinetics of gas reactions.* Acad. Sci. Press, U.S.S.R. In English (1964), Pergamon Press, Oxford.

[§] E. E. Nikitin (1966). *Theory of thermally induced gas phase reactions.* 'Nauka' Press, Moscow. In English (1966), Indiana University Press.

D. Bunker (1966). *Theory of elementary gas reaction rates.* Pergamon Press, Oxford.

time when the book was being prepared for publication. In contrast to the main alphabetical list, the additional list refers to individual topics. This facilitates the finding of original papers, necessary to the reader in so far as there is no reference to the additional literature in the main text.

The author would like to thank Academician V. N. Kondrat'ev and Professor N. D. Sokolov for reading the manuscript and having made a number of important comments. The author is also grateful to co-workers in the laboratory of quantum chemistry of the Institute of Chemical Physics, Academy of Science, U.S.S.R., who took part in the discussion of many problems, with great profit to the author.

CONTENTS

NOTATION   xi

1. THE TRANSITION STATE METHOD
1. The adiabatic approximation. Potential energy surfaces   1
2. Basic assumptions of the transition state method. Deduction of a formula for the rate of an elementary process   7
3. Over-barrier reflection and tunnel corrections   20
4. The quantum partition functions and symmetry numbers   23
5. The isotope effect   28
6. The principle of detailed balance and the transition state method   34

2. THE EXCHANGE OF VIBRATIONAL AND TRANSLATIONAL ENERGY IN MOLECULAR COLLISIONS
7. General observations on the exchange of energy in molecular collisions   41
8. Model of a forced harmonic oscillator   52
9. Vibrational transitions between the lowest levels of diatomic molecules in non-degenerate electronic states   58
10. The vibrational relaxation of oxygen and nitrogen   78
11. Strong coupling of the vibrational and translational motion   83
12. Resonant exchange of vibrational energy   92
13. Exchange of energy in the collisions of polyatomic molecules   96

3. THE EXCHANGE OF ELECTRONIC, VIBRATIONAL, AND TRANSLATIONAL ENERGY IN MOLECULAR COLLISIONS
14. The classification of nonadiabatic transitions   99
15. The linear model. The Landau–Zener formula   107
16. Generalization of the linear model   117
17. Model of non-linear terms   124
18. Nonadiabatic processes in atomic collisions   132
19. Nonadiabatic processes in the collisions of atoms with diatomic molecules   148
20. Vibrational transitions between the lowest levels of a diatomic molecule in a degenerate electronic state   166
21. Vibrational relaxation of nitric oxide   175

4. UNIMOLECULAR REACTIONS
22. The thermal decomposition and isomerization of molecules as unimolecular reactions   179
23. Dependence of the rate constant on the pressure. The mechanism of strongly-activating collisions   181
24. The Slater model   193
25. Kassel's model   208

26. The effect of anharmonicity on the reaction rate	217
27. Nonadiabatic reactions	227

5. THE STATISTICAL THEORY OF REACTIONS

28. The basic assumptions of the statistical theory	237
29. Harmonic model of an active molecule	244
30. An anharmonic model of an active molecule	253
31. Statistical theory with regard for the conservation of angular momentum	258
32. The isotope effect	271
33. The application of the statistical theory to thermal reactions	274

6. DIFFUSION THEORY OF REACTIONS

34. Diffusion in phase space	283
35. Diffusion through energy states	294
36. Relaxation and the transmission of particles across a potential barrier	299
37. Tunnel transitions in a double potential well	305
38. Random walks over the discrete energy levels	310
39. Mechanism of activation and the non-equilibrium distribution function in unimolecular reactions	315

7. DISSOCIATION OF DIATOMIC MOLECULES AND THE RECOMBINATION OF ATOMS

40. Equilibrium theory of decomposition and recombination	322
41. The contribution of various degrees of freedom of a dissociating molecule to the decomposition rate constant	328
42. Variational theory of dissociation and recombination	334
43. The vibrational relaxation of diatomic molecules	342
44. Non-equilibrium theory of dissociation and recombination	350
45. Connection between the rate constants of dissociation and recombination	360
46. Thermal dissociation of oxygen	364

8. BIMOLECULAR REACTIONS

47. Exchange as a bimolecular reaction	367
48. The potential energy surfaces of bimolecular reactions	370
49. Equilibrium theory	386
50. The statistical theory	391
51. Theory of direct reactions	403
52. Dynamics of exchange reactions	412
53. Disturbance of the equilibrium distribution in bimolecular reactions	424

APPENDIX: PROBLEMS OF ELEMENTARY PROCESSES IN THERMAL GAS REACTIONS	434
REFERENCES	446
INDEX	469

NOTATION

Physical constants

c = velocity of light 3×10^{10} cm s^{-1}
e = charge of the electron 4.8×10^{-10} e.s.u.
$\hbar$ = Planck's constant 1.05×10^{-27} erg s
m_e = mass of the electron 9.1×10^{-28} g
a_0 = Bohr radius ($a_0 = \hbar^2/m_e e^2$) 0.53×10^{-8} cm
k = Boltzmann constant 1.38×10^{-16} erg deg^{-1}
N = Avogadro's number 6.023×10^{23} mol^{-1}
R = gas constant ($R = kN$) 1.99 cal deg^{-1} mol^{-1}

Conversion factors for energy units

	cm^{-1}	kcal mol^{-1}	eV
cm^{-1}	1	2.86×10^{-3}	1.24×10^{-4}
kcal mol^{-1}	3.5×10^2	1	4.34×10^{-2}
eV	8.07×10^3	23.06	1

Symbols for physical quantities

X^* = a value X referring to an active molecule.
$X^\star$ = a value X referring to an activated molecule.
X_r = a value X referring to a reaction coordinate.
$X_{\text{trans}}(X_{\text{vib}}, X_{\text{rot}}, X_{\text{el}})$ = a value X, referring to the translational (vibrational, rotational, electronic) degrees of freedom of a molecular system.
$X_{\text{AB,trans}}$ = a value X, referring to the translational degrees of freedom of the molecule AB.
$X_{\text{A–B,trans}}$ = a value X, referring to the translational degrees of freedom describing the relative motion of the fragments A and B.
A = pre-exponential factor in the expression for the reaction rate constant.
a_{nm} = kinematic coefficients of the nuclear Hamiltonian.
b_{nm} = dynamic coefficients of the nuclear Hamiltonian.
b = impact parameter.
D = dissociation energy.
E = total energy of the molecular system.
E_k = an energy level of the molecular system.
E_0 = minimum of the potential energy of the activated molecule.
E_z = energy of zero-point vibrations.
E_a = activation energy.
E_t = kinetic energy of motion along the reaction coordinate.
f = distribution function.
f_0 = equilibrium distribution function.
F = partition function of a molecular system.
g_k = degree of degeneracy of the level E_k.
$g^\star$ = number of equivalent reaction channels.

H = electronic–nuclear Hamiltonian of the system.
$H_e(H'_e)$ = electronic Hamiltonian without (with) regard for the spin–orbit coupling.
H = Hamiltonian of nuclei.
$H^\star = H(p_1, q_1, \ldots, p_r, q_r^\star, \ldots, p_s, q_s)$.
$h(\tau^*)$ = distribution function of the lifetimes of an active molecule.
I = moment of inertia of a molecule.
J = action variable.
$\mathbf{J}$ = total angular momentum of a molecular system.
$\mathbf{j}$ = angular momentum of a fragment of a molecular system.
κ = rate constant of a chemical reaction.
κ_{if} = rate constant of a reaction with given initial and final states i and f.
$\kappa(E)$ = rate constant of a spontaneous unimolecular reaction.
κ_0 = rate constant of a thermal unimolecular reaction of the second order.
κ_∞ = rate constant of a thermal unimolecular reaction of the first order.
K_{eq} = chemical equilibrium constant.
l = relative angular momentum of colliding molecules.
m_A = mass of atom A.
m_r = effective mass corresponding to the reaction coordinate.
M = reduced mass of a diatomic molecule.
n = quantum number of a vibrational energy level of a diatomic molecule.
p_i = generalized momenta.
$P(E, E')$ = transition probability density between levels E and E'.
P_{nm} = transition probability between states n and m.
q_i = generalized coordinates.
$\mathbf{r}$ = set of electron coordinates.
r = distance between nuclei of a diatomic molecule.
$\mathbf{R}$ = set of nuclear coordinates.
$\mathbf{R}_i$ = radius vector of the ith nucleus.
R_{ik} = distance between the atoms i and k.
R = distance between the centroids of two colliding molecules.
R_0 = gas-kinetic radius.
s = number of degrees of freedom of a molecular system.
s = number of oscillators representing a molecule.
$S^\star$ = critical surface.
S = entropy.
T = absolute temperature.
T = kinetic energy of a molecular system.
t = time.
$U(\mathbf{R}), U'(\mathbf{R})$ = adiabatic terms of the Hamiltonian $H_e(H'_e)$.
U_r = potential energy of a diatomic molecule.
V = potential energy responsible for the interaction between the various degrees of freedom of the molecular system.
v = velocity.
$V_{s.o.}$ = spin–orbit coupling.
$W(\mathbf{R})$ = intermolecular interaction.
$W_{\alpha\beta}$ = statistical matrix.
$W(E)$ = number of states of a molecule with energy less than E.

NOTATION

$W_0(E)$ = 'smoothed out' number of states.
$W_{semiclass}(E)$ = semiclassical approximation to the number of states.
Z_0 = gas-kinetic collision number.
Z^* = effective number of deactivating collisions.
x_n = population of the nth state of a molecule.
α = angular coordinate of canonical variables.
α = parameter of the exponential potential.
$\beta = 1/kT$.
γ = angle between internuclear axis of the diatomic molecule AB and the line of collision of AB and C.
γ = mean Massey parameter.
Γ = phase volume of the molecular system.
ε_0 = well depth of the intermolecular potential.
ε_λ = energy of the λth normal vibration.
θ_λ = characteristic vibrational temperature of the λth normal vibration.
λ = de Broglie wavelength.
μ = reduced mass of the colliding molecules.
μ_λ = amplitude factor of the λth normal vibration.
ν_λ = linear frequency of the λth normal vibration.
ξ = Massey parameter.
ρ = density of energy levels.
σ = symmetry number.
$\sigma_{if}(v)$ = cross-section for the transition $i \to f$ with initial velocity v.
τ_0 = collision time.
τ^* = lifetime of a molecular system.
τ_{vib} = vibrational relaxation time.
$\phi_m(\phi'_m)$ = electronic eigenfunction of the Hamiltonian $H_e(H'_e)$.
$\chi_m(\chi'_m)$ = nuclear eigenfunction of the adiabatic vibrational Hamiltonian corresponding to the Hamiltonian $H_e(H'_e)$.
ψ = electronic–nuclear wavefunction of the adiabatic approximation.
Ψ = total wavefunction of the molecular system.
ω_λ = circular frequency of the λth normal vibration.
ω_0 = frequency of small vibrations of a diatomic molecule.

1

THE TRANSITION STATE METHOD

1. The adiabatic approximation. Potential energy surfaces

ONE of the stages in the theoretical calculation of the rate of an elementary process consists of the solution of the wave equation for the electrons and nuclei of the molecular system under consideration. By such a system will be understood the stable or unstable set of atoms, or ions, interaction between which is ultimately responsible for some elementary process or other taking place. The Schrödinger equation for an isolated molecular system has the form

$$i\hbar \frac{\partial \Psi(\mathbf{r}, \mathbf{R})}{\partial t} = H(\mathbf{r}, \mathbf{R})\Psi(\mathbf{r}, \mathbf{R}), \quad (1.1)$$

where Ψ is the wavefunction; $\mathbf{r}$ is the set of coordinates of the electrons; $\mathbf{R}$ is the set of coordinates of the nuclei; and H is the Hamiltonian of the system.

Depending on the type of effect one wishes to describe by eqn. (1.1), the Hamiltonian is written down to some degree of approximation. To sufficient accuracy, H can be represented in the form

$$H = \sum_i T_i + \sum_\alpha T_\alpha + \sum_{i>k} V_{ik} + \sum_{i,\alpha} V_{i\alpha} + \sum_{\alpha>\beta} V_{\alpha\beta} + V_{\text{S.O.}}, \quad (1.2)$$

where T_i and T_α are the kinetic energy operators of the nucleus i and electron α,

$$T_i = -\frac{\hbar^2}{2M_i}\nabla^2_{\mathbf{R}_i}, \qquad T_\alpha = -\frac{\hbar^2}{2m_e}\nabla^2_{\mathbf{r}_\alpha}, \quad (1.3)$$

where $\mathbf{R}_i$ and M_i are respectively the radius vector and mass of the nucleus i; $\mathbf{r}_\alpha$ and m_e are the corresponding quantities for the αth electron; V_{ik}, $V_{i\alpha}$, and $V_{\alpha\beta}$ are potential energy operators representing the Coulomb interaction of the nuclei and electrons,

$$V_{ik} = \frac{Z_i Z_k e^2}{|\mathbf{R}_i - \mathbf{R}_k|}, \qquad V_{i\alpha} = \frac{Z_i e^2}{|\mathbf{R}_i - \mathbf{r}_\alpha|}, \qquad V_{\alpha\beta} = \frac{e^2}{|\mathbf{r}_\alpha - \mathbf{r}_\beta|}, \quad (1.4)$$

and $V_{\text{S.O.}}$ is the operator of the spin–orbit interaction of the electrons due to relativistic effects.

Although, as a rule, the absolute magnitude of the spin–orbit interaction is significantly smaller than the magnitude of the electrostatic interaction (1.4), in a number of cases, it must be taken into account, since the interaction of the spin with the orbital motion leads to new qualitative effects and sometimes to important corrections of a quantitative nature. For $V_{\text{s.o.}}$ we will take the following expression

$$V_{\text{s.o.}} = \sum \frac{1}{2m_e^2 c^2} s_\alpha(\text{grad } V_\alpha p_\alpha), \tag{1.5}$$

where s_α and p_α are the spin and momentum operators of the electron α; V_α is the total electrostatic potential acting on the electron under consideration.

Solution of eqn. (1.1) in its general form is impossible, in view of the mathematical difficulties. Therefore, to determine the function $\Psi(\mathbf{r}, \mathbf{R})$, it is necessary to turn to a series of approximations, taking advantage of the presence of small parameters in the molecular system considered. It is not difficult to persuade oneself that one such small parameter is the ratio of the velocity of the nuclei, limited by the energy of the nuclei, of the order of several electron volts, to the velocity of the electrons in the valence shells of the interacting atoms. Therefore, in the zero-order approximation, it is natural to take the nucleus at rest and first consider the Hamitonian of the electrons for fixed nuclei,

$$H'_e = \sum_\alpha T_\alpha + \sum_{i>k} V_{ik} + \sum_{i,\alpha} V_{i\alpha} + \sum_{\alpha>\beta} V_{\alpha\beta} + V_{\text{s.o.}}. \tag{1.6}$$

The coordinates of the nuclei come into H'_e as parameters.

The solution of the eigenvalue problem with regard to the electronic subsystem which is described by the equation

$$H'_e(\mathbf{r}, \mathbf{R})\phi'_m(\mathbf{r}, \mathbf{R}) = U'_m(\mathbf{R})\phi'_m(\mathbf{r}, \mathbf{R}) \tag{1.7}$$

gives electronic functions ϕ'_m and eigenvalues $U'_m(\mathbf{R})$ depending parametrically on the nuclear coordinates $\mathbf{R}$. The functions $U'_m(\mathbf{R})$ are called *adiabatic electronic terms* of the molecular system. The functions $\phi'_m(\mathbf{r}, \mathbf{R})$ form an *adiabatic* basis corresponding to the Hamiltonian H'_e. Along with the functions ϕ'_m are often introduced functions ϕ_m, which are solutions to the eigenvalue problem with the Hamiltonian

$$H_e = \sum_\alpha T_\alpha + \sum_{\alpha>\beta} V_{\alpha\beta} + \sum_{\alpha,i} V_{\alpha i} + \sum_{i>k} V_{ik}. \tag{1.8}$$

The Hamiltonian (1.8) differs from (1.6), in that the former does not take the spin–orbit interaction into account. It is the usual approximation in present-day calculations on the electronic structure of molecules. The eigenvalues $U_m(\mathbf{R})$, corresponding to the Hamiltonian (1.8), differ from $U'_m(\mathbf{R})$. Nevertheless the ϕ_m are also called adiabatic electronic functions in the literature. In future, we will distinguish these two sets, denoting by a dash the Hamiltonian, wave functions, and energies taking the spin–orbit interaction into account.

The function $\Psi(\mathbf{r}, \mathbf{R})$ of the whole system can be represented in the form of an expansion

$$\Psi(\mathbf{r}, \mathbf{R}) = \left[\sum \chi_m(\mathbf{R})\Phi'_m(\mathbf{r}, \mathbf{R})\right]\exp\left(-\frac{i}{\hbar}Et\right), \qquad (1.9)$$

where E is the total energy of the molecular system; $\chi_m(\mathbf{R})$ are certain functions of the nuclear coordinates.

The functions $\chi_m(\mathbf{R})$ are found from the condition that $\Psi(\mathbf{r}, \mathbf{R})$, written in the form (1.9), satisfies eqn. (1.1). Substituting $\Psi(\mathbf{r}, \mathbf{R})$ into (1.1) and taking it into account that the ϕ'_m satisfy eqn. (1.7), we find the following system of equations for the functions $\chi_m(\mathbf{R})$:

$$-\sum_i T_i \chi_m + [E - C_{mm}(\mathbf{R}) - U'_m(\mathbf{R})]\chi_m = \sum_{m \neq m'} C_{mm'}\chi_{m'}. \qquad (1.10)$$

That the ϕ'_m form a complete normalized orthogonal system of functions,

$$\langle \phi'_m | \phi'_n \rangle \equiv \int (\phi'_m)^* \phi'_n \, d\mathbf{r} = \delta_{mn}, \qquad (1.11)$$

is used specifically in the derivation of (1.10). In (1.10) the operators $C_{mm'}$ are introduced, acting on the nuclear wavefunctions and forming a matrix. The order of the matrix is equal to the number of coupled equations. Although, generally speaking, all the electronic states are taken into account in (1.10), i.e. the number of equations is infinite, in a practical calculation, it has to be limited to a finite system. The sum on the right-hand side of (1.10) applies to the non-diagonal elements $C_{mm'}$, which have the form

$$C_{mm'} = \sum_i \left\{ -\frac{\hbar^2}{2M_i} \langle \phi'_m | \nabla^2_{\mathbf{R}_i} | \phi'_{m'} \rangle - \frac{\hbar^2}{M_i} \langle \phi'_m | \nabla_{\mathbf{R}_i} | \phi'_{m'} \rangle \nabla_{\mathbf{R}_i} \right\}. \qquad (1.12)$$

The diagonal element $C_{mm'}$, which is taken over to the left-hand side of (1.10), may be considered as a correction to the adiabatic energies $U'_m(R)$. It has the same structure as the term $C_{mm'}$ in (1.12).

If the coupling of the equations in (1.10) is neglected and all the elements $C_{mm'}$ are replaced by zero, independent equations are obtained for the χ_m. Each of them reduces to the Schrödinger equation describing the motion of nuclei in a potential $U'_m(\mathbf{R})$,

$$-\sum_i \frac{\hbar^2}{2M_i} \nabla^2_{\mathbf{R}_i} \chi_m(\mathbf{R}, E) + U'_m(\mathbf{R})\chi_m(\mathbf{R}, E) = E\chi_m(\mathbf{R}, E). \qquad (1.13)$$

Depending on the physical conditions, eqn (1.13) corresponds either to an eigenvalue problem (bound states, discrete spectrum E) or to a scattering problem (unbounded states, continuous spectrum E). In the approximation in which (1.10) reduces to a system of separated equations (1.13), the energy $U'_m(\mathbf{R})$ may be interpreted as the potential energy of the nuclei corresponding to the mth electronic state of the system. A criterion for the validity of such an interpretation of U'_m may be ascertained by studying the question of the accuracy with which the system of coupled equations (1.10) may be replaced by independent equations.

This matter is discussed in the literature [4] in application to bound states. It is found that, for the lowest vibrational states of the nuclei, the neglect of the coupling of eqns (1.10) introduces a relative error of order $(m_e/M)^{\frac{1}{4}}$. This ratio appears as a small parameter on account of the coupling of the equations within the limits of perturbation theory.

In the zero-order approximation, the wavefunction $\Psi(\mathbf{r}, \mathbf{R})$ is written in the form

$$\Psi(\mathbf{r}, \mathbf{R}) \approx \Psi_{m,E}(\mathbf{r}, \mathbf{R}) = \chi_{m,E}(\mathbf{R}) \phi'_m(\mathbf{r}, \mathbf{R}) \exp\left(-\frac{iEt}{\hbar}\right). \tag{1.14}$$

The functions $\Psi_{m,E}(\mathbf{r}, \mathbf{R})$ are called the electronic–nuclear functions in the adiabatic approximation. They furnish a complete system in the space of the electronic and nuclear functions. The orthogonality of the functions $\Psi_{m,E}$ corresponding to different electronic states (i.e. to different indices m) is ensured by the factors ϕ'_m. Orthogonality within a single electronic state is ensured by the nuclear functions $\chi_{m,E}$ and $\chi_{m,E'}$, differing in the energy values E. The functions $\chi_{m,E}$ differing only in the index m are not mutually orthogonal in so far as they pertain to different electronic states.

In the study of inelastic atomic and molecular collisions, and also of the intramolecular redistribution of energy, it is usually impossible to limit consideration to the lowest vibrational states. We should therefore consider the possibility of applying the adiabatic approximation to systems of nuclei which are not near a position of equilibrium. To this end, we will simplify the general problem, assuming that the motion of the nuclei may be considered classically. In spite of this, however, a difficulty arises, connected with the ambiguous ascription of the trajectory to any of the adiabatic potentials $U'_m(R)$. In future, in a calculation of the probability of nonadiabatic transitions, we will discuss possible methods of removing these difficulties (Chapter 3). Here we will just give a general formulation of the problem of the adiabatic division of electronic and nuclear motion in the semiclassical approximation. The essence of this approximation consists in the fact that the fast subsystem (the electrons) is considered quantum mechanically and the slow system (the nuclei) is considered classically.

In place of eqn (1.1) with its time independent Hamiltonian, the time dependent electronic Hamiltonian H'_e is considered, together with the corresponding non-stationary wave equation

$$i\hbar \frac{\partial \Psi}{\partial t}(\mathbf{r}, t) = H'_e(\mathbf{r}, \mathbf{R}) \Psi(\mathbf{r}, t). \tag{1.15}$$

The time dependence of the Hamiltonian H'_e is conditioned by the time dependence of the nuclear coordinates $\mathbf{R} = \mathbf{R}(t)$. The choice of some function or other $\mathbf{R}(t)$ is not connected with eqn (1.15). In the solution of (1.15) the trajectory of the motion of the nuclei is considered as given. We will find $\Psi(\mathbf{r}, t)$ in the form of an expansion in terms of the adiabatic functions

$$\Psi(\mathbf{r}, t) = \sum_m a'_m(t)\phi'_m(\mathbf{r}) \exp\left[-\frac{i}{\hbar}\int^t U'_m(\mathbf{R})\,dt\right]. \quad (1.16)$$

This expansion is the semiclassical analogue to (1.9). Substituting (1.16) into (1.15) and using the orthonormal property of the functions ϕ'_m, we obtain the equations for the coefficients a'_m.

$$i\hbar\frac{da'_m}{dt} = \sum_n \left\langle \phi'_m \left| -i\hbar\frac{\partial}{\partial t} \right| \phi_n \right\rangle \exp\left[-\frac{i}{\hbar}\int^t (U'_n - U'_m)\,dt\right] a'_n. \quad (1.17)$$

Eqns (1.17) replace the system (1.10) in the semiclassical approximation. First of all, it can be seen from this that the operator

$$\hat{C} = -i\hbar\frac{\partial}{\partial t}, \quad (1.18)$$

acting on the adiabatic electronic functions, is the semiclassical analogue to the operator of the nonadiabatic coupling $\hat{C}$ whose matrix elements, with respect to the electronic adiabatic functions, are determined by the relationship (1.12). Since ϕ'_m depends on t only through $\mathbf{R}$, a matrix element can be represented in the form

$$\left(i\hbar\frac{\partial}{\partial t}\right)_{mn} = \left\langle \phi'_m \left| i\hbar\frac{\partial}{\partial t} \right| \phi'_n \right\rangle = \sum_i i\hbar\langle \phi'_m | \nabla_{\mathbf{R}_i} | \phi'_n \rangle \dot{\mathbf{R}}_i, \quad (1.19)$$

where $\dot{\mathbf{R}}$ denotes differentiation with respect to t: $\dot{\mathbf{R}} = \partial \mathbf{R}/\partial t$. This expression represents the classical limit of the second term in (1.12), since, in the transition to the classical operator, $(-i\hbar/M_i)\nabla_{\mathbf{R}_i}$ must be replaced by the velocity $\dot{\mathbf{R}}_i$. As regards the first term in (1.12), within the limits of the semiclassical approximation, it is generally neglected, since it is proportional to the highest powers of the expansion of the operator of nonadiabaticity in terms of the ratio of the wavelength to the characteristic dimensions of the potential.

Solving the system of equations (1.17) to first order in perturbation theory, it is not difficult to find the condition under which the expansion sum (1.16) may be limited to one term, i.e. under which the adiabatic approximation is valid.

We will assume that at $t = t_0$ the molecular system was in the electronic state n. Thus the initial conditions for eqns (1.17) are:

$$a'_m = 0, \quad m \neq n, \quad |a'_n| = 1. \quad (1.20)$$

Substituting these values of a'_m in the right-hand side of the system of eqns (1.17) and integrating, we find

$$a'_m(t) = \int_{t_0}^{t} -\langle \phi'_m | \dot\phi'_n \rangle \exp\left[-\frac{i}{\hbar}\int_{t_0}^{t}(U'_n - U'_m)\,dt\right]dt. \tag{1.21}$$

To estimate the order of magnitude of the integral we will assume

$$\langle \phi'_m | \dot\phi'_n \rangle \leqslant \langle \phi'_m | \phi'_n \rangle \frac{v}{a},$$

where v is some mean velocity of the motion of the nuclei in the time interval $t-t_0$; a is a characteristic distance over which the adiabatic electronic function changes substantially.

Assuming then

$$\int_{t_0}^{t}(U'_n - U'_m)\,dt \approx \Delta \overline{U}'_{nm}(t - t_0), \tag{1.22}$$

where $\Delta \overline{U}'_{nm}$ denotes some mean value, we find the following upper bound for the amplitude,

$$|a'_m(t)| \leqslant \frac{v\hbar}{a\Delta \overline{U}'_{nm}}, \qquad m \neq n. \tag{1.23}$$

Thus, if the difference between two electronic energies and the velocity of motion of the nuclei satisfy the condition

$$\xi_{nm} = \frac{\Delta \overline{U}'_{nm} a}{\hbar v} \gg 1, \tag{1.24}$$

then the system, initially in the electronic state n, has the greatest possibility of being found in this state also in the future. Putting this another way, under the condition (1.24) the motion of the nuclei, in practice, does not induce transitions between different electronic states. Instead of the parameter ξ_{nm} (called the Massey parameter) it is convenient in future to introduce an analogous parameter $\xi(\mathbf{R}) = \Delta U(\mathbf{R})a/\hbar v$, in which $\Delta U(\mathbf{R})$ denotes the difference between any two terms (dropping the indices m and n) and which may be attributed to every point of the phase space of the nuclei. In this case, the criterion for the adiabatic approximation may be formulated as follows. In those regions of the phase space of the nuclei in which, for the given electronic state n, the condition (1.24) is fulfilled (for $m \neq n$), the adiabatic electronic energy $U'_n(\mathbf{R})$ may be interpreted as the potential energy of the nuclei. In the system of coordinates $(\mathbf{R}_1, \mathbf{R}_2, \ldots)$ a certain surface, often called the potential energy surface of the nuclei corresponding to the electronic state n, corresponds to the function $U'_n(\mathbf{R})$. The velocity v in (1.24) is determined by the nature of the motion of the representative point of the nuclei on this surface.

In the adiabatic approximation, the wavefunction is written in the form

$$\Psi_n(\mathbf{r}, t) = \phi'_n(\mathbf{r}, \mathbf{R}(t)) \exp\left[-\frac{i}{\hbar} \int^t U'_n(\mathbf{R}) \, dt\right]. \quad (1.25)$$

If the wavefunction $\Psi(\mathbf{r}, t)$ is represented as a superposition of functions of type (1.25) with constant coefficients a'_n and to each of them there corresponds (in the general case) a different trajectory of classical motion, then such a function also corresponds to the adiabatic approximation. The essential feature of nonadiabaticity reflecting the kinematic interaction of the electrons and nuclei becomes apparent precisely in that the coefficients a'_n, and consequently the weighting of the different adiabatic electronic states in the function $\Psi(\mathbf{r}, t)$, change along the trajectories of the classical motion of the nuclei.

2. Basic assumptions of the transition state method. Deduction of a formula for the rate of an elementary process

The first, more or less consistent, method of calculation of the rate constants of various elementary processes, known as the transition state method, was proposed by Eyring [126], [223], Wigner, and Pelzer [400]. For a long time it was the only method allowing the study of specific processes. The special attraction of this method lies in the fact that it avoids the solution of the dynamical part of the problem (i.e., in point of fact, the solution of the many-body problem), and the statistical averaging is carried out on the assumption of a Maxwell–Boltzmann distribution function (i.e. of a Maxwell–Boltzmann probability of locating the representative point of the system in an element of phase space). Instead of representing the rate constants as quantities depending on the physical properties of the reactants, in the transition state method the idea of the activated complex is introduced (see below), whose properties determine the value of the constant and its temperature dependence. The question of the connection between the properties of the activated complex and the properties of the reactants is generally not raised within the framework of this method, which, in fact, offers a formal possibility of avoiding the solution of the dynamical problem. On this basis, the transition state method is sometimes contrasted with collision theory, whose central problem is precisely the investigation of the dynamics of a system of particles. In fact, however, it is always possible to trace the connection between the two methods if the results of collision theory supplement the statistical part, i.e. in the selection of some method of averaging or other.

Different formulations of the basic assumptions of the transition state method are possible [16], [31], [126]. We will formulate them in the following way.

The configuration space of the system of interacting atoms is divided by a so-called critical surface S^* into a series of regions, which are identified with the regions of configuration space corresponding to various stable molecular formations. By the latter, besides stable molecules, are implied also unstable molecules, if their lifetimes exceed by far the characteristic time of intermolecular motion 10^{-12}–10^{-14} s. Such an ascription of various regions to different molecules can be carried out only approximately, where the critical surface itself can be determined to the same degree of approximation. As to the question of the choice of the critical surface, we will return to this further on, and will here assume that such a choice has been made. Near the critical surface S^* the following conditions are assumed to be fulfilled.

1. There exists an adiabatic potential $U(\mathbf{R}_i)$, depending on the coordinates of the nuclei $\mathbf{R}_i$ and determining the dynamics of the motion near S^*.
2. The distribution function for the representative points intersecting S^* in the direction of the reaction products is taken to be the equilibrium distribution function.
3. The rate of an elementary process is identified with the velocity of the representative points across the critical surface in the direction of the reaction products.

In so far as a special role is assigned to systems on the critical surface in the transition state method, the concept of the *activated complex* is introduced into the theory. By definition, by activated complex is understood that state of the molecular system which corresponds to the representative point on the critical surface. In what follows, the activated complex of a system X will be denoted by X^*.

We will now discuss the basic assumptions of the method, limiting ourselves to a classical consideration of the motion of the nuclei. The correctness of the adiabatic approximation implies that the surfaces of potential energy corresponding to different electronic states of the molecular system must be sufficiently separated. Thus condition (1.24), guaranteeing the applicability of the adiabatic approximation, must be fulfilled for any configuration of the nuclei. However, even for the simplest systems, for example an atom or a diatomic molecule, there exist regions of configuration space in which the energies of two electronic states coincide. In particular for a system of three atoms of hydrogen, the splitting ΔU between the two lowest electronic terms tends to zero in the configuration of an equilateral triangle (see Chapter 8). In such a situation, it is possible to neglect nonadiabatic effects of all kinds only in the case when the criterion for the adiabatic approximation $\Delta U/\hbar a v \gg 1$ is fulfilled by the greater part of configuration space of the nuclei accessible to the motion of nuclei with an equilibrium distribution function.

Cases often occur when the adiabatic approximation is known to be violated. Even if one of the initial or final molecules is in a degenerate electronic state then this degeneracy is removed, because of the intermolecular interaction, and the splitting between electronic terms arising from this may be small. For example, in the reactions of an atom of hydrogen or an alkali metal with a halogen molecule Y_2, the atom Y emerges in the ^2P state.† For a linear configuration of the nuclei, a single electronic state of the activated complex is possible, adiabatically correlated with the electronic states of the reactants $H(^2S)$ and $Y_2(^1\Sigma)$, and there are two states $^2\Sigma$ and $^2\Pi$ of the activated complex correlated with the electronic states of $HY(^1\Sigma)$ and $Y(^2P)$. If the reaction proceeds adiabatically, then the representative point will move over the potential surface of the electronic term the whole time, since this term connects the electronic states of the initial and final molecules. It is possible, however, that near S^* a strong nonadiabatic interaction will occur between the states $^2\Sigma$ and $^2\Pi$, so that, in general, one cannot attach the meaning of a potential energy of the system of atoms to the adiabatic terms. For a nonlinear configuration of the nuclei, the situation is complicated even further, since the degeneracy of the term $^2\Pi$ in the linear configuration is split into two components.

We now turn to the second assumption of the theory. In principle, the distribution functions near S^* can be calculated on the basis of a solution of the kinetic equations describing processes of intra- and intermolecular energy exchange. Generally speaking there are no grounds for asserting that the distribution function is near to the equilibrium form. What is more, if it is found to be the exact equilibrium form, then the rate of the flow in both directions is the same, and in fact we are dealing with complete equilibrium. Thus the assumption may also be formulated in the form of an assertion that the distribution function of the representative points near the critical surface differs from its equilibrium form only for a flow opposite to the reaction direction. A simplified one-dimensional model illustrating this assertion is considered in § 36.

To clarify the possibility of identifying the current through the critical surface with the rate of the process, we will consider the simplest dissociation reaction $AB \rightarrow A + B$. In a strict formulation of the problem, the rate of the process might be calculated as the current of atoms A, traversing a sphere of radius R_0 around atom B which divides the noninteracting atoms A and B. Trajectories on which the motion of the representative points describes the decomposition of AB will intersect the sphere $R = R_0$ once, where the current is directed out of the sphere. If now the radius R_0 is decreased, then for certain values, when the interaction of A and B can no longer be considered small, trajectories of a new type appear, intersecting the sphere

---

† Here we neglect the spin–orbit interaction resulting in the splitting of the ^2P term of the halogen atom into the two fine structure components $^2P_{\frac{3}{2}}$ and $^2P_{\frac{1}{2}}$, although this interaction is important for quantitative calculations.

twice. The total current of particles along these trajectories outside the sphere equals zero, though the constituent currents outside and inside the sphere differ from zero. If the sphere $R = R_0$ is now chosen as a critical surface then the current calculated through this surface in the direction of the reaction products will be larger than the true rate. The ambiguity in the choice of $S^\star$ is a fundamental defect of the transition state method, so this method is useful for practical calculations only when reasonable variations in the choice of $S^\star$ affect the final result relatively weakly. In particular, $S^\star$ may be chosen on the basis of requiring a minimum current of representative points in the direction of the reaction products (the so-called variational theory of reactions, see § 42), when the critical surface serves as an 'observation point' for the determination of the reaction rate [118]. The advantage of such a choice of 'observation point' is that the current directed towards the reactants can be neglected.

Classical theory

We will assume that the system of interacting atoms is described by the Hamiltonian

$$H = \tfrac{1}{2}\sum_{ij}^{3n} a_{ij}\dot{q}_i\dot{q}_j + U(q_1, q_2, \dots, q_{3n}), \tag{2.1}$$

where the q_i are generalized coordinates characterizing the position of atoms in the system under consideration, where, for simplicity, we assume that the coefficients a_{ij} do not depend on the q_i.

The centre-of-mass motion may always be removed from the total Hamiltonian, which reduces the number of degrees of freedom which must be considered to $s = 3n - 3$. The rotation of the system as a whole, generally speaking, does not separate out from the internal degrees of freedom. If the coupling between the rotation of the system and other forms of motion of the atoms is neglected, as is often done in the calculation of molecular spectra of stable polyatomic molecules, then the number of independent degrees of freedom in (2.1) is reduced to $s = 3n - 6$. It should be borne in mind, however, that the separation of the rotation and vibrations in polyatomic molecules is due to the smallness of the amplitude of vibration in comparison with the equilibrium interatomic distances [10]. In the problem of the redistribution of particles, generally speaking, there is no analogous small parameter, so the possibility of separating out the rotation, which is usually allowed in the transition state method, demands careful investigation.

Having taken advantage of the definition of the generalized momentum $p_i = \partial H/\partial \dot{q}_i$ we can express (2.1) in the form

$$H = \tfrac{1}{2}\sum_{ij}^{s} g_{ij}p_ip_j + U(q_1, \dots, q_s), \tag{2.2}$$

where the kinematic coefficients a_{ij} are connected with the g_{ik} by the relation $\sum_j a_{ij}g_{jk} = \delta_{ik}$ (δ_{ik} is the Kronecker symbol, equal to unity for $i = k$ and zero for $i \neq k$). An element of phase volume $d\Gamma$ written in terms of the coordinates p_i, q_i, as is well known [38], has the form $d\Gamma = \Pi_i^s dp_i\, dq_i$ (where s is the number of degrees of freedom). In future, however, it is convenient to understand by $d\Gamma$ the dimensionless element of phase volume

$$d\Gamma = \frac{dp_1\, dq_1}{2\pi h} \cdot \frac{dp_2\, dq_2}{2\pi h} \cdots \frac{dp_s\, dq_s}{2\pi h}. \tag{2.3}$$

If the variables in the Hamiltonian (2.2) are separable, then each pair $d\Gamma_i = (dp_i\, dq_i/2\pi h)$ gives (in the classical interpretation) the number of quantum states dn_i corresponding to the element of phase volume.

Having chosen the critical surface in a definite form, one can determine the *reaction coordinate* q_r in such a way that it is normal to the critical surface and on it takes a given value q_r^*. In addition, the motion of a point along this coordinate will not coincide, generally speaking, with a trajectory in phase space. Then the current of particles along q_r will be proportional to the mean rate of change of the phase volume at points of intersection of the critical surface by the reaction coordinate, i.e. at $q_r = q_r^*$,

$$\left\langle \frac{d\Gamma}{dt} \right\rangle = \left\langle \frac{dq_r}{dt} \cdot \frac{dp_r}{2\pi h} \prod_{i \neq r}^s \frac{dp_i\, dq_i}{2\pi h} \right\rangle. \tag{2.4}$$

The averaging, denoted by $\langle ... \rangle$ is accomplished with the help of the distribution function $f(p_i, q_i)$ which gives the probability density for finding the system at the point p_i, q_i of phase space, and includes an integration over all momenta and coordinates except $i = r$ over the interval from $-\infty$ to $+\infty$. As regards the integration over p_r, the corresponding limits are determined by the condition that the current be taken into account only in one direction (see assumption 3, p. 8),

$$\dot{q}_r = \sum_i g_{ri} p_i \geq 0. \tag{2.5}$$

The rate constant κ of the elementary process is defined as the rate of this process per unit concentration of the reactants. Hence for the identification of the mean current (2.4) with the rate constant, the distribution function $f(p_i, q_i)$ must be normalized so that it corresponds to unit density of particles in the region of the reactants. Since the number of particles in phase space is proportional to the partition function, the normalized distribution function on the critical surface must have the form:

$$f^\star(p_i, q_i) = \frac{\exp[-\beta H^\star(p_i, q_i)]}{F}, \tag{2.6}$$

where
$$H^\star(p_i, q_i) = H(p_i, q_i)|_{q_r = q_r^\star}, \qquad \beta = 1/kT,$$
and
$$F = \int \exp[-\beta H(p_i, q_i)] \, d\Gamma, \tag{2.7}$$

where F is the total partition function of the initial system. Thus we obtain the following basic formula for the rate constant of the elementary process

$$\kappa = \int_{\dot q_r = 0}^{\infty} \frac{\dot q_r \, dp_r}{2\pi\hbar} \int_{-\infty}^{\infty} \exp(-\beta H^\star) \, d\Gamma^\star / F, \tag{2.8}$$

where
$$\dot q_r = \frac{dq_r}{dt}, \qquad d\Gamma^\star = \prod_{i \neq r} \frac{dp_i \, dq_i}{2\pi\hbar}.$$

This expression can be simplified when the Hamiltonian of the initial system $H(p_i, q_i)$ and the Hamiltonian of the activated complex $H^\star(p_i, q_i)$ separate into parts corresponding to vibrations and rotations (the translational motion of the centre of mass always separates out). Although as was shown above, the possibility of such a separation in an activated complex is far from evident, nevertheless precisely this approximation is, in fact, used in practical calculations.

Assuming
$$H(p, q) = H_{\text{trans}} + H_{\text{vib}} + H_{\text{rot}}, \tag{2.9}$$
we find
$$F = F_{\text{trans}} \cdot F_{\text{vib}} \cdot F_{\text{rot}}. \tag{2.10}$$

In (2.10) the factors denote the partition functions of the corresponding motion in the initial system. If this is a single molecule, then F corresponds to the molecular partition functions and expression (2.8) gives the unimolecular reaction rate. If the initial system is represented by a pair of reacting molecules, then each of the factors in (2.10) splits into the product of the partition functions for each molecule and κ corresponds to the bimolecular reaction rate.

For the construction of analogous arguments applying to the integral in (2.8), we will assume that, in the expression for the kinetic energy in $H^\star$, the part corresponding to the motion of q_r is separable. We will represent $H^\star$ in the form

$$H^\star = E_t + H_0^\star, \tag{2.11}$$

where

$$E_t = \tfrac{1}{2}m_r\dot{q}_r^2, \qquad H_0^\star = \tfrac{1}{2}\sum_{n,m \neq r} a_{nm}\dot{q}_n\dot{q}_m + U^\star,$$

$$U^\star = U(q_1, q_2, \ldots, q_r^\star, \ldots, q_s).$$

Then the integral in the numerator of (2.8) can be written in the form

$$\int_0^\infty \dot{q}_r \frac{\mathrm{d}p_r}{2\pi\hbar} \int \exp(-\beta H^\star)\,\mathrm{d}\Gamma^\star = \int_0^\infty \frac{\mathrm{d}E_t}{2\pi\hbar} \exp(-\beta E_t) \int \exp(-\beta H^\star)\,\mathrm{d}\Gamma^\star$$

$$= \frac{kT}{2\pi\hbar} F^\star \exp\left(-\frac{E_0}{kT}\right), \qquad (2.12)$$

where E_0 denotes the minimum potential energy of the activated complex.

$$E_0 = \min[U(q_1, \ldots, q_s)]_{q_r = q_r^\star} \qquad (2.13)$$

and $F^\star$ is the total partition function of the activated complex, the energy of which is measured from the level E_0.

$$F^\star = \int \exp[-\beta(H_0^\star - E_0)]\,\mathrm{d}\Gamma^\star. \qquad (2.14)$$

Thus (2.8) takes the form

$$\kappa = \frac{kT}{2\pi\hbar} \frac{F^\star}{F} \exp\left(-\frac{E_0}{kT}\right). \qquad (2.15)$$

It is clear that, after the interaction of the critical surface, the representative point may be reflected from some portion of the energy surface and may intersect $S^\star$ in the reverse direction. It is possible that the contribution of such trajectories is reduced if the form of $S^\star$ is slightly changed, however, then the current in the forward direction is also changed. A calculation of the multiple intersections of the surface cannot be accomplished unless the motion of the point after its transit of the saddle region is considered. However transition state theory avoids just this part of the calculation. As a result, a so-called transmission coefficient $\chi \leqslant 1$ has to be introduced into the theory, effectively taking the multiple transits across $S^\star$ into account. After the introduction of this correction, the basic formula takes the form

$$\kappa = \chi \frac{kT}{2\pi\hbar} \cdot \frac{F^\star}{F} \exp\left(-\frac{E_0}{kT}\right). \qquad (2.16)$$

This expression is useful for the calculation of rate constants only in those cases, when it is possible to find a critical surface $S^\star$ for which the transmission coefficient χ, not determined in the theory, will be of order unity.

The assumption of the separation of the variables in the activated complex is not essential. The integration over the momenta in non-separable variables may be carried out in the general form, since the kinetic energy is always a quadratic in p_i. The integration over the coordinates remains unaccomplished and the arbitrary choice of the critical surface appears only in that the effective mass of the representative point proves to be dependent on the coordinates, which must be taken into account on integration [493], [520]. The expression for the rate constant can also be easily obtained in the case when the critical surface depends not only on the coordinates but on the momenta, which appear as integrals of the motion [347], [350]. A specific example of such a surface is cited in Chapter 8 in connection with a calculation of the rate of ion–molecular reactions.

Quantum theory

The classical theory, presented in the previous section, is correct only in the case when the de Broglie wavelength of the motion of the nuclei is significantly smaller than the characteristic dimensions over which the potential changes by a significant amount. For the vibrations of atoms in molecules, this criterion is fulfilled, on average, by the condition $\hbar\omega_i\beta \ll 1$. Although even in this case, strictly speaking, classical mechanics is inapplicable for the description of atoms in those regions where their velocity is small (close to turning points of the trajectory), this can be ignored, as the contribution from the corresponding quantum effects to the mean reaction rate is small. The characteristic vibrational temperature $\theta_i = \hbar\omega_i/k$ for stable molecules with atomic mass $m \sim 10$ varies in the range 1000–3000 K. Hence, it follows that the relationship $\theta_i/T \ll 1$ is not satisfied in the chief temperature ranges of interest in chemical kinetics, and so a generalization of the transition state theory, taking the quantum mechanical motion of atoms into account, is necessary.

A basic difficulty in the generalization of the theory to quantum systems is that the reaction criterion cannot be formulated as the condition that the representative point intersect the critical surface. This is connected with the fact that, in a quantum-mechanical observation, the coordinates and momenta (and hence the energy) of a system cannot be assigned simultaneously. Moreover, the description of the system of interacting atoms itself changes. In the region of the reactants, for example, it is given by a set of quantum numbers smaller by a factor of two than the number of coordinates and momenta used in classical mechanics for the given state. Above all, the full analogy between the quantum and classical approach is retained when, in classical mechanics, the canonical 'action-angle' variables are used.

We will trace the transition between the classical and quantum approximations in an example with one degree of freedom. As a generalized momentum, we introduce the action variable defined by the relation [36]

$$J = \frac{1}{2\pi}\oint p\,dq, \qquad (2.17)$$

where the integral is taken over all closed trajectories of motion in phase space. The momentum p then depends parametrically on the total energy E, which is conserved along the trajectory. In accordance with the general method of transformation of canonical variables [36], the angular coordinate α conjugate to the action J must be defined by the relation

$$\alpha = \frac{\partial S_0(q, J)}{\partial J}, \tag{2.18}$$

where S_0 the so called 'abbreviated action' equals

$$S_0(q, J) = \int_{q_0}^{q} p \, dq. \tag{2.19}$$

Here the dependence of S_0 on J is obtained in the following way. Initially, S_0 is expressed as a function of q and E, and then the energy is expressed through J by the basic relation (2.17).

With this transformation, the Hamiltonian in the new variables depends only on J, and the equations of motion and the phase volume transform in the following way

$$\dot{p}_k = -\frac{\partial H}{\partial q_k} \rightarrow \dot{J} = -\frac{\partial H}{\partial \alpha} = 0$$

$$\dot{q}_k = \frac{\partial H}{\partial p_k} \rightarrow \dot{\alpha} = \frac{\partial H}{\partial J} = \text{constant}$$

$$d\Gamma = \frac{dp \, dq}{2\pi \hbar} \rightarrow d\Gamma = \frac{dJ \, d\alpha}{2\pi \hbar}. \tag{2.20}$$

In the transformation from classical mechanics to quantum mechanics (within the limits of the quasi-classical approximation) the action variable J plays the role of the quantum numbers (and that is why it takes a discrete set of values) and the angle variable α plays the role of the coordinates on which the wavefunction depends. Being led by the analogy mentioned, the classical problem can approximate the quantum-mechanical problem to a high degree, which is very important in the study of the dynamics of interacting atoms.

We will now study the generalization of the transition state method to the quantum-mechanical case with many degrees of freedom, assuming the variables are separable in the region near S^*. Canonical variables J_k and α_k are introduced for each conjugate pair, coordinate and momentum, describing rotation or vibration. The action variables are considered discrete, in accordance with the quantum rules for oscillators and rotators,

$$J_k \rightarrow \hbar(n_k + \tfrac{1}{2}) \qquad dJ_k \rightarrow \hbar \, dn_k$$
$$J_k \rightarrow \hbar j_k \qquad dJ_k \rightarrow \hbar \, dj_k. \tag{2.21}$$

On integration over the angle variables over the interval 0 to 2π, the factor 2π in the denominator of each pair of variables $dJ_k\, d\alpha_k/2\pi\hbar$ is cancelled and each integral over dn_k should be understood as a sum over all quantum states for the corresponding degree of freedom k:

$$\int (\ldots)\, dn_1\, dn_2 \ldots dn_s \rightarrow \sum_{k_1, k_2, \ldots, k_s} (\ldots). \qquad (2.22)$$

As regards the integration over dE_t, in the classical variant of the transition state method it was attributed to the point q_r^*. On account of the quantum effects, the task of determining the significance of q_r^* is possible only when the potential energy depends on q_r sufficiently weakly.† Therefore the current determined in the classical case by the integral (2.12), in the quantum case must be calculated at a certain point q_r'' sufficiently far from the surface $q_r = q_r^*$ (Fig. 1). The assertion of the transition state method, analogous to this about the

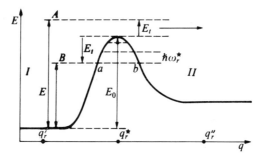

FIG. 1. Penetration of a one-dimensional potential barrier by a particle: state A—above barrier transmission ($E_t > 0$); state B—tunnel transmission ($E_t < 0$).

form of the distribution function with respect to E_t (assumption 2), may refer only to certain points q_r' occuring before the saddle. The condition which determines the points q_r' and q_r'' is the requirement that the current in the direction $q > q_r^*$ should not depend on q_r' and q_r'' (the condition of the sufficient remoteness of q_r' and q_r'' from q_r^*). At the same time, the distance between the points q_r' and q_r'' must be sufficiently small, so that the approximation of separable variables should still be satisfactory. The possibility of simultaneously satisfying these two conditions depends on the specific form of the potential.

In the transit of the representative point over the potential barrier, quantum effects manifest themselves in that there is a probability, differing from zero, of

† The condition of the sufficiently weak dependence of the potential energy on q_r can be formulated in the form of a requirement on the quasi-classical motion of the representative point $d\lambda/dq_r \ll 1$, where λ is the de Broglie wavelength equal to $h/p = h/[2m_r(E-U)]^{\frac{1}{2}}\, dq$. In particular, this condition may be satisfied by all q_r, then the motion is almost classical and $P(E_t)$ in formula (2.23) is close to unity. Since the case $P(E_t) \neq 1$ interests us, we may assume that near the saddle, $d\lambda/dq \geqslant 1$.

over-barrier reflection and tunnel transmission (see § 3). On account of this, in calculating the current along the reaction coordinate in the integral over dE_t, a transmission probability $P(E_t)$ should be introduced, at the same time extending the limits of integration over the whole range of energies for which intersection (over-barrier or tunnelling) of the critial surface is possible. To maintain the maximum analogy between the classical and quantum formulae we will write the expression for the current in the form

$$\chi \frac{kT}{2\pi\hbar} = \int_{-\infty}^{\infty} P(E_t) \exp(-\beta E_t) \frac{dE_t}{2\pi\hbar}. \qquad (2.23)$$

Here, as in formula (2.12) the kinetic energy of the point is measured from the peak of the barrier E_0. However, since in quantum theory the probability of tunnel transmission can be taken into account, integration over E_t also extends into the region of negative energy. The transmission probability $P(E_t)$ must be found as a result of solving the wave equation describing the motion of the representative point in the region of the barrier, in the interval $q' < q_r < q''$ (see § 3).

Thus the expression for the rate constant takes the form

$$\kappa = \frac{\chi kT}{2\pi\hbar} \frac{F^*}{F} \exp\left(-\frac{E_0}{kT}\right), \qquad (2.24)$$

where the partition function is calculated quantum mechanically. In addition, the level of the classical minimum of energy of the initial system is chosen as the zero level for F, but for F^* the level of the saddle E_0 is chosen.

Usually the partition functions refer to the zero energy level of the quantum system. To take this into account in formula (2.24) the substitutions

$$F \to F \exp(-\beta E_z)$$
$$F^* \to F^* \exp(-\beta E_z^*) \qquad (2.25)$$

should be made, where E_z is the zero-point energy of the quantum system.

With regard for relation (2.25), we will rewrite formula (2.24) in the form

$$\kappa = g^* \chi \frac{kT}{2\pi\hbar} \cdot \frac{F^*}{F} \exp\left(-\frac{E_a}{kT}\right), \quad E_a = E_0 + E_z - E_z^*, \qquad (2.26)$$

where χ is the transmission coefficient; F and F^* are calculated by the well-known formula of quantum statistical mechanics [16], [38]. The additional factor g^*, introduced into (2.26), takes into account the possibility of the occurrence of certian identical paths of formation and decomposition of the activated complex. The value of the factor g^* is determined by the symmetry properties of the initial molecules and the activated complex, so it is expedient to consider the question of its calculation together with methods of calculating the quantum partition function.

Although formally (2.16) and (2.26) have a similar aspect, it is necessary to note the different physical meaning of the transmission coefficients χ in essence. The coefficient χ in (2.16) takes into account the fact that the introduction of the critical surface and the reaction coordinates does not give an exact description of the process; for example, reflection of the representative point from a zone of the potential energy surfaces located on the side of the molecules of the reaction products is possible. If the variables in the Hamiltonian were exactly separable, then such a reflection would be absent and consequently $\chi = 1$ would be assumed. In the quantum case (2.26) the coefficient χ differs from unity even when the variables are separable. This is connected with the effects of over-barrier reflection and tunnel transmission of the representative point, along the coordinate q_r. Since (2.16) and (2.26) correspond to one and the same physical model, in regard to the choice of the reaction coordinate, the classical and quantum formulae should be compared on condition $\chi = 1$.

A more precise quantum variant of the transition state method

A fundamental difficulty of the quantum formulation of the transition state method, as already noted, is that the criterion that a reaction is completed cannot be formulated as a condition that some hypersurface S^* be intersected by the representative point. We will assume that we have limited ourselves to an investigation of the dynamics of the representative point near the chosen critical surface, in a certain interval of time $\Delta\tau$. To this time interval corresponds a certain stratum in configuration space, of thickness $\Delta l \approx v\Delta\tau$, bordering the surface S^*. Clearly neither $\Delta\tau$ nor Δl may be too small, as this would result in an increase in the uncertainty in the momentum Δp and energy ΔE in accordance with the basic relations in quantum mechanics,

$$\Delta p \Delta l > \hbar.$$

$$\Delta E \Delta \tau > \hbar. \tag{2.27}$$

Since the transition state method uses the equilibrium distribution function, which changes over an energy interval of order kT, in order that the averaging over energy have a meaning, it is necessary that the relation $\Delta E \ll kT$ be satisfied. But from this it follows that $\Delta\tau$ must be larger than $\hbar/kT$ or what is equivalent, $\Delta l \gg \lambda$, where λ denotes the mean wavelength of the motion along the reaction coordinate. Thus, for the calculation of the current in the quantum case it is necessary to consider explicitly the dynamics of the representative point's motion in the region Δl or to follow the development of the system in time $\Delta\tau$. Therefore in the quantum case, the activated complex cannot be considered as a certain configuration of nuclei in an infinitesimal interval of time. From this it also follows that the local properties of the potential energy surface near the saddle point, which completely determine, for example, the exchange reaction rate in the classical case, are insufficient for the calculation of the mean current in the quantum case. Johnston and Rapp [287] pointed, in particular, to the fact that, for reactions with the transfer of a proton at 300 K, the wavelength of the representative point near the saddle is such that it encloses the whole region of the barrier. Under these conditions, separation of the variables for the activated complex loses any meaning, since the wavefunction of each of the separated equations is defined in a region, whose extent is less than the wavelength.

One of the possible ways out of this difficulty is that, in the saddle region, where the trajectory of the representative point turns sharply, curvilinear coordinates can be introduced, permitting the separation of the variables. In this case, the total wave equation separates into independent equations, each of which describes the motion of curvilinear coordinates q_i. Unlike the simplest case, the total energy may no longer be represented in the form of a sum of the constants of separation, and the potential energy for the motion along the curvilinear reaction coodinate q_r will be dependent, generally speaking, on the constants of separation. Qualitatively, this may be illustrated by the example of the motion in an arc of a circle. In this case, as is well known, the centrifugal term is added to the radial potential, so that the point intersecting S^* experiences centrifugal forces. It is clear too that, in the quantum case, the correction for tunnel transmission will depend not only on the energy E_t corresponding to the reaction coordinate but also on the energies (or more exactly, on the quantum numbers) and other degrees of freedom. Marcus [347], [348], [351] has investigated, in detail, the conditions under which the introduction of curvilinear coordinates of various types into the region of the potential barrier is possible and has derived general expressions for the rate constant.

A second possible way to take account of the curvature of the trajectory is the approximate separation of variables, based on the idea of the adiabatic approximation. As indicated in § 1 the adiabatic approximation is sufficiently accurate only when the masses of the interacting particles differ greatly. For electrons and nuclei, the ratio of masses is of the order 10^{-4}. For particles in a reaction in which heavy atoms and hydrogen atoms participate, this value increases to 10^{-1} to 10^{-2}. It can be seen from this estimate that adiabatic separation of the motion of a proton from the motion of heavy nuclei cannot lay claim to great accuracy. Nevertheless, the adiabatic approximation may serve as a starting point for a discussion of the mechanism of reaction with the transition of a hydrogen atom. This is explained by the fact that S^* permits the approximate separation of variables in a large region, the extent of which considerably exceeds the region found by the quasi-classical approximation (see § 48).

In conclusion, we note that the refinements of the transition state method considered do not solve the basic problem which always arises in the calculation of the rate constant, namely, the problem of the energy exchange between different degrees of freedom. The separation of the variables, simplifying the calculation substantially, brings with it the impossibility of an exchange of energy between the different degrees of freedom. Mathematically, this is expressed by the fact that, for each degree of freedom, there exists an independent integral of motion, the corresponding constant of separation. Therefore the model with the separated variables may serve only as the zero-order approximation to the true Hamiltonian of the system of interacting atoms.

3. Over-barrier reflection and tunnel corrections

Quantum over-barrier reflection and tunnel corrections for the motion of the representative point along coordinate q_r are calculated in formula (2.24) by the introduction of the coefficient χ. This coefficient must be calculated by averaging, over the energy, the transmission probabilities $P(E_t)$ of particles with energy E_t above (or below) the barrier.

In this section we will discuss the problem of the calculation of $\chi(T)$ for a one-dimensional potential barrier with one maximum, remaining (as before) within the limits of the adiabatic approximation. Nonadiabatic effects connected with over-barrier reflections and tunnel corrections are considered in § 27.

The problem of the calculation of $P(E_t)$ can be put in the following way. In the region I (see Fig. 1) there is an incident current of unit amplitude, which is described by the wavefunction

$$\Psi_I^+(q_r) = \exp\left(i\frac{p_r q_r}{\hbar}\right), \qquad q_r < q_r'. \tag{3.1}$$

In the same region there is also the reflected current with unknown amplitude X,

$$\Psi_I^-(q_r) = X \exp\left(-i\frac{p_r q_r}{\hbar}\right), \qquad q_r < q_r'. \tag{3.2}$$

In region II there is only the transmitted current with amplitude Y

$$\Psi_{II}^+(q_r) = Y \exp\left(i\frac{p_r q_r}{\hbar}\right), \qquad q_r > q_r''. \tag{3.3}$$

The explicit assignment of the wavefunctions in regions I and II is equivalent to the imposition of boundary conditions on the solution of the wave equation describing the motion of the representative point in terms of coordinate q_r.

$$-\frac{\hbar^2}{2\mu}\frac{\partial^2 \Psi}{\partial q_r^2} + U(q_r)\Psi = E\Psi. \tag{3.4}$$

The solution of this equation in the region of the potential barrier, with the specified boundary conditions, allows well-defined amplitudes X and Y to be determined, and hence the transmission coefficient $P = |X|^2$ and the reflection coefficient $Q = |Y|^2$.

We will now consider two particular cases, which describe qualitatively the dependence of the transmission coefficient on the energy E_t over a wide range of variation of E_t.

1. Suppose the energy of the particle is insufficient for the barrier to be surmounted classically and suppose the de Broglie wavelength $\hbar/\sqrt{[2\mu(E-U)]}$ in the region under the barrier is appreciably smaller than the width of the barrier at the level $U(q_r) = E$ (i.e. the cut-off line ab in Fig. 1). In this case, in the greater part of the region the conditions of classical motion are fulfilled, from which it is clear that the transmission coefficient will be very small.† The corresponding calculation gives for the transmission coefficient [37]

$$P(E_t) = \exp\left[-\frac{2}{\hbar}\int_a^b \sqrt{[2\mu(U-E)]}\,dq\right] \tag{3.5}$$

which is correct when the exponent of the exponential is appreciably greater than unity.

2. Suppose the energy of the particle is such that, in the region near the peak of the barrier, the conditions for quasi-classical motion are violated but, in the range of variation of E_t under consideration, the potential energy in this region can be represented in the form of a series, about the maximum, in the coordinate q_r, where all the terms in the series, beginning with the cubic, are discarded (a so-called parabolic barrier). In this case $U(q_r) = E_0 - (k/2)q_r^2$ should be substituted into eqn (3.4), after which it can be solved exactly. The expression

$$P(E_t) = \left[1 + \exp\left(-\frac{2\pi E_t}{\hbar \omega_r^\star}\right)\right]^{-1}, \tag{3.6}$$

where

$$\omega_\mu^\star = (k/\mu)^{\frac{1}{2}}$$

is obtained for the transmission coefficient [37].

For an energy sufficiently close to the peak of the barrier, the reflection coefficient will be of order unity, which is connected precisely with the violation of the quasi-classical motion in the barrier region. It is not difficult to see that $\omega_r^\star$ may be interpreted as the frequency of small vibrations of the representative point near the minimum of the 'inverted' potential barrier.

---

† This follows from the fact that in the limit of very small wavelengths, the quasi-classical result must lead to the classical, in which the tunnel effect (the transmission of particles under the potential barrier) is absent.

We will now examine the conditions under which the two limiting cases considered above are sufficient to describe the dependence of $P(E_t)$, for all ranges of variation of the energy. From (3.6), it can be seen that, for $-2\pi E_t/\hbar\omega_r^* \gg 1$, the barrier transmission probability will be exponentially small. The same result can also be obtained from (3.5) if terms of order P^2 in comparison with P are neglected.

Thus the condition of the applicability of the parabolic approximation to the barrier, in the energy region where the transmission coefficient is small, serves as a criterion that the regions of applicability of the expressions (3.5) and (3.6) overlap. In terms of the potential well obtained by turning the barrier upside down, this means that the anharmonic vibrations are small. The condition indicated is sufficiently well satisfied for potential barriers with energy $E_0 > 10\,\text{kcal/mol}^{-1}$ and mass $\mu > 10$.†

For $2\pi E_t/\hbar\omega_r^* \gg 1$, the case-2 model gives a transmission probability close to unity. Although for this the parabolic approximate barrier is incorrect, this does not have practical significance in so far as the limit of the transmission coefficient as $E_t \to \infty$ is, in any case, equal to the classical value, i.e. unity.

It should be borne in mind, however, that all the arguments cited are correct within the limits of applicability of the adiabatic approximation. If nonadiabaticity is taken into account then, for sufficiently high energies, the transmission coefficient may again begin to decrease (see § 27).

The coefficient χ is calculated by formula (2.23). Since, for sufficiently high temperatures, energies $E \sim E_0$ make the main contribution to χ, expression (3.6) can be substituted into (2.23), ignoring the inaccuracy due to the deviation of the form of the barrier from the parabolic. The final result may be approximately represented in the form

$$\chi = \begin{cases} \dfrac{x}{\sin x}, & x = \dfrac{\beta\hbar\omega_r^*}{2} < \pi \\ \exp\left[\beta E_0\left(1 - \dfrac{\pi}{x}\right)\right]\left(1 - \dfrac{\pi}{x}\right)^{-1}, & x = \dfrac{\beta\hbar\omega_r^*}{2} > \pi. \end{cases} \quad (3.7)$$

From the physical significance of the problem of the transmission of the representative point through the potential barrier it is clear that the integrand in (2.23) is the distribution function of the current in the region $q_r > q_r''$. In the classical case the distribution function in this region can be found on the basis of Liouville's equation. The solution of this equation is obvious, since

---

† If the height of the barrier and the mass of the representative point are smaller than these values then, to obtain an interpolation formula between the two limiting cases considered, it is necessary to turn to some model potential. The Eckart potential, analogous to the potential shown in Fig. 1, is often used in such a capacity. The transmission coefficient, which for this type of potential depends on two parameters, is tabulated in the literature [288], [477]. Transmission through a potential barrier of another type is considered in papers [17], [18].

the character of the classical trajectories in the field of a potential $U(q_r)$ is well known (the particle is either transmitted or not transmitted through the barrier).

In the quantum case, instead of the distribution function $f(p,q)$ depending on the momenta and coordinates, the density matrix $\rho(q_1, q_2)$ is introduced, which allows the quantum uncertainty principle to be taken into account [38]. The exact determination of the density matrix and its general properties is discussed in the book [38]. Certain relevant questions are considered below in § 23 and § 37.

In the quasi-classical limit, it is possible to establish a correspondence between the classical distribution function $f(p,q)$ and the function $F(p,q)$ which is obtained by a Fourier transformation of the density matrix. The function $F(p,q)$, called the Wigner distribution function [522], satisfies a certain equation in partial derivatives, which goes over into Liouville's equation in the limit of short wavelengths. From this it follows that the first quantum approximation to the classical distribution function can be obtained if, in this equation, terms are retained containing the small quantity $\lambda/\Delta l$ to the lowest power (here Δl is a quantity characterizing the extent of the potential's influence). Such a problem was solved by Wigner. The distribution function he found, near $q_r = q_r^*$ has the form [552]:

$$F(p_r, q_r) = f(p_r, q_r)\left[1 - \frac{\hbar^2 \beta^2}{8\mu} \cdot \frac{\partial^2 U}{\partial q_r^2} + \frac{\hbar^2 \beta^3}{24\mu^2} \cdot p_r^2 \cdot \frac{\partial^2 U}{\partial q_r^2} + \ldots\right]. \quad (3.8)$$

Integration of this function over all energies gives the expression

$$\chi = 1 + \frac{1}{24}\left(\frac{\hbar\omega_r^*}{kT}\right)^2 + \ldots \quad (3.9)$$

for χ, in which the second term is caused by the contribution of tunnel transmission to the mean current of particles. From the condition in the derivation of (3.8) it follows that this term must be appreciably smaller than unity. It is not difficult to verify that (3.7) and (3.9) coincide in the region $\hbar\omega_r^*/kT \ll 1$.†

4. The quantum partition functions and symmetry numbers

The quantum partition function F of a molecular system is represented by the expression

$$F = \sum_k g_k \exp(-\beta E_k), \quad (4.1)$$

---

† An account of the method of calculating the quantum correction to the classical distribution function is given in the book [38], p. 93.

where E_k is the energy level; g_k is the degree of degeneracy of this level (statistical weight).

The origin of measurement is referred to the lowest level, coinciding with E_z. In what follows, we will assume that the Hamiltonian of the system is approximately separable into parts corresponding to translational, vibrational, and rotational motion of the atoms. Moreover, as a result of the adiabatic approximation, different electronic terms of the system make independent contributions to the partition function. We will assume, for simplicity, that the ground state is non-degenerate and that, in consequence of the large spacing between the electronic terms ($U_m - U_0 \gg kT$), all excited states make negligibly small contributions to the sum (4.1). In this case E_k can be represented in the form of a sum of translational E_trans, vibrational E_vib, and rotational energies E_rot. The statistical weight g_k must take account of the degeneracy of the states and, furthermore, the degeneracy with respect to the spin components of nuclei coming into the molecular system. All states which differ only in the quantum numbers characterizing the spin states of the nuclei are considered degenerate in conformity with the approximation (1.6) adopted in §1, within the limits of which, hyperfine interactions are not taken into account. Knowledge of individual degrees of degeneracy $g_{k,\text{trans}}$, $g_{l,\text{vib}}$, $g_{m,\text{rot}}$, and $g_{n,\text{nucl}}$ is insufficient, however, for the calculation of g_k in (4.1), since in the sum (4.1), not all possible energy states have to be taken into account, but only those which satisfy the demands of Fermi and Bose statistics as regards the permutation of identical particles. The electronic part Ψ'_0 of the total adiabatic function, of course, satisfies the Pauli principle, but its symmetry with respect to permutation of identical nuclei can be arbitrary. At the same time, the total adiabatic function, including the spin functions of the nuclei as well as the electronic and nuclear coordinates factors, must be antisymmetric or symmetric under interchange of identical nuclei with half-integral or integral spin, respectively. This additional requirement (with regard to the wave equation) leads to certain states in the sum (4.1) being forbidden; such states must be omitted from the summation.

Formally this becomes apparent in the fact that g_k in (4.1) is not equal to the product $g_{k,\text{trans}} g_{l,\text{vib}} g_{m,\text{rot}} g_{n,\text{nucl}}$ as might have been expected from basic considerations concerning the independent contributions of each type of motion. Therefore, although E_k can be represented in the form of a sum of contributions of different forms of motion, the partition function F, generally speaking, cannot be written as a product of factors corresponding to different types of motion. This can be formally taken into account by representing F in the form

$$F = F_\text{trans} \cdot F_\text{vib} \cdot F_\text{rot-nucl}. \qquad (4.2)$$

The last factor in (4.2), corresponding to the rotational degrees of freedom and taking the statistics of the nuclei into account, reflects the incomplete separation of the motion in the calculation of F. In the computation of F_{trans} and F_{vib}, the influence of nuclear spin need not be taken into account, since neither the translational motion nor the vibrational motion lead to an exchange of nuclei. The factor F_{trans} of the quantum formula agrees with the corresponding classical expression in so far as the translational motion is not quantized. In contrast to all the other partition functions, only this factor has dimensions (of inverse volume).

For definiteness, we will consider below the calculation of the partition functions of the reactants and the activated complex in dissociation-type reactions

$$AB \to A + B \quad (4.3)$$

and in exchange-type reactions

$$X + Y \to Z + U. \quad (4.4)$$

For process (4.3) the initial system and the activated complex correspond to a single molecule AB, so that from the well-known basic formula [38] we obtain

$$F_{AB,\text{trans}} = F^\star_{AB,\text{trans}} = \frac{(2\pi m_{AB}kT)^{\frac{3}{2}}}{(2\pi\hbar)^3}. \quad (4.5)$$

In the expression (2.26) for the rate constant, these factors cancel and the dimension of κ, provided by the frequency factor $kT/2\pi\hbar$, corresponds to inverse time.

For a reaction of the type (4.4) there are three translational degrees of freedom for the activated complex XY* and six for the reactants X and Y.

$$F^\star_{XY,\text{trans}} = \frac{[2\pi(m_X + m_Y)kT]^{\frac{3}{2}}}{(2\pi\hbar)^3}, \quad (4.6)$$

$$F_{X,\text{trans}} \cdot F_{Y,\text{trans}} = \frac{(2\pi m_X kT)^{\frac{3}{2}}(2\pi m_Y kT)^{\frac{3}{2}}}{(2\pi\hbar)^6}. \quad (4.7)$$

In accordance with this, the dimension of the rate constant is $(\text{time})^{-1}$ $(\text{volume})^{-1}$.

We note that in the product $F_{X,\text{trans}} \cdot F_{Y,\text{trans}}$, we can always separate out the translational partition function corresponding to the motion of the centre of mass. This factor cancels with $F^\star_{XY,\text{trans}}$, so that in future calculations we will make use of this simplification

$$\frac{F^\star_{XY,\text{trans}}}{F_{X,\text{trans}} \cdot F_{Y,\text{trans}}} = \frac{1}{F_{X-Y,\text{trans}}} = \frac{(2\pi\hbar)^3}{(2\pi\mu kT)^{\frac{3}{2}}}, \quad (4.8)$$

where $F_{X-Y,\text{trans}}$ is the translational partition function of the relative motion of the molecules X and Y; μ is the reduced mass of the molecules,

$$\mu = m_X m_Y/(m_X + m_Y).$$

The vibrational sum of states for a system of harmonic oscillators breaks down into a product of the sum of states of different normal vibrations:

$$F_{\text{vib}} = F_{\text{vib}}^{(1)} \cdot F_{\text{vib}}^{(2)} \ldots F_{\text{vib}}^{(\kappa)}, \tag{4.9}$$

$$F_{\text{vib}}^{(k)} = \left[1 - \exp\left(-\frac{\hbar\omega_k}{kT}\right)\right]^{-1}. \tag{4.10}$$

In the calculation of $F_{\text{rot,nucl}}$ the influence of the nuclear spin must be taken into account, since certain rotations of the rigid framework of the molecule are equivalent to a permutation of identical nuclei.† In the general case, the rotational energy of a rigid framework is given by two quantum numbers (J, K) one of which coincides with the angular momentum J. Therefore for $F_{\text{rot,nucl}}$ we have

$$F_{\text{rot,nucl}} = \sum_{J,K} g(J, K, i) \exp[-\beta E_{\text{rot}}(J, K)], \tag{4.11}$$

where $g(J, K, i)$ is the statistical weight of the allowed state (J, K) of the molecule characterized by the set of nuclear spins i.

The rules for the computation of the coefficients $g(J, K, i)$ are formulated in terms of the group theory of the permutations of the identical nuclei and the theory of the point symmetry groups of the rigid framework of the molecule, where the number of irreducible representations of the point symmetry group, realized by the set of functions $\chi_{\text{rot},J,K}$, plays the role of a quantum number [37], [38]. It turns out that the $g(J, K, i)$ change very irregularly with variation of J and K. Nevertheless, if the $g(J, K, i)$ are averaged over an interval $\Delta J, \Delta K \gg 1$, we obtain a sufficiently smoothly-changing function $\bar{g}(J, K, i)$,

$$\bar{g}(J, K, i) = \left[\prod_i (2i+1)\right] \frac{2J+1}{\sigma}, \tag{4.12}$$

where $\prod_i (2i+1)$ is the total statistical weight of the system of nuclei, due to the different orientations of the spins of individual nuclei (identical and different); $(2J+1)$ is the statistical weight of the specified rotational state, due to the degeneracy of the angular momentum with respect to orientations in space; σ is the symmetry number of the rigid framework of the molecule, which is defined as the number of independent rotations of the framework permuting identical nuclei with each other.

The indicated averaging of $g(J, K, i)$ under the summation sign is possible if the difference in the energy levels is small by comparison with kT. In this case, the sum (4.8) is replaced by an integral, which is expressed in terms of the

---

† We remember that the rotational energy is calculated in the approximation of a rigid configuration of the nuclei.

classical rotational partition function

$$F_{\text{rot,nucl}} = \left[\prod_i (2i+1)\right] \frac{1}{\sigma} F_{\text{rot}}, \qquad (4.13)$$

where

$$F_{\text{rot}} = \frac{4\pi(2\pi I kT)}{(2\pi\hbar)^2} \quad \text{for the two-dimensional rotator,} \qquad (4.14)$$

$$F_{\text{rot}} = \frac{8\pi^2(8\pi^3)^{\frac{1}{2}}(I_x I_y I_z)^{\frac{1}{2}}(kT)^{\frac{3}{2}}}{(2\pi\hbar)^3} \quad \text{for the three-dimensional rotator.}$$

Here I, I_x, I_y, I_z are the principal moments of inertia of the rotators.

From (4.13) it can be seen that only the first factor depends on the value of the spins of the nuclei and the symmetry number σ, the appearance of which implies a forbidden series of states for the two possible types of statistics. This simplification occurs only at high temperatures ($kT \gg \Delta E_{\text{rot}}$), when (4.11) may be approximated by an integral. As regards the partition function in the expression for κ, all the factors $\prod_i (2i+1)$ cancel, since the spins of the nuclei i do not change in an elementary event. However, the ratio of the symmetry number for the activated complex and initial system σ^*/σ remains in the formula for the rate constant. Before discussing the contribution of this ratio to the rate constant, we will consider the question of calculating the factor g^* in expression (2.26). Returning to process (4.4), we will determine the statistical factors l and r for the direct and reverse processes, following Bishop and Laidler [163]. The statistical factor l is equal to the number of different complexes XY^*, which may be formed, if all identical atoms in the system X—Y are distinguishable. The statistical factor for the reverse process is determined analogously. For the calculation of g^* we will assume that the representative point does not intersect the critical surface but is reflected from it. The reflection rate is equal to

$$\kappa_{\text{ref}} = \frac{kT}{2\pi\hbar} r[XY^*]_{\text{eq}}, \qquad (4.15)$$

where $[XY^*]_{\text{eq}}$ is the equilibrium concentration of the activated complexes (equilibrium is maintained by the reflection from the surface).

Since the equilibrium concentration is proportional to the corresponding partition functions we have

$$[XY^*]_{\text{eq}} = \frac{F^*_{XY,\text{trans}} F^*_{XY,\text{vib}} F^*_{XY,\text{rot}}}{F_{X,\text{trans}} F_{Y,\text{trans}} F_{X,\text{vib}} F_{Y,\text{vib}} F_{X,\text{rot}} F_{Y,\text{rot}}}$$

$$\times \frac{\sigma_X \sigma_Y}{\sigma^*_{XY}} \exp(-\beta E_a)[X][Y]. \qquad (4.16)$$

Substituting (4.16) into (4.15) we express κ_{ref} in terms of the concentration of the reactants. However, in accordance with the assumptions of the transition state method, the reflection rate must be equal to the reaction rate. Comparing κ_{ref} with κ from (2.26) we find g^*. In addition, we must take into consideration the following theorem [163]: the ratio of the statistical factors equals the ratio of the symmetry numbers of the reaction products, i.e.

$$l/r = \sigma/\sigma^*. \tag{4.17}$$

Taking this relationship into account, we formulate the following rules for the calculation of g^*. The number of independent ways of forming the activated complex is equal to the statistical factor r of its decomposition into the reactants, if the rotational partition function is calculated with regard for the symmetry number (i.e. like nuclei are considered indistinguishable). If the rotational partition function, introduced in (2.26), is calculated without regard for the symmetry number (like nuclei are distinguishable) then g^* is equal to the statistical factor l of formation of the activated complex. Thus we have

$$g^* = \begin{cases} r, & \text{if } \dfrac{F^*}{F} \sim \dfrac{F_{rot}^*}{F_{rot}} \cdot \dfrac{\sigma}{\sigma^*} \\ l, & \text{if } \dfrac{F^*}{F} \sim \dfrac{F_{rot}^*}{F_{rot}}. \end{cases} \tag{4.18}$$

Table 1, taken from the literature [163], illustrates this discussion, with some reactions as examples. The problem of calculating g^* is also considered in the literature [461, 464].

TABLE 1

Statistical factors and symmetry numbers for some reactions

Reactions	l^*	r	σ $(\sigma_X \sigma_Y)$	σ^*
$H + H_2 \to (H \cdots H \cdots H)^*$	2	2	2	2
$H + D_2 \to (H \cdots D \cdots D)^*$	2	1	2	1
$H + HD \to (H \cdots H \cdots D)^*$	1	1	1	1
$H + DH \to (H \cdots D \cdots H)^*$	1	2	1	2
$\begin{array}{c}CH_2\\ \diagup \diagdown \\ CH_2-CH_2\end{array} \to \begin{pmatrix}H\cdots CH_2\\ \vdots \diagup \\ CH-CH_2\end{pmatrix}^* \to$ $\to CH_3-CH=CH_2$	12	2	6	1

5. The isotope effect

By the isotope effect is meant the dependence of the rate constant of an elementary process on the isotopic composition of the molecules. An investigation of the isotope effect can be carried out in a fairly general form, since the isotopic variation in the masses of the atoms does not affect the interatomic interaction potential, i.e. the form of the potential energy surfaces. This conclusion is exact only to the degree that one may speak, in general, of potential energy surfaces, i.e. within the framework of the adiabatic approximation. Since varying the masses of the atoms affects only the kinetic energy of the molecule, it is expedient, to begin with, to consider the classical approximation, when the kinetic and potential energies of the nuclei can be considered separately without coming into conflict with the uncertainty principle.

Turning to the basic formula (2.8) and assuming that the coefficients g_{ij} in (2.2) do not depend on the coordinates, we rewrite the expression for κ in the form

$$\kappa = \langle \dot{q}_r \rangle P_q^\star, \tag{5.1}$$

where $\langle \dot{q}_r \rangle$ is the mean velocity of the reaction coordinate through the critical surface in the direction $q_r > q_r^\star$; $P_q^\star$ is the probability of achieving the critical configuration.

The factors in (5.1) can be represented in the form

$$\langle \dot{q}_r \rangle = \frac{\int \dot{q}_r \exp(-\beta T)\, dp_1 \ldots dp_s}{\int \exp(-\beta T)\, dp_1 \ldots dp_s}, \tag{5.2}$$

$$P_q^\star = \frac{\int \exp(-\beta U^\star) \prod_{i \neq r} dq_i}{\int \exp(-\beta U) \prod_i dq_i}. \tag{5.3}$$

The integral in the numerator of (5.2) is taken over the region

$$(-\infty < \dot{q}_1 < \infty \ldots 0 < \dot{q}_r < \infty \ldots -\infty < \dot{q}_s < \infty),$$

and in the denominator over an infinite range of velocity variables. The integrals in (5.3) extend over the whole region of the coordinate variables.

From (5.1) it can be seen that the ratio of the rate constants of the isotopically-substituted molecules is determined by the ratio of the mean velocities of motion along the reaction coordinate. We substitute the variables p_i in (5.2) by the variables $\dot{q}_i$. The Jacobian of such a transformation is equal to the determinant of the matrix a_{ij}

$$\frac{D(p_1 \ldots p_s)}{D(\dot{q}_1 \ldots \dot{q}_s)} = \|a_{ij}\| = A. \tag{5.4}$$

Next, we go over from the variables $\dot{q}_i$ to new variables $\dot{q}_i'$, with a view to separating out the motion of the reaction coordinate, i.e. to representing the kinetic energy in the form in which it appears in relation (2.11). This transformation with a unit Jacobian is such that

$$\dot{q}_r = \dot{q}'_r$$

$$\dot{q}_m = \dot{q}'_m + \frac{A_{rm}}{A_{rr}} \dot{q}'_r, \qquad (5.5)$$

where A_{rm} is the minor of the determinant A corresponding to the element a_{rm}.

In the variables $\dot{q}'_m$, the expression for the kinetic energy takes the form

$$T = \tfrac{1}{2} m_r (\dot{q}'_r)^2 + \tfrac{1}{2} \sum_{n,m \neq r} a_{nm} \dot{q}'_n \dot{q}'_m; \qquad m_r = \frac{A}{A_{rr}}. \qquad (5.6)$$

Substituting (5.6) into (5.2) and separating out the integration over $\dot{q}'_r$ we find

$$\langle \dot{q}_r \rangle = \frac{\int_0^\infty \dot{q}_r \exp(-\tfrac{1}{2}\beta m_r \dot{q}_r^2)\,d\dot{q}_r}{\int_{-\infty}^\infty \exp(-\tfrac{1}{2}\beta m_r \dot{q}_r^2)\,d\dot{q}_r} = (2\pi m_r \beta)^{-\tfrac{1}{2}}. \qquad (5.7)$$

Thus we obtain for the ratio of the rate constants of two isotopically substituted molecules in the classical case ($\hbar\omega_i \ll kT$)

$$\frac{\kappa'}{\kappa''} = \frac{\langle \dot{q}_r \rangle'}{\langle \dot{q}_r \rangle''} = \left(\frac{m''_r}{m'_r}\right)^{\tfrac{1}{2}}. \qquad (5.8)$$

As an illustration, we will now quote the results of calculating the effective masses for some simple cases. If the change in the distance between atoms A and B is considered as the reaction coordinate q_r, then m_r is equal to the reduced mass of the corresponding pair

$$m_r = \frac{m_A m_B}{m_A + m_B}. \qquad (5.9)$$

For a more complicated case, when q_r can be expressed in the form of a linear combination of the distances between atoms A–B and B–C the result is such that [160], [161]

$$\frac{1}{m_r} = \frac{m_A + m_B}{m_A m_B} + \frac{2}{m_B}\zeta^{\tfrac{1}{2}} + \frac{m_B + m_C}{m_B m_C}\zeta, \qquad (5.10)$$

where the coefficient ζ is defined by the contribution of the distance R_{BC} to the reaction coordinate

$$q_r = R_{AB} - \zeta R_{BC}. \qquad (5.11)$$

The calculation of m_r for the most general case, when q_r is expressed in the form of a linear combination of the bond lengths R_{AB}, R_{BC}, and the angle ϕ_{ABC}, was made by Gatz [233]. In his work Gatz investigated, in particular, the contribution of the different coordinates to the ratio κ'/κ'', for various models of the activated complex.

From (5.9) and (5.10) it can be seen that, on reduction of the mass of any atom whose displacement contributes to the reaction coordinate, the effective mass m_r is decreased and the reaction rate is increased. This effect of isotopic substitution on the reaction rate is called the primary isotope effect.

Before going over to consider the quantum case, we will obtain a relation, useful in future studies, between the effective mass and other properties of the activated complex. Comparing (5.1) with (2.16) and dropping factors independent of the mass, we find

$$m_r^{-\frac{1}{2}} \sim \frac{F_{\text{trans}}^{\star} F_{\text{vib}}^{\star} F_{\text{rot}}^{\star}}{F}. \tag{5.12}$$

The mass dependence of the partition function of the initial system is calculated in the following way. Integration in a Cartesian system of coordinates gives a factor $m_k^{\frac{3}{2}}$ for each component of the kth atom's momentum. Taking it into account that $F_{\text{vib}}^{\star}$, $F_{\text{rot}}^{\star}$ and $F_{\text{trans}}^{\star}$ are proportional respectively to $(\prod_{i=1}^{s-1} \omega_i^{\star})^{-1}$, $(I_x^{\star} I_y^{\star} I_z^{\star})^{\frac{1}{2}}$ and $(\sum_{k=1}^{n} m_k) = M$, we obtain the relation

$$\left(\sum_{k=1}^{n} m_k\right)^{\frac{3}{2}} (I_x^{\star} I_y^{\star} I_z^{\star})^{\frac{1}{2}} \prod_{k=1}^{n} m_k^{-\frac{3}{2}} \sim m_r^{-\frac{1}{2}} \prod_{i=1}^{s-1} \omega_i^{\star}. \tag{5.13}$$

If an analogous derivation is repeated for a stable molecule, separating out the vibrations from the rotation and calculating the partition function, first in Cartesian coordinates and secondly in rotational and vibrational coordinates, then a somewhat different expression is obtained

$$\left(\sum_{k=1}^{n} m_k\right)^{\frac{3}{2}} (I_x I_y I_z) \prod_{k=1}^{n} m_k^{-\frac{3}{2}} \sim \prod_{k=1}^{s} \omega_k, \tag{5.14}$$

in which the ω_k are the frequencies of the normal vibrations of the molecule (the Redlich–Teller formula [10], [15]).

The expressions (5.13) and (5.14) can be represented in a single form, if, to the coordinate q_r is attributed a certain frequency $\omega_r^{\star}$, whose dependence on the effective mass is determined by the relation

$$\omega_r^{\star} \sim m_r^{-\frac{1}{2}}. \tag{5.15}$$

Such a dependence is obtained, for example, for the imaginary frequency characterizing the curvature of a potential barrier of the type shown in Fig. 1. The effective frequency in the expression for the unimolecular reaction rate in the Slater theory (see Chapter 4), also satisfies this relation, although a barrier of the saddle type is generally absent in this model.

Formulae (5.13) and (5.14) enable us to express the ratio of the translational and rotational partition functions, in the expression for the ratio of the constants κ'/κ'', in terms of the vibration frequencies of the initial and activated complexes.

$$\left(\frac{F^\star_{\text{trans}}F^\star_{\text{rot}}}{F_{\text{trans}}F_{\text{rot}}}\right)'\left(\frac{F_{\text{trans}}F_{\text{rot}}}{F^\star_{\text{trans}}F^\star_{\text{rot}}}\right)'' = \left(\frac{\omega^\star_r \prod_{k=1}^{s-1} \omega^\star_k}{\prod_{k=1}^{s} \omega_k}\right)' \left(\frac{\prod_{k=1}^{s} \omega_k}{\omega^\star_r \prod_{k=1}^{s-1} \omega^\star_k}\right)''. \qquad (5.16)$$

The relation gives nothing new from the classical point of view, since it is identical with (5.8). However, from the quantum point of view, it permits a considerable simplification in the investigation of the dependence of κ'/κ'' on the isotopic constitution of the molecules, reducing it to an investigation of the dependence of κ'/κ'' on the frequencies of small vibrations. On the basis of the general formula (2.26), for the quantum case, we have

$$\frac{\kappa'}{\kappa''} = \left(\frac{F^\star_{\text{trans}}F^\star_{\text{rot}}}{F_{\text{trans}}F_{\text{rot}}}\right)' \left(\frac{F^\star_{\text{vib}}}{F_{\text{vib}}}\right)' \left(\frac{F_{\text{trans}}F_{\text{rot}}}{F^\star_{\text{trans}}F^\star_{\text{rot}}}\right)'' \left(\frac{F_{\text{vib}}}{F^\star_{\text{vib}}}\right)'' \exp(-\beta E'_a + \beta E''_a). \qquad (5.17)$$

Considering the rotation as classical, we will express the ratio of the translational and rotational partition functions by formula (5.16), at the same time substituting the quantum vibrational partition functions of the initial system and activated complex. In this way we obtain

$$\frac{\kappa'}{\kappa''} = \frac{\dfrac{\omega^{\star'}_r}{\omega^{\star''}_r} \prod_{k=1}^{s} \dfrac{\omega^{\star'}_k}{\omega^{\star''}_k} \dfrac{[1-\exp(-\beta\hbar\omega^{\star''}_k)]}{[1-\exp(-\beta\hbar\omega^{\star'}_k)]} \cdot \exp(-\beta E^{\star'}_z + \beta E^{\star''}_z)}{\prod_{k=1}^{s} \dfrac{\omega'_k}{\omega''_k} \dfrac{[1-\exp(-\beta\hbar\omega''_k)]}{[1-\exp(-\beta\hbar\omega'_k)]} \cdot \exp(-\beta E'_z + \beta E''_z)}. \qquad (5.18)$$

The ratio of the rate constants determined by this formula has a rather complicated temperature dependence. In the overwhelming majority of cases, however, the situation is such that the reaction rate for light isotopes is higher than the reaction rate for heavy isotopes. Not only the ratio of the frequencies $\omega^\star_r$ contributes favourably to this, but also the ratio of the shifts in the zero-point vibrational energy; the absolute magnitude of the energy difference of the zero-point vibrations of the activated complex and the reactants is usually larger for light isotopes, since, for these, the zero-point energy itself is larger. Nevertheless, in principle cases are possible in which the rate constant for the lighter isotopes is lower (the inverse isotope effect) [161], [162]. The maximum magnitude of the inverse and direct isotope effects can be obtained from the basic equation (5.18), assuming the complete absence of a displacement in the zero-point energy levels in the transition from the reactants to the activated complex. Such an estimate, however, is too rough and gives what is known to be an overestimate of the magnitude of the isotope effect.

It can be shown that the isotopic substitution does not directly change the frequency factor $\omega^\star_r$, but only affects the vibrational partition function.

The corresponding isotope effect is called the secondary isotope effect in the literature. We will consider this in more detail when studying non-equilibrium unimolecular reactions (see Chapter 5).

The relation (5.18) simplifies somewhat in the case when the variation of the frequencies for an isotope substitution $\delta\omega_k$ satisfies the condition $\hbar\delta\omega_k\beta \ll 1$. For definiteness in future we will assume that $\delta\omega_k > 0$, i.e. by ω_k will be understood the frequencies of the lighter molecule. Then each factor of the product in (5.18) can be expanded as a series in $\delta u_k = \beta\hbar\delta\omega_k$ and, if the expansion is limited to the first two terms, represented as

$$\frac{u_k}{u_k + \delta u_k} \exp\left(\frac{\delta u_k}{2}\right) \frac{1 - \exp(-u_k - \delta u_k)}{1 - \exp(-u_k)} = 1 + G(u_k)\delta u_k, \tag{5.19}$$

where $G(u)$ is a function of the form $G(u) = \frac{1}{2} - 1/u + (e^u - 1)^{-1}$, which is tabulated in the literature [159].

Thus the ratio of the constants $\kappa/(\kappa + \delta\kappa)$ for light and heavy isotopes can be written

$$\frac{\kappa}{\kappa + \delta\kappa} = \frac{\omega_r^\star}{\omega_r^\star + \delta\omega_r^\star}\left[1 + \sum_{k=1}^{3n-6} G(u_k)\delta u_k - \sum_{k=1}^{3n-7} G(u_k^\star)\delta u_k^\star\right]. \tag{5.20}$$

On the basis of this formula, it is not difficult to obtain the first correction to the classical formula (5.8) under the condition $\beta\hbar\omega_k \ll 1$. Expanding $G(u)$ in a series, we get

$$G(u) = 1 + \frac{\hbar^2}{24(kT)^2}[\omega^2 - (\omega + \delta\omega)^2]. \tag{5.21}$$

Now advantage can be taken of the fact that the quantity $\sum_{k=1}^{3n-6} \omega_k^2$ is the sum of the eigenvalues of the matrix determining the frequencies of small oscillations, it is therefore equal to the sum of the diagonal elements of that matrix $\sum_{i=1}^{3n-6} \kappa_{ii}/m_i$, where the κ_{ii} are the diagonal force constants. Thus, for the ratio of the partition functions of the reactants in eqn (5.18), in the limit $\beta\hbar\omega_i \ll 1$ we obtain

$$1 + \frac{\hbar^2}{24(kT)^2} \sum_{k=1}^{3n-6} \left(\frac{1}{m_i} - \frac{1}{(m_i + \delta m_i)}\right)\kappa_{ii}. \tag{5.22}$$

Unfortunately, the ratio of the frequencies and partition functions for the activated complex cannot be presented in such a simple form, since the frequencies $\omega_i^\star$ are found by solution of the vibrational problem with regard for the subsidiary boundary conditions and a fixed value of the reaction coordinate. The effect of these limitations can be made clear if, at the same time, the variation of the transmission coefficient on isotopic substitution is taken into account. The connection between the transmission coefficient and the quantum corrections as a result of the shift in the zero-point level is understandable if account is taken of the fact that both these effects are connected with the quantum delocalization of the atoms—with the zero-point vibrations about the position of equilibrium for the stable configuration and with the tunnel penetration of the potential barrier for the unstable configuration.

Staying within the framework of a consistent harmonic model of the activated complex, we will assume that the reaction coordinate describes the motion of the representative point through a parabolic barrier. For this, the transmission coefficient is given by expression (3.9). For the ratio $\chi/(\chi+\delta\chi)$ we obtain

$$\frac{\chi}{\chi+\delta\chi} = 1 + \frac{\hbar^2}{24(kT)^2}[(\omega^*)j^2 - (\omega^*+\delta\omega^*)j^2]. \tag{5.23}$$

If this factor is introduced into the first product in (5.18), then the theory about the sum of the eigenvalues can again be employed, and for the ratio of the partition functions of the activated complexes, the expression

$$\frac{(F^*)'}{(F^*)''} = 1 + \sum_{i=1}^{3n-6} \frac{\hbar^2}{24(kT)^2} \left(\frac{1}{m_i} - \frac{1}{m_i+\delta m_i}\right)\kappa_{ii}^* \tag{5.24}$$

is obtained, in which only the diagonal force constants of the activated complex enter. Thus we finally obtain

$$\frac{\kappa}{\kappa+\delta\kappa} = 1 + \sum \left(\frac{1}{m_i} - \frac{1}{m_i+\delta m_i}\right)\frac{\hbar^2\kappa_{ii}^*}{24(kT)^2} - \sum \left(\frac{1}{m_i} - \frac{1}{m_i+\delta m_i}\right)\frac{\hbar^2\kappa_{ii}}{24(kT)^2} \tag{5.25}$$

for the ratio of the rate constants.

As was shown earlier, a transmission coefficient of the form (3.9) does not take into account the possible reflection of the representative point after intersection of S^*. If this reflection were taken into account, the dependence of χ on the mass of the representative point would be stronger than that given by relation (3.9), but ultimately relation (5.25) would be inapplicable. The process of vibrational excitation of molecules, in which the probability of a transition, formally analogous to the transmission coefficient, depends very strongly on the effective mass of the representative point may serve as a simple example of this type of situation.

6. The principle of detailed balance and the transition state method

The classical equations of motion or the wave equation, if a magnetic field does not enter into them, are reversible in time. This means that, on changing the sign of the time and the initial velocities, the classical representative point will traverse the original trajectory in the reverse direction. In a quantum treatment, the wave equation retains its form if t is changed to $-t$ and, at the same time, the wavefunction Ψ changes to Ψ^*. The latter requirement implies that the transition probability from a state a to a state b is equal to the transition probability from state b^* to state a^*. The symbol * in the case under consideration means that the full sets of quantum numbers a and a^* characterizing the system, differ in the sign of the projection of the angular momentum. This property of the mechanical equations—their microscopic reversibility—enables a useful relation between the macroscopic magnitudes of quantities in the direct and reverse processes to be obtained.

For definiteness, we will return to the bimolecular reaction (4.4). By $\mathbf{i}$ and $\mathbf{f}$, we denote the set of quantum numbers characterizing the internal states of the initial and final molecules. In the set $\mathbf{i}$ (or $\mathbf{f}$) are quantum numbers i (or f) which remain unchanged on changing the sign of the time and numbers i' (or f') which change sign.

Considering the collision of X and Y, we will surround the molecule Y by a closed surface and will follow a certain trajectory L, which intersects the surface and along which X and Y approach one another. For each trajectory L, describing the relative motion of X and Y, we will compare the partial reaction rate $\delta \kappa_L(\mathbf{i}, \mathbf{f})$, which we define as the product of the flux in three-dimensional configuration space and the reaction probability $P_L(\mathbf{i}, \mathbf{f})$ ascribed to this trajectory.†

$$\delta\kappa_L(\mathbf{i}, \mathbf{f}) = P_L(\mathbf{i}, \mathbf{f}) \frac{d\Gamma_{\text{X-Y, trans}}}{dt}. \tag{6.1}$$

It is convenient to express the element of phase volume $d\Gamma_{\text{X-Y,trans}} = dp_x\, dp_y\, dp_z\, dx\, dy\, dz/(2\pi\hbar)^3$ by differential variables, which are integrals of motion in the region, where the interaction of X and Y may be neglected. The magnitudes of the angular momentum l of the relative motion and the component of the momentum l_z along the z axis of the system's coordinates, which is fixed in space, are appropriate. We will denote the angle coordinates conjugate to these two action variables by α and α_z respectively. The angle α describes the orientation of the intermolecular axis $\mathbf{R}$ relative to the line ON, fixed in the plane perpendicular to the vector $\mathbf{l}$, and the angle α_z describes the orientation of the component of the vector $\mathbf{l}$ in a plane xy, fixed in space, with ON. Furthermore, it is sometimes expedient to introduce, instead of the variable l_z, the angle θ defined by the relation $\cos\theta = l_z/l$ (Fig. 2).

---

† We will consider the relative motion of the molecules as classical, since this is usually justified. Such an approach moreover, enables the introduction of a series representation, which will prove to be useful in future in the account of the theory of bimolecular reactions.

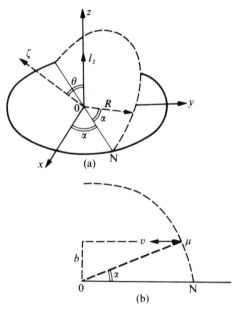

FIG. 2. Angular variables, characterizing the relative motion of two particles: (a) the moving system of coordinates, built on the vectors **l** and **R**, can be considered as a system (x, y, z), whose orientation is defined relative to a stationary system by the three Eulerian angles [36]. Here, the z axis is directed along **l**, the x axis along **R**. The line 0N is the line of nodes; the angles θ, α, α_z coincide with the Eulerian angles θ, ψ, and ϕ; (b) relative motion in the plane perpendicular to **l**.

In the new variables, the element of phase volume can be written in the form

$$d\Gamma_{\text{X-Y,trans}} = \frac{dl\, d\alpha\, dl_z\, d\alpha_z\, dp_R\, dR}{(2\pi\hbar)^3}. \tag{6.2}$$

It is clear that as R increases without limit, the velocity of the representative point is directed along **R**. Therefore the differentiation with respect to time in (6.1) should refer to the coordinate R. In the terminology of the transition state method, this means that R is considered as the reaction coordinate. In the case under consideration such an ascription is completely correct, but is reached by introducing an unknown transition probability. Thus from (6.1) we obtain

$$\delta\kappa_L(\mathbf{i}, \mathbf{f}) = \frac{P_L(\mathbf{i}, \mathbf{f})\, dl\, d\alpha\, dl_z\, d\alpha_z\, dE_{\text{trans}}}{(2\pi\hbar)^3} \tag{6.3}$$

where, as before, dE_{trans} denotes the differential of the relative kinetic energy. Since the assignment of the quantum numbers **i** fixes the internal energy of the molecules X and Y, dE_{trans} can be regarded as the differential of the total energy dE.

We will now go over from the variables l and l_z to the impact parameter b and the angle θ (see Fig. 2(b)), where

$$l = p_R b$$
$$l_z = p_R b \cos\theta. \tag{6.4}$$

The differentiation of l is carried out for constant p_R, the differentiation of l_z for constant magnitude l. In this way, we find the following expression for the element of current of the representative points in the forward direction for the reaction (4.4),

$$\delta\kappa_L(\mathbf{i},\mathbf{f}) = \frac{P_L(\mathbf{i},\mathbf{f})\, d\alpha\, d\alpha_z\, \sin\theta\, d\theta\, b\, db\, p^2\, dE_{trans}}{(2\pi\hbar)^3}. \tag{6.5}$$

In order to find the connection between the currents in the forward and reverse directions, we sum $\delta\kappa(\mathbf{i},\mathbf{f})$ over all variables i', f' occurring in the set $\mathbf{i}, \mathbf{f}$ and changing sign when the time changes sign, and also over the variables α, α_2, θ characterizing the initial relative configuration of the system $X+Y$. To this end, first of all we introduce the mean probability

$$P_{if} = \int \frac{d\alpha\, d\alpha_z\, \sin\theta\, d\theta}{8\pi^2} \frac{1}{g_i} \sum_{i',f'} P_L(\mathbf{i},\mathbf{f}), \tag{6.6}$$

which is obtained from $P_L(\mathbf{i},\mathbf{f})$ by summing over all final states and averaging over initial degenerate states.† The factor $8\pi^2$ is the normalization factor for the set of angles α, α_z, and θ and for the quantum numbers i', the corresponding degrees of degeneracy g_i serves likewise.

The mean probability (6.6) depends on the quantum numbers i and f, which remain from the set $\mathbf{i}, \mathbf{f}$, and also on the initial velocity v_i (or on the initial relative kinetic energy of motion E_{trans}) and on the impact parameter b. We sum (6.5) over all trajectories leading from the initial state i to the final state f under the condition of the constancy of the total energy E. In the notation of eqn (6.5), the total current can be represented in the form

$$\sum_L \delta\kappa_L(\mathbf{i},\mathbf{f}) = \frac{dE}{(2\pi\hbar)^3} g_i 2\pi \int_0^\infty P_{if}(v_i, b) p_i^2 b\, db. \tag{6.7}$$

In the sum on the left, all trajectories are taken into account which lead from the initial state i to the final state f. To each trajectory in (6.7) there corresponds a trajectory which, for the reverse reaction, leads from the state f to the state i, since the set of variables i and f does not change on

---

† It is expedient to carry out the averaging over α_z only in the case when the system of reacting atoms is isotropic. Isotropy is broken, for example, in reactions in beams [258].

changing the sign of the time. Therefore an analogous sum over trajectories for the reverse reaction having the form

$$\sum_L \delta\kappa_L(\mathbf{f}, \mathbf{i}) = \frac{dE}{(2\pi\hbar)^3} g_f 2\pi \int_0^\infty P_{fi}(v_f, b') p_f^2 b' \, db' \tag{6.8}$$

is equal to the sum (6.7). Into formulae (6.7) and (6.8) enter integrals of the mean transition probability over the impact parameters, which have the sense of an effective target area over which a 'hit' occurs, in reactions carrying the system over from an initial state i to a final state f (or vice versa). Hence the expediency is apparent of introducing the idea of effective cross-sections σ of the reaction, defined by the following relations:

$$\sigma_{if}(v_i) = \sigma_{if}(E_{\text{trans}}) = 2\pi \int_0^\infty P_{if}(v_i, b) b \, db, \quad E_{\text{trans}} = \frac{\mu_i v_i^2}{2}, \tag{6.9}$$

$$\sigma_{fi}(v_f) = \sigma_{fi}(E'_{\text{trans}}) = 2\pi \int_0^\infty P_{fi}(v_f, b) b \, db, \quad E'_{\text{trans}} = \frac{\mu_f v_f^2}{2}, \tag{6.10}$$

where μ_i and μ_f denote the reduced masses of the partners in the initial and final states (for the reaction in (4.4), $\mu_i = \mu_{X-Y}$, $\mu_f = \mu_{U-Z}$).

The equality on the left-hand sides of (6.9) and (6.10), following from the microscopic reversibility of the mechanical equations, can be written in the form of a relation between the cross-sections of the direct and reverse processes. Taking (6.9) and (6.10) into account, from the relation

$$\sum_L \delta\kappa(\mathbf{i}, \mathbf{f}) = \sum_L \delta\kappa(\mathbf{f}, \mathbf{i})$$

we obtain

$$g_i p_i^2 \sigma_{if}(v_i) = g_f p_f^2 \sigma_{fi}(v_f). \tag{6.11}$$

The quantities v_i, p_i, v_f, p_f are connected by the conservation law

$$E_{\text{trans}} - E'_{\text{trans}} = \frac{\mu_i v_i^2}{2} - \frac{\mu_f v_f^2}{2} = \frac{p_i^2}{2\mu_i} - \frac{p_f^2}{2\mu_f} = E_f - E_i, \tag{6.12}$$

where μ_i and μ_f are the reduced masses of the initial and final pairs; E_i and E_f are the total internal energies of the systems.

If, in studying the integral relation (6.8), not only the law of conservation is used but also the other integrals of motion, then relations of the type (6.10) can be even more detailed. In particular, if an integral of motion, such as the total angular momentum J, is taken into account, then it is possible to determine a well-defined correlation between the final regions of integration over b and b' in eqns (6.7) and (6.8). Then for each pair of such regions, a relation of the type (6.10) will be fulfilled, where the cross-section carries an additional index J (this is seen in detail in § 31).

We will consider the case where the states i and f are given and the distribution of the relative momentum is the equilibrium distribution. Such a situation arises, for example, in studying the establishment of equilibrium over the internal degrees of freedom in the gaseous phase, which is characterized by the equilibrium distribution over the velocities. For this case the distribution function coincides with the normalized Maxwell distribution:

$$f_{\text{X–Y,trans}} = \frac{\exp(-p^2/2\mu kT)}{F_{\text{trans}}}, \qquad F_{\text{trans}} = \frac{(2\pi\mu kT)^{\frac{3}{2}}}{(2\pi\hbar)^3}. \qquad (6.13)$$

For κ_{if} we obtain

$$\kappa_{if} = \left(\frac{8kT}{\pi\mu}\right)^{\frac{1}{2}} \int \sigma_{if}(E_{\text{trans}}) \exp\left(-\frac{E_{\text{trans}}}{kT}\right) \frac{E_{\text{trans}}\, dE_{\text{trans}}}{(kT)^2}. \qquad (6.14)$$

The rate constant for the reverse process equals

$$\kappa_{fi} = \left(\frac{8kT}{\pi\mu'}\right)^{\frac{1}{2}} \int \sigma_{fi}(E'_{\text{trans}}) \exp\left(-\frac{E'_{\text{trans}}}{kT}\right) \frac{E'_{\text{trans}}\, dE'_{\text{trans}}}{(kT)^2}. \qquad (6.15)$$

In accordance with (6.10) the only differences between the integrals (6.15) and (6.14) consist of a displacement of the energy scale and differences in the factors μ and μ'. Displacing the origin of energy measurement by the amount $E_i - E_f$ in one of the integrals, we obtain the ratio

$$\frac{\kappa_{if}}{\kappa_{fi}} = \left(\frac{\mu_f}{\mu_i}\right)^{\frac{3}{2}} \left(\frac{g_f}{g_i}\right) \exp\left(\frac{E_i - E_f}{kT}\right), \qquad (6.16)$$

$$\mu_i = \mu, \qquad \mu_f = \mu'.$$

This relation expresses the principle of detailed balance for the elementary rate constants of the direct and reverse processes, under the condition that the translational degrees of freedom are characterized by the equilibrium distribution function. Discussion of the principle of detailed balance in its application to specific processes is given in the review by Light, Ross, and Shuler [333].

If the distribution function $f_{\text{X,Y}}(i)$ is known for the internal initial states, then the total rate constant for the bimolecular reaction can be obtained by averaging the normalized current over all initial states i and summing over all final states f. Turning to the definition of the reaction cross-section (6.9) and taking account of relation (6.8) we obtain

$$\kappa = \sum_f \sum_i f_{\text{X,Y}}(i) g_i \int \sigma_{if}(v_i) f_{\text{X–Y,trans}}(v_i) v_i \frac{4\pi p_i^2\, dp_i}{(2\pi\hbar)^3}. \qquad (6.17)$$

This formula establishes the general connection between the rate constant and the cross-section of any process.

If the initial distribution is the equilibrium distribution, then by $f_{X-Y,\text{trans}}$ and $f_{X,Y}(i)$ are understood the Boltzmann distribution functions for the relative velocities and for the internal states of the reactants X and Y.

In conclusion, we note that the conservation condition for the current along the set of trajectories leading from the region of the reactants into the region of the reaction products enables us to obtain a useful relation between the reaction cross-section and certain characteristics of the activated state [333], [353]. The integral equality which was obtained from (2.26) and (6.17) serves as a basis for this relation

$$g^\star \chi \frac{kT}{2\pi\hbar} \frac{F^\star_{XY}}{F_X F_Y} \exp\left(-\frac{E_a}{kT}\right)$$
$$= \sum_f \sum_i \int \sigma_{if}(v_i) v_i \exp\left(-\frac{\mu_i v_i^2}{2kT} - \frac{E_{X,i}}{kT} - \frac{E_{Y,i}}{kT}\right) g_{X,i} g_{Y,i} F_{XY,\text{trans}} \frac{4\pi p_i^2 \, dp_i}{(2\pi\hbar)^3 F_X F_Y}, \quad (6.18)$$

where $E_{X,i}$ is the internal energy of the molecule X in the state i; $g_{X,i}$ is the corresponding statistical weight.

If, for example, the reaction cross-section is known, then the expediency of introducing the concept of the activated state and the possibility of isolating a certain reaction coordinate can be judged by the precision with which these relations are satisfied. The solution to the problem in reverse, namely, the recovery of the reaction cross-section in terms of temperature-dependent rates, calculated by the transition state method, is also of interest. Generally speaking, the solution is not unique, because to one variable T on the left of (6.18) there corresponds a set of variables i and f on the right. If, however, an additional assumption is made concerning the dependence of σ_{ij} on the energy of the different degrees of freedom, then the integral in (6.18) can be inverted and the cross-section thus expressed as a function of the energy.

In particular, advantage can be taken of the fact that, in the exponent on the right-hand side of (6.18), only the total energy of the molecules X and Y occurs (the sum of the internal and relative translational energies), and a relation can be obtained between the total cross-section and features of the energy spectrum of the activated complex. This relation, which is found by applying the inverse Laplace transform to both sides of the equality (6.18), enables the connection between the transition state method and collision theory to be traced.

2

THE EXCHANGE OF VIBRATIONAL AND TRANSLATIONAL ENERGY IN MOLECULAR COLLISIONS

7. General observations on the exchange of energy in molecular collisions

A SYSTEMATIC formulation of problems on the calculation of cross-sections in inelastic molecular collisions accompanied by the excitation or deactivation of vibrations must be based on the solution of a wave equation in which all possible inelastic processes (change of rotational, vibrational, and electronic states of the particles) are taken into account. It is clear that, mathematically, the problem is exceptionally complicated, even if the potential energy surface is known for the system of interacting atoms. As a rule, however, we do not have sufficient information available about the intermolecular interaction which would have to be used as a starting point for the solution of the dynamical problem on the redistribution of the energy in collisions. Hence, a large variety of models are used in the interpretation of the experimental data on vibrational relaxation. At the same time, it is essential that the model representations relate not only to the potentials but also to a sufficiently arbitrary division of the different degrees of freedom into active and inactive, as regards some specific process. The premeditated exclusion of the possibility of electronic or rotational excitations, in studying the problem of vibrational energy transfer, can lead to very substantial errors. Therefore, first it is necessary to ascertain to what degree it is possible to divide these processes, since just such a division enables concrete results to be obtained.

Electronic excitation

The basic parameter determining the probability of the electronic excitation of molecules in collisions is the Massey parameter

$$\xi(R, r, \gamma) = \frac{\Delta U(R, r, \gamma)}{\hbar \alpha v}, \tag{7.1}$$

where ΔU is the adiabatic splitting of the electronic terms; $1/\alpha$ is the characteristic radius of action of the potential; v is the velocity of the nuclei.

In contrast to the one-dimensional case (the problem of inelastic atomic collisions) even the most simple type of collision (between an atom and a diatomic molecule) demands the introduction of three coordinates: the internuclear distance r_{AB} of the diatomic molecule, the distance R_{AB-C}

between the molecule AB and the atom C, and the angle γ characterizing the relative orientation of **R** and **r**. In the coordinate space of three dimensions, there is always a region where the parameter $\xi(R, r, \gamma)$ is known to be small (see § 14 and § 48) and where, in consequence, transitions between different electronic states of the molecular system AB–C are probable. Therefore it is possible to ignore consideration of the electronic excitation only in those cases where the nuclei are energetically forbidden to reach the region where the terms approach one another. Such a situation is possible if the splitting of the electronic terms in the isolated molecule ΔU_{AB} and in the isolated atom ΔU_C are sufficiently large, so that they fulfill the adiabatic criteria

$$\Delta U_{AB}, \quad \Delta U_C \gg \hbar \alpha v. \tag{7.2}$$

In practice, this means that the partners in a collision must be in a non-degenerate electronic ground state. The theory set forth in this chapter is correct for this type of collision.

Vibrational excitation

To estimate the effectiveness of vibrational excitation in molecular collisions, we will consider the simplest example of the collision of a diatomic molecule AB with an atom C. In the adiabatic approximation, the classical Hamiltonian of such a system can be represented in the form

$$H = T + U(R, r, \gamma), \tag{7.3}$$

where T is the kinetic energy of the atoms; U is the energy of interaction of the three atoms.

For future use, it is convenient to represent $U(R, r, \gamma)$ in the form

$$U(R, r, \gamma) = U(r) + W(R, r, \gamma), \tag{7.4}$$

where $U(r)$ is the interaction energy of the atoms A and B in the isolated molecule AB; $W(R, r, \gamma)$ is the interaction energy of the molecule AB with the atom C.

In particular, if the rotation of the molecule AB and the relative motion of AB and C are described in a system of coordinates fixed in space, then T can be written in the form

$$T = \frac{p_r^2}{2M} + \frac{j^2}{2Mr^2} + \frac{p_R^2}{2\mu} + \frac{l^2}{2\mu R^2}, \tag{7.5}$$

where **j** is the angular momentum of the isolated molecule AB; **l** is the relative angular momentum of the colliding pair.

The vector **l** is defined by assigning the modulus l, the magnitude of the component l_z along a z axis, fixed in the space of the coordinate system, and the angle α_z characterizing the direction of the projection of **l** onto the xy plane (see Fig. 2). To specify the orientation of **R**, it is necessary to introduce

§7   TRANSLATIONAL ENERGY IN MOLECULAR COLLISIONS   43

another angle α characterizing its deflection in a plane normal to the vector **l**. The vector **j** and the rotation of the molecular axis (modulus j, component j_z, angles α' and α'_z) are specified in an analogous way. For future use, it is expedient to represent the Hamiltonian H in the form

$$H = H_{\text{AB,vib.rot}} + T_{\text{AB-C}} + W(R, r, \gamma), \tag{7.6a}$$

$$H_{\text{AB,vib.rot}} = H_{\text{AB,vib}} + H_{\text{AB,rot}} = \left[\frac{p_r^2}{2M} + U(r)\right] + \frac{j^2}{2Mr^2}, \tag{7.6b}$$

$$T_{\text{AB-C}} = \frac{p_R^2}{2\mu} + \frac{l^2}{2\mu R^2}. \tag{7.7}$$

In the adopted coordinates, the angle γ is a complicated function of the angles α, α_z, α', α'_z so that the equation of motion takes on a very complex form. This is partly connected with the fact that, in the Hamiltonian (7.6), no account has been taken of the obvious integral of motion, namely the total angular momentum **J** of the system of three particles. Explicit regard for the conservation of **J** would lead to a simplification of the equation of motion, since a number of independent variables could be excluded from the Hamiltonian (7.6). If W does not depend on γ, the conservation of **J** implies the independent conservation of the angular momenta **j** and **l**.

In particular, the Hamiltonian of three particles in the form (7.5) is convenient for the calculation of the partition functions of the free molecules AB and C, when the interaction between AB and C can generally be neglected. The element of phase volume in these coordinates has the form

$$d\Gamma = d\Gamma_{\text{AB,vib}}\, d\Gamma_{\text{AB,rot}}\, d\Gamma_{\text{AB-C,trans}}\, d\Gamma_{\text{AB-C,rot}}, \tag{7.8}$$

where

$$d\Gamma_{\text{AB,vib}} = \frac{dp_r\, dr}{(2\pi\hbar)}, \qquad d\Gamma_{\text{AB,rot}} = \frac{dj\, d\alpha'\, dj_z\, d\alpha'_z}{(2\pi\hbar)^2},$$

$$d\Gamma_{\text{AB-C,trans}} = \frac{dp_R\, dR}{(2\pi\hbar)}, \qquad d\Gamma_{\text{AB-C,rot}} = \frac{dl\, d\alpha\, dl_z\, d\alpha_z}{(2\pi\hbar)^2}.$$

In conditions typical of gas kinetics, the classical approximation, in practice, turns out to be applicable to all degrees of freedom save the vibrational. The calculation of the quantum vibrations can be accomplished easily if by $H_{\text{AB,vib.rot}}$ is understood the quantum Hamiltonian of a rotating molecule and $d\Gamma_{\text{AB,vib}}$ is replaced by Δn, the number of vibrational quantum states.

In order to express H in the relative coordinates R, r, and γ and to take the law of conservation of the total angular momentum **J** into account explicitly, it is necessary to transform to a moving system of coordinates (see Fig. 3). The component J_x of the total angular momentum **J** along the x axis of the

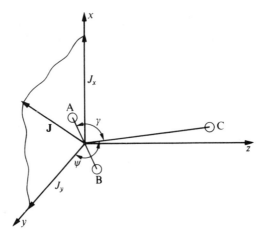

FIG. 3. Angular variables characterizing the motion of three particles.

moving system of coordinates is determined by the motion of the atoms in a fixed plane. We will define the orientation of **R** and **r** in the yz plane by the angles ψ and $\psi+\gamma$ measured relative to some stationary y axis, lying in the same plane. Then we have

$$J_x = A\dot\psi + a(\dot\psi + \dot\gamma), \qquad A = \mu R^2, \qquad a = Mr^2, \qquad (7.9)$$

where M is the reduced mass of AB; μ is the reduced mass of AB and C.

The kinetic energy T_1 of the motion of the atoms in the plane has the form

$$T_1 = \frac{\mu\dot R^2}{2} + \frac{M\dot r^2}{2} + \frac{A\dot\psi^2}{2} + \frac{a(\dot\psi+\dot\gamma)^2}{2}. \qquad (7.10)$$

We now assume that the y axis coincides with the line of intersection of the plane ABC with the plane passing through the x axis and the vector **J**. The corresponding component J_y describes the rotation of the plane ABC. The kinetic energy of this rotation is T_2, i.e. the contribution of the nonplanar motion of the atoms to the total kinetic energy T equals

$$T_2 = \tfrac{1}{2}(J_y)^2(\boldsymbol{I}^{-1})_{yy}, \qquad (7.11)$$

where $\boldsymbol{I}$ is the inertia tensor of the system ABC and $\boldsymbol{I}^{-1}$ is the inverse tensor.

In the chosen system of coordinates, the non-diagonal components of the tensor I_{ik} are absent, if one of the indices refers to the x axis. The inertia tensor

§7   TRANSLATIONAL ENERGY IN MOLECULAR COLLISIONS

I has the form

$$I = \begin{bmatrix} I_{xx} & 0 & 0 \\ 0 & I_{yy} & I_{yz} \\ 0 & I_{zy} & I_{zz} \end{bmatrix}$$

$$= \begin{bmatrix} A+a & 0 & 0 \\ 0 & A\sin^2\psi + a\sin^2(\psi+\gamma) & -A\sin\psi\cos\psi - a\sin(\psi+\gamma)\cos(\psi+\gamma) \\ 0 & -A\sin\psi\cos\psi - a\sin(\psi+\gamma)\cos(\psi+\gamma) & A\cos^2\psi + a\cos^2(\psi+\gamma) \end{bmatrix}$$

(7.12)

The components of the inverse tensor $(I^{-1})_{yy}$ equal

$$(I^{-1})_{yy} = \frac{I_{zz}(A+a)}{\det(I)} = \frac{A\cos^2\psi + a\cos^2(\psi+\gamma)}{Aa\sin^2\gamma}.$$ (7.13)

Substituting (7.13) into (7.11) and taking the law of conservation of the total angular momentum into account,

$$J^2 = (J_x)^2 + (J_y)^2,$$ (7.14)

we obtain

$$T = T_1 + T_2 = \frac{p_R^2}{2M} + \frac{p_R^2}{2\mu} + \frac{1}{2A}(P_\gamma - P_\psi)^2 + \frac{1}{2a}P_\gamma^2 +$$
$$+ \left[\frac{\cos^2\psi}{2a} + \frac{\cos^2(\psi+\gamma)}{2A}\right](J^2 - P_\psi^2)\bigg/\sin^2\gamma,$$ (7.15)

where

$$P_\psi = J_x = A\dot\psi + a(\dot\psi + \dot\gamma), \qquad P_\gamma = a(\dot\psi + \dot\gamma).$$ (7.16)

In such a system of coordinates, the Hamiltonian depends on the total angular momentum J (an integral of motion), on the angle ψ, describing the rotation of the system in the plane, and on the corresponding conjugate momentum P_ψ in addition to the internal coordinates and momenta (p_r, P_R, P_γ, r, R, γ).

An element of phase volume in the relative coordinates has the form

$$d\Gamma = d\Gamma_{AB,vib}\, d\Gamma_{ABC,rot}\, d\Gamma_{AB-C,trans},$$ (7.17)

where

$$d\Gamma_{AB,vib} = \Delta n \approx \frac{dp_r \, dr}{2\pi\hbar},$$

$$d\Gamma_{ABC,rot} = dJ \, d\alpha \, dJ_z \, d\alpha_z \frac{dP_\psi \, d\psi}{(2\pi\hbar)^3},$$

$$d\Gamma_{AB-C,trans} = \frac{dP_\gamma \, d\gamma \, dP_R \, dR}{(2\pi\hbar)^2}.$$

These variables are convenient for calculating the partition function of the activated complex, since the critical surface (and reaction coordinate) is determined only by the relative coordinates (R, γ) which enter $d\Gamma_{AB-C,trans}$.

We will estimate the Massey parameter for a transition between the vibrational levels of the molecule AB. We will assume that the relative energy of AB and C, in the region of their interaction, is appreciably smaller than the dissociation energy D of the molecule AB. In conformity with this, for a thermal collision we will assume.

$$D \gg kT \qquad (7.18)$$

Moreover, the interaction of AB and C cannot markedly change the structure of the vibrational terms in AB, so that the energy difference ΔU must be very close to the difference in the vibrational terms ΔE_{vib} for the free molecule AB. As regards the magnitude of ΔE_{vib}, for single-quantum transitions $\Delta E_{vib} = \hbar\omega$, and the parameter α in (7.1) can be connected with the frequency ω_0 of the fundamental vibrations of the molecule AB, its reduced mass M and the dissociation energy D. With this aim, we will consider the Morse potential, which gives the qualitatively correct dependence of the electronic ground term of the diatomic molecule on the internuclear distance:

$$U(r) = D[1 - \exp(-\alpha' r + \alpha' r_e)]^2 \qquad (7.19)$$

In expression (7.19), the potential energy is measured from its minimum at $r = r_e$, close to which $U(r)$ can be represented in the form

$$U(r) = U(r_e) + D(\alpha')^2(r - r_e)^2, \qquad \alpha'(r - r_e) \ll 1. \qquad (7.20)$$

Close to the minimum, $U(r)$ is replaced by the potential of a harmonic oscillator with a force constant $\kappa = 2D(\alpha')^2$. Hence we find the connection between the frequency of small vibrations ω_0 and the potential parameters D and α' to be

$$\omega_0 = \sqrt{\left(\frac{\kappa}{M}\right)} = \alpha'\sqrt{\left(\frac{2D}{M}\right)}. \qquad (7.21)$$

For increasing amplitude of oscillation, anharmonic corrections appearing in the dependence of the vibrational frequency on the total energy E_{vib} become

§7 TRANSLATIONAL ENERGY IN MOLECULAR COLLISIONS

important. To clarify the dependence of $\omega = \omega(E)$, we will study, firstly, the classical problem of the vibrations of a Morse oscillator. The equation of motion has the form

$$\frac{M\dot{x}^2}{2} + D[1 - \exp(-\alpha' x)]^2 = E, \qquad x = r - r_e. \tag{7.22}$$

The solution to eqn (7.22) is conveniently presented in a form such that maximum compression of the bond in the molecule AB occurs at time $t = 0$, i.e. the time is measured from the turning point of the motion of the nuclei under the repulsive part of the potential. In this case we have

$$\exp(\alpha' x) = \begin{cases} \dfrac{(1 - \cos\theta \cos\omega t)}{\sin^2\theta}, & \sin\theta = \left(\dfrac{D-E}{D}\right)^{\frac{1}{2}}, \quad E \leqslant D. \quad (7.23a) \\ \omega = \omega_0 \sin\theta \\[4pt] \dfrac{(\cosh\theta \cosh\omega t - 1)}{\sinh^2\theta}, & \sinh\theta = \left(\dfrac{-D+E}{D}\right)^{\frac{1}{2}}, \quad E \geqslant D. \quad (7.23b) \\ \omega = \omega_0 \sinh\theta \end{cases}$$

The first equality is correct for bound states ($E \leqslant D$), consequently E has the meaning of a vibrational energy $E = E_{\text{vib}}$; the second equality applies to dissociation. In the first case, the motion is periodic with a frequency equal to

$$\omega = \omega_0 \sqrt{\left(\frac{D - E_{\text{vib}}}{D}\right)}. \tag{7.24}$$

In the second case, the atoms separate to infinity and the parameter $1/\omega$ has the meaning of a collision time, during the course of which the atoms A and B effectively interact. In addition, the quantity E can be identified with the relative translational energy of the pair scattering to infinity, i.e. it is assumed that $E = E_{\text{trans}}$.

The solution to the quantum problem of the Morse oscillator gives the expression for the energy levels [37]

$$E_n = \hbar\omega_0(n + \tfrac{1}{2}) - \hbar\omega_0 x_e(n + \tfrac{1}{2})^2, \qquad x_e = \frac{\hbar\omega_0}{4D}, \tag{7.25}$$

in which n runs from 0 to $n_{\max} = N$, determined by the condition $E(N+1) \leqslant E(N)$. Under the condition $D/\hbar\omega_0 \gg 1$ the value of N can be found from (7.25) by considering E_n as a continuous function of n and finding its maximum

$$\left.\frac{dE_n}{dn}\right|_{n=N} = 0, \qquad N = \frac{2D}{\hbar\omega}. \tag{7.26}$$

Formula (7.26) gives the total number of bound states of the Morse oscillator. The transition frequency $n+1 \to n$ is determined by the expression

$$\omega_{n+1,n} = \frac{E_{n+1} - E_n}{\hbar} = \omega_0(1 - 2x_e n). \qquad (7.27)$$

It is not difficult to verify that if n in (7.27) is expressed in terms of E by formula (7.25), then the expression obtained agrees with that in (7.24) for $n \gg 1$.

This substitution illustrates the transition from the quantum formula for the energy and frequency to the classical formula for large vibrational numbers n.

Thus, for single-quantum vibrational transitions we find

$$\xi(E_{\text{vib}}) = \frac{\omega}{\alpha \bar{v}} \sim \left(\frac{D - E_{\text{vib}}}{kT}\right)^{\frac{1}{2}} \left(\frac{\mu}{M}\right)^{\frac{1}{2}}, \qquad (7.28)$$

where μ is the reduced mass of AB+C.

Hence it follows that for low vibrational levels ($E_{\text{vib}} \ll D$) under the condition of not too high temperatures ($kT \ll D$) and in the case when the mass of the atom C is comparable to or greater than the mass of AB ($\mu \geqslant M$), the collisions bear an adiabatic character in relation to the excitation of vibrations of AB.

In the problem of the vibrational excitation of molecules, there is a small parameter which can be employed in the construction of a theory of excitations. This parameter is the ratio of the amplitude x of the vibrations of the nuclei to the radius of action of the intermolecular force $1/\alpha$. Since the amplitude of vibration x is directly proportional to $n^{\frac{1}{2}}$, where n is the vibrational quantum number and is inversely proportional to the quantity $(M\omega_0)^{\frac{1}{2}}$, with given parameters of the electronic term, from dimensional considerations it follows that

$$\alpha x \sim n^{\frac{1}{2}} \left(\frac{m_e}{M}\right)^{\frac{1}{4}}. \qquad (7.29)$$

The ratio $(m_e/M)^{\frac{1}{4}}$ is the usual parameter in the adiabatic approximation [4]. Since, near the dissociation limit, it is evident that $\alpha x \sim 1$, from (7.29) we find the connection between the total number of vibrational quanta n_{max} and the mass M of the nuclei for molecules whose dissociation energy is of the order of the ionization potential to be

$$n_{\text{max}} \approx \left(\frac{M}{m_e}\right)^{\frac{1}{2}}. \qquad (7.30)$$

§7 TRANSLATIONAL ENERGY IN MOLECULAR COLLISIONS

Rotational excitation

The Massey parameter for the rotational excitation of a diatomic molecule can be defined as the ratio of the collision time to the period of rotation

$$\xi = \frac{\omega_{\text{rot}}}{\alpha \bar{v}} \sim \frac{1}{r_e \alpha} \left(\frac{\mu}{M}\right)^{\frac{1}{2}}. \tag{7.31}$$

For the majority of molecules, this ratio is of the order of unity for all ranges of temperature important in gas dynamics and chemical kinetics.

From this estimate and also directly from the experimental data, it follows that in the collision of a molecule of average atomic weight, the rotational quantum number changes by several units.† This implies a strong interaction between the rotation of the molecules and their translational motion, which cannot be correctly taken into account within the limits of perturbation theory. Since the zero-order approximation, which would correspond to a function describing the rotational excitation of a molecule with unchanging vibrational and electronic quantum numbers, is generally unknown, it is difficult to use the parameters of the interaction of the electronic and vibrational motion of the molecules with the translational and rotational motion in the solution to the problem of the electronic and vibrational transitions. In connection with this, the theory employs simplified models of various kinds, which are often constructed specifically with the aim of expressing some particular effect. As a rule one and the same model cannot be used for a satisfactory interpretation of experiments on the rotational and vibrational excitation of molecules.

Although great mathematical difficulties hinder the development of the general theory of molecular collisions, nevertheless, it is possible to indicate processes for which the problem is appreciably simplified. The vibrational excitation of molecules in weak shock waves is an example. A typical situation in the transmission of weak shock waves (the temperature behind the front being up to 5000–7000 K) in a mixture of monatomic and diatomic gases is such that, immediately behind the front, at distances of the order of a few free path lengths of the molecules, the Boltzmann distribution of the molecules, with regard to velocities and rotational states, is established. There then follows a wide zone in which exchange between translational and vibrational energies of the molecules takes place and where the Maxwell distribution of the molecules with respect to the velocities is not violated during the course

---

† The ratio of the rotational relaxation rate to the translational relaxation rate can serve as a parameter characterizing the average interaction of the rotating molecule with the translational motion. This ratio for molecules of the N_2 or O_2 type is approximately equal to 0·1 [211]. Different models describing the rotational transitions and relaxation of diatomic molecules are considered in the review of Takayanagi [512] and in the book by Stupochenko, Losev, and Osipov [114].

of the characteristic time of exchange. This fact makes it possible to consider vibrational relaxation as transitions between the vibrational states of the molecules under the effect of the interaction of the system with a heat reservoir (heat bath), to which must be ascribed not only translational degrees of freedom of the molecules and atoms but also rotational degrees of freedom. With this approach, the rate constant of an elementary process can be calculated within the framework of the transition state method, where, for its applicability, the following conditions must be fulfilled.

1. In some region L of the coordinates R and γ, where the chief contribution to the inelastic transition probability occurs, the distribution of all the states, with the exception of the vibrational or electronic states of AB, must be the equilibrium distribution. For this, it is clearly necessary that the vibrational relaxation time should be appreciably longer than the rotational relaxation time and the relaxation time with regard to the translational degrees of freedom.
2. The dynamical variables entering the set $\mathbf{r}$, $\mathbf{R}$ must be separable in the region L, and the conditions for quasi-classical motion in the coordinates R and γ must be fulfilled in this region. The dimensions of the region must be appreciably smaller than the gas kinetic radius R_0, so that the definition of the transition state should have meaning.

In an application of the transition state method to classical systems, the latter condition is always satisfied formally, since the dimensions of the region L can be regarded as vanishingly small (see Chapter 1). For real systems, the minimal dimensions of L are limited at least to the de Broglie wavelength of the rotational or translational motion.

It is not difficult to see that the first condition is satisfied when the interaction of the vibrations of the molecules AB with the heat reservoir is small. This small coupling is a consequence of the occurrence of the small parameter in (7.29). Fulfillment of the second condition is connected with specific assumptions as to the form of the intermolecular interaction. It is essential that the assumption about the separation of the variables be fulfilled in a region whose dimensions $1/\alpha$ are substantially smaller than the gas-kinetic radius R_0. Nowadays, for molecular collisions under the above-mentioned conditions, the fact that the dimensions of the region of nonadiabatic transitions are appreciably smaller than R_0 is firmly established. Typical values of the parameter αR_0 vary in the range 5–15 [262], [468], [469]. Thus the assumption of the separability of the variables in the Hamiltonian describing the relative motion of AB and C in some region L is the only assumption of the theory which lays claim to a quantitative estimate of the rate constants of the elementary processes of vibrational or electronic excitation of molecules in thermal collisions. Under these conditions, the problem of a real collision breaks down into a series of simpler problems for one-dimensional motion, and the rate constant κ can be represented as a certain average value of the

§7 TRANSLATIONAL ENERGY IN MOLECULAR COLLISIONS

transition probability for the one-dimensional motion. For atomic collisions, such a reduction can be made exactly if advantage is taken of the smallness of the parameter $1/\alpha R_0$ and the spherical symmetry. For molecular collisions, the reduction is approximate and depends on the possibility of carrying out the separation of the variables in a small region L. Unfortunately, at present we do not have sufficient information available about the intermolecular potential to test this assumption. Sometimes (see, for example, [511]) it is assumed that the interaction potential can be written in the form of a sum of pair potentials referring to different atoms. In this case, clearly, the radius of action of the pair potential gives both the effective breadth of the region of nonadiabaticity and a characteristic dimension over which it is already impossible to separate the variables in the Hamiltonian. However, theoretical calculations for simple models of the systems $H-H_2$ [354], H_2-H_2 [220], and H_2-He [314], [445] clearly show that the approximation of a pair interaction is totally inadequate to represent the true potential in those regions of the coordinate R which are important in slow collisions. In particular, a calculation of the exchange potential of the H_2-He [314] system shows that, in the isosceles triangle configuration, an increase in the internuclear distance of H–H, with R_{He-H_2} = constant, increases the repulsion. A model with a pair interaction which qualitatively gives an exchange interaction for the collinear configuration of the nuclei leads to the opposite result. Bearing in mind that approximate pair interactions limit to a high degree the possibility of separating the variables, the assumption of the separation of the variables in the region L may be thought sufficiently reasonable. Moreover it should be taken into account that in a different sort of application (for example, an estimate of the width of the non-equilibrium zone behind a shock wavefront), where not such high accuracy is required, it often turns out to be sufficient to estimate the order of magnitude of the vibrational relaxation time and obtain its general dependence on temperature, parameters, potential, etc.

Since, in a calculation of the rate of vibrational excitation, definite assumptions have to be made about the form of the potential, it is necessary to estimate, in a qualitative form, the possible contribution to the nonadiabatic transition probability of the short-range (exchange) and long-range (polarization) parts of the potential. For this, it is sufficient to notice that R_0 should be substituted in the Massey parameter (7.1) as the characteristic dimension of the long-range part of the potential. In adiabatic collisions, for which the transition probability depends exponentially on ξ_{vib}, the long-range part of the potential does not make a substantial contribution to the transition probability because ξ_{vib} (polarization) $> \xi_{vib}$ (exchange). This makes possible the approximation of the potential by a monotonic function with a small radius of action; for example, an exponential could be chosen. The parameters in the potential can be estimated, in principle, on the basis of independent experiments on the elastic scattering of molecules and atoms, and also on the basis of calculating a number of properties of gaseous mixtures in which the intermolecular interaction is important, with a consequent comparison of theory with experiment. Such a method was used in a series of papers (see reviews [42], [98]) where definite rules were obtained correlating the parameter α with the Van der Waals radius R_0 of the molecules and the depth ε_0 of the potential well, which were determined from experiments on transport phenomena. On the other hand, experiments on vibrational relaxation can sometimes be regarded as a source of information on the short-range intermolecular interactions, for which the interaction energy is of the order 0·1–1 eV [12].

One of the long-term lines of research into intermolecular interactions is a study of the form of the optical band spectrum under known conditions, enabling important information about the nature of the potential to be obtained. Of all forms of spectra, those which, in the absence of an interaction, are forbidden by the selection rules are found to be the most sensitive to the intermolecular interaction (see review [120]). In particular, the infrared absorption and emission spectra of nonpolar molecules are relevant here. The absorption is due to the dipole moment of the colliding pair which arises at the time of collision.

The theory of induced spectra is constructed in a manner analogous to the theory of inelastic collisions. It turns out that the form of the induced absorption bands depends to a greater degree than the integral intensity on the details of the molecular collisions. In particular, the band halfwidth is connected with the radius of action of the exchange force, the determination of which is of great interest in connection with the study of various relaxation processes in gases.

8. Model of a forced harmonic oscillator

One of the simplest models describing the energy exchange between the translational and vibrational degrees of freedom in molecular collisions qualitatively correctly corresponds to a harmonic oscillator under the effect of an external force. The Hamiltonian of the colliding partners, the diatomic molecule AB and the particle C, which we will consider devoid of internal degrees of freedom is written in the form:

$$H_{AB-C} = H_{AB} + T_{AB-C} + U_{AB-C}(x, R), \tag{8.1}$$

where H_{AB} is the Hamiltonian of the isolated molecule AB; T_{AB-C} is the relative kinetic energy of AB and C; U_{AB-C} is the potential energy of the system AB and C.

On the assumption that x is small, U_{AB-C} can be approximated to sufficient accuracy by the first two terms in the expansion

$$U_{AB-C}(x, R) = W(R) - xF(R), \tag{8.2}$$

where

$$W(R) = U_{AB-C}(0, R)$$

$$F(R) = -\left[\frac{\partial U(x, R)}{\partial x}\right]_{x=0}.$$

With this approximation, (8.1) is transformed into

$$H_{AB-C} = H_{AB} + [T_{AB-C} + W(R)] - xF(R). \tag{8.3}$$

The vibrational Hamiltonian H_{AB}, by consideration of the isolated molecule AB, yields its vibrational functions $\phi_n(x)$. The second term in (8.3), the Hamiltonian of the relative motion of AB and C, written down on the assumption that the molecule AB does not vibrate, contributes, in a classical solution to the problem, the trajectory $R(t)$, which depends parametrically on the energy E of the relative motion. The third term in (8.3), responsible for the interaction and dependent on time through the function $R(t)$, leads to transitions between the stationary states of the molecule AB. The non-stationary vibrational function Ψ of the molecule AB satisfies the wave equation

$$i\hbar\frac{\partial \Psi}{\partial t} = H_{AB}\Psi - F(t)x\Psi(t). \tag{8.4}$$

We will seek a solution Ψ in some interval of time during which there is a single collision. Since, in a gas, the duration τ of a collision is far smaller than the time between successive collisions $1/Z_0$ this interval can be considered infinitely great in comparison with τ. Representing the molecule by a one-dimensional harmonic oscillator, we will rewrite (8.4) in the form

$$i\hbar\frac{\partial \Psi}{\partial t} = -\frac{\hbar^2}{2M}\cdot\frac{\partial^2 \Psi}{\partial x^2} + M^2\frac{\omega^2 x^2}{2}\Psi - F(t)x\Psi. \tag{8.5}$$

The solutions Ψ_m to this equation, which tend to the stationary solutions $\phi_m(x)$ of the free harmonic oscillator as $t \to -\infty$ have the form [19], [137], [302]

$$\Psi_m = N_m \exp\left\{\frac{i}{\hbar}\left[M\dot{u}x - \int_{-\infty}^t (L + E_m)\, dt\right]\right\}\phi_m(x - u). \tag{8.6}$$

Here u is a function of time, describing the vibrations of a classical oscillator

$$M\ddot{u} + M\omega\, u = F(t) \tag{8.7}$$

and is found from the solution to this equation under the initial conditions $u(-\infty) = \dot{u}(-\infty) = 0$. The function L represents the Lagrangian of the harmonic oscillator:

$$L(t) = \tfrac{1}{2}M\dot{u}^2 - \tfrac{1}{2}M\omega^2 u^2. \tag{8.8}$$

The transition probability $n \to m$ per collision equals

$$P_{nm} = \left| \int_{-\infty}^{\infty} \Psi_m^*(x, t \to +\infty) \phi_n(x) \, dx \right|^2. \tag{8.9}$$

Calculation of this integral gives

$$P_{nm} = \frac{n!}{m!} \exp(-\varepsilon) \varepsilon^{m-n} |L_n^{m-n}(\varepsilon)|^2$$

$$= n! m! \exp(-\varepsilon) \varepsilon^{n+m} \left\{ \sum_{k=0}^{\min(n,m)} \frac{(-1)^k \varepsilon^{-k}}{(n-k)! k! (m-k)!} \right\}^2, \quad \varepsilon = \frac{E_{0,\text{class}}}{\hbar \omega}, \tag{8.10}$$

where L_n^{m-n} is the Laguerre polynomial; $E_{0,\text{class}}$ is the energy transferred to the classical non-vibrating oscillator.

In particular, for the transition $0 \to m$ we obtain

$$P_{0m} = \frac{\varepsilon^m}{m!} \exp(-\varepsilon). \tag{8.11}$$

Using (8.11) it is easy to obtain the result that the mean value of the energy $\bar{E}_{0,\text{qu}}$, transferred to the quantum harmonic oscillator, is the same as the corresponding classical quantity:

$$\bar{E}_{0,\text{qu}} = \sum_{m=0}^{\infty} \hbar \omega m P_{0m} = \hbar \omega \sum_{m=0}^{\infty} \frac{\varepsilon^{m+1}}{m!} \exp(-\varepsilon) = E_{0,\text{class}}. \tag{8.12}$$

This result is correct for any (not only the zero-order) initial state, and also for the mean square energy transfer [19], [423], [427].

The classical quantities E_{class} and $(\bar{E}_{\text{class}})^2$ are not difficult to calculate, on solving the problem of the forced vibrations of a harmonic oscillator. If the initial phase and amplitude of the vibrations are equal to ϕ and A, then the increment in energy ΔE_{class} during the time the force F acts, is given by the relation [511]:

$$\Delta E_{\text{class}} = \frac{1}{2M} \left| \int_{-\infty}^{\infty} F(t) \exp(i\omega t) \, dt \right|^2 - \omega A \cos \phi \int_{-\infty}^{\infty} F(t) \exp(i\omega t) \, dt. \tag{8.13}$$

The mean value obtained from this by averaging ΔE_{class} and $(\Delta E_{\text{class}})^2$ over the initial phase ϕ:

$$\Delta \bar{E}_{\text{class}} = \Delta E_{0,\text{class}}$$
$$\overline{(\Delta E_{\text{class}})^2} = (\Delta E_{0,\text{class}})^2 + 2 E_{\text{vib}} \Delta E_{0,\text{class}}, \tag{8.14}$$

where E_{vib} is the initial vibrational energy of the oscillator.

Assuming $A = 0$ in (8.13) we find

$$\varepsilon = \frac{\Delta E_{0,\text{class}}}{\hbar \omega} = \frac{1}{2 \hbar \omega M} \left| \int_{-\infty}^{\infty} F(t) \exp(i\omega t) \, dt \right|^2. \tag{8.15}$$

We will now consider in more detail the case of adiabatic collisions, when the Massey parameter $\omega/\alpha v$ is large.

If the collision time $\tau \sim 1/\alpha v$ is sufficiently large in comparison with the period of the molecular vibrations $2\pi/\omega$, then the exponential factor in the integrand in (8.15) changes sign many times during the time for $F(t)$ to change appreciably. In this connection, it is convenient to use the following method to calculate the integral [320]. We will displace the path of integration with respect to t into the upper half plane of the complex variable t. In a parallel displacement of the contour by the value $i\Delta\tau$ the integral acquires a factor $\exp(-\omega\Delta\tau)$. Such a displacement may be carried out until the contour reaches the singularity of the integrand closest to the real axis $t_c = \tau' + i\tau''$, where $F(t_c)$ tends to infinity. The contour can be displaced further and deformed only in such a way that it does not intersect the point t_c. In particular, it can be chosen so that the integrand is an exponentially decreasing function of the variables of integration, so that the whole integral can be evaluated without difficulty. In this way we obtain

$$\varepsilon = B(\omega)\exp(-2\omega\tau''), \qquad (8.16)$$

where the pre-exponential term B depends only weakly on ω in comparison with the exponential factor. An explicit expression for B and τ'' in terms of the constants in the potential can be obtained, of course, only for a definite choice of interaction. From physical considerations, it is clear that $\tau'' = l/v$, where l is a certain characteristic length depending on the type of potential. Therefore (8.16) gives the basic dependence of the parameter ε on the velocity v of the colliding molecule:

$$\varepsilon = B\exp\left(-\frac{v_0}{v}\right), \qquad v_0 = 2\omega l. \qquad (8.17)$$

Since in the limiting case $\omega l/v \gg 1$ the parameter ε is found to be small, the formula (8.10) can be simplified. Retaining only the terms linear in ε in the sum in (8.10), we obtain

$$P_{n,n+1} = P_{n+1,n} = (n+1)\varepsilon,$$
$$P_{n,n+k} = 0, \qquad |k| > 1. \qquad (8.18)$$

These expressions agree with the result which can be obtained by first-order perturbation theory with an interaction $xF(t)$:

$$P_{n,n+1} = \left|\frac{1}{\hbar}\int_{-\infty}^{\infty} x_{n,n+1} F(t)\exp(i\omega t)\,dt\right|^2. \qquad (8.19)$$

The tending to zero of the transition probabilities for non-neighbouring states and also the linear dependence of $P_{n,n+1}$ on the vibrational quantum number for transitions between neighbouring levels are consequences of the smallness of the interaction and its linearity in the amplitude of vibration x.

The mean transition probability $\langle P_{n,n+1}\rangle$ is obtained from (8.17) and (8.18) by integrating $P_{n,n+1}$ over all velocities with a one-dimensional current distribution function, normalized to unity:

$$\langle P_{n,n+1}\rangle = \int_0^\infty P_{n,n+1}(v)\exp(-\beta E)\beta\,dE, \qquad E = \frac{\mu v^2}{2}. \quad (8.20)$$

It is not difficult to ascertain that the integrand has a sharp maximum near a certain value $v = v^*$, due to the minimum in the function $f = -v_0/v + \beta\mu v^2/2$ in the exponential. Therefore the integration over dv can be limited to a not very large range Δv near to v^*, representing $f(v)$ in this region by an expansion $f(v^*) + f''(v^*)(v-v^*)^2/2$ and formally extending the limits of integration from $-\infty$ to $+\infty$ (the so-called saddle-point method).† The calculation gives:

$$\langle P_{n,n+1}\rangle = (n+1)B(\omega, v^*)2\sqrt{\left(\frac{\pi}{3}\right)}\left(\frac{\gamma}{2}\right)^{\frac{1}{3}}\exp\left[-3\left(\frac{\gamma}{2}\right)^{\frac{2}{3}}\right]$$

$$\gamma = v_0\left(\frac{2kT}{\mu}\right)^{-\frac{1}{2}}, \quad (8.21)$$

where the condition $\omega l/v \gg 1$ now takes the form $\gamma \gg 1$.

Landau and Teller [320] approximated the interaction potential W and perturbation F by exponential functions $W(R) = A\exp(-\alpha R)$ and $F = [m_B/(m_A+m_B)]\alpha A\exp(-\alpha R)$. For such a model, the solution to the classical equation of motion gives

$$\exp(-\alpha R) = \cosh^{-2}\left(\frac{\alpha vt}{2}\right), \quad (8.22)$$

from which we immediately find $t_c = i\pi/\alpha v$ or $l = \pi/\alpha$. In this case (8.21) takes the form

$$\langle P_{n,n+1}\rangle = \langle P_{n+1,n}\rangle = (n+1)Z_{10,\text{vib}}Z_{\text{trans}} \quad (8.23a)$$

$$Z_{10,\text{vib}} = \frac{1}{2}\cdot\frac{m_A^2+m_B^2}{(m_A+m_B)^2}\cdot\frac{\mu}{M}\cdot\frac{2\pi\theta^2}{\theta'}, \quad \theta' = \frac{4\pi^2\omega^2\mu}{\alpha^2 k}, \quad \theta = \frac{\hbar\omega}{k}, \quad (8.23b)$$

$$Z_{\text{trans}} = \left(\frac{\theta'}{\pi\theta}\right)^2\sqrt{\left(\frac{2\pi}{3}\right)}\left(\frac{\theta'}{T}\right)^{\frac{1}{6}}\exp\left[-\frac{3}{2}\left(\frac{\theta'}{T}\right)^{\frac{1}{3}}\right], \quad (8.23c)$$

where the standard parameters θ' and θ usually employed in calculations of vibrational transition probabilities are introduced. The factor $Z_{10,\text{vib}}$ is obtained by averaging the probability over the two possible realizations of a linear collision: AB–C and C–AB.

† A discussion of the exactness of this approximation is contained in the literature, [480], [533].

The basic features of expression (8.23), known in the literature as the Landau–Teller formula, are the following: the already-mentioned linear dependence of the transition probability on the vibrational quantum number and the approximate linear dependence of $\ln\langle P_{n,n+1}\rangle$ on $T^{-\frac{1}{3}}$. From (8.23), it can be seen that the mean probability of the direct $n \to n+1$ and reverse $n+1 \to n$ transitions are equal:

$$\frac{\langle P_{n,n+1}\rangle}{\langle P_{n+1,n}\rangle} = 1. \tag{8.24}$$

This relation contradicts detailed balance, which establishes the general connection between the probabilities of the direct and reverse processes by the basic requirement of the exact compensation of the two elementary processes:

$$\frac{\langle P_{n,n+1}\rangle}{\langle P_{n+1,n}\rangle} = \frac{x_{n+1,\mathrm{eq}}}{x_{n,\mathrm{eq}}}, \tag{8.25}$$

where $x_{n,\mathrm{eq}}$ is the equilibrium population of the level n. Taking it into account that $x_{n,\mathrm{eq}} \approx \exp(-n\hbar\omega\beta)$ we find that relation (8.24) is obtained from (8.25) only under the condition $\hbar\omega/kT \ll 1$. This contradiction is connected with the inconsistency in the derivation of (8.19), in which no account has been taken of the fact that the excitation or deactivation of the vibrations of a molecule AB, in which it receives or gives out energy $\hbar\omega$, is accompanied by a corresponding decrease or increase in the energy of the relative motion. Everywhere earlier it had been assumed that the relative motion of AB and C could be characterized by a definite trajectory. The velocity v of the molecules AB and C at infinity, the same for both the initial and final states, occurs in formula (8.22) as a characteristic of the motion along this trajectory.

Taking the restrictions which are imposed on $P_{n,n+1}$ by the choice of model into account, the following lines of improvement of the Landau–Teller formula are indicated:

1. Extension of the model to the three-dimensional case.
2. Accounting for the change in the relative velocity of the molecules as a result of the inelastic nature of the process.
3. Accounting for the effect of anharmonicity on the vibrational transition probabilities.
4. Investigation of the dependence of the vibrational transition probability on the form of the intermolecular potential.
5. Clarification of the connection between the semiclassical approximation and the solution to the corresponding quantum problem.

The following paragraphs are devoted to these problems.

9. Vibrational transitions between the lowest levels of diatomic molecules in non-degenerate electronic states

The theory, set forth below, of the vibrational transitions in diatomic molecules induced by collisions presumes that, on collision, the electronic state of the colliding partners is not changed and that the vibrational transition probability is so small that, for its calculation, perturbation theory can be enlisted. The latter assumption enables the model considered in § 8 to be improved.

In place of the Hamiltonians (8.1) and (8.2), we will consider the more general Hamiltonian

$$H = H_{AB} + [T_{AB-C} + W(R, \gamma)] + V(R, \gamma, x), \qquad (9.1)$$

where H_{AB} is the vibrational–rotational Hamiltonian of the molecule AB; $[T_{AB-C} + W(R, \gamma)]$ is the Hamiltonian of the relative motion of AB and C, depending on the relative distance and orientation of the colliding molecules; $V(R, \gamma, x)$ is the interaction responsible for the vibrational transitions in AB.

If the final term in (9.1) is neglected, then the variables are still not separable in the remaining Hamiltonian, since it describes collisions accompanied by rotational transitions. If W does not depend on γ, then the rotational state of the molecule AB does not change as a result of the collision, but if $W(R, \gamma)$ is described by the form $W(R) + \delta W(R, \gamma)$, where δW is sufficiently small, then the collisions induce rotational transitions with a small probability. In this case the interaction of the translational motion with the rotational motion can be described within the framework of perturbation theory.

We will now go over to the refinements of the Landau–Teller model mentioned at the end of § 8, using the Hamiltonian (9.1). They are all in the nature of corrections to the basic formula (8.23). The problem is simplified, because attention can be concentrated on some particular point and then all the improvements can be summed up in one general formula (see formula (9.39)).

Generalization of the oscillator model to the three-dimensional case

The basic difficulty of this generalization is that it is necessary to take into account the anisotropic intermolecular potential and the vibrational–rotational transitions produced by it, at the same time. In such a formulation, the problem of calculating the vibrational–rotational transition probabilities is so complicated that only by introducing sufficiently simplified models can real success be achieved. One such is the model of a radially vibrating diatomic molecule, within the limits of which the potential responsible for the elastic scattering is assumed independent of the mutual orientation of the colliding partners. In addition, of course, the exchange of energy between the rotational and translational degrees of freedom of the colliding molecules

§9 TRANSLATIONAL ENERGY IN MOLECULAR COLLISIONS

is completely neglected (the model of Schwartz, Slawsky, and Herzfeld [468], [469]).

For an intermolecular potential depending only on the radial coordinate R, the dynamical problem of a three-dimensional collision reduces to a one-dimensional problem, where the true potential $W(R)$ is changed to an effective potential $W_{eff}(R) = W(R) + l^2/2\mu R^2$. Introducing the impact parameter b, instead of the relative angular momentum l, the law of conservation of energy, correct for elastic collisions, can be written in the form

$$\mu \frac{\dot{R}^2}{2} + W(R) + E_{trans} \frac{b^2}{R^2} = E_{trans}. \qquad (9.2)$$

The essential feature of the interaction $W(R)$ is that the radius of action $1/\alpha$ of the exchange potential is appreciably smaller than the gas-kinetic radius R_0 of the molecule. This enables the centrifugal potential $b^2 E_{trans}/R^2$, in the region of closest approach of the molecules, to be approximated by the constant $b^2 E_{trans}/R_0^2$, which is obtained by substituting the value R_0 for R. Such an approximation is more valid for collisions close to head-on, when the centrifugal energy forms a small fraction of the total energy. But it is precisely these collisions which make the largest contribution to the mean transition probability, on condition $\omega\tau \gg 1$, when collisions with large radial velocity are the most effective for the transfer of energy.

This standard approximation is known in the literature as the method of modified wavenumber, first introduced by Takayanagi [506] in the calculation of rotational transition probabilities and widely employed in the study of vibrational transitions [507], [511], [513]. In this approximation, in place of the relative energy E_{trans}, it is possible to introduce the radial energy E_R, rewriting (9.2) in the following form

$$\mu \frac{\dot{R}^2}{2} + W(R) = E_R, \qquad E_R = E_{trans}\left(1 - \frac{b^2}{R_0^2}\right). \qquad (9.3)$$

Now the problem is reduced to the one-dimensional case considered in §8. The only difference is that the velocity in the formulae of §8 must be replaced by the radial velocity $v_R = v(1 - b^2/R_0^2)^{\frac{1}{2}}$.

In this way, the three-dimensional vibrational transition probability P_{nm}, which, generally speaking, depends on the two variables v and b, can be expressed in terms of just the one variable v_R, within the limits of the approximation considered. The cross-section σ_{nm} for a non-elastic collision, as a result of which the vibrational transition $n \to m$ occurs, is given by the general expression (6.9), taking the form

$$\sigma_{nm}(v) = 2\pi \int_0^\infty P_{nm}(v, b) b \, db \qquad (9.4)$$

in the given case.

In accordance with the definition adopted for v_R and E_R, the integration over b in (9.4) extends only over the interval $0 \leqslant b \leqslant R_0$, in fact, since for $b > R_0$ the transition probability tends to zero. Introducing the corresponding energies as arguments in (9.4) in place of the velocities and replacing the integration over b by an integration over E_R, we rewrite (9.4) in the form

$$\sigma_{nm}(E_{\text{trans}}) = \frac{\pi R_0^2}{E_{\text{trans}}} \int_0^{E_{\text{trans}}} P_{nm}(E_R)\, dE_R. \tag{9.5}$$

The rate constant for the vibrational transition $n \to m$ can be expressed in terms of the cross-section by relation (6.14):

$$\kappa_{nm}(T) = \left(\frac{8kT}{\pi\mu}\right)^{\frac{1}{2}} \int_0^\infty \left[\frac{\pi R_0^2}{E} \int_0^E P_{nm}(E_R)\, dE_R\right] \exp\left(\frac{-E}{kT}\right) \frac{E\, dE}{(kT)^2}. \tag{9.6}$$

Integrating (9.6) by parts, we obtain the following expression for κ_{nm}:

$$\kappa_{nm} = \left(\frac{8kT}{\pi\mu}\right)^{\frac{1}{2}} \pi R_0^2 \langle P_{nm} \rangle, \tag{9.7}$$

in which the value $\langle P_{nm} \rangle$ is defined as the mean transition probability for a one-dimensional collision, to conform with (8.20). Thus (9.7) establishes the connection between the three-dimensional and one-dimensional models.

We note that formula (9.7) can be obtained easily by the transition state method. We choose the radius R as the reaction coordinate, and a sphere of radius R^* as the critical surface, assuming that the vibrational transition $n \to m$ is accomplished with probability P_{nm}, when R attains the critical value R^*. Then the general expression for κ_{nm} is described by the form

$$\kappa_{nm} = \frac{kT}{2\pi\hbar} \frac{F^\star_{\text{AB--C}}}{F_{\text{AB}} F_{\text{C}}} \int_0^\infty P_{nm}(E_t) \exp\left(-\frac{E_t}{kT}\right) \frac{dE_t}{kT}. \tag{9.8}$$

The integral in (9.8) already has the form of a one-dimensional mean transition probability. In calculating the factors in front of the integral, the partition functions of the initial molecules AB and C are identified with the translational partition functions, but the partition function of the activated complex is represented as a product of $F^\star_{\text{AB--C,trans}}$ and $F^\star_{\text{AB--C,rot}}$, where the latter quantity is identified with the partition function of a rigid two-dimensional rotator with a moment of inertia $\mu(R^*)^2$. A simple calculation gives

$$\frac{kT}{2\pi\hbar} \frac{F^\star_{\text{AB--C}}}{F_{\text{AB}} F_{\text{C}}} = \left(\frac{8kT}{\pi\mu}\right)^{\frac{1}{2}} \pi(R^*)^2. \tag{9.9}$$

Thus (9.8) leads to (9.7) if the critical radius R^* is identified with the gas-kinetic radius R_0. The choice of critical surface $R = R^*$ is not strictly determined, since, in reality, the transition between the two vibrational states takes place in the region of interaction of the molecules, i.e. over an interval of radius of order $1/\alpha$. This means that the transition state method is as accurate as the method of the modified wavenumber.

However eqn (9.7) is obtained as a particular case of the general formula of the transition state theory with the simplest choice of critical surface. It may be that another choice of critical surface might succeed in taking the non-central nature of the intermolecular interaction into account, thus appreciably extending the region of applicability of the three-dimensional model.

It is not difficult to understand why the vibrational–rotational transitions may make an appreciable contribution to the vibrational relaxation rate. It is connected with the fact that only the Fourier component $V(\omega)$ of frequency ω derived from the potential $V(t)$ by

$$V(\omega) = \int_{-\infty}^{\infty} V(R, \gamma) \exp(i\omega t) \, dt \qquad (9.10)$$

is effective in the vibrational transitions. The time dependence of V, moreover, is determined by the two-dimensional trajectory of the relative motion of the colliding molecules, i.e. by the two functions $R = R(t)$ and $\gamma = \gamma(t)$. Since, usually, the weight of the high-frequency components in the resolution of $V(t)$ along the trajectory of the rotational motion is greater than the weight of the high-frequency components, for the same potential, in its resolution along the trajectory of translational motion, even a weak coupling of the vibrations with the rotation in the region can affect the vibrational transition probability appreciably. In an attempt to take the vibrational–rotational interaction into account, one has, of necessity, to go beyond the bounds of the one-dimensional model, which leads to great mathematical difficulties. The simplifications mentioned above, introduced with the aim of treating the three-dimensional model, neglect this interaction, in point of fact, since an assumption has been made about the isotropic nature of the intermolecular potential. Only in the case of a very strongly anisotropic potential, characteristic, for example, of chemically active molecules, can a number of results be obtained [63], on the assumption that the relative rotational motion of the colliding molecules is changed into vibrational motion. Just how important the interaction is between the intermolecular vibrations and the rotation of the pair as a whole remains completely obscure. It is possible to speak of the physical effect of the interaction of the intramolecular vibrations with the intermolecular only when it lies outside the errors introduced by the model itself.

The mechanisms of certain processes of this type are considered in the literature [63].

Great difficulties arise when the intermolecular interaction in the region of closest approach is comparable to the hindered internal rotation in the molecules.

Here, in a certain sense, the two limiting cases of weak and strong coupling of the molecular rotation and the vibrations should be distinguished. Under conditions of weak coupling (δW small) only single quantum (for symmetric molecules, double quanta) rotational transitions occur with appreciable probability, whose effect, in fact, is to reduce the energy transfer in comparison with the energy of the vibrational quantum. Since the vibrational transition probability in the adiabatic region depends very strongly on the magnitude of the energy transfer, even a small contribution of the rotation to the value of the energy transfer changes $P_{nn'}$ substantially.

Approximate estimates, only taking account of single-quantum rotational transitions, show [68] that the interaction of the vibrations and the rotation increases the vibrational relaxation rate 3–5 times in comparison with the value obtained for the simplest models taking no account of this interaction. Experimental data recently obtained for systems which most closely approach the simplified theoretical models confirm this result [373]. It should be borne in mind that the factor mentioned above is of the same order of magnitude as the error which arises in the attempt to estimate the vibrational relaxation time with a potential whose constants are determined by independent means. Therefore the contribution of the vibrational–rotational transitions can be found experimentally by comparing the experimental results for different systems, whose only difference consists of the possible occurrence of similar effects (for example, the vibrational relaxation of oxygen in O_2+O_2 or O_2+Ar [373]) rather than by comparing theory with experiment.

In the presence of strong coupling between the rotation and the vibration, some results can be derived based on the above-mentioned ideas of the transition state method. To take account of the vibrational–rotational interaction in the basic exponential factor in the expression for $P_{nn'}$, we will assume that the rotational and translational degrees of freedom of the molecule AB are characterized by the equilibrium distribution in the interaction region L and that an AB + C collision which does not lead to a change in the vibrational state of AB can be described within the framework of classical mechanics. To each such collision there corresponds a classical trajectory which describes, generally speaking, the rotational excitation or deactivation of the rotator AB. The vibrational excitation probability corresponding to relative motion along this trajectory is calculated according to the standard formula (8.19), which must now be written in the form

$$P_{nn'} = \left| \frac{1}{\hbar} \int_{-\infty}^{\infty} [V(R,\gamma,x)]_{nn'} \exp\left[\frac{i}{\hbar}\int^{t}(E_n - E_{n'} + V_{nn} - V_{n'n'})\,dt\right] dt \right|^2. \quad (9.11)$$

§9 TRANSLATIONAL ENERGY IN MOLECULAR COLLISIONS 63

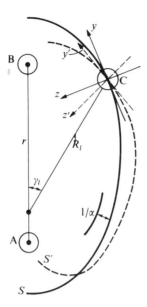

Fig. 4. Local system of coordinates on the curve of intersection of the critical surface with the plane of the three atoms: S is the curve of intersection for vibrational transitions; S' is the curve of intersection for electronic–vibrational transitions.

In contrast to (8.19), account is taken here of the possible non-linear dependence of V on x and also of the diagonal matrix elements of the interaction (an exponential term) in the transition probability. The time dependence of $[V(R, \gamma, x)]_{nn'}$ is determined by the dependence of R and γ on the time, assumed known for the given trajectory of the relative motion.

It is important, for later on, that only a small portion of the trajectory near the turning point, where the width of the transition region L is of order $1/\alpha$ (Fig. 4), makes a contribution to the integral in (9.11). Therefore, in the description of the relative motion of the molecules in this region, being given, for example, by the two functions $R = R(t)$ and $\gamma = \gamma(t)$, it is convenient to introduce a new system of Cartesian coordinates x, y, z. The origin of each of these systems is referred to a fixed point R_1, γ_1, and the orientation of the axes is assigned in the following way: the x axis is directed perpendicular to the plane of the three atoms; the y axis is along the tangent to the equipotential line (the level line) of the interaction $W(R, \gamma)$; and the z axis is along the vector grad W. The level lines $W(R, \gamma) = E$ determine the geometrical position of the turning points of different trajectories, characterized by the total energy E. Turning to the implicit function $W(R, \gamma) = E$, we find the equation for the lines of constant energy $R = R_E(\gamma, E)$. One such line is denoted by the curve S in Fig. 4. The condition $R_1\alpha \gg 1$ enables the separation of the x, y, z

variables in the equation of motion to be carried out in a small region near the point R_l, γ_l and the classical trajectories

$$x = x(t, v_x), \quad y = y(t, v_y), \quad z = z(t, v_z) \tag{9.12}$$

to be found. Here v_x, v_y, v_z denote the velocity components of the relative motion along the corresponding axes, where they take on the meaning of velocities outside the interaction region, i.e. in point of fact, the velocities at infinity. Moreover, the velocity components play the roles of parameters over which the transition probability $P_{nn'}$ in a small region near R_l, γ_l must be averaged.

We will consider a trajectory L whose turning point is at R_l, γ_l, and we will make the natural assumption that the transition probability for $n \to n'$ depends only on the direction of the velocity along the vector grad W, i.e. $P_{nn'} = P_{nn'}(v_z)$. From the results of § 8, it follows that the averaged transition probability depends very strongly on the reduced mass of the partners. In the case under consideration, the reduced mass is replaced by an effective mass μ_l characterizing the motion of the representative point along the z coordinate. The value of μ_l depends on the position of the coordinate system x, y, z, i.e. on R_l, γ_l. The relative configuration of the partners for which μ_l is minimal makes the largest contribution to the averaged transition probability. Therefore, amongst all possible regions L characterized by the given R_l and γ_l, that which contains the trajectories L making the chief contribution to the mean transition probability can be separated out. It is natural to attach the meaning of coordinates to the corresponding coordinates R^* and γ^* of this region, characterizing the structure of the activated complex ABC^*, and, to the curve S, the meaning of a cross-section of the critical surface by the plane of the three atoms ABC. Thus from the point of view of the transition state method, a vibrational transition is regarded as taking place in the immediate neighbourhood of S^*. The coordinates R^* and γ^* are determined from the two equations

$$W(R^*, \gamma^*) = E^*$$

$$\mu_l(R^*, \gamma^*) = \min[\mu_l(R_l, \gamma_l)], \tag{9.13}$$

where E^* is the relative energy of AB and C making the maximum contribution to the mean transition probability; the minimal value μ_l acquires the meaning of the mass m_r, which enters the formula of the transition state method.

By analogy with (2.8) we will write

$$\kappa_{nn'} = \frac{\int P_{nn'}(v_z) \exp(-\beta H^*) \frac{v_z \, dp_z}{2\pi\hbar} d\Gamma^*}{F_{AB,rot} F_{AB-C,trans}}, \tag{9.14}$$

where $d\Gamma^*$ is an element of phase volume of the four degrees of freedom: the three rotations of the system ABC^* and the small displacement y along the curve S around the point R^*, γ^*.

Since the v_k in (9.12) are identified as the velocities at infinity, only the kinetic energy T^*, conveniently represented in the form (7.15), which, of course, does not take the kinetic energy of vibration into account, should be substituted in place of H^*.

In contrast to (2.12), the integral in the numerator of (9.14) cannot be factorized into the partition function of the activated complex and an integral of a one-dimensional average over velocities, since μ_l depends on y. Integration over $d\Gamma^*_{rot}$ gives the rotational partition function $F^*_{ABC,rot}$ and integration over dp_y gives the Maxwellian distribution function of a one-dimensional current with an effective mass μ_l depending on y. The corresponding calculations give

$$\kappa_{nn'} = \frac{F^*_{ABC,rot}}{F_{AB,rot}F_{AB-C,trans}} \cdot \frac{(2\pi kT)^{\frac{3}{2}}}{(2\pi\hbar)^3} \cdot \left[\frac{\mu^2 M}{m_r(M+\mu\rho^2)}\right]^{\frac{3}{2}}$$

$$\times \int_0^\infty P_{nn'}(v_z)v_z \exp\left[-\beta\frac{\mu_l(y)}{2}v_z^2\right]\frac{dp_z}{2\pi\hbar}dy, \quad (9.15a)$$

$$\mu_l = \frac{\mu\lambda[\rho^2+(\rho')^2]}{\lambda[\rho^2+(\rho')^2]+(\rho\rho')^2}, \quad m_r = \mu_l(0) = \min \mu_l \quad (9.15b)$$

$$\rho = \frac{R^*}{r_e}, \quad \rho' = \frac{(dR^*/d\gamma)}{r_e}. \quad (9.15c)$$

In expression (9.15a) the dependence on $\mu_l(y)$ is preserved only in the main exponential factor. In the remaining cases $\mu_l = m_r$ is assumed. The average over v_z in the integral (9.15a) gives the mean transition probability for the one-dimensional motion, in which the effective reduced mass depends on y. Substituting the explicit expression for the partition function into (9.15a) we obtain

$$\kappa_{nn'} = 2\pi R^* \sin \gamma^* \left(\frac{kT}{2\pi m_r}\right)^{\frac{1}{2}} \int \langle P_{nn'}[\mu_l(y)]\rangle\, dy, \quad (9.16)$$

where the y-dependence of the mean transition probability is recorded and the region of integration over y is assumed small in comparison with R^*. It is not difficult to trace the connection between eqns (9.16) and (9.7). If it is assumed that μ_l does not depend on y, then the angle γ^* is not fixed by conditions (9.13) and the factor $\sin \gamma^*$ in (9.16) must therefore stay under the integral sign. Next, using $dy = R\, d\gamma$ and integrating over γ, we obtain (9.7). When μ_l depends markedly on y, so that a small variation of y near $y = 0$ leads to an appreciable diminution of the exponential factor, in the expression for the mean transition probability, in an evaluation of the integral in (9.16), it is sufficient to represent $\mu_l(y)$ in the form of an expansion $\mu_l(y) = m_r + \frac{1}{2}m_r'' y^2$, substituting this expansion in the exponent of the exponential. The first term of the expansion gives the mean transition probability for the effective mass m_r and the second gives a certain characteristic length δR^*.

In particular, for the Landau–Teller model, where the probability of one-dimensional collisions is expressed by formulae (8.23), we obtain

$$\kappa_{n,n+1} = 2\pi R^* \delta R^* \sin \gamma^* \sqrt{\left(\frac{kT}{2\pi m_r}\right)} \langle P_{n,n+1}(\theta_r') \rangle, \qquad (9.17a)$$

where

$$\theta_r' = \frac{4\pi^2 \omega^2 m_r}{\alpha^2 k}$$

$$\delta R^* = 2\sqrt{\left(\frac{\pi m_r}{m_r''}\right)}\left(\frac{T}{\theta_r'}\right)^{\frac{1}{6}} R^*. \qquad (9.17b)$$

On substituting R^* and γ^* from (9.13) into formula (9.17), it should be borne in mind that these quantities depend very weakly on E^*, which determines the equipotential curve identified with the critical surface S^*. In particular, it can be assumed that $E^* = kT$, it is hardly advisable to carry out a more precise specification of E^* within the approximation of the transition state method.

The correction found, which takes account of the contribution of the rotation of the molecule to the vibrational transition probability is of significance when the ratio μ_l/μ differs from unity by a value exceeding $1/\alpha R^*$ (the parameter characterizing the accuracy of the model itself). If R depends on γ very weakly, so that $R_\gamma \approx \Delta R = R_{\max} - R_{\min} \ll 1/\alpha$ then it should be assumed that $\mu_l = \mu$ and the temperature dependence of $\langle P_{10} \rangle$ will be determined by the exponential factor, which is obtained for a model of a radially vibrating sphere. The contribution of the rotational transitions to $\langle P_{10} \rangle$ can be calculated, in this case, by perturbation theory, in which the smallness of the parameter $1/\alpha R^*$ is utilized.

Another limiting case, which can be obtained from (9.16), corresponds to the model of a roughly spherical molecule. For the case under consideration, for given R, R_γ takes any value and, for the most part, $|R_\gamma| \gg R$. In this limit we obtain

$$\mu_l = \frac{\mu \lambda r_e^2}{[\lambda r_e^2 + (R^*)^2]}, \qquad (9.17c)$$

where for $\lambda \ll 1$ the effective mass μ_l can be appreciably smaller than μ. This implies that, on the whole, vibrational deactivation takes place at the expense of a loss of rotational energy of the same molecule, and the role of the partner

C is reduced to satisfying the law of conservation of total angular momentum (the momentum of AB itself being lost, carried away in the form of the relative momentum of the AB–C pair.) The rough molecule model is probably adequate for polyatomic molecules [376]. For diatomic molecules, it is doubtful whether it can be assumed that the level curves $R = R(\gamma)$ are 'cut' so that $|R_\gamma| \gg |R|$ on average.

At the present time we do not have available sufficient results of theoretical calculations of the potential $W(R, \gamma)$ of multi-electron atomic–molecular systems. The simple model system H_2–He, mentioned in §7, enables some qualitative deductions to be made concerning the interaction potential of a molecule with like nuclei and the atom of an inert gas. For the H_2–He system, the radius of action of the exponential repulsion $1/\alpha$ does not depend on γ and E, in practice, and equals ~ 0.5 Å. Moreover the asymmetry parameter ΔR is of the order 0.2 a.u. and the radius $R^\star$ is close to 3.3–3.5 a.u. (for $E_{\text{trans}} \approx 1$ eV). If the same relation between $R^\star$, $1/\alpha$, and ΔR is also preserved for other qualitatively analogous systems, then the contribution of the rotation to the vibrational relaxation rate must be comparatively small, and the vibrational relaxation time can be calculated in accordance with the model of the radially vibrating molecules.

The largest contribution of rotation to the vibrational relaxation rate is to be expected for heteroatomic molecules, for which the centre of gravity is strongly displaced relative to the centroid of the charge distribution. In this respect, the recently found relaxation of HI molecules in an Ar atmosphere behind shock wavefronts [198] is of interest. If the HI molecule is represented by a sphere rotating about an axis which passes through the I atom then $\gamma^\star$ is found to be close to $\pi/4$, and $\mu_l = \mu/8$. An estimate of the contribution of the rotation of HI to the vibrational transition rate $n = 1 \to n = 0$, carried out on the basis of the general formula (9.16), shows that the vibrational rotational energy exchange increases $\langle P_{10} \rangle$ by a factor of 10^2 ($T = 2300$ K) and 10^3 ($T = 1400$ K) in comparison with values calculated for the model of the radially vibrating sphere. This correction factor is close to the experimental values (30 and 1.2×10^3 at $T = 2300$ K and 1400 K respectively), but to what extent this agreement is significant is difficult to say, because the adiabatic potentials are unknown. Nevertheless, if $\langle P_{10} \rangle$ is calculated for rough molecules (i.e. formula (9.15) is used, which gives $\mu_l = \mu/32$), then the values found for $\langle P_{10} \rangle$ exceed the experimental values by an order of magnitude 2–3 [376].

Calculation of the change in the relative kinetic energy of molecules in inelastic collisions

In a consistent quantum-mechanical solution of the problem of an inelastic collision, the energies of the initial and final states of the partners enter explicitly into the expression for the transition probability, since the wavefunctions of the states depend on them. In the case of adiabatic collisions under consideration, it is only necessary to take account of the change in relative energy ΔE of the molecule in the exponential term of formula (8.16), which gives the main velocity dependence of the transition probability. Corrections to the pre-exponential factor, as will be seen from an estimate to be made, are usually so small that they can be neglected. This means that it is sufficient to determine the exponential factor in the quasi-classical matrix element of the interaction $\langle \chi(E_i)|V|\chi(E_f)\rangle$, where E_i and E_f are the energies of the initial and final states. The calculation is carried out by a method analogous to the method of distorting the contour of integration in the classical variational calculation. The quasi-classical wavefunction contains an oscillatory factor of the form [37]

$$\chi(E) \sim \exp\left\{\frac{i}{\hbar}\int^R [2\mu(E-W(R))]^{\frac{1}{2}}\,dR\right\}. \qquad (9.18)$$

By distorting the contour of integration by dR in the complex plane, the oscillatory character of the function

$$\chi^*(E_i)\chi(E_f) \sim \exp\left[\frac{i}{\hbar}\int (P_i - P_f)\,dR\right]$$

can be transformed into a decaying factor, which makes possible the exact calculation of the exponential factor. Calculations for an exponential potential $W(R) \sim \exp(-\alpha R)$ give (see [37])

$$P_{if} \sim \exp\left(-\frac{\pi\mu}{\alpha\hbar}|v_i - v_f|\right). \qquad (9.19)$$

Thus this factor replaces the exponential factor of expression (8.17). For changes of energy ΔE small in comparison with the initial energy E_i, formula (8.16) can be represented in the form

$$P_{if} \sim \exp\left[-\frac{v_0}{v_i}\left(1 \pm \frac{|\Delta E|}{4E_i}\right)\right]. \qquad (9.20)$$

We will now quote another derivation of (9.20), correct for any potential $W(R)$ [56], [383].

The limiting expression is a result of the semiclassical calculation and would be obtained from the exact quantum-mechanical formula for P under the condition that the difference in the velocities $(v_i - v_f)$ is small, and for $\Delta E \to 0$, the parameter v of the semiclassical theory is some symmetric function of v_i and v_f. As $v_f \to v_i$, this function must transform into v. These conditions are sufficiently accurate, up to the term linear in ΔE, to express v as a function of v_i and ΔE independent of the form of the exact function

§9   TRANSLATIONAL ENERGY IN MOLECULAR COLLISIONS   69

$v = v(v_i, v_f)$. In fact

$$v = v(v_i, v_f) = v(v_i, v_i) + \frac{\partial v}{\partial v_f}(v_f - v_i) + \ldots = v_i \pm \frac{|\Delta E|}{2\mu v_i} + \ldots \qquad (9.21)$$

because, by symmetry, we have $(\partial v/\partial v_f)_{v_f=v_i} = \tfrac{1}{2}$; this agrees with the correction in (9.20). The additional factor $\exp[\pm(v_0/v_i)(|\Delta E|/4E_i)]$, arising as a correction in (9.20), should now be taken into account in the averaging of the transition probability in accordance with (8.20). Since this factor has the nature of a correction to the main factor $\exp(-v_0/v_i)$ it can be taken outside the integral in (8.20) at the point $v_i = v^*$ where the integrand is a maximum. Simple calculations, which we will omit here, show that the result of such an approximation amounts to replacing the unaveraged factor by the averaged factor

$$\exp\left(\pm \frac{v_0}{v_i} \frac{|\Delta E|}{4E_i}\right) \rightarrow \exp\left(\pm \frac{|\Delta E|}{2kT}\right). \qquad (9.22)$$

Thus this correction removes the defect of the semiclassical approximation, of which relation (8.24) is a consequence. Taking this factor into account, the probabilities of the direct and reverse transitions satisfy the equation of detailed balance (8.25) which can be written in the form

$$\frac{\langle P_{n,n+1}\rangle}{\langle P_{n+1,n}\rangle} = \exp\left(-\frac{|\Delta E_{n,n+1}|}{kT}\right). \qquad (9.23)$$

The next terms in the expansion of the exponential in (9.20) must be of order $(\Delta E/E_i)^2$ or of order $(\beta \Delta E/4)^2$ after averaging over the velocities. The possibility of limiting the calculation of the probability by the semiclassical method to the first correction of the type (9.22) is determined by two conditions: the relative smallness of the expansion terms of the function $v = v(v_i, v_f)$ in the parameter $\Delta E/E_i$ and the small absolute contribution, quadratic in ΔE, of the correction to the exponentially small transition probability. On averaging over the equilibrium distribution, these two conditions must be fulfilled for those velocities which give the chief contribution to the integral. It is not difficult to verify that they can be put in the form

$$\left(\frac{\theta'}{T}\right)^{\frac{1}{3}} \gg \max\{1, (\beta\Delta E)^2\}. \qquad (9.24)$$

If P is not exponentially small and the inequality $(\theta'/T)^{\frac{1}{3}} \gg 1$ is consequently not fulfilled, then the condition $\beta\Delta E \ll 1$ will always be satisfied. Moreover the relation (9.23) can always be considered as satisfied, since, by the addition to $\langle P_{n,n+1}\rangle$ or to $\langle P_{n+1,n}\rangle$ of quantities of the first order of smallness (for example, by the replacement of $\langle P_{n,n+1}\rangle$ by $\langle P_{n,n+1}\rangle(1-\beta\Delta E/2)$ and $\langle P_{n+1,n}\rangle$ by $\langle P_{n+1,n}\rangle(1+\beta\Delta E/2)$, the relation (9.23) can always be satisfied. As regards the probabilities $\langle P_{n,n+1} \rangle$ and $\langle P_{n+1,n} \rangle$ themselves, they can be considered equal to within an accuracy of small terms of the order $\beta \Delta E$. This 'rectification' procedure of the semiclassical approximation for $\beta \Delta E \ll 1$ is widely employed in the deduction of kinetic equations for systems with a small separation between the levels (see, for example, [275]).

Calculation of the anharmonicity of the vibrations

In an attempt to take account of the effect of anharmonicity in the vibrations on the vibrational transition probabilities, two problems arise: the manifestation of anharmonicity in the interaction potential ('intermolecular anharmonicity') and the effect of the anharmonic corrections on the magnitude of the vibrational quanta ('intramolecular anharmonicity'). To begin with, we will consider the first problem.

Owing to the adiabatic nature of the interaction of the vibrations with the translational motion, even a small change in the adiabatic terms appreciably changes the exponential factor in expression (8.21) Therefore the anharmonic correction to the diagonal elements V_{nn} of the interaction potential must be taken into account, to the degree of accuracy which would secure sufficient accuracy for the whole exponential factor. Without taking into account the anharmonicity, the adiabatic vibrational terms $E_n + V_{nn}$ are almost parallel, and the separation between them does not depend on R. If, however, the anharmonicity is taken into account, then this leads to an additional drawing together or drawing apart by the value $\delta V(R) = V_{n+1,n+1} - V_{n,n}$. Under the condition $\delta V(R) \ll \hbar\omega$ the effect of anharmonicity on the general position of the terms will be small. However, in the region of maximum approach at $R \sim R^\star$, the exponential term is changed by the factor $\exp[-v_0/v \cdot \delta V(R^\star/\hbar\omega)]$, which may be rather large as a result of the adiabatic nature of the collision ($v_0/v \gg 1$). Therefore the condition for neglecting the anharmonicity in the interaction potential can by written in the form:

$$\left(\frac{v_0}{v}\right)\left(\frac{\delta V}{\hbar\omega}\right) \ll 1. \tag{9.25}$$

It is evident that this condition is more restricting than that usually formulated as the condition for the applicability of the harmonic approximation to the description of molecular vibrations

$$\frac{\delta V}{\hbar\omega} \ll 1. \tag{9.26}$$

To calculate the anharmonicity, we will proceed from a one-dimensional model, for whose description we will employ the classical trajectory $R(t)$ of the relative motion in a potential $W(R)$ and the function $V(t, x)$ obtained from $V(R, x)$ by the substitution of a given function of time in place of R.

§9   TRANSLATIONAL ENERGY IN MOLECULAR COLLISIONS   71

For comparison with the Landau–Teller formula, we will approximate the potential by an exponential function and will write the matrix element $V_{nn'}$ in the form

$$\delta V = V_{nn} - V_{n'n'} = \delta V^* \cosh^{-2}\left(\frac{\alpha v t}{2}\right)$$

$$V_{n,n+1} = \frac{Bm_B}{m_A + m_B} \alpha \cosh^{-2}\left(\frac{\alpha v t}{2}\right) \qquad (9.27)$$

where α and δv^* are independent parameters of the theory.

Generally speaking, the non-diagonal matrix elements and the difference between the diagonal elements have a different dependence on time, so that the approximation (9.27) imposes certain restrictions on the final results. At present, however, we do not have sufficient information available about the intermolecular potential to justify the introduction of two independent functions. In particular, the relations (9.27) are exact if V is approximated by a function of the form $V = V_1(R)V_2(x)$. Substituting (9.27) into (9.1), we obtain an expression in which it is convenient to separate out the factor obtained in the Landau–Teller approximation [383], [370]:

$$\langle P_{n,n+1} \rangle = A_{n,n+1}(n+1) Z_{10,\text{vib}} Z_{\text{trans}}, \qquad (9.28)$$

where $Z_{10,\text{vib}}$ and Z_{trans} are defined by the relations (8.23b) and (8.23c). The corresponding calculations give [72]

$$A_{n,n+1} = \left| \Phi\left(1 + \frac{i\omega}{\alpha v^*}, 2, \frac{4\delta V^*}{\hbar \alpha v^*}\right) \right|^2, \qquad (9.29)$$

where Φ is the confluent hypergeometric function for which the following asymptotic expansion is correct [217]:

$$|\Phi(1 + i\xi, 2, i\,\delta\xi)|^2 = \begin{cases} 1, & \delta\xi \ll 1 \quad (9.30\text{a}) \\ \dfrac{\exp(2\pi|\xi|)}{\pi(\delta\xi)^2 \xi}, & 1 \ll \xi \ll \delta\xi \quad (9.30\text{b}) \\ \dfrac{1}{(\xi\,\delta\xi)} \cdot J_1^2[2\sqrt{(\xi\,\delta\xi)}], & \delta\xi \ll \xi \quad (9.30\text{c}) \end{cases}$$

where J_1 is a Bessel function.

Up till now, no limits have been imposed on the parameters ξ and $\delta\xi$. Therefore (9.29) is correct also for the case when the difference between the diagonal terms $V_{nn} - V_{n'n'}$ is so great that the terms of the first approximation $E_n + V_{nn}$ and $E_{n'} + V_{n'n'}$ cross. In this case, the correction factor $A_{nn'}$ is exponentially large, so that the expression (9.28), in fact, does not contain the characteristic exponential of Landau–Teller. Such a situation arises for vibrational transitions in molecules with degenerate electronic states (see § 20). Here we will note especially case (9.30c), which is characteristic for adiabatic collisions and for which the correction factor A only depends on the parameter $\xi \, \delta\xi$. The specific evaluation of this factor for vibrational transitions in N_2 will be given below.

We will now go over the second problem, which concerns the effect of anharmonicity on the magnitude of the energy transfer. For the lowest vibrational levels of molecules, the vibrational energy can be represented by a quadratic function of n (formula (7.25)) and the vibrational quanta by a linear function of n (formula (7.27)). Substituting these expansions into the Landau–Teller formula (8.23), we find

$$\langle P_{n+1,n} \rangle = (n + 1)\langle P_{10} \rangle \gamma^n, \qquad (9.31)$$

where

$$\gamma = \exp\left[\left(\frac{\theta'}{T}\right)^{\frac{1}{3}} x_e\right]. \qquad (9.32)$$

Turning to the evaluation of the parameter x_e for the nitrogen molecule (see below) we find that, for the fifth or sixth vibrational level, the anharmonic correction to the magnitude of the vibrational quantum increases the single-quantum vibrational transition probability, at $T \approx 300$ K, by an order of magnitude. A specific estimate of this effect in the application to the vibrational relaxation of oxygen is mentioned in the literature [58].

In connection with what has been said, it is necessary to note that taking account of anharmonicity in the pre-exponential could lead to the possibility of vibrational transitions across one, two, etc. levels. The corresponding pre-exponential factors decrease in this sequence as powers of the anharmonicity coefficient. This decrease is, however, accompanied by an additional sharp decrease in the exponential factor, and, as a result, an increase in the energy transfer and a growth in the exponent of the exponential of $\Delta E_{nm}\tau''/\hbar$. This is the principal difference between the appearance of anharmonicity in vibrational transitions in collisions and in spontaneous optical transitions. In the latter, the intensities of the overtones are actually reduced in proportion to the anharmonicity constant. Therefore the proportionality, sometimes assumed, between the lifetimes of the excited vibrational states in connection with emission and deactivation in collisions (see [196]) is not in fact satisfied.

The dependence of the vibrational transition probability on the intermolecular potential

The problem of the dependence of the transition probability on the intermolecular potential is very important, since interpretation of the experimental data is usually conducted in terms of some model or other of the interaction, chosen in advance. In previous discussion, it was everywhere assumed (just as it was accepted in the work of Landau and Teller [320]) that the intermolecular potential is a monotonically decreasing function of the distance R, and can therefore be approximated by an exponential. In connection with this, it is necessary to ascertain the correction to the Landau–Teller formula dependent on that part of the intermolecular potential which is responsible for the attraction of the molecules at great distances. This means that, in addition to the basic parameter of the potential namely the characteristic length $1/\alpha$, a second parameter should be introduced, the depth of the potential well ε_0.

Since the direct effect of the long-range part of the potential on the transition probability, estimated by the Massey parameter (see § 7), is negligibly small, Herzfeld and coworkers [262], [469] assumed that the attraction was apparent only in an additional acceleration of the molecules in the region of their interaction. The corresponding correction to the transition probability may be found by the same method as was used to take account of the change in velocity of the molecules as a result of the inelastic nature of the collisions. Namely, by substituting $[2(E_{\text{trans}}+\varepsilon_0)/\mu]^{\frac{1}{2}}$ into (8.17) for $v = (2E_{\text{trans}}/\mu)^{\frac{1}{2}}$ and expanding $(E_{\text{trans}}+\varepsilon_0)^{\frac{1}{2}}$ in powers of the ratio ε_0/E, we find the first correction factor to the main factor $\exp(-v_0/v)$. In complete analogy with (9.22), on averaging, this additional factor changes in the following way

$$\exp\left(\frac{v_0}{v}\cdot\frac{\varepsilon_0}{2E_{\text{trans}}}\right) \rightarrow \exp\left(\frac{\varepsilon_0}{kT}\right). \tag{9.33}$$

For molecules with saturated chemical bonds, the depth of the Van der Waals minimum usually works out at $\varepsilon_0/k = 200\text{–}600$ K. Thus, according to this method of calculation, the attraction may increase the transition probability several times.

In subsequent work, it was made clear that the attraction does not lead to just an acceleration of the molecules and hence cannot be taken into account simply by replacing E by $E+\varepsilon_0$. To investigate this problem, we will consider the Lennard–Jones potential, widely employed in the interpretation of the properties of gas transfer. All calculations will be carried out within the framework of a one-dimensional model, since its generalization to the three-dimensional case can be developed in accordance with the method set out above [56].

We will assume that the interaction responsible for the elastic scattering of AB and C can be described by the Lennard–Jones potential

$$W(R) = 4\varepsilon_0\left[\left(\frac{R_0}{R}\right)^{2n} - \left(\frac{R_0}{R}\right)^n\right], \tag{9.34}$$

where ε_0 is the depth of the attractive potential well; R_0 is the Van der Waals radius.

Concerning the interaction $V(R, x)$ responsible for the excitation of the vibrations, we will only assume that it depends on R in the same way as the potential $W(R)$. To first-order perturbation theory we obtain

$$P_{nm} = \left| \frac{1}{\hbar} \int_{-\infty}^{\infty} [V(R, x)]_{nm} \exp\left(-\frac{i\Delta E_{nm} t}{\hbar}\right) dt \right|^2. \qquad (9.35)$$

The main problem is estimating the collision time τ'', which comes into the exponential of expression (8.16). For the Lennard–Jones potential, this time turns out to be

$$\tau'' = \int_0^{R_t} \left[\frac{4\varepsilon_0}{E_{\text{trans}}} \left(\frac{R_0^{2n}}{R^{2n}} - \frac{R_0^n}{R^n} \right) - 1 \right]^{-\frac{1}{2}} \frac{dR}{v}, \qquad (9.36)$$

where the upper limit R_t corresponds to the turning point where the radicand tends to zero. If $2n/R_0$ is identified with the value α and it is taken into consideration that $2n$ is practically always sufficiently large in comparison with 1, so that we can put $(4\varepsilon_0/E_{\text{trans}})^{1/2n} \sim 1$, we find that the characteristic time $\tau''_{\text{L-J}}$ for the Lennard–Jones potential is decreased in comparison with the corresponding quantity for a model with an exponential potential

$$\tau''_{\text{L-J}} = \tau'' \left[1 - \frac{2}{\pi} \left(\frac{\varepsilon_0}{E} \right)^{\frac{1}{2}} \right]. \qquad (9.37)$$

The calculation of this correction, on averaging, indicates the substitution

$$\exp\left[\frac{v_0}{v} \cdot \frac{2}{\pi} \sqrt{\left(\frac{\varepsilon_0}{E} \right)} \right] \rightarrow \exp\left[\frac{2}{\pi} \left(\frac{\theta'}{T} \right)^{\frac{1}{6}} \left(\frac{\varepsilon_0}{kT} \right)^{\frac{1}{2}} \right]. \qquad (9.38)$$

Thus we see that the effect of the attraction is to increase the vibrational transition probability to a greater degree than might have been expected on the basis of (9.33). This follows from the fact that for conditions typical of a molecular collision, the inequality $\chi \gg \varepsilon_0/kT$ is fulfilled, so that the quantity $(\chi\varepsilon_0/kT)^{\frac{1}{2}}$ is appreciably larger than ε_0/kT. Although in the calculation of τ'', the analytical properties of the function $W(R)$ in the complex plane were used, which are, generally speaking, different for potentials constructed from powers and exponentials; the final expressions are the same in the approximation in which a power can be approximated by an exponential close to the turning point on the real t axis. In this respect, the theory justifies the definition of the constant α according to the Lennard–Jones potential, which was assumed in a number of papers [206], [262]. To what extent a real potential may be approximated by a Lennard–Jones potential remains, of course, an open question.

For the Morse intermolecular potential and for the potential proposed by Parker [394], the correction to the attraction is also given by expression (9.38). This is connected with the fact, that in all these models, the radius of action of the repulsive part of the potential is chosen smaller by a factor of two than the radius of action of the attractive part.

Further corrections to the transition probability and also some general problems of its dependence on the form of the potential are discussed in the literature [478], [479], [481], [549]. Since refinements of this type go beyond the limits of accuracy of the approximate method of calculating $P_{n,n+1}$ under consideration, we will not dwell on them here.

We will only mention here the papers in which the transition probabilities are calculated for the Morse intermolecular potential [127], [205], [510], [572] and for a special model in which the centre of the attractive force does not coincide with the centre of the repulsive force and each is approximated by an exponential function [394]. Parker's model [394] lays claim to a simultaneous description of vibrational and rotational transitions, and is discussed in the literature [396] from this point of view. In the Parker potential, besides the parameters α and ε_0, is introduced the distance d^* between the two centres of repulsion, on which only the pre-exponential factor (8.23c) depends. At present, however, we do not have information available regarding the magnitude of this parameter.

We will now write down the general expression for the single-quantum vibrational transition probability, taking into account all the corrections found above:

$$\langle P_{n,n+1} \rangle = (n+1) Z_{10,\text{vib}} Z_{\text{trans}} \left[\exp\left(-\frac{\theta}{2T}\right) A \gamma^n Z_{\text{rot}} B \right],$$

$$\theta = \frac{\hbar \omega}{k}. \tag{9.39}$$

The factor in front of the brackets is obtained in the Landau–Teller approximation (8.23).

The factor inside the square brackets is the correction to the Landau–Teller formula found in this section. The first factor $\exp(-\theta/2T)$ is due to the change in the kinetic energy of the colliding molecules. The second and third factors are corrections for anharmonicity (9.29) and (9.32). The fourth factor Z_{rot} takes account of asymmetry in the interaction potential, and finally the fifth is caused by the long-range part of the intermolecular potential not taken into account in (8.19). For the Morse and Lennard–Jones potentials this factor equals

$$B = \exp\left[\frac{2}{\pi}\left(\frac{\theta'}{T}\right)^{\frac{1}{6}}\left(\frac{\theta^*}{T}\right)^{\frac{1}{2}}\right], \qquad \theta^* = \frac{\varepsilon_0}{k}. \tag{9.40}$$

In the presence of a strong vibrational–rotational interaction, the parameter θ' determined by the relation (8.23) must be replaced by θ'_r from (9.17). In addition, the factors $\delta R^*/R^*$ should be introduced into Z_{rot}.

The mean transition probability (9.39) refers to a single kinetic collision with mean velocity $v = (8kT/\pi\mu)^{\frac{1}{2}}$ and cross-section πR_0^2. That the attraction may increase the cross-section somewhat at the expense of deforming the trajectory of the relative motion has not been taken into account here. However, this effect is very small and can be taken into account formally by the introduction of a weak temperature dependence in the radius R_0 [262].

Connection between the semiclassical approximation and the quantum solution

For the great majority of important cases, the semiclassical approximation is found to be satisfactory in describing single-quantum vibrational transitions if, in addition, the symmetrization of the probability relative to the initial and final states is carried out by the method considered in the preceding section. Nevertheless, it is interesting to trace the connection between the quantum-mechanical solution of a problem and the classical approximation, since the method of symmetrization itself is substantiated by the possibility of a transition from one to the other. Moreover a clarification of the connection between the two approaches shows how the exponential dependence of the cross-section on the velocity for slow collisions (a dependence of type (9.2)) goes over into the $\sigma \sim 1/v$ dependence, which must be satisfied in the limiting case of small velocities [37].

The quantum transition probability is calculated by solving the system of coupled wave equations to first order in the parameter $\alpha x_{nn'}$ (the so-called method of distorted waves [50]). Analysis of the solution found by Jackson and Mott [284] for the exponential potential shows that the semiclassical expression for the transition probability is obtained from the quantum solution by the substitution

$$\frac{\sinh \pi q_i \sinh \pi q_f}{(\cosh \pi q_i - \cosh \pi q_f)^2} \to \exp\left(-\frac{2\pi\omega}{\alpha v}\right), \quad q_m = \frac{2k_m}{\alpha}. \quad (9.41)$$

Such a substitution is correct on condition that the wavevectors of the initial and final states are large in comparison with the inverse radius of action of the force:

$$\pi q_i \gg 1, \quad \pi q_f \gg 1. \quad (9.42)$$

In this case the transition probability can be described by the form

$$P \sim C \exp\left(-\frac{2\pi|E_i - E_f|}{\hbar \alpha v}\right), \quad (9.43)$$

where C is a constant independent of the velocity.

§9   TRANSLATIONAL ENERGY IN MOLECULAR COLLISIONS   77

The total cross-section is shown to be of order

$$\sigma \sim \pi R_0^2 C \frac{\hbar \alpha v}{2\pi |E_i - E_f|} \exp\left(-\frac{2\pi |E_i - E_f|}{\hbar \alpha v}\right). \tag{9.44}$$

If the quasi-classical condition is broken in the initial state and it satisfies a relation contrary to (9.42), then we obtain

$$\frac{\sinh \pi q_i \sinh \pi q_f}{(\cosh \pi q_i - \cosh \pi q_f)^2} \approx 2\pi q_i \exp(-\pi q_f). \tag{9.45}$$

The chief contribution to the cross-section comes from just the s-wave scattering, so that we can write

$$\sigma \approx \sigma(l=0) = \frac{\pi}{k_i^2} P = \frac{2\pi^2}{k_i \alpha} C \exp\left[-\frac{2\pi |E_i - E_f|}{\hbar \alpha}\sqrt{\left(\frac{2\mu}{|E_i - E_f|}\right)}\right]. \tag{9.46}$$

For vibrational transitions between the lowest levels of the molecule, the quantity $(|E_i - E_f|/2\mu)^{\frac{1}{2}}$ is of the order of the velocity of the nuclei v_{nucl}. The ratio $|E_i - E_f|/\hbar \alpha$ can also be represented in the form $v_{\text{nucl}}(M/m_e)^{\frac{1}{4}}$, if the connection between the magnitude of the vibrational quantum with the radius of action of the exchange force and the energy of dissociation of the molecules is taken into account. In this way, we obtain the following conversion of the exponentially small cross-section in the quasi-classical region ($\pi q_i, \pi q_f \gg 1$) for slow collisions ($v \ll v_{\text{nucl}}(M/m_e)^{\frac{1}{4}}$) to the s-wave cross-section ($\pi q_i \ll 1$) in the limit of small velocities:

$$\pi R_0^2 \frac{Cv\hbar \alpha}{\pi |\Delta E|} \exp\left(-\frac{2\pi |\Delta E|}{\hbar \alpha v}\right) \rightarrow \pi \alpha^{-2} C \frac{2\pi \alpha \hbar}{\mu v} \exp\left[-\frac{2\pi \sqrt{(2\mu \Delta E)}}{\hbar \alpha}\right]. \tag{9.47}$$

We will also note that a quantum variant of a model with diverging or converging terms in the first approximation was considered by Mies [369]. In this case, the confluent hypergeometric function in relation (9.29) must be replaced by the hypergeometric function, which in the limit of small wavelengths $\lambda \alpha \ll 1$ tends to the confluent form of this function.

As an illustration of the quantum treatment of a collision, we mention the model calculation of the vibrational transition probability in a hydrogen molecule in collision with a helium atom, carried out by Mies [371] for an interaction potential calculated theoretically [314] in the self-consistent field approximation. The greatest deviation from the results of a semiclassical calculation should be expected for the given system. In fact, the results of this work clearly show the insufficient accuracy of the semiclassical approximation, and also the limited nature of the model of a spherically symmetrical molecule. At present, however, it is not clear to what extent the results of this work are applicable to other systems.

10. The vibrational relaxation of oxygen and nitrogen

Information concerning the probability of individual transitions $n \rightleftarrows n+1$ in diatomic molecules can be obtained on the basis of experimental data on the vibrational relaxation of molecules. In this section, as an illustration we will briefly discuss the experimental data on the relaxation of oxygen and nitrogen molecules and will quote theoretical estimates of various parameters entering expression (9.39).

Experimentally, the quantity under observation, namely the vibrational relaxation time τ_{vib}, must be calculated theoretically on the basis of a solution to the system of kinetic equations for the non-equilibrium vibrational distribution function, which depends on the probabilities of all transitions. For the harmonic oscillator model such an investigation gives (see § 38):

$$\frac{1}{\tau_{vib}} \equiv \frac{Z}{Z_{vib}} = Z_0 \langle P_{10} \rangle \left[1 - \exp\left(-\frac{\hbar\omega}{kT}\right) \right], \qquad (10.1)$$

where Z_0 is the gas-kinetic number of collisions of the molecule AB undergoing relaxation with the molecules C of the reservoir per unit time; Z_{vib} is the number of collisions effective in the relaxation.

The last factor in (10.1) arises as a result of the reduction of the many-level problem ($n = 0, 1, 2, \ldots$) to a single relaxation equation for the mean vibrational energy. Since such a reduction is possible only for the harmonic oscillator, subsequently in the interpretation of $(\tau_{vib})_{exp}$ by way of $\langle P_{10} \rangle_{theor}$, we will assume everywhere that $\gamma = 1$.

Experimental data on the relaxation of oxygen in O_2–O_2 collisions are discussed in a number of papers [13], [42], [373]. As an illustration of the sensitivity of τ_{vib} to the parameters of the theory, various formulae are quoted below which describe the temperature dependence of τ_{vib} over sufficiently wide ranges of temperature [396], [542].

1. If only the basic temperature dependence, namely, the exponential term of the Landau–Teller formula is left in expression (9.39) for $\langle P_{10} \rangle$ and in formula (10.1), then the experimental data can be described by the relations

$$\frac{1}{\tau_{vib}} = 0.67 \times 10^{10} \exp(-133 T^{-\frac{1}{3}}) \text{ atm s}^{-1}, \qquad 600 \text{ K} < T < 2600 \text{ K [542]}$$
$$\qquad (10.2)$$
$$\frac{1}{\tau_{vib}} = 0.58 \times 10^{10} \exp(-133.39 T^{-\frac{1}{3}}) \text{ atm s}^{-1}, \qquad 600 \text{ K} < T < 6000 \text{ K [396]}$$

2. If it is assumed that the temperature dependence is determined by expressions (10.1) and (9.39), without taking account of the correction to the

§ 10   TRANSLATIONAL ENERGY IN MOLECULAR COLLISIONS   79

attraction, then for τ_{vib} we have [542]

$$\frac{1}{\tau_{vib}} = 61 \times 10^{10} T^{\frac{1}{2}} \exp\left(-198 T^{-\frac{1}{3}} + \frac{2239}{2T}\right)\left[1 - \exp\left(-\frac{2239}{T}\right)\right] \text{ atm s}^{-1} \quad (10.3)$$

where $\hbar\omega_{O_2}/k = 2239$ K.

3. With the same accuracy as in (10.3), the experimental data are correlated with the relation obtained in the work of Landau and Teller [320]:

$$\frac{1}{\tau_{vib}} = 45 \times 10^{10} T^{\frac{1}{6}} \exp(-172 T^{-\frac{1}{3}})\left[1 - \exp\left(-\frac{2239}{T}\right)\right] \text{ atm s}^{-1}. \quad (10.4)$$

A detailed consideration of the experimental data on the relaxation of oxygen confirms the basic temperature dependence of the relaxation time $\ln \tau_{vib} \sim T^{-\frac{1}{3}}$, which follows from the Landau–Teller formula [114], [542]. Moreover, from relations (10.2), (10.3), and (10.4) it can be seen that a comparatively small variation of the temperature dependence of the pre-exponential term noticeably changes the value of the constant in the exponent of the exponential. This means that if the temperaure dependence τ_{vib} is used to obtain information about the constant α of the intermolecular potential, then the theoretical temperature dependence of the pre-exponential must be known sufficiently exactly. If the accuracy is sufficient and τ_{vib} is determined over some region ΔT near the temperature T, then we can obtain information about that part of the intermolecular potential which corresponds to the energy $W \sim kT(\theta'/T)^{\frac{1}{3}}$, since a collision with energy $E^*_{trans} = kT(\theta'/T)^{\frac{2}{3}}$ makes the chief contribution to the vibrational transition probability.

Such a method of estimation in connection with the mean interaction of O_2–O_2, was employed by Generalov and Losev [12]. As to an *a priori* estimate of α, existing methods, chiefly based on the extrapolation of the intermolecular potential in the region of small internuclear separations, generally speaking give an error 2–3 times out. This means that the parameter $(\theta'/T)^{-\frac{1}{3}}$ is determined with an accuracy up to a factor 1·6–2·1. For oxygen, however, the extrapolation of the potential and its representation in the form of an exponential function near the point $R = R^*$ enables the magnitude of $\langle P_{10} \rangle$ to be predicted with an accuracy up to a factor 3–5, over a wide range of temperature [117], [542].

The vibrational relaxation of nitrogen, like oxygen, has been sufficiently well studied [372]. In particular, in the temperature range 1900–5600 K, the experimental value of τ_{vib} can be described by the formula

$$\frac{1}{\tau_{vib}} = 1.74 \times 10^{11} \exp(-236 T^{-\frac{1}{3}}) \text{ atm s}^{-1}. \quad (10.5)$$

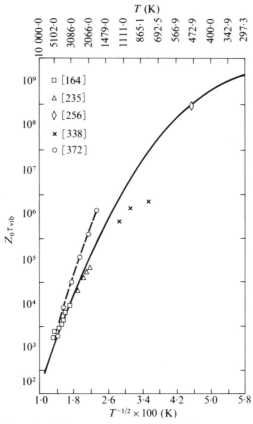

FIG. 5. Vibrational relaxation of nitrogen: comparison of theory with experiment (Parker [396]).

The value $\alpha = 3\cdot127\,\text{A}^{-1}$ corresponds to the exponential factor entering this expression. In Parker's work [396] another value $\alpha = 4\cdot073\,\text{A}^{-1}$ is proposed, for which the theory reproduces also the results of the low-temperature data ($T = 470\,\text{K}$) obtained by the method of ultrasonic dispersion. In Fig. 5, taken from the paper [396], are represented the experimental results of various authors (points), the correlation with formula (10.5) (dashed curve) and the calculation in accordance with the Parker model [394] (continuous curve).

Within the limits of this model, the temperature dependence of $\langle P_{10} \rangle$ is in fact determined by formula (9.39), in which a correction for the attraction is included, but a correction taking account of the change in kinetic energy on collision (i.e. by the factor $\exp(\hbar\omega/2kT)$ is neglected. The continuous curve corresponds to the parameter $\alpha = 4\cdot073\,\text{A}^{-1}$ and $\theta^* = 81\,\text{K}$. Without considering the reliability of the experimental data, we will only note here

§ 10   TRANSLATIONAL ENERGY IN MOLECULAR COLLISIONS   81

that the probable reason for the great variation in the experimental data in all the experiments on relaxation is the unchecked presence in the gas of small concentrations of impurities, which are very effective in the exchange of vibrational and translational energy.

Another check of the theory might be carried out on the basis of an investigation of the dependence of the exponent of the exponential on the reduced mass of the partners. Here, however, the difficulty occurs that the parameter α is different for different pairs and there are no-well founded laws correlating the parameter for the partners $A+B$ with the corresponding parameters for the collisions of $A+A$ and $B+B$. Therefore one can talk about the appearance of a mass effect only in those cases where it is clear beforehand that the variation of θ for a change in μ far exceeds the variation of θ due to a change in α. Precisely this situation arises in the comparison of the relaxation times of oxygen in an atmosphere of O_2 and He. In the latter case, the relaxation time in the temperature range 300–1600 K can be described by the formula [372], [395]

$$\frac{1}{\tau_{vib}} = 0.19 \times 10^9 \exp(-60 T^{-\frac{1}{3}}). \tag{10.6}$$

Assuming

$$\alpha_{O_2-O_2} \approx \alpha_{O_2-He}$$

we find

$$\left(\frac{\theta_{O_2-O_2}}{\theta_{O_2-He}}\right)^{\frac{1}{3}}_{theor} \approx \left(\frac{\mu_{O_2-O_2}}{\mu_{O_2-He}}\right)^{\frac{1}{3}} = 1.65. \tag{10.7}$$

Comparing (10.6) and (10.2) we obtain

$$\left(\frac{\theta_{O_2-O_2}}{\theta_{O_2-He}}\right)^{\frac{1}{3}}_{exp} = 2.2. \tag{10.8}$$

The discrepancy between (10.7) and (10.8) should be attributed to the violation of the assumption about the equality of the parameters α. An independent evaluation of α can be carried out in those cases where the different partners in the collision possess electron shells characterized by similar properties. Thus one might expect that the value of α in the series of O_2–He, O_2–D_2, O_2–H_2 would be approximately constant, so that all the differences in the values of $\langle P_{10} \rangle$ must be attributed to the effect of changes in the reduced mass μ. Theoretical estimates of the ratios

$$\tau_{vib,O_2-He}/\tau_{vib,O_2-D_2}, \quad \tau_{vib,O_2-D_2}/\tau_{vib,O_2-H_2},$$

carried out in accordance with (9.39) with different methods for determining α, give the values 1.30–1.64 and 6.2–7.7 at $T = 303$ K. The experimental values are 1·49 and 16·1 [269]. The discrepancy between theory and experiment for O_2–H_2 and O_2–D_2 cannot be explained by the inaccuracy in the estimate of α, and must be attributed to the different effect of the rotation of H_2 and D_2 on the vibrational transitions in O_2.

The examples presented above illustrate sufficiently well the general situation regarding the semi-empirical treatment of the experimental results on the vibrational relaxation of diatomic molecules on the basis of the theory set forth in § 9.

We will now consider to what extent the assumption about the harmonicity of the vibrations of the molecule undergoing relaxation is fulfilled. Formula (9.39) contains two corrections for anharmonicity. The first of them, the factor A, takes account of the distortion of the molecular terms as a result of the molecular anharmonicity. The factor A does not depend on the vibrational quantum number n, and therefore the condition for the harmonic approximation (the linear dependence of $P_{n,n+1}$ on $n+1$) essential for the derivation of (8.18) is not broken. Nevertheless, this correction affects the value of P_{10} substantially. The change in the separation between neighbouring vibrational terms of the molecules is found from the relation

$$\delta V(R) = \frac{\partial V}{\partial x}(x_{n+1,n+1} - x_{nn}) + \frac{1}{2}\frac{\partial^2 V}{\partial x^2}(x^2_{n+1,n+1} - x^2_{nn}), \qquad (10.9)$$

in which the matrix element x_{nn} should be calculated taking account of the anharmonicity in the vibration; for the calculation of x^2_{nn} the harmonic approximation is sufficient. Since in the region of interaction $\partial V/\partial x > 0$, $\partial^2 V/\partial x^2 > 0$ and each of the brackets in (10.9) is positive, it can be seen that, as the molecules approach, the separation between the vibrational terms increases. At the turning point, δV is at a maximum and equals

$$\delta V^* = \frac{\mu v^2 \Gamma}{2},$$

$$\Gamma = \lambda\alpha(x_{n+1,n+1} - x_{nn}) + \frac{(\lambda\alpha)^2}{2}(x^2_{n+1,n+1} - x^2_{nn}). \qquad (10.10)$$

The explicit form of the factor Γ is given, on the assumption of an exponential intermolecular interaction $U(R, x) \sim \exp(-\alpha R + \lambda\alpha x)$, where Γ does not depend on n in the approximation under consideration. Limiting ourselves to the case $\omega/\alpha v \gg 1$, $\delta V^*/\hbar\alpha v \leqslant 1$ and expressing these parameters in terms of θ, θ', and T we find, from (9.30c),

$$A = \frac{1}{y}J_1^2(2\sqrt{y}), \quad y = \frac{\Gamma}{2\pi}\cdot\frac{\theta'}{\theta}. \qquad (10.11)$$

We note that the factor A does not depend on the temperature and is therefore not essential in the interpretation of the temperature dependence of $\langle P_{10} \rangle$. Evaluation of the factor A for N_2–N_2 collisions gives the value 0·06–0·07 [370]. This correction can be amalgamated with the factor $Z_{10,\text{vib}}$ which in this case equals 0·0065 [425].

We now turn to the factor γ in formula (9.39). Although γ is close to unity, for sufficiently large n the exponential dependence of γ^n begins to prevail over the linear Landau–Teller dependence. It is of interest to find the value of the quantum number n^* for which the rates of the linear and exponential growth of $P_{n,n+1}$ become equal. Taking account of the relation

$$\frac{\langle P_{n,n+1} \rangle}{\langle P_{n-1,n} \rangle} = \frac{n+1}{n}\gamma \tag{10.12}$$

and expanding γ, defined by eqn (9.32), in a series, we find

$$n^* \approx \frac{T^{1/3}}{(\theta')^{\frac{1}{3}} x_e}. \tag{10.13}$$

Turning to (10.5) we find that for N_2–N_2 collisions, the parameter $(\theta')^{\frac{1}{3}}$ is approximately equal to $150\,(\text{K})^{\frac{1}{3}}$. Then from (10.13) it follows that for $T = 1000$ K, for transitions between levels with $n > n^* = 5$, the linear dependence of the Landau–Teller law is markedly violated. This violation, however, does not, in practice, affect the kinetic vibrational relaxation in shock wave conditions. In fact, for such a type of relaxation, the quantum numbers

$$\langle n(T) \rangle \sim \exp(-\hbar\omega/kT)[1-\exp(-\hbar\omega/kT)]^{-1}$$

are important. Comparing $\langle n(T) \rangle$ and n^*, it is not difficult to verify the fulfillment of the condition

$$\langle n(T) \rangle \ll n^*, \tag{10.14}$$

which ensures the applicability of the harmonic approximation in the calculation of the vibrational relaxation time in shock waves.

The situation is essentially changed when the 'cooling' of highly-excited molecules ($n \gg \langle n(T) \rangle$) in a heat reservoir at a sufficiently low temperature (a condition typical of relaxation in experiments with flash photolysis) is considered. In this case, transitions between strongly-excited states are important, and it is necessary, of course, to take the anharmonic correction into account, owing to the lessening of the vibrational quanta with increase in n. The influence of this effect on the vibrational relaxation of oxygen is discussed in the work [58].

11. Strong coupling of the vibrational and translational motion

The theory of inelastic molecular collisions set forth in §9 assumes that the transition probability $n \to n'$ can be calculated by first-order perturbation theory. The condition

$$\sum_{n' \neq n} P_{nn'} \ll 1 \tag{11.1}$$

serves as a criterion for the validity of this assumption, and its violation necessitates taking higher orders of perturbation theory into account. To clarify the limits of applicability of the perturbation theory, we will calculate the transition probability $P_{nn'}$, no longer assuming an interaction linear in x and not going beyond the limits of the adiabatic approximation. We will assume that $U_{\text{AB-C}}(R, x)$ can be represented in the form

$$U_{\text{AB-C}}(R, x) = A \exp(-\alpha R_{\text{BC}}) = A \exp[-\alpha(R - \lambda x)],$$

$$\lambda = \frac{m_{\text{A}}}{m_{\text{A}} + m_{\text{B}}}. \tag{11.2}$$

In addition, we will take the anharmonicity of the vibrations into account, which, for example, may be done within the framework of the Morse oscillator model. If calculated formally to first-order perturbation theory, the transition probability $P_{nn'}$ equals

$$P_{nn'} = Z_{nn',\text{vib}} Z_{\text{trans}}, \tag{11.3}$$

where

$$Z_{nn',\text{vib}} = \left| \int_{-\infty}^{\infty} \phi_n(x) \exp(\alpha \lambda x) \phi_{n'}(x) \, dx \right|^2 \tag{11.4a}$$

$$Z_{\text{trans}} = \left\langle \left| \int_{-\infty}^{\infty} \frac{A}{\hbar} \exp[-\alpha R(t)] \exp\left(\frac{i \Delta E_{nn'} t}{\hbar}\right) dt \right|^2 \right\rangle. \tag{11.4b}$$

All further calculations are substantially simplified if consideration is limited to molecules with identical nuclei. In this case $\lambda = \frac{1}{2}$ and the coefficient $\lambda \alpha$ can be shown to be close to the coefficient α' of the Morse potential. Physically, this is because the parameters characterizing the repulsive parts of the intermolecular and intramolecular potentials must be close, but in the Morse potential the repulsive part is proportional to $\exp[-2\alpha'(r - r_e)]$. The condition $\alpha = 2\alpha'$ enables the integral of (11.4a) to be calculated comparatively easily,

$$Z_{n+1,n,\text{vib}} = Z_{n,n+1,\text{vib}} = \frac{(n+1)}{2N} \cdot \frac{N^{\frac{3}{2}}(N - n/2)^{\frac{1}{2}}}{(N-n)^2}, \tag{11.5}$$

where

$$Z_{n,n+k,\text{vib}} = 0, \quad k > 1.$$

Here, N denotes the number of levels in the Morse potential $N = 2D/\hbar \omega_e$, and the first factor agrees with $Z_{n,n+1,\text{vib}}$ for the harmonic oscillator. We will now estimate the factor Z_{trans}. To begin with, it is necessary to calculate the integral in (11.4b) with the functions (8.22), then to average the square of its modulus over the equilibrium distribution of a one-dimensional current. The calculation gives

$$Z_{\text{trans}} = \frac{8\mu kT}{\alpha^2 \hbar^2} f(\gamma), \tag{11.6}$$

where

$$f(\gamma) = \int_0^\infty \exp(-y) \frac{y^2 \, dy}{\sinh^2(\gamma/\sqrt{y})}, \qquad \gamma = \frac{\pi \Delta E_{nn'}}{\hbar \alpha} \sqrt{\left(\frac{\mu}{2kT}\right)}. \tag{11.7}$$

For $\gamma \gg 1$ and $n \ll N$, (11.7) goes over into formula (8.21), valid for transitions between the lower vibrational levels of the molecule.† It is easy to verify this, if it is taken into account that, for $\gamma \gg 1$, the integral (11.7) can be calculated by the saddle-point method:

$$f(\gamma) = 8\sqrt{\left(\frac{\pi}{3}\right)} \gamma^{\frac{7}{3}} \exp(-3\gamma^{\frac{2}{3}}), \qquad \gamma \gg 1. \tag{11.8}$$

The adiabacity function f increases sharply with increase of quantum number and, in the limit $\gamma \to 0$, the function $f(\gamma) \to 1$. The simple interpolation formula [300] can be used for an approximate estimate of $f(\gamma)$

$$f(\gamma) \approx \frac{1}{2}\left[3 - \exp\left(-\frac{2\gamma}{3}\right)\right] \exp\left(-\frac{2\gamma}{3}\right) \tag{11.9}$$

giving an accuracy of about 20 per cent over the range of the variable $0 < \gamma < 20$.

The function $f(\gamma)$ reaches its maximum value $f \approx 1$ for an impulsive collision, when the mean collision time proves to be small in comparison with the period of vibration (i.e. for $\gamma \ll 1$). Moreover,

$$Z_{\text{max,trans}} = \frac{4\mu}{M} \cdot \frac{kTD}{\hbar \omega_0} \qquad \text{for } \alpha = 2\alpha'. \tag{11.10}$$

Since transitions between non-neighbouring states are forbidden, for the model considered, the condition for the applicability of perturbation theory to the calculation of the transition probability from the level n takes the particularly simple form $P_{n,n+1}, P_{n,n-1} \ll 1$. If this condition is broken, it becomes necessary to resort to a more accurate calculation, which takes the higher-order corrections of perturbation theory into account.

† If the trajectory of the colliding molecules is calculated taking the attraction into account, then the function $f(\gamma)$ must be replaced by a function of two variables $f(\gamma, x^*)$ where x^* characterizes the effect of the attraction on the probability of energy exchange. The function $f(\gamma, x^*)$ is calculated for the Morse potential in reference [396]. We note also papers [508], [509], in which the values of Z_{trans} are tabulated corresponding to the quantum transition probabilities (9.41).

The simplest model enabling the multiple quantum transitions, which occur in the higher orders of perturbation theory, to be considered corresponds to a harmonic oscillator under an external force. The region of applicability of this model is limited by the condition $(\hbar\omega) \ll (kT)$. On taking account, additionally, of the variation in the velocity of the molecules, the range of applicability can be extended to lower temperatures $(\hbar\omega \geqslant kT)$ also (see § 9). Numerical calculations within the framework of this model were carried out by Osipov [99] for the $O_2 + Ar$ system. The results of the calculation show that the transition probabilities are close to values for which the semiclassical approximation, which does not take into account the reaction of the oscillator on the relative motion of $AB + C$, is completely inapplicable. Hence the necessity of developing a theory which, already in the zero-order approximation, is able to take account of the loss of relative energy of AB and C becomes evident. Such a calculation can be carried out, within the strong coupling approximation, only under certain simplifying assumptions. In this connection, mention should be made of the work of Shuler and Zwanzig [486] and Mies and Shuler [368], in which a model was investigated free from the shortcomings of the model with an external force. In these papers, the transition probabilities for the impulsive interaction of an incident atom with an atom of the molecule were calculated numerically, and both the vibrations of the oscillator and also the relative motion were considered quantum mechanically, with full regard for the exchange of energy.

In the papers mentioned above, the impulsive nature of the interaction (an instantaneous collision) was secured by the particular choice of potential, a potential with rigid walls.

The mathematical part of the problem here reduces to the solution of a system of equations, giving the boundary condition of the incident and scattered waves on the absolutely rigid surface of the vibrating atom of the molecule. In spite of the fact that, within the framework of this model, the interaction of the incident atom with the atoms of the molecule is approximated by a sum of pair potentials, the solution of the problem in the general form cannot be formulated in terms of a pair interaction. This is connected with the contribution of the so-called multiple collisions. To explain this effect, we will consider a classical linear system A–B ... C, in which the atoms A and B are initially at rest. As a result of the collision, the atom C transfers some momentum to atom B, which clearly begins to move in the direction of A, whilst atom C changes its direction of motion. If, after the collision, C moves sufficiently slowly, then atom B, having rebounded from A may 'catch C up' and transfer to it part of the momentum received in the first collision. As the result of such repeated collisions, the full effectiveness of the energy transfer is diminished. Calculation of the second, third, etc. collisions forms the main difficulty in the theory of impulsive interactions.

§11 TRANSLATIONAL ENERGY IN MOLECULAR COLLISIONS

The physical effects, to which repeated collisions give rise, are considered in papers [153], [545].

It is clear that the problem is appreciably simplified if repeated collisions can be neglected. For this to be possible, it is necessary that the velocity of particle C, before and after the collision, exceed by far the velocity of B. Since for an oscillator, the velocity v is of the order of the ratio of the amplitude of vibration to the period T, the condition will be satisfied for

$$\frac{\omega}{\alpha v} \ll 1, \tag{11.11}$$

where it has been taken into account that $\alpha' < \alpha$.

At the same time condition (11.11) may be interpreted in another sense. We will revert to a real potential, which can be characterized by a finite (non-zero) radius of action $1/\alpha$. In this case, (11.11) means that the collision time $\tau_{\text{coll}} \sim 1/\alpha v$ is appreciably smaller than the period of vibration $1/\omega$. It is then clear physically that the result of a collision can be characterized completely by the momentum transferred from C to B, considering the interaction of C and B as a collision of two free particles and completely neglecting the effect of the potential $U_{AB}(r)$, which does not manage to show up in so short a time τ. Thus we see that relation (11.11), though imposing limits on the rigid sphere model, at the same time broadens the class of potentials to which the impulse approximation can be applied. Returning to the estimate of the product $\omega\tau$ by formula (7.28), we find that for real potentials and on condition $\alpha' \sim \alpha$, relation (11.11) can be satisfied only for small values of the ratio μ/M. The ratio μ/M is small for the molecules of a heavy distomic gas interacting with light particles. If the vibrational energy of the molecules AB undergoing relaxation and the translational energy of the molecule C are of the same order, then the following inequalities are simultaneously satisfied:

$$v_{AB} \ll v_{AB-C} \sim v_C$$

$$p_{AB} \gg p_{AB-C} \sim p_C. \tag{11.12}$$

They imply that the collision of C with B may be considered as a collision of C with a stationary wall. In addition, the momentum of C changes sign and B gains momentum $\Delta p_B \sim 2\mu v_{AB-C} \approx 2m_C v_i$. The momentum gained by atom B is distributed between the momentum p of the centre of gravity motion of the molecule and the momentum χ of the relative motion. Only the latter component is effective in the excitation of vibrations in AB.

We will consider for simplicity the one-dimensional case and will assume that, before the collision, the molecule is characterized by the vibrational functions ϕ_n. In our case it is convenient to use functions in the momentum representation which depend on the relative momentum p conjugate to the coordinate x. After the collision, the momentum p changes by the value χ, so that the result of a single interaction of B and C leads to an instantaneous (condition $\omega\tau \ll 1$) change [24], [49]

$$\phi_n(p) \to \phi_n(p-\chi), \tag{11.13}$$

where the minus sign on the right of (11.13) is explained in that the transfer of momentum tends to contract the molecule. Expanding $\phi_n(p-\chi)$ in terms of the stationary functions $\phi_m(p)$ we find the transition probability

$$P_{nm} = \left| \int_{-\infty}^{\infty} \phi_m^*(p)\phi_n(p-\chi)\,dp \right|^2. \tag{11.14}$$

The expression (11.14) can be rewritten in the form

$$P_{nm} = \left| \int_{-\infty}^{\infty} \phi_m^*(p)\hat{F}(-\chi)\phi_n(p)\,dp \right|^2, \tag{11.15}$$

where $\hat{F}(\chi)$ is an operator displacing the momentum by the value χ.

It is not difficult to verify that the formal expression for $\hat{F}(-\chi)$ has the form

$$\hat{F}(-\chi) = \exp\left(-\chi\frac{\partial}{\partial p}\right) = \exp\left(-i\frac{\chi}{\hbar}\hat{x}\right), \quad \hat{x} = i\hbar\frac{\partial}{\partial p}. \tag{11.16}$$

The exponential on the right is written in operator form, correct in any representation. Rewriting (11.15) with regard for (11.16) in the coordinate representation, we obtain†

$$P_{nm} = \left| \int_{-\infty}^{\infty} \phi_m^*(x) \exp\left(-i\frac{\chi}{\hbar}x\right)\phi_n(x)\,dx \right|^2. \tag{11.17}$$

Expressions (11.14) and (11.17) are completely equivalent: one can be obtained from the other as the result of a Fourier transform applied to the wavefunction. Therefore a calculation can be carried through using either of these expressions, basing the choice on grounds of convenience.

Condition (11.12), satisfied by the velocities and momenta both before and after the collisions, in fact implies that the reaction of the vibrations of the oscillator on the translational motion of atom C is neglected (atom B being represented by a stationary wall). Hence it is clear that the simplified variant of the impulse approximation considered is, in this sense, analogous to representing the actual collision process as the effect of an external force. Therefore, for the harmonic oscillator model, expression (11.17) must agree with (8.10), in which ε is calculated on the assumption that $\omega = 0$. Since (11.17) can also be extended to the model of a highly-anharmonic oscillator,

---

† Another derivation of this formula is given in the papers of Osipov [85], [86], [96].

this enables the exchange of vibrational–translational energy at high levels to be investigated, in certain cases.

We will now estimate the energy transfer for a system to which the impulse approximation is applicable. Considering a linear collision of AB ... C under conditions (11.12) (for $m_B = m_A$) we have:

$$\Delta E_{AB} = \Delta E_{AB,kin} = \frac{p_B \Delta p_B}{m_B} = \frac{p_B}{m_B}(2\mu v_C) = v_B(2\mu v_C)$$

$$\langle \Delta E_{AB}^2 \rangle = 4 E_{AB,kin} \left(\frac{\mu}{m_B}\right) \left\langle \frac{\mu v^2}{2} \right\rangle. \qquad (11.18)$$

Taking the contribution of the configuration C ... AB into account and calculating the average of $\langle \mu v^2/2 \rangle$ with respect to a normalized current, we obtain

$$\langle \Delta E_{AB}^2 \rangle = 4 \frac{\mu}{M} E_{AB,kin} kT. \qquad (11.19)$$

The impulse approximation has been used in a number of papers investigating the relaxation of heavy molecules in a light gas atmosphere [25], [89], [93] and in the calculation of the moderating power of a gas caused by the excitation of vibrations [110].

Finally, we will consider a numerical method of solution of the system of equations for the transition amplitude, with an intermolecular potential reproducing the true interaction more realistically. Such calculations were carried out by Rapp and Sharp [425], [428], [431], [474] and also by Marriott [355]–[358] for an exponential potential within the limits of the semiclassical and quantum methods. Up to the present time, however, not enough experience has accumulated in the solution of problems with many equations and strong coupling, to enable more or less definitive observations to be made. The numerical calculations succeed in taking into account the combined contribution of the direct transitions $n \to n'$ via one, two, etc. levels (allowed even in first order as a result of the anharmonicity of the intermolecular interaction) and transitions $n \to n'' \to n'$ through an intermediate level (allowed in higher orders for the harmonic potential). In this connection, it would be interesting to find a model which would permit the role of these effects to be made clear in a comparatively simple form. The model of a quantum oscillator in the field of a given force $F(t)$ takes account of the contribution of the transitions via an intermediate level, since the assumption of the linearity of the interaction operator in vibrational coordinates implies the complete neglect of direct multiple-quantum transitions. For the harmonic oscillator, with an interaction of arbitrary form $U(x, R)$, the analogy between the quantum and classical problems is lost. Nevertheless, as a satisfactory zero-order approximation, it turns out to be expedient

[527] to use the functions (8.6), in which $u(t)$ is determined by the solution of the equation

$$M\ddot{u} + M\omega^2 u = -\frac{\partial U(u, R)}{\partial u}. \tag{11.20}$$

Substitution of the function $u(t)$ into the final quantum result (8.9), taking account of the non-linear coupling of the oscillator with the external force in the classical approximation, gives, in some measure, the contribution of the direct multiple-quantum transitions to the probability $P_{n \to n'}$. Calculation shows, for example, that, for an oscillator with an exponential interaction $U \sim \exp(-\alpha\lambda x) f(t)$, the probability of multiple-quantum transitions at small velocities is somewhat smaller than for the oscillator with a linear interaction $-xf(t)$, which takes only the first term in the expansion of $U(x, R)$ [527], [528] into account. In this case, the qualitative effect of non-linearity agrees with the result found within the framework of perturbation theory (see §9): calculation of the higher terms in the expansion of $U(x, R)$ leads principally to a small additional separation of the vibrational terms of the oscillator, which reduces the probability of the single-quantum transitions. It seems to be impossible to predict the qualitative contribution of direct transitions for an arbitrary type of interaction.

However, it is clear that the intermediate levels n'' play an essential role in the determination of the transition probabilities $n \to n'$. Fig. 6 illustrates this case. Although the final population of states with $n = 2, 3$ is found to be small, after the end of the action of the perturbation on the system, which is initially in the level $n = 0$, the probability amplitudes $a_2(t)$ and $a_3(t)$ appreciably affect the amplitude $a_1(t)$, whose asymptotic value $|a_1(+\infty)|^2$ gives

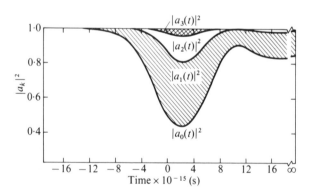

FIG. 6. Variation in the population of the nth level $|a_n|^2$ of a molecule of the N_2 type during the time of linear collision with an atom of the Ar type. Collision velocity 6×10^5 cm s^{-1} (Sharp and Rapp [474].

P_{01} [474]. In the given case, the states $n'' = 2$ and $n'' = 3$ serve as intermediate states for the transition $0 \to 1$.

In connection with the difficulties of solving the quantum problem on the excitation of vibrations, the calculations carried out within the limits of classical mechanics, with an exact solution of the equations of motion, are of interest. Such studies can only be carried out, of course, with the aid of computers. To begin with, the numerical investigation of the simplest system, namely the one-dimensional collision of a harmonic oscillator AB with the atom C, enables the limits of applicability of the model of a forced oscillator to be clarified. The calculations of Kelley and Wolfsberg [301] show that the possibility of replacing the interaction $xV(R)$ by a function of time $-xF(t)$ is basically determined by the value of the single parameter μ/M. As already mentioned, for small values of the ratio μ/M, the energy transfer is small, so that the deformation of the trajectory of the relative motion of AB and C, caused by the change in relative energy, can be neglected. The energy transfer differs from the ΔE_{class}, determined by expression (8.13), as μ/M increases. For example, for a molecule with identical nuclei and an exponential interaction between B and C, the actual value of the energy transfer ΔE turns out to be 1·7 or 3 times smaller than ΔE_{class} for values of μ/M equal to 1·3 and 3 respectively.

As another example, we cite the papers [154]–[157], [254] and the work of Alterman and Wilson [128], in which the classical equations of motion are integrated for the collision of a linear Morse oscillator with an atom, where the intermolecular interaction (taken into consideration only between the nearest atoms) is approximated by the Lennard–Jones potential. The masses and parameters of the potential for such a calculation were chosen appropriate to a model of the Br_2–Se system. A result of the investigation is an estimation of the kinetics of vibrational relaxation, calculated on the basis of a statistical averaging of the dynamical characteristics of different trajectories. In particular, the work contains a calculation of the classical transition probability $P(E-E')$ between levels E and E' of the molecule on collision, which is analogous to the quantum transition probability $P_{nn'}$. An example of such a function is shown in Fig. 7, where the energy transfer ΔE_{class} is plotted as the argument. For this case, the initial vibrational energy E (Br_2) is equal to $0·216D$ and the temperature of the heat reservoir $T = 4000$ K ($\approx 0·202D$). From the diagram can also be seen the preferred contribution from deactivation $E' < E$ in comparison with activation $E' > E$, which is due precisely to the change in the energy of the relative motion of the partners, not taken into account in the model with the external force.

Classical calculations of this nature enable the change in the mechanism of exchange of vibrational and translational energies with increase in vibrational energy and temperature to be followed [35]. For heavy molecules, a classical description must be completely satisfactory.

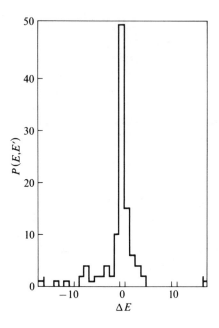

FIG. 7. Unnormalized transition probability $P(E, E')$ for the linear collision $Br_2 + Xe$ (Altermann and Wilson [128]). The energy transfer ΔE is plotted along the abscissa in units 0.276×10^{-13} ergs $= 0.435\hbar\omega_0$.

12. Resonant exchange of vibrational energy

In the collision of diatomic molecules AB and CD, simultaneous vibrational transitions are possible

$$AB(n) + CD(m) \rightarrow AB(n') + CD(m'). \quad (12.1)$$

If the energy difference $\Delta E_{vib} = |E_n + E_m - E_{n'} - E_{m'}|$ is so small that it can be supplied by the relative kinetic energy of the partners, with appreciable probability, then process (12.1) is referred to as a resonant exchange of vibrational energy. From the foregoing, it is clear that the condition for resonance can be written in the form

$$\frac{\Delta E_{vib}}{\hbar \alpha v} \leqslant 1. \quad (12.2)$$

The calculation of the probability of the reaction (12.1) to first order in perturbation theory can be simply carried out. To this end, in the Hamiltonian of the two interacting molecules, the interaction term $V(R, x_1, x_2)$ containing the vibrational coordinates of the molecules AB and CD should be separated

§12 TRANSLATIONAL ENERGY IN MOLECULAR COLLISIONS 93

out. Next, the classical trajectory $R = R(t)$ is introduced, and the standard formula is used to calculate $P_{nn'}^{mm'}$.

$$P_{nn'}^{mm'} = \left| \frac{1}{\hbar} \int [V(R(t), x_1, x_2]_{nn'}^{mm'} \exp\left(\frac{i\Delta E_{nn'}^{mm'} t}{\hbar}\right) dt \right|^2, \quad (12.3)$$

in which the exponential factor can be replaced by unity as a result of the inequality (12.2).

If the interaction $V(R, x_1, x_2)$ is approximated by the lowest term in its expansion in powers of x_1 and x_2 and the molecules are represented by harmonic oscillators, having assumed

$$V(R, x_1, x_2) = f(R)\alpha_1 x_1 \alpha_2 x_2, \quad (12.4)$$

then for the mean transition probability $n \to n+1$, $m \to m-1$ we obtain the expression

$$\langle P_{n,n+1}^{m,m-1} \rangle = (n+1)m Z_{01}^{AB} Z_{01}^{CD} Z_{trans}, \quad (12.5)$$

where

$$Z_{01}^{AB} = \left| \int_{-\infty}^{\infty} \phi_0^{AB}(x_1) \alpha_1 x_1 \phi_1^{AB}(x_1) dx_1 \right|^2 \quad (12.6)$$

$$Z_{01}^{CD} = \left| \int_{-\infty}^{\infty} \phi_0^{CD}(x_2) \alpha_2 x_2 \phi_1^{CD}(x_2) dx_2 \right|^2$$

and

$$Z_{trans} = \left\langle \left| \frac{1}{\hbar} \int_{-\infty}^{\infty} f(R) dt \right|^2 \right\rangle \approx \frac{kT \cdot D}{(\hbar\omega)^2}. \quad (12.7)$$

On the basis of these formulae, we will estimate the probability for the exchange of the lowest vibrational quantum between two identical molecules. The factor Z_{01} for a molecule of the N_2 type has a value of the order 0·01 [425], [426], [429]. Hence we find that for parameters $T \approx 300$ K, $\omega \approx 3000$ cm^{-1}, and $D \approx 5$ eV, the value of $\langle P_{01}^{10} \rangle$ lies in the range 10^{-4}–10^{-3}.

An important characteristic of resonant exchange by vibrational quanta is that the mean exchange probability grows with increase in temperature. In this, the process under consideration is distinguished from, for example, the resonant transfer of excitation or charge in atomic or molecular collisions; for these processes the probability of transfer (or the corresponding cross-section) diminishes with increasing temperature. Physically, the reason for the difference is the following. If there are two systems capable of energy interchange at resonance, then the frequency of the exchange Ω is determined by the matrix element of the interaction between the two states a and b, whose wavefunctions are stationary solutions of the Schrödinger equation with the interaction switched off. The nature of the resonant interaction depends essentially on the ratio between the period of exchange and the interaction time τ (the collision time). For $\Omega\tau \gg 1$, the excitation passes from one system to the other many times during the time of interaction, so that, on average, after the interaction is switched off, the probability of finding the excitation on one system is equal to $\frac{1}{2}$. Just such a situation occurs in the transfer of excitation or in charge exchange, for which the cross-section is equal to the effective area, with such impact parameters b as satisfy the condition $\Omega(b)\tau(b) \geqslant 1$. For $\Omega\tau \ll 1$, the transition probability is small and amounts to a value of the order $(\Omega\tau)^2$. This case occurs in the transfer of vibrational energy, where the small value of the product $\Omega\tau$ is secured by the small value of the matrix element $(\alpha x)_{01}$, i.e., in the final analysis, by the smallness of the amplitude of vibration in comparison with the radius of action of the force. The temperature dependence of the transition probability is directly determined by the temperature dependence of $(\Omega\tau)^2$, which follows from the physical meaning of Ω and τ: $\Omega \sim \overline{W} \sim kT$, $\tau \sim 1/v \sim T^{-\frac{1}{2}}$.

In order to trace the smooth passage between these two cases, it is necessary to solve the problem without the use of perturbation theory. This can be done for a model system of two oscillators, coupled by an interaction bilinear in the displacements x_1 and x_2. The Hamiltonian of such a system has the form

$$H = \left(\frac{p_1^2}{2M} + \frac{M\omega^2 x_1^2}{2}\right) + \left(\frac{p_2^2}{2M} + \frac{M\omega^2 x_2^2}{2}\right) + f(t)x_1 x_2. \qquad (12.8)$$

Comparing (12.8) with (8.4), it can be seen that this model for the resonant exchange of energy is very close to the model for the excitation of a single oscillator by an external force. The function $f(t)$ in (12.8), like the $F(t)$ in (8.4), has the form of discrete impulses, each of which describes the switching on of the interaction in a single collision. To calculate the transition probability due to a single collision, we only need to know the function $f(t)$ for a single switching on of the interaction.

The beats of two oscillators, introduced by the time dependence of the interaction $f(t)$, correspond to the classical solution of the problem with the Hamiltonian (12.8). To find a solution to the quantum problem, we will make certain assumptions about the values of ω, τ, and Ω, namely: we will assume the inequalities $\omega\tau \gg 1$ and $\omega \gg \Omega$. The first of these corresponds exactly to the physical situation in the study of the resonant transfer of energy, when we can neglect the conversion of vibrational energy into translational motion. It also signifies that terms of the type $F_1(t)x_1$ and $F_2(t)x_2$, which in principle might have been added to (12.8), can be neglected. The second condition allows us to disregard the change in the fundamental frequency of the oscillator as a result of the interaction due to terms of the type $f_1(t)x_1^2$ and $f_2(t)x_2^2$ omitted in (12.8). The transition probabilities found under these

§12 TRANSLATIONAL ENERGY IN MOLECULAR COLLISIONS

assumptions equal [71]:

$$P_{n,n+\Delta}^{m,m-\Delta} = \frac{(1-\mu)^\Delta(1+\mu)^{n-m+\Delta}(m-\Delta)!}{2^{n+m}n!m!(n+\Delta)!} \left\{ \frac{d^{n+\Delta}}{d\mu^{n+\Delta}}[(1-\mu)^n(1+\mu)^m] \right\}^2, \quad (12.9)$$

where

$$\mu = \cos 2\theta, \qquad \theta = \Omega\tau = (Z_{01}^{AB}Z_{01}^{CD})^{\frac{1}{2}} \int_{-\infty}^{\infty} \frac{f(t)\,dt}{\hbar}. \quad (12.10)$$

In particular, for transitions $0 \to \Delta$, $N \to N-\Delta$ we obtain

$$P_{0,\Delta}^{N,N-\Delta} = \frac{N!}{2^N \Delta!(N-\Delta)!}(1-\mu)^\Delta(1+\mu)^{N-\Delta}. \quad (12.11)$$

Formulae (12.9) and (12.11) are the analogues to formulae (8.10) and (8.11) describing the excitation of an oscillator under the action of a random force. For small interactions, the parameter $(1-\mu)$ is small, and in (12.9) we need retain only the lowest power of θ. The result of first-order perturbation theory is then obtained

$$P_{n,n+1}^{m,m-1} = (n+1)m\theta^2 \quad (12.12)$$

corresponding to (12.5). This replaces formula (8.18) which describes a single-quantum transition in the conversion of vibrational energy into translational.

The change-over from the resonant to the non-resonant case can be followed only for weak interactions, where perturbation theory is applicable to the calculation of probabilities. The transition probability $n \to n'$, $m \to m'$ takes the form

$$P_{nn'}^{mm'} = Z_{nn',\text{vib}}^{AB} Z_{mm',\text{vib}}^{CD} \frac{8\mu kT}{\hbar^2 \alpha^2} f(\gamma), \quad (12.13)$$

where $f(\gamma)$ is the function determined by the integral (11.7); the argument γ is given by the expression

$$\gamma = \frac{\pi|E_n + E_m - E_{n'} - E_{m'}|}{\hbar\alpha} \sqrt{\left(\frac{\mu}{2kT}\right)}. \quad (12.14)$$

On taking anharmonicity into account, transitions between one, two, or more levels become possible. For molecules AB and CD with frequencies differing appreciably, such processes must be taken into account, generally speaking, since the big reduction in the vibrational factors Z_{vib} for transitions $n \to n\pm 2$ in comparison with transitions $n \to n\pm 1$ may well be compensated by an increase in the function f due to the collision conditions approaching those for resonance. In this sense, the manifestation of anharmonicity in the collision of diatomic molecules differs essentially from the case of a collision of a diatomic molecule with an atom, when multiple quantum transitions can be neglected under adiabatic conditions.

Calculations for a number of molecular systems, in which the effect of the proximity to resonance on the value of the quantum exchange probability is investigated, are contained in paper [504]. In particular, for the approximate evaluation of P_{10}^{01} for a molecule of the N_2 type, Rapp [430] suggested employing the formula

$$P_{10}^{01} = 3.7 \times 10^{-6} T \cosh^{-2}\left(\frac{0.174 \Delta E}{T^{\frac{1}{2}}}\right), \qquad (12.16)$$

where ΔE is expressed in cm^{-1} and T in degrees K.

13. Exchange of energy in the collisions of polyatomic molecules

In contrast to the collision of diatomic molecules, in collisions of polyatomic molecules, the exchange of the vibrational and translational energies is accompanied by a redistribution of the vibrational energy within each molecule. In considering transitions between the lowest vibrational states of the molecules, we can limit ourselves to the harmonic approximation and represent the molecule by a system of independent oscillators, each of which corresponds to a definite normal vibration. However, we should keep in mind that such a representation is only valid when the difference $E_n - E_m$ in the vibrational terms, calculated in the harmonic approximation, greatly exceeds the matrix element of the anharmonic interaction between these states. If, by chance, two harmonic terms coincide or turn out to be close, then the anharmonicity must be taken into account by constructing correct vibrational functions, which should be written as a linear combination of the functions of a system of harmonic oscillators. Nevertheless, if the difference between the energies of the two terms is neglected and they are regarded as degenerate, then it is not possible to take the anharmonic interaction between the normal vibrations into account to the same degree of accuracy. The vibrational relaxation times and the transition probabilities between the lower vibrational levels of polyatomic molecules are customarily considered to precisely this approximation [262], [263], [560].

Although the intramolecular anharmonicity can often be nelected, the anharmonicity due to the intermolecular interaction must certainly be taken into account. The higher powers in the expansion of the interaction $V(R, Q_1, Q_2, \ldots, Q_s)$ in terms of the normal coordinates Q_i are responsible for the vibrational transitions by which some of the vibrational quantum numbers suddenly change. Here, it is possible to trace the analogy with the case of a collision between two diatomic molecules with almost resonant exchange of vibrational energy. However, the existence of several normal vibrations in polyatomic molecules ensures a greater chance of satisfying the conditions for resonance when only a small fraction of the relative energy of the partners is converted into vibrational energy.

Within the framework of the model of a radially vibrating sphere, the mean probability $P_{nn'}^{mm'}$ of a vibrational transition in which the quantum number of one partner change from $n \to n'$, whilst that of the other changes from $m \to m'$, can be calculated by formula (12.13). It differs from the case of diatomic molecules only in the definition of the vibrational factors $Z_{nn',\text{vib}}$, which must now be attributed to specific normal vibrations. Since vibrations of the surface atoms of the molecule along the radius r are important for the model being considered, the factors $Z_{nn',\text{vib}}(\omega_i)$ are defined in terms of the derivatives of r with respect to the ith normal coordinate. In practice, generally, transitions with Δn equal to 0, 1, and 2 are of interest. The corresponding expressions for such $Z_{nn',\text{vib}}$ are [356], [357], [504]

$$Z_{nn,\text{vib}} = 1, \tag{13.1a}$$

$$Z_{n,n+1,\text{vib}}(\omega_i) = \frac{\alpha^2 \hbar}{2 M_i \omega_i}(n+1) \sum_{k=1}^{p} \frac{(\partial r_k/\partial Q_i)^2}{p}, \tag{13.1b}$$

$$Z_{n,n+2,\text{vib}}(\omega_i) = \frac{\alpha^4 \hbar^2}{4 M_i^2 \omega_i^2} \frac{(n+1)(n+2)}{4} \sum_{k=1}^{p} \frac{(\partial r_k/\partial Q_i)^4}{p}, \tag{13.1c}$$

where M_i and ω_i are the effective masses and frequencies respectively of the normal vibrations; r_k is the radial coordinate of the surface of the kth atom.

As a result of the large contributions of the combinational transitions, the relaxation of each normal vibration is not an independent process. The set of relaxation times $\tau_{i,\text{vib}}$, which are obtained as the eigenvalues of the kinetic matrix of the system of equations describing the exchange of energy between the different normal vibrations and between the vibrational and translational motion, serves as an analogue to the relaxation time τ_{vib}. If the characteristic times differ sufficiently in magnitude, then each can be roughly attributed to some elementary process, in the same sense as the relaxation time τ_{vib} of the diatomic molecules is attributed to the 1–0 transition.

Such an ascription, together with the results of calculating the transition probabilities for certain combinational processes for methane, are given in Table 2.

The slowest stage corresponds to the deactivation of the first low-frequency vibrational state. Deactivation (and activation) of other normal vibrations proceeds as a result of combinational processes and not as a result of direct excitation. In going over to deuterated methane, the transition probability $P_{10}(v_4)$ must increase, since the magnitude of the vibrational quantum $\hbar\omega_4$ is decreased and the reduced mass M_4 changes insignificantly. The corresponding increase of the exponential factor exceeds the decrease of the

TABLE 2

Vibrational transitions for a molecule of methane at $T = 298$ K

Symmetry of the normal vibration and frequency (cm^{-1})	Combinational transition responsible for relaxation of the corresponding normal vibration
Model of radially-vibrating molecules [262], [504]	
$v_1(a_1) = 2915$ (CH$_4$)	$P_{01,10}(v_1, v_3) = 1.9 \times 10^{-2}$
$v_2(e) = 1534$ (CH$_4$)	$P_{01,10}(v_2, v_4) = 7.8 \times 10^{-3}$
$v_3(f_2) = 3019$ (CH$_4$)	$P_{01,20}(v_3, v_2) = 1.3 \times 10^{-3}$
$v_4(f_2) = 1306$ (CH$_4$)	$P_{10}(v_4) = 0.8 \times 10^{-4}$
$v_4(f_2) = 996$ (CD$_4$)	$P_{10}(v_4) = 2.1 \times 10^{-4}$
Model of rough molecules [504]	
$v_4(f_2) = 1306$ (CH$_4$)	$P_{10}(v_4) = (0.8-0.5) \times 10^{-4}$
$v_4(f_2) = 996$ (CD$_4$)	$P_{10}(v_4) = (0.4-0.2) \times 10^{-4}$

pre-exponential term determined by the reduction in the amplitude of vibration on isotopic substitution.

Experiment does not confirm this deduction of the theory: the vibrational relaxation rate of CD$_4$ at $T = 298$ K proves to be approximately twice smaller (and not greater) than the relaxation rate for CH$_4$.

A reasonable explanation of this contradiction is that the rotation of either the molecule being deactivated or of the partner in the collision participates in the vibrational deactivation process. For example, for the rough sphere models, the exponent of the exponential in the expression for $P_{10}(v_4)$ is reduced in going from H to D. This is connected with the fact that, although the effective mass of the normal vibration ω_4 is increased in the ratio $|\omega_4(CH_4)/\omega_4(CD_4)|^2 \approx 1.72$, the effective reduced mass m_r increases even more—by a factor of two. The over-all change in the probability is such that the vibrational deactivation probability for CH$_4$ proves to be greater than for CD$_4$.

An analogous situation occurs for other polyatomic molecules containing light atoms, where the model of rough molecules enables a satisfactory interpretation to be made of the experimental results on vibrational transitions induced by collisions [207]–[210], [270], [276].

3

THE EXCHANGE OF ELECTRONIC, VIBRATIONAL, AND TRANSLATIONAL ENERGY IN MOLECULAR COLLISIONS

14. The classification of nonadiabatic transitions

IN this chapter, in an account of the theory of nonadiabatic transitions, we will limit ourselves to the semiclassical approximation, within whose bounds, as was shown in § 1, the motion of the nuclei is considered as classical. Such a division of the full problem of the interaction of electrons and nuclei into two parts—the quantum problem of the calculation of the adiabatic electronic terms and the semiclassical problem of the calculation of the probability of nonadiabatic transitions induced by the motion of the nuclei along the given classical trajectory $R(t)$—considerably simplifies the theory. At the same time, the assumption of the existence of a single trajectory $R(t)$, attributed straight away to several adiabatic terms, introduces a certain indefiniteness into the theory, since, generally speaking, the parameters of this trajectory cannot be fixed precisely. This problem has already been discussed, in application to vibrational transitions, in § 6, where a special procedure for the symmetrization of the transition probabilities with respect to the initial and final relative velocities of the colliding molecules was introduced. The analogous problem of the symmetrization of the electronic transition probability always arises, in some form or other, within the framework of the semiclassical approximation. Moreover, it is found that, in a number of the most important cases, a well-defined correspondence can be established between the semiclassical approximation and the exact quantum calculation. It is precisely this which finally warrants the use of the semiclassical approximation. Later on, we will return to a discussion of this problem.

If the adiabatic terms of the molecular system are known, then the time dependence of the probability amplitudes a'_n of the different adiabatic functions ϕ'_n, which contribute to the nonadiabatic function Ψ, is expressed by a system of equations (1.17) In principle, this system is infinite, since it couples the amplitudes of all the electronic states. To obtain concrete results, it is necessary to resort to some form of simplification leading to the replacement of the infinite system by a finite one. However, the following problem arises. The adiabatic electronic functions are usually written down in the form of a linear combination of functions with respect to a certain basis. Therefore the expansion (1.16) in fact contains two expansions—with respect to the adiabatic functions and with respect to the functions of the chosen basis. In such a situation, it is often simpler to find the nonadiabatic function Ψ, from the start, in the form of an expansion in terms of the functions of the chosen basis, determining the expansion coefficients from a system of auxiliary equations.

For the derivation of the corresponding system of equations and their comparison with (1.17) we will consider the somewhat more general problem of nonadiabatic transitions, assuming that in a certain molecular system there exist two types of subsystem—the fast and the slow. On average, the former is characterized by higher velocities than the latter. We will denote the set of coordinates of the first subsystem by **r** and of the second by **R**. If the Hamiltonian of the fast subsystem, which depends on **r** as on a set of dynamical variables and on **R** as on a set of parameters, is denoted by $H(\mathbf{r}, \mathbf{R})$, then the adiabatic functions of this subsystem $\phi_n(\mathbf{r}, \mathbf{R})$ are determined by the eigenvalue equations

$$H(\mathbf{r}, \mathbf{R})\phi_n(\mathbf{r}, \mathbf{R}) = U_n(\mathbf{R})\phi_n(\mathbf{r}, \mathbf{R}). \tag{14.1}$$

If by **r** is understood the coordinates of the electrons and by **R** the coordinates of the nuclei, then $H(\mathbf{r}, \mathbf{R})$ agrees with the electronic Hamiltonian $H_e(\mathbf{r}, \mathbf{R})$ or $H'_e(\mathbf{r}, \mathbf{R})$ and $\phi_n(\mathbf{r}, \mathbf{R})$ with the corresponding electronic adiabatic functions (§ 1). Sometimes the coordinates of the electrons and the vibrational coordinates of the nuclei of the molecular system can be attributed to the set **r**. Then $H(\mathbf{r}, \mathbf{R})$ will represent the electronic–vibrational Hamiltonian, parametrically dependent on the set of coordinates **R**, characterizing the rotation of the molecular system and the slow relative motion of its separate parts (see §§ 20, 21). Introducing the classical motion, which is described by the trajectory $\mathbf{R} = \mathbf{R}(t)$ in the space of the coordinates **R**, we obtain eqn. (1.15) for the nonadiabatic function Ψ of the fast subsystem, whence we arrive at the system of equations (1.17), after the introduction of the expansion (1.16). In accordance with the notation assumed in formulae (1.15)–(1.17), here, H'_e, ϕ'_n and U'_n should be replaced by $H(\mathbf{r}, \mathbf{R})$, $\phi_n(\mathbf{r}, \mathbf{R})$, and $U(\mathbf{R})$. In place of the expansion (1.16) we can introduce an expansion in terms of a certain basis set $\phi_n^0(\mathbf{r}, \mathbf{R})$ whose components are defined as the eigenfunctions of the operator $H^0(\mathbf{r}, \mathbf{R})$.

$$H^0(\mathbf{r}, \mathbf{R})\phi_n^0(\mathbf{r}, \mathbf{R}) = U_n^0 \phi_n^0(\mathbf{r}, \mathbf{R}). \tag{14.2}$$

In the representation ϕ_n^0, the matrix of the Hamiltonian $H(\mathbf{r}, \mathbf{R})$ is non-diagonal, so that the functions ϕ_n are expressed in the form of a linear combination of the functions ϕ_n^0:

$$\phi_n = \sum_m c_{nm} \phi_m^0. \tag{14.3}$$

Formula (14.3) agrees formally with the expression which gives the connection between the molecular and atomic orbitals in the one-electron theory of

§ 14   TRANSLATIONAL ENERGY IN MOLECULAR COLLISIONS   101

molecules. Therefore in problems on the nonadiabatic transitions, the functions ϕ_n are often spoken of as the molecular basis and the functions ϕ_n^0 as the atomic basis.

We will represent the nonadiabatic function Ψ in the form of an expansion

$$\Psi(\mathbf{r}, t) = \sum_n b_n(t) \exp\left[-\frac{i}{\hbar}\int^t H_{nn}(\mathbf{R})\,dt\right]\phi_n^0(\mathbf{r}, \mathbf{R}), \quad (14.4)$$

where H_{nn} is the diagonal element of the Hamiltonian $H(\mathbf{r}, \mathbf{R})$ with respect to the basis ϕ_n^0.

Substituting $\Psi(\mathbf{r}, t)$ in the non-stationary equation

$$i\hbar\frac{\partial\Psi(\mathbf{r}, t)}{\partial t} = H[\mathbf{r}, \mathbf{R}(t)]\Psi(\mathbf{r}, t), \quad (14.5)$$

we obtain the following system of equations for the coefficients:

$$i\hbar\dot{b}_n = \sum_m \left[H_{nm}(t) - i\hbar\left(\frac{\partial}{\partial t}\right)^0_{nm}\right]\exp\left[-\frac{i}{\hbar}\int^t (H_{mm}-H_{nn})\,dt\right]b_m, \quad (14.6)$$

where the matrix elements are defined by the relations

$$H_{nm}(\mathbf{R}) = \langle\phi_n^0|H(\mathbf{r},\mathbf{R})|\phi_m^0\rangle$$

$$\left(\frac{\partial}{\partial t}\right)^0_{nm} = \langle\phi_n^0|\frac{\partial}{\partial t}|\phi_m^0\rangle$$

$$\mathbf{R} = \mathbf{R}(t). \quad (14.7)$$

From (14.6), it can be seen that the states of the Hamiltonian $H^0(\mathbf{r}, \mathbf{R})$ are connected not only by the nonadiabaticity operator $C = -i\hbar(\partial/\partial t)$, but also by the adiabatic interaction, which arises as a result of the arbitrary choice of the basis ϕ_n^0. Of course, in many problems the choice of the basis ϕ_n^0 is determined by some physical condition or other, where the interaction which leads to the difference between H^0 and H can be demonstrated and, what is more important, evaluated. In a formal derivation of the system (14.6), however, the choice of the basis ϕ_n^0 remains arbitrary.

In an exact solution to the problem of finding the nonadiabatic function Ψ, the systems of equations (1.17) and (14.6) are completely equivalent. If, however, the systems are restricted to a finite number of N equations, then (1.17) and (14.6), generally speaking, will give different results. It is precisely at this stage that the choice of the basis ϕ_n^0 proves to be important, since different sets of N functions ϕ_n^0 will give different approximations to the exact function. In particular, it can be shown that the finite set ϕ_n^0, giving the best approximation to the adiabatic functions in the sense of the expansion (14.3), will not be the best in the sense of the expansion (14.4).

In future, for simplicity, we will limit ourselves to the so-called two-state approximation, where the nonadiabatic coupling between only two electronic terms is taken into account. The grounds for this approximation are that, at sufficiently low nuclear energies (in the scale of atomic units), the Massey parameter (1.24) will be large for the majority of pairs of states, so that, for them, the dynamic coupling of the electrons and nuclei can be neglected. However, this parameter may be small for states whose energies turn out to be close for some reason or other. It is not very likely that the energies of several states will prove to be close. The simplest situation is one in which only two electronic terms converge. This enables us to retain only two terms out of the total sum in (14.6), i.e. to approximate the nonadiabatic function by an expression of the form

$$\Psi(\mathbf{r}, t) = a_1(t)\exp\left[-\frac{i}{\hbar}\int^t U_1(\mathbf{R})\,dt\right]\phi_1(\mathbf{r}, \mathbf{R})$$
$$+ a_2(t)\exp\left[-\frac{i}{\hbar}\int^t U_2(\mathbf{R})\,dt\right]\phi_2(\mathbf{r}, \mathbf{R}), \quad (14.8)$$

where ϕ_k are the adiabatic electronic functions.

The possibility that terms U_1 and U_2 converge in certain regions of the configuration space of the nuclei is physically connected with the fact that the total adiabatic Hamiltonian can be represented in the form of a certain zero-order Hamiltonian and a comparatively small perturbation $H = H^0 + V$. The choice of the zero-order Hamiltonian H^0 is not unique. We only require that the corresponding eigenfunctions $U_k^0(\mathbf{R})$ be sufficiently smooth functions of the coordinates $\mathbf{R}$, i.e. they should change substantially only over distances of the order of atomic dimensions. In this case, the strong convergence of the two terms U_1^0 and U_2^0 at a certain point $\mathbf{R}_p$ inevitably implies that they cross over at a point near to $\mathbf{R}_p$. We use this fact to investigate the forms of the potential energy surfaces in the region of their convergence. The matrix of the Hamiltonian H with respect to a basis of functions ϕ_1^0 and ϕ_2^0 has the form

$$H = \begin{pmatrix} H_{11}(\mathbf{R}) & H_{12}(\mathbf{R}) \\ H_{21}(\mathbf{R}) & H_{22}(\mathbf{R}) \end{pmatrix}, \quad (14.9)$$

where the eigenvalues of the matrix are

$$U_{1,2} = \bar{U}(\mathbf{R}) \pm \Delta U(\mathbf{R}),$$
$$\bar{U}(\mathbf{R}) = \tfrac{1}{2}[H_{11}(\mathbf{R}) + H_{22}(\mathbf{R})],$$
$$\Delta U(\mathbf{R}) = \tfrac{1}{2}\{[H_{11}(\mathbf{R}) - H_{22}(\mathbf{R})]^2 + 4|H_{12}(\mathbf{R})|^2\}^{\frac{1}{2}}. \quad (14.10)$$

The difference in the adiabatic energies $\Delta U(\mathbf{R})$ is determined by the value of the radical in (14.10). As proposed, ϕ_n^0 changes smoothly for a change in the nuclear coordinates, therefore the matrix elements H_{ik} and the eigenvalues

U_n^0 too change appreciably only over an interval of the order of an atomic unit a_0. Therefore in a study of the dependence of ΔU on the configuration of the nuclei, after their small displacement by $|\delta \mathbf{R}| \ll a_0$, close to a certain point $\mathbf{R}_c$, the matrix elements under the radical can be expanded in a series in $\delta \mathbf{R} = \mathbf{R} - \mathbf{R}_c$, which is then limited to the first few terms. Of course, to what extent such an analysis of terms in a small neighbourhood is sufficient for a calculation of the nonadiabatic transition probabilities remains an open question. Leaving a proper discussion to the following sections, we will investigate the general conditions under which the adiabatic terms may cross over. To this end, we will subject the choice of points $\mathbf{R}_c$ to the condition for intersection, i.e. to the requirement that $\Delta U(\mathbf{R}_c) = 0$. The tending to zero of the expression (14.10) implies the fulfillment of the relations

$$H_{11}(\mathbf{R}_c) - H_{22}(\mathbf{R}_c) = 0, \qquad (14.11\text{a})$$

$$|H_{12}(\mathbf{R}_c)| = 0. \qquad (14.11\text{b})$$

If the matrix element H_{12} is complex (which occurs, for example, when the spin–orbit coupling is considered as a perturbation) then (14.11b) in fact represents the two conditions

$$\text{Re } H_{12}(\mathbf{R}_c) = 0, \qquad (14.12\text{a})$$

$$\text{Im } H_{12}(\mathbf{R}_c) = 0. \qquad (14.12\text{b})$$

Thus, in the most general case, there are three additional conditions ((14.11a) and (14.12)) imposed on a possible choice of the points of intersection of the adiabatic terms. Hence, it follows that, for a system with s degrees of freedom, the s-dimensional hypersurfaces of the adiabatic electronic terms (s-dimensional manifolds) intersect in a manifold of dimension $s-3$.

The following particular cases deserve special attention. If $H(\mathbf{r}, \mathbf{R})$ is identified with the electronic Hamiltonian $H_e(\mathbf{r}, \mathbf{R})$ in which the spin–orbit coupling is disregarded, then the matrix elements H_{ik} can be considered real (since the electronic functions ϕ_n can be taken to be real). Then condition (14.12b) is satisfied identically, and there remain only two additional limitations on the choice of points of intersection. Therefore the dimension of the manifold of intersection increases to $(s-2)$. In this case, terms of a system with one degree of freedom do not intersect; with two, they intersect at one point; with three, they intersect along a curve, and so on.

If, finally, the off-diagonal matrix element H_{12} vanishes identically, then the dimension of the manifold of intersection equals $s-1$, i.e. the intersection of the terms of a one-dimensional system in a point is possible, the terms of a two-dimensional system intersect along a curve and so on. The tending to zero of the off-diagonal matrix element H_{12} in a basis of functions of the zero-order approximation is usually connected with the fact that ϕ_1^0 and ϕ_2^0 refer to different types of symmetry of the total adiabatic Hamiltonian. In particular, for a system of two atoms, the Hamiltonian $H^0(\mathbf{r}, \mathbf{R})$ can be chosen so that it does not contain part of the electrostatic interaction between the electrons and nuclei (for example, the exchange interactions of the electrons belonging to different atoms) and the spin–orbit coupling. In this case, into the set of quantum numbers n of the functions ϕ_n^0 enter the components of the angular momentum of the electrons Λ along the molecular axis and the component of the total spin along this axis Σ. The matrix elements of the electrostatic part of the interaction, coming into $H(\mathbf{r}, \mathbf{R})$ are diagonal with respect to the quantum numbers Λ and Σ, but matrix elements of the spin–orbit coupling are non-diagonal with respect to Λ and Σ but diagonal with respect to the exact quantum number of the adiabatic functions of the Hamiltonian $H(\mathbf{r}, \mathbf{R})$, namely the component Ω of the total angular momentum of the electrons along the axis ($\Omega = \Lambda + \Sigma$).

Thus the adiabatic molecular terms with the same value of Ω will not intersect. If, however, the spin–orbit coupling is considered as negligibly small, then the rule for intersection can be formulated separately in relation to the quantum numbers Λ and Σ. In such a situation, intersection of terms with the same value of Ω is possible only if these terms differ in the values of Λ and Σ. In considering specific problems, the imprecision which exists in the literature over the definition of the rules for the non-intersection of the terms with the same symmetry in a system of two atoms should be borne in mind. The possibility of the intersection of the adiabatic terms of the same symmetry, in the multi-dimensional case, enables the surfaces $U_1(\mathbf{R})$ and $U_2(\mathbf{R})$ to be considered as two sheets of the same surface. In particular, for systems with two degrees of freedom, the double-sheeted nature of the surface becomes apparent in the following way. For motion over the lower sheet along a closed path embracing the point of intersection (or contact) of the surface, the electronic wavefunctions ϕ_1 of the lower state changes sign as it travels round (and similarly for the upper state).

Thus, even for motion on one sheet only, there is an effect which indicates the existence of another sheet.

If the classical trajectory $\mathbf{R} = \mathbf{R}(t)$ is assigned then, generally speaking, it will not pass through the locus of points corresponding to the intersection of the potential energy surfaces. Therefore along this trajectory $\Delta U|\mathbf{R}(t)|$ does not tend to zero, but possibly goes through a minimum. The functions ϕ_n, whose time dependence for motion along the given trajectory induces non-adiabatic transitions, can be expressed in terms of the functions ϕ_n^0. It is convenient to do this because in view of the properties of ϕ_i^0 assumed (a substantial variation of ϕ_i^0 over lengths of order a_0) an estimate is known for the matrix operator of the nonadiabaticity between these functions:

$$\left\langle \phi_1^0 \left| -i\hbar \frac{\partial}{\partial t} \right| \phi_2^0 \right\rangle \sim -i\hbar v \left\langle \phi_1^0 \frac{\partial \phi_2^0}{\partial R} \right\rangle \sim -i\hbar v/a_0. \qquad (14.13)$$

As regards the matrix elements of the operator $-i\hbar\partial/\partial t$ with respect to the functions ϕ_i, which enter into the basic equations (1.17) for the nonadiabatic transition amplitudes, they depend on the expansion coefficients of the ϕ_i with respect to the ϕ_i^0, i.e., in the end, on how closely the trajectory passes to the point of intersection of the terms.

Diagonalizing the matrix (14.9), it is easy to find the following formulae, which express the adiabatic functions of the exact Hamiltonian (here, by the exact Hamiltonian is to be understood the Hamiltonian of the two states considered) in terms of the adiabatic functions of the zero-order Hamiltonian:

$$\phi_1 = \phi_1^0 \cos\chi + \phi_2^0 \sin\chi$$
$$\phi_2 = -\phi_1^0 \sin\chi + \phi_2^0 \cos\chi, \qquad (14.14)$$

where

$$\chi = \tfrac{1}{2}\arctan\frac{2V_{12}}{E_1^0 + V_{11} - E_2^0 - V_{22}}.$$

The equations for the probability amplitudes a_1 and a_2, which are obtained from the system (1.17) if all its states are discarded except the two under consideration and the transformation (14.14) is substituted, have the form

$$i\hbar\dot{a}_1 = i\hbar(\dot{\chi} + \langle\phi_1^0|\dot{\phi}_2^0\rangle)\exp\left[-\frac{i}{\hbar}\int^t (U_2 - U_1)\,dt\right]a_2$$

$$i\hbar\dot{a}_2 = -i\hbar(\dot{\chi} + \langle\phi_1^0|\dot{\phi}_2^0\rangle)\exp\left[+\frac{i}{\hbar}\int^t (U_2 - U_1)\,dt\right]a_1, \qquad (14.15)$$

where it is taken into account here that ϕ_1^0 and ϕ_2^0 are real, orthogonal functions.

Expressing Ψ in the form of an expansion in the functions ϕ_n^0

$$\Psi = b_1 \exp\left(-\frac{i}{\hbar}\int^t H_{11}\,dt\right)\phi_1^0 + b_2 \exp\left(-\frac{i}{\hbar}\int^t H_{22}\,dt\right)\phi_2^0, \qquad (14.16)$$

it is not difficult to obtain the system of equations for the coefficients

$$i\hbar\dot{b}_1 = (H_{12} - i\hbar\langle\phi_1^0|\dot{\phi}_2^0\rangle)\exp\left[-\frac{i}{\hbar}\int^t (H_{22} - H_{11})\,dt\right]b_2$$

$$i\hbar\dot{b}_2 = (H_{21} + i\hbar\langle\phi_1^0|\dot{\phi}_2^0\rangle)\exp\left[\frac{i}{\hbar}\int^t (H_{22} - H_{11})\,dt\right]b_1. \qquad (14.17)$$

Returning to eqns (14.15), it is easy to find parameters whose values conveniently classify the nonadiabatic transitions. First we note that a typical formulation of the problem is such that, in the initial and final portions of a trajectory $R_\alpha = R_\alpha(t)$, the nonadiabatic coupling of the states ϕ_1 and ϕ_2 can be neglected. This means that the wavefunction is represented by relation (1.16) with constant coefficients a_i. Therefore, as one parameter, is is expedient to introduce the difference in the energies $\Delta U_\infty = (U_1 - U_2)_\infty$, which characterizes the splitting of the terms far from the region of nonadiabatic coupling. In particular, here, the U_i may be the energy levels of the colliding atoms or molecules at large internuclear separations; this justifies the introduction of the notation ∞. In the motion of the representative point along the given trajectory, nonadiabatic transitions occur in the region where the terms converge. In this region, the nonadiabatic coupling can be characterized by the frequency (or inverse time) of the variation of the coefficients in eqn (14.15). To each of them ($d\chi/dt$ and $\langle \phi_1^0 | d\phi_2^0/dt \rangle$) there corresponds a characteristic time τ_1 and τ_2.

The solution to eqns (14.15) is determined, in the main, by the frequency of oscillation of the exponential terms. For a high frequency of oscillation and a slow variation of the pre-exponential terms, which are proportional to the velocity of the nuclei, the total effect of the integration is close to zero. In order to separate out the resultant small effect of an increase (or decrease) of the transition amplitude from terms, generally speaking, large at all times but often oscillating, it is convenient to go over from an integration over the real time axis to an integration over a path in the complex time plane. This method of integration of a system of differential equations may be regarded as a generalization of the method of calculating definite integrals by deforming the contour of integration in the complex plane of the variables of integration, which was employed to calculate the vibrational transition probabilities (see § 8). If the path of integration is chosen so that it passes through the point t_c, where the frequency $\Delta U(t)/\hbar$ tends to zero, the effect of the oscillating factors is reduced to a considerable degree. Therefore a special significance is attached to the moment t_c, determined by the condition

$$\Delta U(t_c) = 0. \tag{14.18}$$

It is not difficult to see that this is the condition for the intersection of the terms U_1 and U_2. Since, however, the chosen trajectory, generally speaking, does not pass through the point of intersection, the time t_c must be complex,

$$t_c = \tau' + i\tau''. \tag{14.19}$$

On the basis of these qualitative considerations, we will now define τ' as the time corresponding to the moment of passing through the centre of the region of nonadiabaticity, τ'' as the effective time of passage through this region, and $\Delta U_p = \Delta U(\tau')$ as the splitting of the terms in the region of nonadiabaticity. A strict proof that for small velocities ($\Delta U_p \tau''/\hbar \gg 1$) the transitions between the electronic terms are actually localized in a region of width of order τ'' near the point τ' is given in adiabatic perturbation theory [37], [106].

Thus there are three parameters with the dimension of a frequency, namely, $\Delta U_\infty/\hbar$, $\Delta U_p/\hbar$, and $1/\tau''$.

From these, two dimensionless parameters can be formed, whose values chiefly determine the nonadiabatic transition probabilities. In particular, the relative splitting of the terms $\Delta U_p/\Delta U_\infty$ serves as one of the parameters and the Massey parameter, referred either to the centre of the region of nonadiabaticity ($\xi_p = \Delta U_p \tau''/\hbar$) or taken far away from it ($\xi_\infty = \Delta U_\infty \tau''/\hbar$), is a second.

15. The linear model. The Landau–Zener formula

The two-state approximation for the linear one-dimensional model

As was noted earlier, the nonadiabatic coupling is strong only between two states whose terms converge strongly. We will consider in detail the case when it is possible to consider just two electronic states and motion along a single coordinate R. The strong convergence of three or more terms is not unlikely, but, all the same, for simplicity we will consider the so-called two-state approximation.

We will assume that there are two adiabatic functions ϕ_1 and ϕ_2, corresponding to the two adiabatic terms U_1 and U_2, which are represented in Fig. 8 by the continuous curves, denoted by indices 1 and 2. If the nonadiabatic coupling is appreciable then the functions ϕ_1 and ϕ_2 must depend strongly on R. Returning to transformation (14.14), we will represent ϕ_i as a linear combination of the other pair of functions ϕ_i^0, which depend weakly on R. We now choose the ϕ_i^0 in such a way that on leaving the region of strong

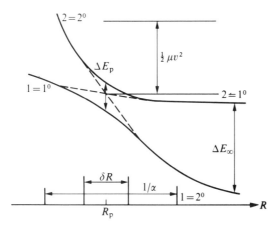

FIG. 8. Adiabatic terms (solid curves) and terms of the zero-order approximation (broken curves) in the region of nonadiabatic coupling.

coupling, of width δR, the functions ϕ_i^0 will go over into the ϕ_i. This means that, for $|R - R_p| \to \infty$, the condition

$$\frac{H_{12}(R)}{[H_{11}(R) - H_{22}(R)]} \to 0 \tag{15.1}$$

must be fulfilled. At this stage, it is necessary to approximate the dependence of the matrix elements on R. Limiting ourselves to the one-dimensional case, we will expand H_{12} and $H_{11} - H_{22}$ as a series in $R - R_p$ in the neighbourhood of the point R_p, which is still not determined. Retaining only the first two terms, we obtain

$$H_{12} = H_{12}(R_p) + H'_{12}(R_p)(R - R_p)$$
$$H_{11} - H_{22} \equiv \Delta H(R) = \Delta H(R_p) + \Delta H'(R_p)(R - R_p). \tag{15.2}$$

An estimate of the effect of the higher terms in the expansion will be given below. Identifying R_p with the point R_c, determined from eqn (14.11a), we have $\Delta H(R_p) = 0$. Condition (15.1) then implies that, at this point, $H'_{12}(R)$ must tend to zero. The choice of the functions ϕ_i^0 is determined by this, and the matrix of H_e with respect to the basis ϕ_i^0 now takes the form

$$H_e(\phi_0) = \begin{bmatrix} E_0 + \dfrac{\Delta F}{2} x & a \\ a & E_0 - \dfrac{\Delta F}{2} x \end{bmatrix}, \tag{15.3}$$

where

$$x = R - R_p; \qquad a = H_{12}(R_p);$$

$$E_0 = H_{11}(R_p) = H_{22}(R_p); \qquad \Delta F = \frac{\partial}{\partial R}(H_{11} - H_{22})\bigg|_{R = R_p}.$$

The diagonal matrix elements H_{11} and H_{22} are shown in Fig. 8 by dashed lines, labelled 1^0 and 2^0. In terms of these parameters, the eigenvalues U_{12} are

$$U_{12} = \bar{U} \pm \Delta U,$$
$$\Delta U = \tfrac{1}{2}[(\Delta F x)^2 + 4a^2]^{\frac{1}{2}}, \tag{15.4}$$

and the mixing parameter χ is given by the expression

$$\chi = \tfrac{1}{2} \arctan \frac{2a}{\Delta F x}, \tag{15.5}$$

where x is a function of the time which still has to be chosen in some way.

§15 TRANSLATIONAL ENERGY IN MOLECULAR COLLISIONS

The nonadiabatic function Ψ can be expanded in terms of the set ϕ_1, ϕ_2 or ϕ_1^0, ϕ_2^0, and the expansion coefficients determined from eqns (14.15) and (14.17).

Neglecting the terms $\langle \phi_1^0 | \dot\phi_2^0 \rangle$ for the present, it can be seen that each pair of equations is uncoupled as $|x| \to \infty$. In (14.15) in this limit $\dot\chi \to 0$, and in (14.17) the rapid oscillations of the exponential factor effectively averages the interaction. Near $x = 0$, i.e. for $R = R_p$, the coupling of the equations is, generally speaking, large in a region δR, whose magnitude must be estimated.

We will consider the restriction which is imposed on the model by the linear approximation (15.2). We revert to the limit of small velocities. The uncoupling in (14.15) is chiefly caused by the decrease in the nonadiabatic coupling, i.e. by condition (15.1). For a rough estimate, we will assume that the slowly varying functions ϕ_i^0 are changed over a length $1/\alpha$ of the order of the Bohr radius. Then $\Delta F \approx \Delta U_\infty \alpha$. From (15.1) we see that the uncoupling takes place over distances x exceeding δR,

$$x \gg \delta R \sim \frac{a}{\Delta F} = \frac{a}{\Delta U_\infty \alpha}. \tag{15.6}$$

On the other hand, δR must be much less than $1/\alpha$, in order that the higher terms in the expansion in (15.2) be insignificant. To meet this demand, we arrive at the first restriction of the theory

$$\frac{2a}{\Delta U_\infty} = \frac{\Delta U_p}{\Delta U_\infty} \ll 1 \tag{15.7}$$

We will now consider the limit of large velocities. The exponential term in (14.17) has the form

$$\exp\left[-\frac{i}{\hbar}\int (H_{22} - H_{11})\,dt\right] = \exp\left[-\frac{i}{\hbar}\Delta F \int x(t)\,dt\right]. \tag{15.8}$$

Near $R \approx R_p$ we will assume

$$x(t) = vt, \tag{15.9}$$

where v is the velocity of the nuclei at the point of intersection of the zero-order terms. Substituting (15.9) into (15.8) we find that the exponential oscillates weakly in a time δt determined by the condition

$$\frac{\Delta F (\delta t)^2 v}{\hbar} \approx 1. \tag{15.10}$$

Thus, for δR we have $\delta R \sim v\,\delta t$, and the condition $\delta R \ll 1/\alpha$ takes the form

$$\left(\frac{v\hbar}{\Delta F}\right)^{\frac{1}{2}} \ll 1. \tag{15.11}$$

Assuming here that $\Delta F \approx \Delta U_\infty \alpha$, we find

$$\frac{\Delta U_\infty}{\hbar a v} \gg 1. \tag{15.12}$$

This is the second restriction of the theory.

For particle velocities typical in chemical reactions and values of ΔU_∞ of the order of 1–3 eV, the conditions (15.7) and (15.12) do not limit the applicability of the linear model to any marked degree. The only serious difficulty concerns the choice of the trajectory and considerable attention will be paid to precisely this problem later on.

The Landau–Zener formula

The problem of a nonadiabatic transition in a system of two linear terms was first considered by Landau [319] and Zener [571]. They assumed that the higher terms in the expansion in (15.9) were insignificant. Thus the velocity v at the point of intersection of the zero-order terms served as the only parameter of the trajectory. We will now obtain the results of Landau and Zener in two limiting cases. We will assume that a is sufficiently small and that it is possible to solve (14.17) by perturbation theory. Denoting the transition probability $\phi_1^0 \to \phi_2^0$ by P_{12}^0 and the transition probability $\phi_1 \to \phi_2$ by P_{12}, it can be seen from Fig. 8 that $P_{12} = 1 - P_{12}^0$, since the adiabatic and zero-order terms change places after passage through the region of nonadiabaticity.

The initial conditions of the problem are

$$b_1(-\infty) = 1, \qquad b_2(-\infty) = 0. \tag{15.13}$$

To the first order in perturbation theory we obtain

$$b_2(+\infty) = \frac{a}{i\hbar} \int_{-\infty}^{\infty} \exp\left[-\frac{i}{\hbar} \int^t (H_{11} - H_{22}) \, dt\right] dt = \frac{a}{i\hbar} \left(\frac{2\pi\hbar}{i\Delta F v}\right)^{\frac{1}{2}}. \tag{15.14}$$

Thus

$$P_{12}^0 = \frac{2\pi a^2}{\Delta F \hbar v}, \quad \text{if } P_{12}^0 \ll 1. \tag{15.15}$$

If a is sufficiently large, then P_{12}^0 may be greater than unity and perturbation theory unsuitable.

Landau [319] first showed that the main factor in the expression for the transition probability for large adiabatic coupling (large a) can be calculated by integrating eqns (14.17) over the complex plane of t. Omitting the details of the calculation we quote the final result:

$$a_2(+\infty) \approx \exp\left[-\frac{i}{\hbar} \int^{t_c} \Delta U(t) \, dt\right], \tag{15.16}$$

where t_c is the solution to eqn (14.18), i.e. $t_c = -2ai/\Delta F v$.

Thus substituting (15.4) into (15.16) we obtain

$$P_{12} = |a_2(+\infty)|^2 \approx \exp\left[\frac{2}{\hbar} \operatorname{Im} \int^{t_c} \Delta U(t)\, dt\right] = \exp\left(-\frac{2\pi a^2}{\Delta F \hbar v}\right). \quad (15.17)$$

The correct pre-exponential factor in this expression was calculated by Zener [571] by means of an exact solution to eqn (14.17). The transition probability, not only for large a but for any case, was shown to be equal to

$$P_{12} = \exp\left(-\frac{2\pi a^2}{\Delta F \hbar v}\right) = 1 - P_{12}^0. \quad (15.18)$$

If here, $2\pi a^2/\Delta F \hbar v \ll 1$, then (15.18) goes over into (15.15).

In atomic collisions, nuclei usually pass twice through the region of strong coupling. Then the total probability can be obtained by summing the currents for each term, if the interference between the incident and reflected waves is neglected,

$$P = 2P_{12}(1 - P_{12}) = 2P_{12}^0(1 - P_{12}^0). \quad (15.19)$$

Substituting from (15.17) we find

$$P = 2\exp\left(-\frac{2\pi a^2}{\Delta F \hbar v}\right)\left[1 - \exp\left(-\frac{2\pi a^2}{\Delta F \hbar v}\right)\right]. \quad (15.20)$$

This is the well-known formula of Landau and Zener. Analogous results were obtained by Stueckelberg [505], by considering two coupled wave equations under the same boundary conditions. The condition for the applicability of (15.20), apart from the restrictions (15.7) and (15.12), is that it should be possible to neglect the higher terms in (15.9). The velocity can be considered constant if the interval δR is small in comparison with the interval over which the energy of the nuclei changes.

$$F_i\, \delta R \ll \frac{\mu v^2}{2}. \quad (15.21)$$

This criterion, together with the criterion for neglecting the interference, can be written in the form

$$S_i = \frac{1}{\hbar}\int_{R_t}^{R_p}\left\{2\mu\left[\frac{\mu v^2}{2} - U_i(R)\right]\right\}^{\frac{1}{2}} dR \gg 1, \qquad \Delta S = |S_1 - S_2| \gg 1. \quad (15.22)$$

where R_t denotes the turning point of the motion of the nuclei for each of the adiabatic terms.

It is precisely condition (15.22) that substantially narrows the region of applicability of the Landau–Zener formula for low energies. Later on, we will discuss a recent development of the theory, which removes restriction (15.22). At higher energies (several hundred electron volts) other limitations of the

Landau–Zener model arise, which have been considered in detail by Bates [1], [139].

We will now estimate the contribution to the transition probability of the second term in brackets in eqn (14.17). Neglecting the first term, i.e. the off-diagonal matrix element H_{12} of the adiabatic (static) interaction, we find by first-order perturbation theory

$$P^0_{12} = 2\pi\hbar^2 \frac{|\langle \phi^0_1 | \dot{\phi}^0_2 \rangle|^2}{\Delta Fhv}. \tag{15.23}$$

Returning to the estimate of (14.13) we find $P^0_{12} \sim \hbar a v/\Delta U_\infty$. Comparing this result with (15.12) it is easy to verify that, within the limits of the linear model considered, i.e. under the restrictions (15.7) and (15.12), the non-adiabatic coupling between atomic functions can be taken into account in first-order perturbation theory. Thus the second term in brackets in (14.17) must be retained only when H_{12} is very small or equal to zero. As an important example of such a situation we mention the nonadiabatic transitions between terms of different symmetry in a system of two atoms, induced by the rotation of the axis of the quasi-molecule. In this case, $H_{12} = 0$ and the nonadiabaticity operator $\hat{C}$ is expressed in the following way by the angular momentum operator $\mathbf{J}_x$ acting on the functions ϕ^0_i in the molecular system of coordinates

$$\hat{C} = -i\hbar \frac{\partial}{\partial t} = -i\hbar \dot{\phi} \left(\frac{\partial}{\partial \phi} \right) = -\omega_x J_x, \tag{15.24}$$

where ϕ and $\dot{\phi} = \omega_x$ are the angle of rotation and angular velocity of rotation of the molecular axis in a plane perpendicular to the vector $\mathbf{l}$ of the relative angular momentum of the atoms; J_x is the component of the angular momentum of the electrons along the vector $\mathbf{l}$.

Substituting (15.24) into (15.23) we find

$$P^0_{12} = \frac{2\pi \omega_x^2 (J_x)^2_{12}}{\Delta Fhv}. \tag{15.25}$$

In problems on collisions it is often convenient to introduce two other parameters—the relative energy E of the atoms at infinity and the impact parameter b—instead of v and ω. Taking note that v in (15.25) denotes the radial velocity, on the basis of the laws of energy and angular momentum conservation we find

$$\frac{\mu v^2}{2} = E \left(1 - \frac{b^2}{R_p^2} \right) - U_p, \tag{15.26a}$$

$$\mu \omega R_p = \mu v_\infty b = (2\mu E)^{\frac{1}{2}} b. \tag{15.26b}$$

Substituting (15.26) into (15.25) we obtain the following expression for the transition probability between terms of different symmetry in atomic collisions [37]:

$$P = 2P_{12}^0(1-P_{12}^0) \approx 2P_{12}^0 = 4\pi \frac{2Eb^2/R_p^4}{\mu \Delta F \hbar \left[\frac{2}{\hbar}\left(E - U_p - \frac{Eb^2}{R_p^2}\right)\right]^{\frac{1}{2}}}. \qquad (15.27)$$

From expression (15.18), it is evident that the nonadiabatic transition probability is small for large splitting of the adiabatic terms. Therefore, in induced nonadiabatic transitions at low (thermal) energies, usually only the comparatively weak spin–orbit coupling, which connects terms of the same symmetry, is effective. Terms of different symmetry are connected by the Coriolis interaction (15.24), also called the orbital–rotational interaction.

Effect of the turning point

If condition (15.21) is not satisfied then, to calculate the transition probability correctly, the acceleration of the representative point on passing through the region of strong nonadiabatic interaction should be taken into account. An investigation of the quantum problem with the Hamiltonian (15.3) shows [5] that, in place of the approximation (15.9) for the trajectory $x(t)$, we should take

$$x = vt + \frac{Ft^2}{2\mu}, \qquad (15.28)$$

corresponding to uniformly accelerated motion of the representative point. The force F, which comes into (15.28), is determined by the gradients of the zero-order terms at the point R_p:

$$F = \left[\left(\frac{dH_{11}}{dR}\right)\left(\frac{dH_{22}}{dR}\right)\right]_{R=R_p}^{\frac{1}{2}} = (F_1 F_2)^{\frac{1}{2}}. \qquad (15.29)$$

With such a definition of force, the semiclassical problem with the trajectory (15.28) turns out to be exactly equivalent to the quantum problem, in which the motion of the point along the two electronic terms figures explicitly. The definition (15.29) has meaning when F_1 and F_2 have the same sign. The case of different signs will be considered briefly in the discussion of nonadiabatic reactions. Changing the origin of the time coordinate, we will represent (15.28) in the form

$$x = \frac{F}{2\mu}t^2 - \frac{\mu v^2}{2F}. \qquad (15.30)$$

It can be seen from this that the moment $t = 0$ corresponds to the turning of the trajectory. Therefore, the additional term in (15.28) (by comparison with

(15.9)) in the calculation of the transition probability reflects the effect of the turning point. Omitting the details of the calculation, presented in papers [5], [64] and the review [390], we note only the basic formulae which replace expressions (15.5), (15.17), and (15.20).

If a is sufficiently small, then the transition probability P, calculated with trajectory (15.30) which takes account of the double passage through the region of nonadiabaticity (up to and after the turn), is shown to be equal to

$$P = \pi^2 b^{\frac{2}{3}}[Ai(-\varepsilon b^{\frac{1}{3}})], \qquad (15.31)$$

where Ai is the Airy function; $\varepsilon = \mu v^2/2\Delta U_p \cdot \Delta F/F$; $b = (\Delta U_p/\varepsilon^* \cdot F/\Delta F)^{\frac{2}{3}}$; $\varepsilon^* = (\hbar^2 F^4/2\mu\Delta F^2)^{\frac{1}{3}}$.

In comparison with (15.15), an additional parameter ε^* appears here, characterizing the effect of the turning point. In the limiting cases of high and low energies, the following approximations are valid:

$$P = \begin{cases} \dfrac{\pi b}{\sqrt{\varepsilon}} \sin^2\left(\dfrac{2}{3}\varepsilon^{\frac{3}{2}}b + \dfrac{\pi}{4}\right) = \dfrac{8\pi a^2}{\Delta F h v}\sin^2\left[\dfrac{2}{3}\left(\dfrac{\mu v^2}{2\varepsilon^*}\right)^{\frac{3}{2}} + \dfrac{\pi}{4}\right], & \dfrac{\mu v^2}{2\varepsilon^*} \gg 1 \\[2ex] \dfrac{\pi b}{4\sqrt{|\varepsilon|}} \exp\left(-\dfrac{4}{3}|\varepsilon|^{\frac{3}{2}}b\right) = \dfrac{2\pi a^2}{\Delta F h v}\exp\left(-\dfrac{4}{3}\left|\dfrac{\mu v^2}{2\varepsilon^*}\right|^{\frac{3}{2}}\right), & \dfrac{\mu v^2}{2\varepsilon^*} \ll -1. \end{cases} \quad (15.32)$$

Two peculiarities of P should be noted. First, for positive energies P oscillates, this is connected with the effect of interference between the incident and reflected waves. If the varying factor $\sin^2[\frac{2}{3}(\mu v^2/2\varepsilon^*)^{\frac{3}{2}} + \pi/4]$ is replaced by its mean value $\frac{1}{2}$ then we again revert to formula (15.15). Secondly, for negative energies P decreases monotonically, since the transition probability differs from zero only when the system does not reach the point of intersection of the zero-order terms. This is a result of the nonadiabatic tunnel effect.

For large values of a, the exponential factors in the expression for P can be calculated by formula (15.17), in which the trajectory (15.30) is substituted. Qualitatively, it is not difficult to understand how the proximity of the turning point to the region of the nonadiabatic transition influences the transition probability. Since P in (15.17) grows rapidly with increase in velocity, the acceleration, caused by the action of the force F, must lead to an increase in the transition probability in comparison with the value calculated under the approximation of uniform motion. The calculation confirms this deduction; the formula for the transition probability, in which the first correction for acceleration is taken into account, has the form [64], [390]

$$P = \exp\left\{-\dfrac{2\pi a^2}{\Delta F h v}\left[1 - \dfrac{3}{32}\left(\dfrac{4a^2}{\mu v^2} \cdot \dfrac{F}{\Delta F}\right)^2 + \ldots\right]\right\}. \qquad (15.33)$$

This expression, which replaces (15.17), is valid when the exponent of the exponential appreciably exceeds unity and the correction term in the brackets is small.

For intermediate values of a, generally speaking, the transition probability cannot be expressed analytically. If, however, the conditions are such that the phase ΔS is large, then formula (15.20) turns out to be valid. Thus the Landau–Zener formula can be used to calculate nonadiabatic transition probabilities in conditions where the possibility of a classical description of the motion of the nuclei is ruled out. In fact, for the validity of this formula, it is necessary to satisfy not two inequalities (15.22) but only one:

$$S_F = \frac{1}{\hbar}\int_{R_t}^{R_p}(2\mu)^{\frac{1}{2}}\left[\frac{\mu v^2}{2} - Fx\right]^{\frac{1}{2}}dx \gg 1. \tag{15.34}$$

A more detailed investigation of different limiting cases of the linear model is given in the review [390].

The mean transition probability

For thermal processes, the mean transition probability $\langle P \rangle$ is of interest in practice. It can be calculated by the transition state method if x is identified with the reaction coordinate q_r. On the basis of the formulae in § 2, we have

$$\langle P \rangle = \int_{-\infty}^{\infty} P(E_t) \exp(-\beta E_t)\beta\, dE_t. \tag{15.35}$$

By analogy with (2.23) here, the lower limit of integration over the energy E_t, measured from the point of intersection of the terms of the zero-order approximation, is extended to $-\infty$, with a view to taking into account possible tunnel effects. Calculated in accordance with (15.35), the mean transition probability can be interpreted as a transmission coefficient in the general formula of the transition state method, in which the critical surface $S^\star$ is given by the equation $x = 0$.

For *small values of a*, the calculation of (15.35) gives [384]:

$$\langle P \rangle = \frac{2\sqrt{\pi}\,\pi a^2}{\hbar \Delta F(2kT/\mu)^{\frac{1}{2}}}\exp\left[\frac{1}{12}\left(\frac{\varepsilon^*}{kT}\right)^3\right]. \tag{15.36}$$

Here $\langle P \rangle$ consists of two factors. The first (pre-exponential) factor expresses the mean probability of over-barrier transmission (with relative energy $\mu v_\infty^2/2 > U_p$), the second (exponential) factor gives the correction for the tunnel effect ($\mu v_\infty^2/2 < U_p$). This can lead to a considerable lowering of the activation energy U_p. Thus, for sufficiently low temperatures, $\langle P \rangle$ may exceed unity. However, this does not contradict the obvious condition

$$\kappa < \pi R_p^2 (8kT/\pi\mu)^{\frac{1}{2}},$$

because the tunnel effect is only significant when the activation energy is comparatively large and the total exponential (Arrhenius factor and tunnel correction) is small, i.e.

$$\exp\left[-\beta U_p + \frac{(\beta\varepsilon^*)^3}{12}\right] \ll 1.$$

If the matrix element a (which, by assumption, does not depend on R) depends on other coordinates, then an additional temperature dependence may appear in (15.36). It can be found by averaging the transition probability, with respect to the other coordinates, independently of the averaging represented by (15.35). The transition between two molecular terms of different symmetry, coupled by the operator (15.24), serves as an example. Averaging the square of the angular velocity for a two-dimensional rotator, we find

$$\langle a^2 \rangle = \langle \omega^2 \rangle (J_x)_{12}^2 = \frac{2kT}{\mu R_p^2}(J_x)_{12}^2. \tag{15.37}$$

With *strong coupling*, the Landau–Zener formula can be used to calculate $\langle P \rangle$ over a wide range of the parameters ε and b. The transition probability P then depends on only one parameter $2\pi a^2/\Delta Fhv$ which enables $\langle P \rangle$ to be calculated comparatively easily. Substituting (15.20) into (15.35) we find

$$\langle P \rangle = 2F(\gamma) - 2F(2\gamma), \tag{15.38}$$

where

$$F(\gamma) = \int_0^\infty \exp\left(-\frac{\gamma}{x} - x^2\right) 2x \, dx,$$

$$\gamma = \frac{2\pi a^2}{\Delta F\hbar}\left(\frac{\mu}{2kT}\right)^{\frac{1}{2}}. \tag{15.39}$$

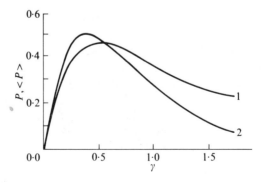

Fig. 9. The Landau–Zener transition probability (curve 2) as a function of the parameter $\gamma = 2\pi a^2/\Delta Fhv$. The mean transition probability (curve 1) as a function of the parameter $\gamma = 2\pi a^2/\Delta F\hbar\sqrt{(2kT/\mu)}$.

A graph of the function $\langle P \rangle$ is presented in Fig. 9. The function is characterized by the expansions [7]:

$$\langle P \rangle = \begin{cases} 2\sqrt{\pi\gamma}, & \gamma \ll 1 \quad (15.40a) \\ \langle P \rangle_{\max} = 0.427, & \gamma = 0.50 \quad (15.40b) \\ 4\sqrt{\left(\dfrac{\pi}{3}\right)\left(\dfrac{\gamma}{2}\right)^{\frac{1}{3}}} \exp\left[-3\left(\dfrac{\gamma}{2}\right)^{\frac{2}{3}}\right], & \gamma \gg 1. \quad (15.40c) \end{cases}$$

It is evident that (15.38) goes over into (15.36) on fulfillment of the conditions

$$\gamma \ll 1, \qquad \beta\varepsilon^* < 1, \qquad (15.41)$$

which imply that first-order perturbation theory is valid and tunnel corrections are neglected.

For *large values of a*, the probability $\langle P \rangle$ is exponentially small, and (15.35) can be calculated easily by the saddle-point method. The integrand has a sharp maximum at the velocity v^* determined by the condition

$$\operatorname{Im} \int^{t_c} \Delta U(t_c) \, dt - \frac{\mu v^2}{2kT} = \text{minimum}. \qquad (15.42)$$

In practice, it is sufficient to use the first term in the expansion in (15.33), so that the minimization condition takes the form

$$\frac{\partial}{\partial v}\left(\frac{2\pi a^2}{\Delta F\hbar v} + \frac{\mu v^2}{2kT}\right)\bigg|_{v=v^*} = 0. \qquad (15.43)$$

Then v^* from (15.43) can be introduced into the higher terms of the expansion (15.33) to calculate the tunnel corrections. This method gives

$$\langle P \rangle = 4\sqrt{\left(\frac{\pi}{3}\right)\left(\frac{\gamma}{2}\right)^{\frac{1}{3}}} \exp\left[-3\left(\frac{\gamma}{2}\right)^{\frac{2}{3}} + \frac{3}{16}(\varepsilon_T)^{-2}\left(\frac{\gamma}{2}\right)^{-\frac{2}{3}} + \ldots\right], \quad (15.44)$$

where

$$\varepsilon_T = \frac{kT\Delta F}{\Delta U_p F}.$$

This formula is correct when $\gamma \gg 1$ and the second term is small in comparison with the first (nevertheless it may be of the order of, or even larger than, unity). If the tunnel correction is neglected (the second term in (15.44)) then (15.44) reduces to (15.40c).

16. Generalization of the the linear model

In a number of cases, there arises the necessity for taking into account the nonadiabatic coupling between certain electronic states and even between states belonging to the continuous spectrum. In a similar situation, the exact solutions to certain model problems are of great value, enabling the details of some particular mechanism or other, connected with nonadiabatic processes, to be clarified. In this paragraph, we will consider certain generalizations of the linear model, permitting comparatively simple analytical solutions. Specific examples of elementary processes, corresponding to these processes, are cited in §§ 18 and 19.

Extension of the Landau–Zener model to a multi-term system

We will assume that a single term H_{00} of the zero-order approximation intersects the set of terms H_{kk} ($k = 1, 2, \ldots, n$) with derivatives $dH_{kk}/dx = F$ independent of k. In the semiclassical approximation, it may be assumed (without loss of generality) that $F = 0$, since the transition probability depends only on the difference between the diagonal matrix elements. We will assume that the only off-diagonal matrix elements different from zero are H_{k0} ($k = 1, 2, \ldots, n$), and that they do not depend on x (which is a feature of the linear model). Fig. 10 illustrates this situation and the points indicate the

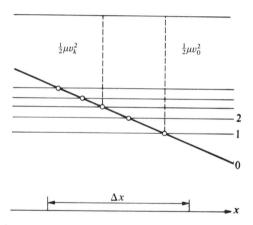

FIG. 10. The interaction of the single term H_{00} with the many parallel terms H_{kk} ($k = 1, 2, \ldots$).

positions of pseudo-intersections which arise if the adiabatic interaction (i.e. H_{k0}) is taken into account. We will now find the nonadiabatic wave function in the form

$$\Psi(t) = B_0 \phi_0^0 + \sum_{k=1}^{n} B_k(t) \phi_k^0, \tag{16.1}$$

where, for the trajectory, we will take the approximation of Landau–Zener $x = v(t - t_0)$, where v denotes some average value amongst the v_k. Substituting (16.1) into the wave equation, we obtain the following system of equations for

§ 16 TRANSLATIONAL ENERGY IN MOLECULAR COLLISIONS 119

the $B_k(t)$;

$$i\hbar \frac{dB_0}{dt} = -\Delta F v t B_0 + \sum H_{0k} B_k, \qquad (16.2a)$$

$$i\hbar \frac{dB_k}{dt} = H_{kk} B_k + H_{k0} B_0, \qquad (16.2b)$$

where $\Delta F = dH_{00}/dx$; $t_0 = -H_{00}|_{x=0}/\Delta F v$ and all H_{k0} are assumed real, for simplicity.

Eqns (16.2) were solved by Ovchinnikova [81], [84] (for $n = 2$) and Osherov [104] and Demkov [22] (for arbitrary n). We will not discuss the details of calculation here but will present just the main results: the transition probabilities P_{lk} from the initial state l to the final state k can be calculated as the probability of an independent pair of transitions, each of which is localized at the point of intersection of the terms H_{00} and H_{jj}. The probability of a pair transition is given by the Landau–Zener formula

$$P_j = \exp\left(-\frac{2\pi |H_{0j}|^2}{\Delta F \hbar v}\right). \qquad (16.3)$$

Thus, for a single passage through the region of nonadiabatic coupling, we obtain

$$P_{kl} = (1 - p_k) p_{k+1} p_{k+2} \cdots p_{l-1} (1 - p_l), \qquad (16.4a)$$

$$P_{0l} = p_1 p_2 \cdots p_{l-1} (1 - p_l). \qquad (16.4b)$$

The transition probability from the state ϕ_0^0 to all states ϕ_n^0 with indices $n \leq k$ equals

$$P(k) = \sum_{n=1}^{k} P_{0n} = 1 - p_1 p_2 \cdots p_k = 1 - \exp\left(\frac{2\pi}{\Delta F \hbar v} \sum_{n=1}^{k} |H_{0n}|^2\right). \qquad (16.5)$$

The expression (16.5) enables the transition from a discrete state ϕ_0^0 into a state with a continuous energy spectrum ϕ_k^0 to be described. To this end, in the sums over n we should go over to the limit $H_{nn} - H_{n'n'} \to 0$, replacing the summation by an integration. By the functions ϕ_n^0 are to be understood functions of the continuous spectrum ϕ_ε^0, normalized over the energy by the δ-function. Taking it into account that dn and $d\varepsilon$ are connected by the relation $dn = \rho(\varepsilon) d\varepsilon$, where $\rho(\varepsilon)$ is the density of states with energy ε, we find

$$P(\varepsilon) = 1 - \exp\left[-\frac{2\pi}{\Delta F \hbar v} \int_0^\varepsilon |H_{0\varepsilon'}|^2 \rho(\varepsilon') d\varepsilon'\right], \qquad (16.6)$$

where $P(\varepsilon)$ denotes the transition probability from the state ϕ_0^0 into a state ϕ_ε^0, whose energy lies in the band $0 \leq \varepsilon' \leq \varepsilon$ adjacent to the lower boundary of the continuous spectrum.

The expression (16.6) permits of a simple interpretation in terms of the decay of the quasi-stationary state $\Phi_0(t)$, arising out of ϕ_0^0 due to its interaction with the states of the continuous energy spectrum. In place of the variable ε, we will introduce t, the time the system stays in states with energies $\varepsilon > 0$,

$$\varepsilon = \Delta F x, \qquad x = (t-t_1)v. \tag{16.7}$$

Then, with regard for (16.7), (16.6) can be rewritten in the form:

$$P(t) = 1 - \exp\left[-\frac{1}{\hbar}\int_{t_1}^{t}\Gamma(\varepsilon')\,dt'\right], \tag{16.8}$$

where

$$\Gamma(\varepsilon') = 2\pi|\langle\phi_0^0|H|\phi_{\varepsilon'}^0\rangle|^2\rho(\varepsilon'). \tag{16.8a}$$

Here $P(t)$ denotes the transition probability into a state with a continuous energy spectrum at time t; the time dependence of $\Gamma(\varepsilon')$ is obtained by substituting the function $\varepsilon' = \varepsilon'(t')$ from (16.7). As regards the decay probability of the state ϕ_0 at time t, it is obviously given by the exponential term in (16.8).

The function $\Gamma(\varepsilon)$ coincides with the width of the electronic term introduced to describe a quasi-stationary state. Within the limits of this description, the quasi-stationary functions Φ_0 are sought as eigenfunctions of the Hamiltonian H_e, corresponding to complex eigenvalues of the term

$$\mathfrak{U}_0(\mathbf{R}) = U_0(\mathbf{R}) - \frac{i}{2}\Gamma(\mathbf{R}). \tag{16.9}$$

We will assume that the Hamiltonian H can be represented approximately in the form of a sum of two terms, one of which is responsible for the formation of bound states (in particular, of the state ϕ_0^0) and the second for the formation of states of the continuous spectrum ϕ_ε^0. We will separate out these two terms in the Hamiltonian H^0, considering the difference $H - H^0$ as a perturbation which gives rise to an interaction between states of the discrete and continuous spectra. Then the imaginary part of the term $\mathfrak{U}_0(\mathbf{R})$ can be calculated by the formula†

$$\tfrac{1}{2}\Gamma(\mathbf{R}) = \pi|\langle\phi_0^0|H|\phi_\varepsilon^0\rangle|^2\rho(\varepsilon), \tag{16.10}$$

in which $\mathbf{R}$ and ε are connected by the relation

$$\varepsilon = H_{00}(\mathbf{R}). \tag{16.10a}$$

† We will not discuss here the conditions under which the introduction of complex eigenvalues of $\mathfrak{U}_0$ is possible and the level width is given by formula (16.10). On this question see book [37].

The adiabatic approximation for the quasi-stationary states is obtained by extending the basic formula (1.25) to the case of complex terms $U_0(\mathbf{R})$:

$$\Phi_0(\mathbf{r}, t) = \phi_0^0 \exp\left[-\frac{i}{\hbar}\int^t \mathfrak{U}_0(\mathbf{R})\,dt\right] = \phi_0^0 \exp\left[-\frac{i}{\hbar}\int^t U_0\,dt - \frac{1}{2\hbar}\int^t \Gamma\,dt\right], \tag{16.11}$$

where the integrals in the exponential are taken along some classical trajectory $\mathbf{R} = \mathbf{R}(t)$. From (16.11) it follows that the decay of the state Φ_0, as the result of interaction with the states of the continuous spectrum, is described by the relation

$$|\Phi_0(\mathbf{r}, t)|^2 = |\phi_0^0(\mathbf{r})|^2 \exp\left[-\frac{1}{\hbar}\int_{t_1}^t \Gamma(\mathbf{R})\,dt\right], \tag{16.12}$$

for which (16.8), in particular, follows.

The semiclassical approximation in its application to the linear model is valid in the case when all the velocities v_k can be approximated by just one, and the total length Δx of the region of nonadiabaticity is very much less than a characteristic dimension $1/\alpha$. This, as previously, determines the limits of applicability of the model on the side of high and low energies. For processes with low energies, the neglect of the effect of the turning point is a very serious limitation. To extend the region of applicability of the linear multi-term model, it is necessary to go over to the quantum equations. If we put $F = 0$ in these equations then, after a Laplace transformation, a system analogous to (16.5) is obtained. This suggests that the quantum system of equations, with the additional restriction $F = 0$, may also be solved in an analytical form. However, we will not dwell on this problem.

Extension of the Landau–Zener model to a system with two degrees of freedom

The simplest generalization of the Landau–Zener model to a multi-dimensional system can be represented by the case of two variables. For a two-level approximation, the adiabatic electronic terms U_i are given by the eigenvalues of the matrix $H_e(\phi^0)$ (15.3). Just as in § 15 we will assume that the H_{ik} are weakly-varying functions of two coordinates x and y, so that the linear approximation will be sufficient. The expansion analogous to (15.2) now takes the form

$$H_{12}(x, y) = H_{12}^c + \frac{\partial H_{12}^c}{\partial x}(x - x_c) + \frac{\partial H_{12}^c}{\partial y}(y - y_c) \tag{16.13a}$$

$$H_{11} - H_{22} = \Delta H(x, y) = \Delta H^c + \frac{\partial \Delta H^c}{\partial x}(x - x_c) + \frac{\partial \Delta H^c}{\partial y}(y - y_c), \tag{16.13b}$$

where the index c denotes that the functions are calculated at fixed points x_c and y_c (e.g. $H_{12}^c = H_{12}(x_c, y_c)$).

Noting that, in the general case, H_{12} is a complex quantity but ΔH is real, we will determine x_c and y_c from a system of two real equations

$$\Delta H^c = \Delta H(x_c, y_c) = 0,$$
$$\operatorname{Re} H_{12}^c = \operatorname{Re} H_{12}(x_c, y_c) = 0. \qquad (16.14)$$

Measuring x_c and y_c from the origin of coordinates, we can now simplify the expansion (16.13) by two linear transformations. The first is a linear transformation of the coordinates (a rotation of the coordinate system), and the second is a linear transformation of the functions of the zero-order approximation. The latter does not depend on the former if the matrix of the transformation does not depend on x and y. Introducing the new transformed variables X and Y referred to x_c and y_c as the origin of coordinates and the new transformed functions of the zero-order approximation ϕ_1^0 and ϕ_2^0, we obtain the matrix [387]:

$$H_e(\phi^0) = \begin{bmatrix} E_0 + \tfrac{1}{2}\Delta F_1 X & iA + \tfrac{1}{2}\Delta F_2 Y \\ -iA + \tfrac{1}{2}\Delta F_2 Y & E_0 - \tfrac{1}{2}\Delta F_1 X \end{bmatrix}, \qquad (16.15)$$

where $A = \operatorname{Im} H_{12}^c$; ΔF_2 is assumed real.†

As regards the trajectory, we will take the Landau–Zener approximation

$$X = V_1(t-t_1),$$
$$Y = V_2(t-t_2). \qquad (16.16)$$

It is evident that condition (15.7) is not satisfied for any trajectory of this type, therefore the interaction between functions ϕ_1^0 and ϕ_2^0 does not vanish at the limits $t \to \pm\infty$. But since the ratio $H_{12}/\Delta H_{12}$ tends in this limit to a constant value, the interaction between the functions of the zero-order approximation as $t \to \pm\infty$ can be eliminated by the choice of a corresponding linear combination, different for each trajectory, characterized by its ratio V_1/V_2. For the fixed trajectory, $H_e(t)$ can be simplified and represented in the form

$$H_e(t) = \begin{bmatrix} E_0 + \tfrac{1}{2}\mathfrak{N}t & \mathfrak{M} \\ \mathfrak{M} & E_0 - \tfrac{1}{2}\mathfrak{N}t \end{bmatrix}, \qquad (16.17)$$

where

$$\mathfrak{N} = [(\Delta F_1 V_1)^2 + (\Delta F_2 V_2)^2]^{\tfrac{1}{2}},$$

$$\mathfrak{M} = \left[A^2 + \frac{1}{4} \frac{(\Delta F_1 V_1)^2 (\Delta F_2 V_2)^2 (t_1 - t_2)^2}{(\Delta F_1 V_1)^2 + (\Delta F_2 V_2)^2} \right]^{\tfrac{1}{2}}. \qquad (16.18)$$

† Usually the term $\tfrac{1}{2}\Delta F_2 Y$ expresses the Coulomb or exchange contribution to the interatomic or intermolecular interaction, so that this assumption is justified. The term iA is chiefly due to the spin–orbit coupling, which can be considered as unchanged for small displacements of the nuclei.

§16   TRANSLATIONAL ENERGY IN MOLECULAR COLLISIONS   123

The matrix (16.17) is formally equivalent to (15.3) if the trajectory (15.9) is substituted into the latter. Then the transition probability P^0_{12} for this fixed trajectory can be written down immediately without solution of eqn (14.15) or eqn (14.7):

$$P^0_{12} = \exp\left(-\frac{2\pi \mathfrak{M}^2}{\hbar \mathfrak{N}}\right). \tag{16.19}$$

We will now discuss this expression in terms of the adiabatic splitting ΔU. From (15.4) we find

$$\Delta U = (A^2 + \Delta F_1^2 X_1^2 + \Delta F_2^2 Y^2)^{\frac{1}{2}}. \tag{16.20}$$

Hence it can be seen that the adiabatic potential energy surfaces in the region of maximum convergence have the form of paraboloids if $A \neq 0$. For $A = 0$, the adiabatic potential energy surfaces are intersected at the point $X = 0, Y = 0$, which is the vertex of a double cone. It is in precisely such a case (Im $H_{12} = 0$) that the well-known theorem is valid; the number of dimensions of the manifold, in which the intersection of the adiabatic terms occurs equals $s - 2$, where s denotes the number of degrees of freedom (see [37]). The shortest distance l between the point of maximum convergence of the terms and the straight line trajectory

$$l = V_1 V_2 |t_1 - t_2|(V_1^2 + V_2^2)^{-\frac{1}{2}} \tag{16.21}$$

may be introduced instead of the time difference $t_1 - t_2$ in (16.18). If, moreover, the change in the splitting is studied for motion along the trajectory (16.16) then $\mathfrak{M}$ can be represented in the form

$$\mathfrak{M} = \min_L \Delta U(X, Y), \tag{16.22}$$

where the index L denotes that the minimum of ΔU is sought along the given trajectory. Written down in this form $\mathfrak{M}$ is analogous to the splitting $2a$ in the one-dimensional case. This was used by Teller [516] to calculate the mean two-dimensional transition probability for the particular case $A = 0, V_2 = 0$. In contrast to the one-dimensional case, $\mathfrak{M}$, in addition, depends on the parameters l and V_1/V_2 given by the trajectory in the region of strong nonadiabatic coupling.

In the two-dimensional case, as in the one-dimensional, the repeated transit of the represenative point in the region of strong coupling is possible. In contrast to the one-dimensional case, in which the second transit of the region of nonadiabaticity always takes place along the same trajectory, in the two-dimensional case there is small probability that a point which once goes near the position of maximum convergence (or near the vertex of the double cone) should get into the same region again. Since the transition probability P_{12} is small for large l, for a double transit $P \approx P^0_{12}$, (i.e. (16.19)) should be taken in place of formula (15.20).

We will now consider the case when the magnitude of H_{12} is so small that, in eqns (14.17), as well as the first terms, the second terms in the brackets should be taken into account. In the limiting case of $H_{12} = 0$, eqns (16.14) determine the line of intersection of the two adiabatic terms in whose neighbourhood the nonadiabatic transitions induced by the operator $\hat{C}$ take place. The operator $\hat{C}$, acting on the zero-order functions along some trajectory L, can be expressed in the form of a linear combination of momentum operators, by analogy with (15.24). We will separate out that part $\hat{C}_{\text{rot}}$ of the operator $\hat{C}$ which induces transitions between states due to the rotation of the system of atoms being considered. In the general case, there are three instantaneous axes of rotation, so that $\hat{C}_{\text{rot}}$ can be represented in the form

$$\hat{C}_{\text{rot}} = -\omega_x J_x - \omega_y J_y - \omega_z J_z, \qquad (16.23)$$

where the J_i are the components of the electronic angular momentum in the molecular (rotating) system of coordinates; ω_i are the components of the angular velocity of the nuclei in the same system.

The difference between this expansion and (15.24) is that even for fixed relative distances between the atoms, the ω_i change with time, since the components of the angular velocity are not integrals of motion in the moving system of coordinates. If, however, the characteristic time for a variation in the ω_i is appreciably shorter than the nonadiabatic transition time δt, then the Landau–Zener formula, in which the matrix element of the interaction is regarded as a constant, can be employed to calculate the probability P_{12}.

17. Model of non-linear terms

If the difference between the electronic terms of the non-interacting system is not large then condition (15.12) for the applicability of the linear term model may prove to be violated. The small value of the Massey parameter $\xi = \Delta U_\infty a/\hbar v$ implies that nonadiabatic transitions between terms will commence at distances where the intermolecular interaction will turn out to be of the same order as the initial splitting of the terms. In such a case, the size of the region where the nonadiabatic coupling is large can be of the order of a_0. Therefore the probability of a nonadiabatic transition will essentially depend on the details of the adiabatic terms and not on their general characteristics which, in the previous example, were the size of the off-diagonal matrix element H_{12} and the gradients of the zero-order terms F_1 and F_2 at the point of intersection.

Below, we quote the results of a solution to a model problem on the nonadiabatic transitions which gives qualitative information about the terms of many real systems. We will limit ourselves to the two-state approximation. Moreover, we will neglect the second term in eqns (14.15) in comparison with the first. In accordance with this, we will assume that for $R \gg R_p$

the atomic functions ϕ_1^0 and ϕ_2^0 are exact adiabatic functions. For $R \ll R_p$, the molecular functions ϕ_i, which are constructed in the form of linear combinations of the atomic functions, are adiabatic functions. The corresponding expansion coefficients must not depend on R because in this region, by assumption, the nonadiabatic coupling must disappear. Therefore the ϕ_i can be represented in the form

$$\phi_1 = \phi_1^0 \sin\frac{\theta}{2} - \phi_2^0 \cos\frac{\theta}{2}$$

$$\phi_2 = \phi_1^0 \cos\frac{\theta}{2} + \phi_2^0 \sin\frac{\theta}{2}. \qquad (17.1)$$

It is now necessary to write down the Hamiltonian $H = H_0 + V$ with the condition that ϕ_i^0 and ϕ_i be eigenfunctions of H for $R \gg R_p$ and for $R \ll R_p$. In the intermediate region ($R \approx R_p$) the perturbation must be chosen so that it reproduces qualitatively the dependence of the adiabatic terms U_i on the interatomic separation R. To take into account the exchange interaction, which depends approximately exponentially on distance, we will take the following approximation for the matrix of H_e with respect to the basis ϕ_i^0 [67], [69], [383]:

$$H_e(\phi^0) = \begin{bmatrix} E_0 + \dfrac{\Delta\varepsilon}{2} - \dfrac{A}{2}\cos\theta\exp(-\alpha R) & -\dfrac{A}{2}\sin\theta\exp(-\alpha R) \\ -\dfrac{A}{2}\sin\theta\exp(-\alpha R) & E_0 - \dfrac{\Delta\varepsilon}{2} + \dfrac{A}{2}\cos\theta\exp(-\alpha R) \end{bmatrix}.$$

$$(17.2)$$

Here, $\Delta\varepsilon$, A (on the assumption $A > 0$), θ, and α are free parameters of the problem, which can be varied in order to approximate a real Hamiltonian by a matrix with elements (17.2). Instead of the parameter A, the coordinate of the centre of the region of nonadiabaticity R_p can be introduced, on the basis of the relation $A = \Delta\varepsilon \exp(-\alpha R_p)$. Measuring the internuclear separation from the point R_p, we find the following expression for the splitting of the adiabatic terms:

$$\Delta U_{12} = \Delta\varepsilon[(1 - \cos\theta\exp(-\alpha x))^2 + \sin^2\theta\exp(-2\alpha x)]^{\frac{1}{2}},$$

$$x = R - R_p. \qquad (17.3)$$

The adiabatic functions which correspond to this splitting are of the form

$$\phi_1 = \phi_1^0 \sin\chi - \phi_2^0 \cos\chi,$$

$$\phi_2 = \phi_1^0 \cos\chi + \phi_2^0 \sin\chi,$$

$$\tan 2\chi = \frac{\sin\theta\exp(-\alpha x)}{\cos\theta\exp(-\alpha x) - 1}. \qquad (17.4)$$

It is not difficult to verify that, in the two limiting cases when $\exp(-\alpha x) \ll 1$ and when $\exp(-\alpha x) \gg 1$, the adiabatic functions ϕ_i (17.4) agree with the functions ϕ_i^0 and ϕ_i from (17.1). The solutions to the equations $\Delta U_{12} = 0$ determine the point of intersection of the adiabatic terms. The two points closest to the real axis of R are given by the relation

$$R_c^\pm = \frac{1}{\alpha} \ln \left| \frac{A}{\Delta \varepsilon} \right| \pm \frac{i\theta}{\alpha}. \qquad (17.5)$$

All other points lie on the straight line going through the points R_c^+ and R_c^- and are $\pm n\pi$ from them. In accordance with the definitions given in § 14, the centre of the region of nonadiabaticity R_p and its width δR are

$$R_p = \frac{1}{\alpha} \ln \frac{A}{\Delta \varepsilon},$$

$$\delta R = \frac{\theta}{\alpha}. \qquad (17.6)$$

The splitting of the adiabatic terms at the centre of the region of non-adiabaticity equals

$$\Delta U_p = 2 \Delta \varepsilon \sin\left(\frac{\theta}{2}\right). \qquad (17.7)$$

As is seen from (17.2) the parameter θ varies in the range $0 \leqslant \theta \leqslant \pi$.

It is not difficult now to explain the way in which the different ranges of the variable θ correspond to the behaviour of the adiabatic terms. For $\theta \ll 1$ a quasi-intersection of terms takes place, since the ratio $\Delta U_p / \Delta U_\infty$ turns out to be small, as follows from (17.7). If, in addition, it be demanded that the ratio $\Delta \varepsilon / \alpha \hbar v$ be small then, in fact, we arrive at conditions (15.7) and (15.12), i.e. at the model of linear terms.

For $\theta = \pi/2$, as can be seen from (17.2), parallel terms of the zero-order approximation are obtained, with an exponential coupling between them. This situation is typical of non-resonant single-stage charge-exchange reactions [21]

$$A^+ + B \to A + B^+. \qquad (17.8)$$

For $\theta > \pi/2$ both the terms of the zero-order approximation and the adiabatic terms diverge as R decreases. A specific example of such a process will be considered in a subsequent paragraph.

From what has been said above, it can be seen that the model described by the Hamiltonian (17.2) is more general than the model of linear terms. However, the Landau–Zener model permits of an analytical solution to the quantum problem of the motion of the nuclei though, just as for the model with an exponential interaction, a solution can only be found in the semiclassical approximation. Therefore the model with linear terms permits investigation of the case of very low atomic velocities, and in this sense it encompasses a much wider class of problems.

Both models successfully supplement each other, and in a common region of applicability go over into each other. This region is determined by conditions (15.7) and (15.12) and by the additional requirement that the splitting ΔU_p be small in comparison with the energy of the atoms $\mu v^2/2$. This latter requirement enables us to ignore the uncertainty in the parameters of the trajectory of the motion of the nuclei, which gives rise to the principal drawback of the semiclassical approximation, namely, the replacement of the true motion of the nuclei along the two terms by motion along a certain average term.

The system of equations for the probability amplitudes is obtained by substituting ΔU_{12} and χ from (17.3) and (17.4) into (14.15). In addition, it is still necessary to choose the trajectory for the motion of the nuclei $R = R(t)$ in the region where the atomic functions change into molecular functions. If $R(t)$ depends comparatively weakly on t, in the interval $\Delta t \sim 1/\alpha v$, then a series expansion can be used for $R(t)$. Attributing $t = 0$ to the point R_p, we obtain

$$R = R_p + v_R t + \dots . \qquad (17.9)$$

For a diatomic system (atomic collisions) by v is here understood the radial velocity $v_R = v(1-b^2/R_p^2)^{1/2}$. For a system with many degrees of freedom v_R represents the velocity of the motion along the reaction coordinate. Omitting the details of the calculation, we will quote only the final expression for the nonadiabatic transition probability in a system of two terms with the Hamiltonian (17.2) and the classical trajectory (17.9) [67], [69], [383]:

$$P_{12} = \exp(-\lambda + \lambda \cos\theta)\frac{\sinh(\lambda + \lambda \cos\theta)}{\sinh(2\lambda)}, \qquad \lambda = \frac{\pi \Delta \varepsilon}{2\hbar \alpha v_R}. \qquad (17.10)$$

The formula for the transition probability for a double transit of the region of nonadiabaticity is obtained by substituting (17.10) into (15.19):

$$P = 2\exp(2\lambda \cos\theta)\frac{\sinh(\lambda - \lambda \cos\theta)\sinh(\lambda + \lambda \cos\theta)}{\sinh^2(2\lambda)}. \qquad (17.11)$$

A diagram of the transition probability levels, calculated in accordance with (17.11) is presented in Fig. 11. In the formula (17.11) the interference of the probability amplitudes in a double transit across the region of nonadiabaticity has not been taken into account. This disadvantage can be removed, whilst remaining within the limits of the semiclassical approximation; to do this, it is necessary to observe the change of phase difference $\Delta S = (1/\hbar)\int \Delta U\, dt$ in the region $R < R_p$. The corresponding calculations,

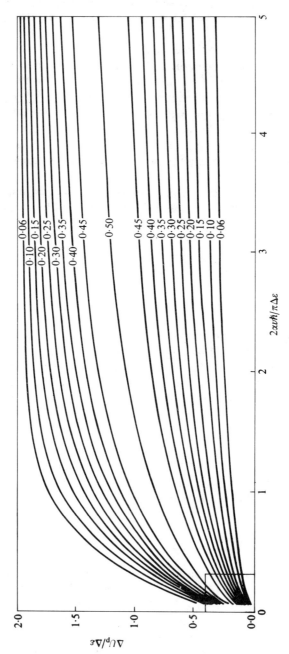

Fig. 11. Graph of the transition probability levels (17.11), corresponding to the non-linear model (17.2). The region of validity of the linear model in the approximation of a constant velocity is shown near the origin of coordinates.

based on the theory of sudden perturbations [23], show that the interference can be taken into account by introducing into (17.11) an additional factor B equal to

$$B = 2\sin^2\left[\int_{t_1}^{t_2} \frac{\Delta U(t)}{2\hbar} dt\right], \qquad (17.12)$$

where by $\Delta U(t)$ is understood the adiabatic splitting of the terms in the region $R < R_p$, and the integral is taken along the chosen trajectory during the time the system moves in this region.

It is essential to notice that, in the region $R < R_p$, the splitting of the terms ΔU depends on R in a completely arbitrary way and is in no way connected with expression (17.3) chosen to approximate ΔU in the transit region. Therefore the factor B is an additional parameter of the model.†

The expression (17.11) for $\theta \ll 1$ and $\lambda \gg 1$ goes over into the Landau–Zener formula which, in these variables, has the form

$$P = 2\exp(-\lambda\theta^2)[1 - \exp(-\lambda\theta^2)]. \qquad (17.13)$$

A detailed investigation of the transition probability in two limiting cases, for large and small values of the Massey parameter $\xi = \Delta\varepsilon/\alpha\hbar v$ in the transition region is of interest.

Slow transit of the region of nonadiabaticity

For large values of the parameter $\lambda(1 - \cos\theta)$ (slow transit), formula (17.11) can be simplified. Replacing all the hyperbolic functions by exponentials, we obtain

$$P = 2\exp\left[-\frac{\pi\Delta\varepsilon}{\hbar\alpha v}(1 - \cos\theta)\right]. \qquad (17.14)$$

This expression agrees with the result of calculating the transition probability by formula (15.17), within the limits of the adiabatic perturbation theory.

Since the semiclassical approximation may introduce a small error, in the limiting case of small velocities, in practice, it is of interest to find a quantum correction to the main term in the exponent of the exponential. This can be done exactly because in the case of an exponentially small transition probability, the expression for the exponent of the exponential

---

† For linear terms (see § 16) the interference term is expressed in terms of the same parameters which determine the transition probability in a single transit. This is connected simply with the fact that the linear approximation to the terms is considered valid in all the regions of motion of the nuclei from which contributions arise to the transition probability P and to the phase ΔS.

in the quantum case is known [37]:

$$P = 2\exp\left\{-\frac{2\sqrt{(2\mu)}}{\hbar}\operatorname{Im}\left[\int^{R_c}\left(E-\bar{U}+\frac{\Delta U_{12}}{2}\right)^{\frac{1}{2}}dR\right.\right.$$
$$\left.\left. -\int^{R_c}\left(E-\bar{U}-\frac{\Delta U_{12}}{2}\right)^{\frac{1}{2}}dR\right]\right\}. \qquad (17.15)$$

In (17.15) in addition to the splitting of the adiabatic terms ΔU_{12} the function $\bar{U}(R)$ is introduced, in terms of which are expressed the adiabatic terms under consideration:

$$U_1 = \bar{U} + \frac{\Delta U_{12}}{2},$$

$$U_2 = \bar{U} - \frac{\Delta U_{12}}{2}. \qquad (17.16)$$

In formula (17.15) the pre-exponential factor 2 is written down by analogy with (17.14), and we will not consider the possible quantum correction to this factor. For (17.15) to reduce to (17.14), the energy E of the nuclei should be considered as greatly exceeding $\bar{U} \pm \Delta U_{12}/2$ and each of the expressions under the integral sign should be expanded in a series in powers of $1/E$. Calculation of the first terms in the expansion and the corresponding integrals gives

$$P = 2\exp\left[-\frac{\pi\Delta\varepsilon}{\hbar\alpha v}(1-\cos\theta)\left(1 \pm \frac{\Delta\varepsilon}{4E} + \frac{E_0}{4E} + \ldots\right)\right]. \qquad (17.17)$$

In this expression v and E now denote the velocity and energy of the atoms, referred to the asymptotic value of the initial electronic term, and the correction term $\pm\Delta\varepsilon/4E$ takes account of the fact that the electronic transition is accompanied by a change in the energy of the nuclei (the positive sign referring to excitation, the negative sign to deactivation). The correction $E_0/4E$ takes account of the curvature of the trajectory of the relative motion of the atoms, disregarded within the framework of the classical approximation. The value of E_0 is given by the formula

$$E_0 = \operatorname{Im}\frac{2}{\pi}\int^{R_c}U(R)\frac{\Delta U_{12}(R)\,dR}{\Delta\varepsilon(1-\cos\theta)} \qquad (17.18)$$

and, as will be seen below, has the meaning of an effective activation energy.

The mean transition probability $\langle P \rangle$ is calculated by the general formula (15.35), where it may be interpreted as a transmission coefficient in the formulae of the transition state method, in which the critical surface S^* is determined by the condition $R^* = R_p$. Since the width of the region of nonadiabaticity is of the order $1/\alpha$, it is clear that the position of the critical surface is determined to within the same error. Therefore in the rate constant

§17 TRANSLATIONAL ENERGY IN MOLECULAR COLLISIONS

of the process, in its calculation by the transition state method, a relative error of the order of $1/\alpha R_p$ is introduced. Calculation of the integral (15.35) with P from (17.17) gives [72]

$$\langle P \rangle = 4\sqrt{\left(\frac{\pi}{3}\right)}\gamma^{\frac{1}{3}}\exp\left[-3\left(\frac{\gamma}{2}\right)^{\frac{2}{3}} \pm \frac{\beta\Delta\varepsilon}{2} - \beta E_0\right],$$

$$\gamma = \frac{\pi(1-\cos\theta)\Delta\varepsilon}{\hbar\alpha(2kT/\mu)^{\frac{1}{2}}}. \tag{17.19}$$

The first term in the exponential can be obtained in the semiclassical approximation, and the other two, which are assumed small by comparison with the first, represent quantum corrections. In particular, the term $\pm\beta\Delta\varepsilon/2$ secures the fulfillment of detailed balance between the direct and reverse processes.

In Fig. 12, the mean transition probabilities $\langle P \rangle$ calculated with functions (17.11) for the values $\cos\theta = \frac{1}{3}$ and $\cos\theta = -\frac{1}{3}$ are plotted against

$$\gamma' = \frac{\pi\Delta\varepsilon}{\hbar\alpha}\left(\frac{2kT}{\mu}\right)^{-\frac{1}{2}}.$$

For $\gamma' \gg 1$, the probability $\langle P \rangle$ is described by formula (17.19), in which only the first term in the exponential should be taken into account.

Fast transit through the region of nonadiabaticity

For fast transit through the region of nonadiabaticity (condition $\pi\Delta\varepsilon/\hbar\alpha v \ll 1$) we find, from expression (17.11),

$$P = \sin^2\left[\int_{t_1}^{t_2}\frac{\Delta U(t)}{2\hbar}dt\right]\sin^2\theta. \tag{17.20}$$

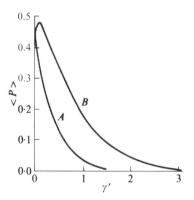

FIG. 12. The dependence on $\gamma' = \pi\Delta\varepsilon\sqrt{(\mu/2kT)}/\hbar\alpha$ of the mean transition probability for the non-linear model (17.2). Curves A and B correspond to values $\cos\theta = -\frac{1}{3}, \cos\theta = \frac{1}{3}$ respectively.

It is evident that, if the interference is neglected and the first oscillating factor replaced by the mean value of $\frac{1}{2}$ then, in contrast to the Landau–Zener model, the transition probability does not tend to zero in the limit of high velocities. This difference is wholly due to the fact that the adiabatic functions on the two sides of the transition region are connected by the general linear relation (17.1) and do not simply go over into one another. If in (17.20) the limit of high velocities is considered then, with regard to the oscillating factor, for a small phase difference $\int_{t_1}^{t_2} \Delta U(t)/2\hbar \, dt$ the probability turns out to be inversely proportional to the velocity. In this limit, the results for the exponential model can be obtained within the limits of the generalized Born approximation [37].

In conclusion, we note that the possibility of calculating the transition probability for a general form of nonadiabatic coupling in the two limiting cases (fast and slow transit) enables various types of interpolation formulae to be constructed; put forward as general approximate solutions of the system of temporal equations, they are widely employed nowadays in the interpretation of experimental data on charge exchange and transfer of excitation. The formulae of Rosen and Zener [447], Gurnee and Magee [246], Rapp and Francis [424], L. Vainshtein, L. Prenyakov, and I. Sobel'man [9] belong to this category. All these formulae give exact results for the resonant energy exchange and a correct answer for a weak interaction when perturbation theory is valid.

18. Nonadiabatic processes in atomic collisions

The only non-elastic process in atomic collisions is when the electronic state of one partner (or two simultaneously) changes on collision. As a result of the laws of conservation of total energy and total angular momentum, the problem of calculating the cross-section for some process reduces to the solution of a system of coupled equations, for one-dimensional motion, whose number equals the number of possible channels. The problem consists of calculating the scattering amplitudes into the various channels and is considered in general form in the general theory of atomic collisions [50]. In many cases, however, interference effects can be neglected and consideration limited to transition probabilities proportional to the squared moduli of the corresponding scattering amplitudes. In this approximation, the results of §§ 15–17 are immediately applicable to the calculation of rate constants of atomic processes connected with processes of electronic excitation and deactivation, transfer of excitation, and charge exchange.

All such types of process separate conveniently into three groups:
(a) Resonant processes for which the change in the electronic state of the colliding partners is not accompanied by a change in the internal energy ($\Delta\varepsilon = 0$).

(b) Non-resonant processes with small energy transfer (in the scale of energies of electronic excitation of atoms).

(c) Non-resonant processes with large energy transfer (of the order of the energy of electronic excitation of atoms).

We will now pass on to a discussion of the mechanism of some of such processes on the basis of the theory of nonadiabatic transitions set forth above, but will first pause to consider the adiabatic characteristics of the states of the two interacting atoms.

Adiabatic terms of a system of two atoms

The adiabatic terms of a system of two atoms or ions are calculated in the usual way by diagonalizing the matrix of the electronic Hamiltonian H'_e, written down with respect to a basis consisting of a finite number of the wavefunctions of the free atoms. The matrix elements of H'_e are expressed in terms of the matrix elements of the Hamiltonian H_e, which excludes the spin–orbit coupling and by the matrix elements of the spin–orbit coupling operator $V_{\text{s.o.}}$. For slow atomic collisions, when the adiabatic distortion of the atomic functions in the collision process is comparatively small, to a sufficiently good approximation the matrix elements of $V_{\text{s.o.}}$ between functions localized on different nuclei can be neglected, and for functions localized on one and the same nucleus, the matrix elements of the interaction $V_{\text{s.o.}}$ can be expressed in terms of the parameters of the free atom.

As regards the Hamiltonian H_e, its matrix elements are written down in terms of the excitation energies of the free atoms and the matrix elements of the interatomic interaction. Diagonalization of the matrix H_e yields the adiabatic electronic terms of the system of two atoms, disregarding the spin–orbit coupling. These terms are classified according to the total spin S of all the electrons, the total component Λ of the angular momentum of the electrons along the line joining the atoms (the intermolecular axis), and the parity (g or u) of the electronic state (in the case of identical partners) [37]. All remaining quantum numbers which are necessary for the full description of a term are usually denoted by a character in front of the term's symbol $^{2S+1}\Lambda_\omega$ ($\Lambda = \Sigma, \Pi, \Delta \ldots$ for components 0, 1, 2, ...; $\omega =$ g or u).

At sufficiently large interatomic distances R, the adiabatic terms of the system of atoms A and B can be represented in the form of a superposition of the energies of the zero-, first- and second-order approximations

$$U(^{2S+1}\Lambda_\omega, R) = E_A + E_B + W^{(1)}(^{2S+1}\Lambda_\omega, R) + W^{(2)}(^{2S+1}\Lambda_\omega, R). \quad (18.1)$$

Here, the first and second terms are the excitation energies of atoms A and B, the third term is the interaction energy of the first-order approximation, and the fourth is the interaction energy of the second-order approximation. All these contributions are calculated using atomic functions, taken in a suitable combination to secure the correct symmetry of the term $^{2S+1}\Lambda_\omega$.

As R decreases, the representation of $U(R)$ in the form of a series in perturbation theory becomes invalid so that, for the general properties of the terms, it is more convenient to consider not the contribution of the different interactions in $U(R)$ but the matrix elements of the Hamiltonian H_e directly, in terms of which these contributions are expressed.

Concise information concerning the dependence of the matrix elements of H_e on the interatomic separation will be presented below in connection with the discussion of particular processes.

If the terms $U(^{2S+1}\Lambda_\omega, R)$ are known, they may be used to calculate the terms of the system of two atoms with regard for the spin–orbit coupling by diagonalizing the matrix H'_e. Moreover, generally speaking, terms with different values of Λ and S are mixed, and from the two quantum numbers Λ and S only one quantum number Ω can be formed, characterizing the adiabatic state. This quantum number is defined as the component of the total spin of the electrons along the intermolecular axis. As regards the parity, it does not change on account of the spin–orbit coupling. Thus the adiabatic terms are denoted by Ω_ω. If only one term with $S' = S$ and $\Lambda' = \Lambda$ makes the chief contribution to the state function of the term Ω_ω, which is being represented by a linear combination of functions of different terms $^{2S+1}\Lambda'_\omega$, then the corresponding term is denoted by $^{2S+1}\Lambda_{\Omega,\omega}$.

We note that the adiabatic terms $U'(\Omega_\omega, R)$ can be determined by the direct diagonalization of the matrix of H'_e, without an intermediate calculation of the terms $U(^{2S+1}\Lambda_\omega)$. The significance of the intermediate stage is that the approximation (18.1) may sometimes be used for $U(^{2S+1}\Lambda_\omega)$, which achieves considerable simplification in subsequent calculations.

Resonant processes

As an illustration of resonant processes we will consider the charge exchange of an atom and an ion in an S-state and the transfer of excitation in an S–P transition.

The resonant charge exchange

$$A^+ + A(S) \rightarrow A(S) + A^+ \tag{18.2}$$

is described in a crude approximation by two functions corresponding to the localization of the electron on one centre or other. We will suppose that these functions coincide with the atomic functions and will neglect the nonadiabatic coupling between the atomic functions induced by the motion of the nuclei. This approximation implies that the mean momentum of the electron, localized on the moving atom A, is considered to be equal to zero, and that the transition of the electron between the two centres, moving in opposite directions, is not accompanied by momentum transfer. With these approximations, eqns (14.15) separate, and U_1 and U_2 refer to the adiabatic $^2\Sigma_g$ and $^2\Sigma_u$ terms of the A_2^+ system.

The system of equations (14.17) can be used in the same way for the amplitudes b_1 and b_2, discarding the second terms in the brackets of the pre-exponential factors. The connection between the adiabatic splitting ΔU and the matrix elements H_{ik} is given by relation (14.10) whence we find for the resonant process being considered

$$\Delta U(R) = U(^2\Sigma_u) - U(^2\Sigma_g) = 2|H_{12}|. \tag{18.3}$$

The solution to the system (14.17) can easily be obtained for the arbitrary function $H_{12}(t)$. The transition probability 1–2 giving the charge-exchange probability turns out to equal

$$P = \sin^2\left(\int_{-\infty}^{\infty} \frac{H_{12}}{\hbar} dt\right), \tag{18.4}$$

which of course agrees with (17.20). The cross-section for charge exchange is obtained by integrating P over all impact parameters b.

This integration can be carried out approximately in the following way. Noting that P is a function oscillating about its mean value $\bar{P} = \frac{1}{2}$, we replace P by $\bar{P}$ for all values of the impact parameter b smaller than some critical value $b = R^*$, and for $b > R^*$ we assume $P = 0$. Thus we have

$$\sigma(v) = 2\pi \int_0^\infty \sin^2\left(\int_{-\infty}^\infty \frac{H_0(t)}{\hbar} dt\right) b\, db = \frac{\pi}{2}[R^*(v)]^2, \tag{18.5}$$

where R^* is determined by the condition that P should tend to zero for $b > R^*$ sufficiently rapidly. Following Firsov[121], we will represent this condition in the form

$$\int_{-\infty}^\infty \frac{H_{12}(t)}{\hbar} dt = \frac{1}{\pi}, \tag{18.6}$$

where the integral on the left is calculated along the trajectory with impact parameter $b = R^*$. We notice that the introduction of the critical impact parameter $b = R^*$, which effectively separates the region of charge exchange with probability $\frac{1}{2}$ from the region of non-interacting partners, is analogous to the introduction of the critical surface S^* in the transition state method. In the case under consideration, however, this critical surface turns out to be dependent on the velocity v.

Now it is necessary to make definite assumptions concerning the dependence of H_{12} on R and on the form of the trajectory. As regards the matrix element of the interaction H_{12} (or of the adiabatic splitting ΔU), for large interatomic distances it can be represented in the form

$$\Delta U(R) = 2H_{12}(R) = BR^m \exp(-\alpha R). \tag{18.7}$$

Here B and m are certain constants and α is given by the relation

$$\alpha = \frac{\sqrt{(I/I_H)}}{a_0}, \tag{18.8}$$

where I is the ionization potential of the outer electron of the atom A; I_H is the ionization potential of an atom of hydrogen; a_0 is the Bohr radius.

The parameter α defined by relation (18.8) agrees with the orbital exponential for the outer electron of atom A and the expression (18.7), valid for $(Ra_0)\alpha^2 \gg 1$, yields an approximation to the one-electron resonance integral.

If the curvature of the trajectory is completely neglected then in calculating (18.6) we should put

$$R(t) = \sqrt{[(R^*)^2 + (vt)^2]}. \tag{18.9}$$

Substituting (18.9) into (18.7) and carrying out the integration (18.6) under the condition $(R^*a_0)\alpha^2 \gg 1$, we find

$$\frac{B(R^*)^m}{\hbar v} \exp(-\alpha R^*) \sqrt{\left(\frac{\pi R^*}{2\alpha}\right)} = \frac{1}{\pi}. \tag{18.10}$$

This expression gives an equation for the function $R^* = R^*(v)$, which determines the dependence of the cross-section on the velocity. Since the dependence of the left-hand side of (18.10) on R^* is chiefly exponential, the reverse dependence will be approximately logarithmic, so that (18.5) can be represented in the form

$$[\sigma(E)]^{\frac{1}{2}} = C_1 - C_2 \ln E, \tag{18.11}$$

where E is the relative energy of A^+ and A; C_1 and C_2 are constants.

A more exact expression for the cross-section can be obtained as a result of integrating the oscillating factor (18.4) directly over the impact parameter [109]. However, this destroys the possibility of interpreting the cross-section as a rate constant in terms of the concepts of the transition state as a state corresponding to a definite value of R^*.

The validity of expression (18.10) is limited by two conditions imposed above and below on the energy E (or the velocity). The restriction below is connected with the fact that, as the energy decreases, the trajectory of the relative motion becomes more curved by the polarization potential $W(R)$, so that (18.9) becomes inapplicable. For the collision between an atom and an ion, the long-range part of the potential $W(R)$ corresponds to a polarizing interaction

$$W(R) = -\frac{\alpha_p e^2}{2R^4}, \tag{18.12}$$

where α_p is the polarizability of the neutral partners.

Examination of the trajectories of motion of particles in such a potential, conducted in § 31, shows that they all divide into two groups: trajectories of the first group with impact parameters b exceeding a certain value b_c are weakly curved; trajectories of the second group with impact parameters $b < b_c$ are strongly curved. The trajectories of the second group describe phenomena, known in mechanics as the 'fall' of a particle onto the centre (see [36]). In fact, however, due to the close-range part of the potential, the convergence of the pair only proceeds to a certain finite separation so that, until the moment of maximum convergence, the partners may perform some revolutions. In such a situation, one speaks of the formation of a polarization complex. The expression for the cross-section for the formation of a polarization complex has the form (see § 31 formula (50.16)):

$$\sigma^c = \pi b_c^2 = \frac{2\pi e}{v} \sqrt{\left(\frac{\alpha}{\mu}\right)}. \tag{18.13}$$

From a comparison of (18.11) and (18.13), it can be seen that for resonant charge exchange in the region of sufficiently low velocities, the cross-section for formation of the complex σ^c exceeds the cross-section σ, calculated within the approximation of an undeflected trajectory. In this case, the cross-section for charge exchange should be calculated simply as the cross-section for the formation of a complex with the probability factor $P = \frac{1}{2}$, taking account of the two equiprobable ways for the complex to break up.

The restrictions on v from above are connected with the repeatedly-used condition $(R^\star a_0)\alpha^2 \gg 1$. Thus, finally, the conditions for the validity of relation (18.11) take the form

$$R^\star(v) \gg \frac{1}{\alpha^2 a_0}$$

$$R^\star(v) \gg b_c(v) = \left(\frac{2e}{v}\right)^{\frac{1}{2}} \left(\frac{\alpha_p}{\mu}\right)^{\frac{1}{4}}, \tag{18.14}$$

where the function $R^\star$ is determined from (18.10).

We will now pass on to the process of resonant excitation exchange in S–P transitions:

$$A^*(P) + A(S) \rightarrow A(S) + A^*(P). \tag{18.15}$$

We notice the following interesting difference between this process and the process of resonant charge exchange (18.2), namely, the degeneracy of the initial and final states and the long-range nature of the interaction. The latter is connected with the fact that the matrix element of the interaction is taken between states which are coupled by a dipole transition.

The long-range nature of the interaction in the adiabatic states of the A_2^* system may be understood on the basis of the following qualitative reasoning. Since the process (18.15) is a resonant process, the eigenfunctions of the Hamiltonian H_e are described in the form of a linear combination of functions of the excited and non-excited atoms. In a mixed (i.e. hybrid) S–P state, an atom may possess a dipole moment, directed along that component of the P-function which is admixed with the S-function. As a result we obtain a dipole–dipole interaction decreasing as R^{-3}.

For definiteness, we will consider atoms with one outer electron, so that each of the S- and P-atomic terms will be a doublet. Simple calculations give the following system of adiabatic molecular terms

$$\left.\begin{array}{l}U(^1\Sigma_g, R)\\ U(^3\Sigma_u, R)\end{array}\right\} = 2e^2\frac{(z_{SP}^A)^2}{R^3} \qquad (18.16a)$$

$$\left.\begin{array}{l}U(^1\Pi_u, R)\\ U(^3\Pi_g, R)\end{array}\right\} = e^2\frac{(z_{SP}^A)^2}{R^3} \qquad (18.16b)$$

$$\left.\begin{array}{l}U(^1\Pi_g, R)\\ U(^3\Pi_u, R)\end{array}\right\} = -e^2\frac{(z_{SP}^A)^2}{R^3} \qquad (18.16c)$$

$$\left.\begin{array}{l}U(^1\Sigma_u, R)\\ U(^3\Sigma_g, R)\end{array}\right\} = -2e^2\frac{(z_{SP}^A)^2}{R^3}, \qquad (18.16d)$$

where z_{SP}^A is the matrix element of the dipole moment

$$z_{SP}^A = \langle S^A|z|P_z^A\rangle.$$

The relative motion of the partners does not give rise to transitions between terms of different multiplets and parity. If the transitions between the Σ and Π states are neglected, then it is found that only the pairs of states $^{2S+1}\Lambda_g$ and $^{2S+1}\Lambda_u$ interact, and the adiabatic splitting between them is given by the expression

$$\Delta U(\Sigma, R) = 4e^2\frac{(z_{SP}^A)^2}{R^3}, \qquad (18.17a)$$

$$\Delta U(\Pi, R) = 2e^2\frac{(z_{SP}^A)^2}{R^3}. \qquad (18.17b)$$

Now the problem has been reduced to the preceding one with one difference: the splitting given by (18.7) has been replaced by (18.17). A calculation analogous to the one carried out above gives the following expression for the cross-section σ for process (18.15) under the condition that, before the collision, all orientations of the angular momentum of atom A*(P)'s electron are equiprobable:

$$\sigma = \frac{4}{3}\frac{\pi^2 e^2(z_{SP}^A)^2}{\hbar v}. \qquad (18.18)$$

§18   TRANSLATIONAL ENERGY IN MOLECULAR COLLISIONS   139

It might be thought, however, that this value of the cross-section is an overestimate by comparison with the true value, owing to the neglect of the Coriolis interaction between the Σ- and Π-states. In fact, the mixing of the Σ- and Π- states of the same multiplet and parity leads to a reduction in the splitting between the new, improved terms, and thus to a reduction in the cross-section, owing to a reduction in the critical value R^*, which is determined from condition (18.6). A correct calculation of the Coriolis interaction, however, demands a numerical solution of the system of equations for non-adiabatic coupling.

Non-resonant processes with small energy transfer $\Delta\varepsilon$

As an illustration of such processes, we will discuss non-resonant charge exchange with the participation of the S-states and the modification of the fine structure of an atom in a P-state.

We will describe the charge-exchange process

$$A^+ + B \to A + B^+ + \Delta\varepsilon \tag{18.19}$$

in a basis of the functions S_A and S_B. We have a matrix of the second order for the determination of the adiabatic terms. Its essential feature is that the diagonal elements of the interaction decrease, as R increases, faster than the off-diagonal elements. As regards the off-diagonal term H_{12}, for it approximation (18.7) is valid, where $\alpha = \frac{1}{2}(\alpha_A + \alpha_B)$.

We now note that, under the condition $(Ra_0)\alpha^2 \gg 1$, the pre-exponential factor in (18.7) can be regarded as constant by comparison with the exponential, if R varies in the range $\delta R \approx 1/\alpha$ near some fixed value R_0. We choose R_0 to be at the centre of the region of nonadiabaticity, assuming $R_0 = R_p$ and determining R_p from the relation

$$BR_p^m \exp(-\alpha R_p) = \Delta\varepsilon. \tag{18.20}$$

After these approximations, the problem of non-resonant charge exchange reduces to the model problem with Hamiltonian (17.2), with the following values for the parameters

$$\cos\theta = 0$$
$$A = \Delta\varepsilon \exp(\alpha R_p)$$
$$\alpha = \tfrac{1}{2}(\alpha_A + \alpha_B)$$
$$E_0 + \tfrac{1}{2}\Delta\varepsilon = E_A, \qquad E_0 - \tfrac{1}{2}\Delta\varepsilon = E_B, \tag{18.21}$$

where R_p is determined by relation (18.20).

The charge-exchange probability is given by formula (17.11) which, for $\theta = \pi/2$ and with regard for (17.12), takes the form

$$P = \sin^2\left(\int_{-\infty}^{\infty} \frac{H_{12}}{\hbar} dt\right) \cosh^{-2}\left(\frac{\pi\Delta\varepsilon}{2\alpha\hbar v_R}\right),$$

$$v_R = v\sqrt{\left(1 - \frac{b^2}{R_p^2}\right)}. \tag{18.22}$$

The cross-section, calculated with the probability (18.22), is tabulated as a function of $\lambda = \pi\Delta\varepsilon/2\alpha\hbar v$ in the book by Smirnov mentioned in the preface. Here we will only note the limiting expressions, which are obtained for large and small values of the parameter λ.

For $\lambda \gg 1$, the transition probability is exponentially small and rapidly decreases for increasing b. In all significant regions of integration over b the oscillating factor is replaced by the average value $\frac{1}{2}$, so that we finally obtain

$$\sigma = 2\pi R_p^2 \frac{\exp(-2\lambda)}{\lambda}. \tag{18.23}$$

For $\lambda \ll 1$, the second factor in (18.22) can be replaced by 1 in all the significant regions of integration, so that the problem becomes analogous to the problem of calculating the cross-section for resonant charge exchange. Thus we obtain

$$\sigma = \frac{\pi}{2}[R^\star(v)]^2, \tag{18.24}$$

where $R^\star(v)$ is determined from (18.10).

Comparison of the limiting expressions (18.23) and (18.24) shows that the cross-section σ of the non-resonant process, considered as a function of velocity, goes through a maximum whose position is determined by the condition that the Massey parameter be close to unity.

We note finally, that the smallness condition on $\Delta\varepsilon$ is satisfied in the approximation of H_{12} by the asymptotic expression (18.20).

We will now pass on to the second process: the transitions between the components of the fine structure of atoms in collision. For definiteness, we will consider the collision of excited atoms of an alkali metal M with the atoms of an inert gas X:

$$M^*(^2P_{\frac{1}{2}}) + X(^1S_0) \to M^*(^2P_{\frac{3}{2}}) + X(^1S_0) + \Delta\varepsilon. \tag{18.25}$$

The most interesting feature of these reactions (the so-called intermultiplet mixing) is the very large variation of the average cross-section with changes in $\Delta\varepsilon$ ($\Delta\varepsilon$ changes from 16 cm^{-1} to 548 cm^{-1} in the series from Na to Cs) and in the temperature.

We will first discuss qualitatively the mixing mechanism. To this end, we will consider the relative motion of the partners M* and X along some trajectory. Before the collision, the orbital angular momentum **l** and the spin **s** of the optical electron in M* are coupled to give the total momenta $j = \frac{1}{2}$ and $j = \frac{3}{2}$. At a certain interatomic distance, the electrostatic (polarization or exchange) forces become so intense that they lead to axial symmetry in the field acting on the electron. In the rotating molecular system of coordinates, this sets in when the electrostatic interaction is larger than the Coriolis interaction. At even shorter distances partial breaking up of the spin–orbit coupling occurs; this takes place at a distance R_p where the electrostatic interaction is comparable with the spin–orbit coupling. In addition, initially **l** is coupled with the molecular axis **n**, and the spin **s** is polarized in a magnetic field directed along **n**. The temporary breaking down of the coupling, caused by the competition between the electrostatic and magnetic interactions and described in terms of the nonadiabatic interactions between terms of like symmetry, is the first mixing mechanism [523].

At shorter distances still, **l** is strongly coupled with **n** and rotates together with the axis. The rotating magnetic field may lead to the reorientation of **s** under favourable competition between the magnetic and Coriolis interactions. This effect, which is described in terms of the nonadiabatic coupling between terms of different symmetry, is the second mixing mechanism.

To obtain quantitative information regarding the probability of mixing, we will start from the adiabatic approximation which is obtained on diagonalizing the energy matrix of the two atoms for a fixed separation and a fixed axis. At this stage, as a reasonable approximation, the spin–orbit coupling operator may be considered independent of the distance between the atoms. Among the molecular states, there are six ($A\frac{1}{2}$, $A\frac{3}{2}$, and $B\frac{1}{2}$, each doubly degenerate) which correlate with the six atomic states $^2P_{\frac{1}{2}}$ and $^2P_{\frac{3}{2}}$ as $R \to \infty$. These six states can be chosen as the basis in the calculation of the nonadiabatic coupling, i.e. the coupling induced by the motion of the atoms. The advantage of such a representation of the process (18.25) is that the interaction turns out to be localized and the transitions between the molecular states can be ascribed to a definite interatomic separation R_p.

Disregarding the spin–orbit coupling, the electronic Hamiltonian of the system M*(P)+X(S) yields two terms $U(\Sigma, R)$ and $U(\Pi, R)$ which are used to calculate the true adiabatic terms [20], [78], [80] $U'(R)$ of the above-mentioned three states $A\frac{1}{2}$, $A\frac{3}{2}$, and $B\frac{1}{2}$:

$$U'(A\tfrac{1}{2}, R) = \tfrac{1}{2}[U(\Sigma) + U(\Pi) + \Delta\varepsilon - \Delta U'],$$

$$U'(B\tfrac{1}{2}, R) = \tfrac{1}{2}[U(\Sigma) + U(\Pi) + \Delta\varepsilon + \Delta U'],$$

$$U'(A\tfrac{3}{2}, R) = \tfrac{1}{2}[U(\Pi) + \Delta\varepsilon],$$

$$\Delta U' = \{(\Delta\varepsilon)^2 + \tfrac{2}{3}\Delta\varepsilon[U(\Sigma) - U(\Pi)] + [U(\Sigma) - U(\Pi)]^2\}^{\frac{1}{2}}. \quad (18.26)$$

where, here, the origin of the energy measurement is referred to the state $^2P_{\frac{1}{2}} + {}^1S_0$ of the separated partners.

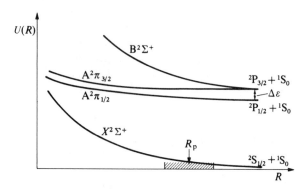

Fig. 13. Qualitative form of the adiabatic terms of the alkali metal M and inert gas X system. Weak polarizing interaction.

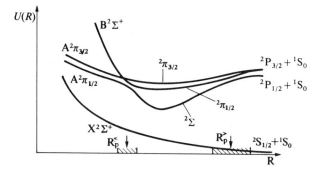

Fig. 14. Qualitative form of the adiabatic terms of the M + X system. Strong polarizing interaction.

In connection with the sign of the difference $E_\sigma - E_\pi$, the two different situations represented in Fig. 13 and Fig. 14 are possible.† At very large distances, the polarization forces, for which the Σ-terms lie lower than the Π-terms, make the chief contribution to the splitting $U(\Sigma) - U(\Pi)$. The exchange forces play a main role at short distances and the difference $E_\sigma - E_\pi$ is positive. The distance $R = R_p$ close to the centre of the region of nonadiabaticity is of interest. The value R_p is found from the condition

$$|U(\Sigma) - U(\Pi)| \approx \Delta\varepsilon. \tag{18.27}$$

† In these diagrams, the terms of the excited states are not taken to small distances, since the asymptotic formulae, on whose basis the diagrams are constructed, are correct only for sufficiently large distances.

The larger root $R_p^>$ of this equation determines the position of the outer region of nonadiabaticity. If for $R = R_p^>$ the difference $U(\Sigma) - U(\Pi)$ is negative, then, at some point $R < R_p^>$, it changes sign and yet another root $R_p^<$ will correspond to eqn (18.27). This root (if it exists) gives the position of the second region of nonadiabaticity. A diagram of the distribution of the terms is presented in Fig. 14.

An investigation of the system of equations which describes the non-adiabatic transitions between the molecular states (which we will not present here) enables simple expressions to be obtained for the mean transition cross-sections in two limiting cases: for $\Delta\varepsilon/\alpha h v \ll 1$, when an approximate solution to the three equations can be found, and for $\Delta\varepsilon/\alpha h v \gg 1$, when transitions between terms of like symmetry make the chief contribution, so that the system of three equations reduces to a system of two equations [20], [78], [80].

Under the condition $\Delta\varepsilon/\alpha h v \ll 1$, the cross-section for reactions (18.25) is found by the method of sudden perturbations: three atomic functions are joined onto the molecular functions at a certain value of R equal to R^*. In addition, in the outer region $(R > R^*)$, the intermolecular interaction is completely neglected and, in the inner region $(R < R^*)$, the spin–orbit coupling is neglected. In this way we obtain

$$\sigma = \tfrac{2}{3}\pi[R^*(v)]^2. \tag{18.28}$$

This expression is analogous to formula (18.5) in which the factor $\tfrac{1}{2}$ has been replaced by $\tfrac{2}{3}$. The specific form of the function $R^*(v)$ depends on the intermolecular potential in the region $R \approx R^*$. The asymptotic dependence of the Σ- and Π-terms of the M*–X system at large distances is given by [78], [80]:

$$U(\Sigma) = AR^m \exp(-\alpha R) - \tfrac{7}{5}\alpha_X e^2 \langle r^2 \rangle R^{-6}, \quad m = \frac{4}{\alpha} - 2, \tag{18.29a}$$

$$U(\Pi) = -\tfrac{4}{5}\alpha_X e^2 \langle r^2 \rangle R^{-6}. \tag{18.29b}$$

In (18.29a) the first term gives the contribution of the exchange interaction, where, with an accuracy up to a constant factor, it agrees with the asymptotic value of the squared wavefunction of the outer electron in the excited atom M*. Therefore the constant α can be expressed in terms of the ionization potential I^* of the atom M*(^{2}P):

$$\alpha = \frac{2}{a_0}\left(\frac{I^*}{I_H}\right)^{\tfrac{1}{2}},$$

where I_H is the ionization potential of a hydrogen atom.

The constant A is determined by the nature of the interaction between the electron M* and the atom X, and in certain cases may be expressed in terms of the scattering length of a slow electron on atom X [82]. Since the cross-section of the reaction (18.25) depends very weakly on A, we will not discuss the question of the calculation of A. In the approximation under consideration, the exchange interaction in the state Π should be neglected, since the π-orbital of the outer electron is close to zero in the region of maximum electron density of atom X. The second term in (18.29a) corresponds to the Van der Waals interaction of M* and X. Here, α_X is the polarizability of atom X and $\langle r^2 \rangle$ is the mean square distance of the excited p-electron in atom M*. Thus all the parameters of the adiabatic terms of the quasi-molecule are expressed in terms of the free-atom constants. Later on, in a discussion of the reactions (18.25), we will assume that (18.29) is valid in the region of a nonadiabatic transition for $R = R_p$, although, in a number of cases, higher terms of the multipole expansion may be important [20].

The critical value R^* is found on the basis of the ideas expressed above. In the given case, the equation for the determination of R^* has the form

$$\frac{1}{2\hbar} \int_{-\infty}^{\infty} [U(\Sigma, \sqrt{\{(R^*)^2 + (vt)^2\}}) - U(\Pi, \sqrt{\{(R^*)^2 + (vt)^2\}})] \, dt = \frac{1}{\pi}. \quad (18.30)$$

If only the polarization interaction is substituted here for the difference $U(\Sigma) - U(\Pi)$, then we obtain

$$\sigma_{\frac{1}{2} \to \frac{3}{2}} = 2 \cdot 18 \left(\frac{\alpha_X e^2 \langle r^2 \rangle}{\hbar v} \right)^{\frac{2}{5}}. \quad (18.31)$$

The result found for $\sigma_{\frac{1}{2} \to \frac{3}{2}}$ on the basis of an approximate solution to the system of three coupled equations, under the condition $\Delta \varepsilon / \alpha v \hbar \ll 1$, is

$$\sigma_{\frac{1}{2} \to \frac{3}{2}} = 1 \cdot 65 \left(\frac{\alpha_X e^2 \langle r^2 \rangle}{\hbar v} \right)^{\frac{2}{5}}. \quad (18.32)$$

The difference between the numerical coefficients of eqns (18.31) and (18.32) reflects the effect of the different approximations in the calculation of R^*, and thus demonstrates the accuracy which may generally be expected from the method of joining states in a system of three or more states.†

If it is assumed that in the joining region the exchange interaction makes the chief contribution, then a transcendental equation of the type (18.10) is obtained for R^*, the solution to which is the function $R^* = R^*(v)$, which decreases approximately logarithmically with increase in velocity. With sufficient accuracy we may assume [78], [388]:

$$\sigma_{\frac{1}{2} \to \frac{3}{2}} = \frac{2}{3} \pi (R^*)^2 = \frac{2}{3} \frac{\pi}{\alpha^2} \left(\ln \frac{A R_p^m \sqrt{\pi}}{\hbar v \sqrt{(2\alpha R_p)}} \right)^2, \quad (18.33)$$

where R_p is determined by eqn (18.27).

† In the case under consideration, the joining method is, in point of fact, equivalent to the transition state method if the existence of some critical distance, separating the initial and final states, is assumed.

§ 18   TRANSLATIONAL ENERGY IN MOLECULAR COLLISIONS   145

In the general case both the exchange and the polarization forces contribute to the cross-section. However, as an approximate estimate of $\sigma_{\frac{1}{2}\to\frac{3}{2}}$ the larger of the two calculated values σ_{pol} (18.31) and σ_{exch} (18.33) can be chosen. It is clear that in the limit of large velocities, the relation $\sigma_{\text{exch}} \gg \sigma_{\text{pol}}$ will be ultimately satisfied, since the exchange forces make the main contribution to the splitting of the molecular terms at sufficiently short distances. Of course, these distances, all the same, must appreciably exceed the gas-kinetic radius R_0, in order that the exchange interaction should be reproduced by an exponential function.

As the velocity decreases, the cross-section σ_{pol} increases, however, expression (18.31) ultimately becomes invalid since the assumed condition $\Delta\varepsilon/\hbar\,\alpha v \ll 1$ is broken. In the region of velocities $\Delta\varepsilon/\hbar\,\alpha v \approx 1$ the transition cross-section reaches a maximum

$$\sigma_{\frac{1}{2}\to\frac{3}{2}}^{\max} \approx \tfrac{2}{3}\pi R_p^2, \tag{18.34}$$

and then beings to fall for a further increase in the parameter $\Delta\varepsilon/\hbar\alpha v$. This region corresponds to almost-adiabatic collisions. Here, we may confine consideration to the coupling of the two electronic terms of symmetry $\Omega = \tfrac{1}{2}$. If X is an atom with a small polarizability, for example He, then in the region of nonadiabatic coupling, in general, the polarization correction can be dropped, and we obtain:

$$U(\Sigma) - U(\Pi) = A R_p^m \exp(-\alpha R). \tag{18.35}$$

Within the limits of this approximation for the terms $U(\Sigma)$ and $U(\Pi)$, which determine the adiabatic terms U' by formulae (18.26), the calculation of the transition probability and the cross-section for the reaction (18.25) reduces to the model problem with the Hamiltonian (17.2). Identification of the parameters of this Hamiltonian gives

$$\cos\theta = -\tfrac{1}{3}$$

$$A = \Delta\varepsilon \exp(\alpha R_p)$$

$$\alpha = \frac{2}{a_0}\sqrt{(I^*/I_H)}$$

$$E_0 + \frac{\Delta\varepsilon}{2} = E(^2P_{\frac{3}{2}}) + E(^1S_0)$$

$$E_0 - \frac{\Delta\varepsilon}{2} = E(^2P_{\frac{1}{2}}) + E(^1S_0), \tag{18.36}$$

where R_p is determined from (18.27) on substituting relation (18.35).

The mean transition cross-section is found by integrating P from (17.11) over the impact parameters and averaging over the velocities. In addition, in place of the parameter $\lambda = \pi\Delta\varepsilon/2\hbar\alpha v$, a new parameter $\gamma = \tfrac{4}{3}\pi\Delta\varepsilon/\alpha\sqrt{(\mu/kT)}$

arises, whose magnitude determines the adiabatic ($\gamma \gg 1$) or nonadiabatic ($\gamma \ll 1$) nature of the process. In particular, for $\gamma \gg 1$, the average cross-section of the reaction (18.25) has the form [80]

$$\langle \sigma_{\frac{1}{2} \to \frac{3}{2}} \rangle = \pi R_p^2 4 \sqrt{\left(\frac{\pi}{3}\right)\left(\frac{\gamma}{2}\right)^{\frac{1}{3}}} \exp\left[-3\left(\frac{\gamma}{2}\right)^{\frac{2}{3}} - \frac{\beta \Delta \varepsilon}{2}\right]. \qquad (18.37)$$

In Table 3, calculations using this formula [20] and experimental cross-sections of various processes are compared.

TABLE 3

Mean transition cross-sections $j = \frac{1}{2} \to j = \frac{3}{2}$ for various partners

Partners	T(K)	R_p (atomic units)	γ	$-\delta\varepsilon/\Delta\varepsilon$	$\sigma_{\frac{1}{2}} \to \sigma_{\frac{3}{2}}$ cm² Theoretical	Experimental†
K+He	368	14	2.05	0.02	3×10^{-15}	6.6×10^{-15}
K+Ne	368	14	3.82	0.04	0.7×10^{-15}	1.3×10^{-15}
K+Ar	368	14	3.79	0.15	0.6×10^{-15}	3.6×10^{-15}
K+Kr	368	14	4.21	0.23	0.7×10^{-15}	5.9×10^{-15}
K+Xe	368	14	3.49	0.36	1.2×10^{-15}	10.3×10^{-15}
		$\Delta\varepsilon = 58$ cm⁻¹,		$\alpha = 0.90$ atomic units		
Rb+He	340	13	9.25	0.008	10×10^{-18}	7.9×10^{-18}
Rb+He	373	13	8.8	0.008	12×10^{-18}	10×10^{-18} [149]
Rb+Ne	340	13	19.00	0.016	0.8×10^{-19}	1.4×10^{-19}
Rb+Ne	373	13	18.15	0.016	1.3×10^{-19}	10^{-19}–10^{-18} [149]
		$\Delta\varepsilon = 238$ cm⁻¹,		$\alpha = 0.87$ atomic units		
Cs+He	315	12	23.4	0.005	5×10^{-21}	5.7×10^{-21}
		$\Delta\varepsilon = 554$ cm⁻¹,		$\alpha = 0.85$ atomic units		
Na+Ar	450	Calculation of exchange forces only			1.2×10^{-14}	10^{-14}
Na+Ar	450	Calculation of polarization forces only			1.32×10^{-14} [183]	10^{-14}
		$\Delta\varepsilon = 16$ cm⁻¹,		$\alpha = 0.94$ atomic units		

† The experimental data, quoted without a reference, are copied from the papers [195], [313].

In the calculation [20], a small correction to the contribution of the polarization was taken into account, which slightly reduces the value of $\Delta\varepsilon$ which must be substituted into γ by comparison with the value of the splitting between the fine structure components of M*. This correction is given in Table 3 in the form of the ratio $\delta\varepsilon/\Delta\varepsilon$. The data in Table 3 show that the theory enables a qualitative explanation to be given of the variation in the transition cross-sections over a factor of 10^7 from light to heavy atoms in the series Na, K, Rb, Cs. At the same time it can be seen that there is a marked discrepancy between theory and experiment. It is possible that this lies in the insufficient accuracy of the adiabatic potentials. Moreover, it is very probable that there is yet another mechanism present in transitions. Calculations on the model system Li–He [458], [466] show that the energy of the $^2\Pi$-term remains constant to within comparatively short distances, where it rapidly begins to increase. In the region of increasing potential, the molecular axis of M*–X rotates comparatively rapidly. This may induce transitions between the $^2\Pi_{\frac{1}{2}}$ and $^2\Pi_{\frac{3}{2}}$ components, which ultimately appear as transitions between the doublet components of atom M*. Calculations of the cross-section for this mechanism cannot be carried out without knowledge of the terms at short interatomic separations.

Non-resonant processes with large energy transfer $\Delta\varepsilon$

The electronic deactivation reaction of an excited atom M* in collision with an atom X, whose internal states do not change in the collision process, is the simplest example of this type of process.

$$M^* + X \rightarrow M + X + \Delta\varepsilon. \qquad (18.38)$$

The energy given off usually amounts to a few electron volts.

The Massey parameter in this case turns out to be very large, not only in thermal conditions but even for relative energies of a few tenths of an electron volt.

If, nevertheless, process (18.38) proceeds at an appreciable rate under thermal conditions, this means that the minimum convergence $\Delta U_{\min}$ of the adiabatic terms, which correlate with the initial and final states of the partners, is much smaller than $\Delta\varepsilon$. Thus, in the cases of interest to us, we can expect the conditions

$$\frac{\Delta\varepsilon}{\hbar\alpha v} \gg 1, \qquad \frac{\Delta U_{\min}}{\Delta\varepsilon} \ll 1 \qquad (18.39)$$

to be satisfied. But these are precisely the conditions for the applicability of the linear model for which the transition probability is given by the Landau–Zener formula (15.20). Unfortunately, the parameters of the linear model (the activation energy E_a, the matrix element of the interaction, etc.), in terms of which the transition probability is expressed, are usually unknown, and for their determination the asymptotic method of estimating the interatomic interaction cannot be used. This is connected with the fact that the convergence of terms whose separation in the free partners is great continues up to comparatively short interatomic distances, and information is very sparse on adiabatic potentials at such distances.

As an example of a process of type (18.38) which has been frequently studied experimentally we mention the quenching of the resonance fluorescence of sodium by argon:

$$\text{Na}^*(^2P) + \text{Ar}(^1S_0) \rightarrow \text{Na}(^2S) + \text{Ar}(^1S_0) + 2 \cdot 1 \text{ eV}. \qquad (18.40)$$

The adiabatic electronic terms of this system are presented qualitatively in Fig. 13. At the present time, non-empirical calculations exist only for the Li–He system [458], [466], whose terms agree with Fig. 13. The excited Π-term, on being extrapolated to short internuclear distances, possibly crosses the ground Σ-term. Unfortunately, such an extrapolation cannot be justified because, at short distances, the interaction of the electrons of He with the electrons of the Li shell begins to appear.

In the paper [7] an attempt was made to interpret the experimental data on the quenching of the resonance fluoresence of sodium by argon on the basis of a simplified variant of the linear model. Experimental data was enlisted, which indicated very small values of the activation energy E_a and a comparatively large nonadiabatic transition probability, approaching the maximum value $P_{\max} = \frac{1}{2}$. In the light of new experimental data [256], [531], indicating a comparatively small value of the reaction cross-section, the interpretation given in reference [7] cannot be considered final. With a view to developing a general scheme for the interpretation of the temperature dependence of the rate constant of a process of type (18.40), a calculation was carried out [389] on the average quenching probability, taking into account the spin–orbit and Coriolis interactions of the ground Σ-term and the excited Π-term. The basic assumption, which is open to criticism, resulted in the presence of an intersection between these terms. Unfortunately, at the present time it is not at all clear to what extent this is valid.

19. Nonadiabatic processes in the collision of atoms with diatomic molecules

As was shown above, the spin–orbit coupling is one of the causes of nonadiabatic transitions in atomic collisions; at high temperatures, the non-adiabatic coupling of the electron momentum with the rotation of the quasi-molecule makes an appreciable contribution. These interactions connect terms of different coordinate symmetry relative to the collision axis. The splitting between terms of like symmetry is often very large in the system of two atoms, so that nonadiabatic transitions take place between them with very small probability. In this sense, the mechanism of nonadiabatic transitions in the collision of an atom with a diatomic molecule XY differs in essence from the transition mechanism in atomic collisions, since in a system where the number of atoms is greater than two, terms of like symmetry may cross and the probability of nonadiabatic transition in the neighbourhood of the point of intersection must be considerable (see § 16). A theoretical calculation for similar processes is complicated by the necessity to consider the multi-dimensional motion of the nuclei over the adiabatic potential energy surfaces. The case of two-dimensional motion would be the simplest after the one-dimensional. Such a situation occurs, under certain restrictions, in the collision of an atom with a diatomic molecule or in a nonadiabatic transition in triatomic non-linear molecules. Here, as in the case of atomic collisions, it is expedient to consider different situations depending on the nature of the electronic states between which the transitions occur. If these electronic states belong to different electronic configurations of the atom, then the corresponding adiabatic potential energy surfaces of the triatomic system $XY + A$ will differ very appreciably in form and, in consequence, their sharp convergence, and even intersection, is probable. A theoretical estimate of the position of the regions of the nonadiabatic transition in such a case is very unreliable, as a rule, since it must be founded on a calculation of the adiabatic terms in which account is taken of a sufficiently large number of atomic and molecular orbitals. Even for the simplest system of an alkali metal and an inert gas, considered earlier, the position of the point of intersection (if there is one) of the terms $X^2\Sigma$ and $A^2\Pi$ is unknown, in the general case.

If the electronic states of the atom which participate in the transition relate to different terms of one and the same configuration, then it is possible to state more-or-less definite assumptions about the structure of the surfaces. Here, the simplification is connected with the fact that the splitting between the surfaces is determined chiefly by the correlation energy of one and the same configuration, which changes comparatively weakly for a change in the nuclear coordinates. An example of a simple estimate of the relative distribution of the adiabatic terms is given in § 21, in connection with the study of electronic vibrational transitions.

Finally, if the electronic terms under consideration concern the fine structure components of the atomic terms, then nonadiabatic transitions between them usually take place at intermolecular distances sufficiently large that the XY molecule only weakly perturbs the wavefunction of atom A. In this case, in an approximate calculation, the unperturbed electronic functions of atom A can be used which, of course, appreciably simplifies the calculation of the adiabatic terms. One such process occurring in atomic collisions was considered in a previous section (the transitions between the fine structure components of alkali metal atoms). Below, we will consider the characteristics of analogous processes in atomic–molecular collisions by the example of transitions between the fine structure components of an excited mercury atom.

Electronic excitation and deactivation in the system $M-X_2$ leading to the excitation of vibrations in X_2

To be precise, we will consider a nonadiabatic transition causing the deactivation of an atom of an alkali metal M in the electronic ^2P state in collision with a diatomic molecule X_2 having a closed electronic shell (state $^1\Sigma_g^+$):

$$M^*(^2P) + X_2(^1\Sigma_g^+, n) \to M(^2S) + X_2(^1\Sigma_g^+, n') + \Delta E. \tag{19.1}$$

The energy ΔE liberated in (19.1) may be distributed in different ways between the relative translational energy of M and X_2 and the rotational and translational energies of X_2. We will assume, to begin with, that the adiabatic potentials acting on the X nuclei in the triatomic complex MX_2, in the neighbourhood of the region of nonadiabaticity, differ so little that, in accordance with the Franck–Condon principle, the vibrational quantum number n of the X_2 molecule will not change (i.e. $n' = n$) in an adabatic transition.

The general characteristics of the adiabatic potentials of the system under consideration can be obtained on the basis of a correlation diagram for the linear ($C_{\infty v}$) and symmetric triangle (C_{2v}) configurations of MX_2, considered at a fixed distance r_n which corresponds to the nth vibrational state of X_2. For the configuration $C_{\infty v}$, the terms found, without taking account of the spin–orbit coupling, can be classified in the same way as for the diatomic molecule, and their dependence on the intermolecular distance R is analogous to the dependence of the terms of the M–X system. Intersection of the ground term $^2\Sigma(^2S + {}^1\Sigma_g)$ and first excited term $^2\Pi(^2P + {}^1\Sigma_g)$ is possible (cf. Fig. 13). For the configuration C_{2v} for large R, the structure of the terms of MX_2 is analogous to the structure of the terms of MX. However, as the intermolecular distance decreases, the Π-term is split in two: into a symmetrical term B_1 and an anti-symmetrical term B_2, with respect to the plane of the three nuclei. The state B_1 does not interact with the ground state A_1, since they differ in symmetry with respect to a rotation by an angle Π about the altitude of the triangle MX_2 which joins M to the centre of gravity of X_2. The elements of the symmetry group C_{2v}, the irreducible representations, and their characters are given in Table 4. In addition, the representations by which the coordinates x, y, z, which are proportional to the components of the p-function of the valence atom M and the rotations R_x, R_y, R_z about these axes, transform, are shown in the table. For uniformity, the quantity ω is introduced, transforming like the product xy.

For small deformations of the configuration of an isosceles triangle (symmetry group C_s) the states B_1 and A_1 interact, since they are both symmetric with respect to the only symmetry operation in an arbitrary triatomic system, namely, reflection in the plane of the three atoms. These terms are represented in Fig. 15, where the intersection of the surfaces in a cone is shown in large scale. Here $\pi/2 - \gamma$ denotes the angle between the axis **n** of the X_2 molecule and **R**. It is not difficult to verify that the spin–orbit coupling separates any terms of the triatomic system under consideration. Therefore the terms of the Hamiltonian H'_e will have the form of a paraboloid in this region (see § 16) instead of contact at the vertex of a cone. Moreover,

TABLE 4

Symmetry of irreducible representations of the point groups C_{2v} and C_s†

	Group C_{2v}				Group C_s		
Single-valued representation	E	C_2	σ_v	σ'_v	Single-valued representation	E	σ_v
$A_1(z)$	1	1	1	1	$A'(y, z, R_x)$	1	1
$B_1(y, R_x)$	1	-1	1	-1			
$A_2(w = xy, R_z)$	1	1	-1	-1			
$B_2(x, R_y)$	1	-1	-1	1	$A''(x, R_y, R_z)$	1	-1
Two-valued representation	E	Q	C_2, C_2Q	σ_v, σ_vQ	σ'_v, σ'_vQ	Two-valued representation	E	σ_v
$E''\begin{pmatrix}\alpha\\\beta\end{pmatrix}$	2	-2	0	0	0	$E''\begin{pmatrix}\alpha\\\beta\end{pmatrix}$	$\begin{pmatrix}1\\1\end{pmatrix}$	$\begin{pmatrix}i\\-i\end{pmatrix}$

† The standard notation is used here: E is the identity operation; C_2 is a rotation of π about an axis of symmetry; σ_v is a reflection in the plane of the three atoms; σ'_v is a reflection in a perpendicular plane passing through the axis of symmetry; Q denotes a rotation of 2π about the axis of symmetry which, for the two-valued representation, is formally considered different from the operation E. The system of coordinates is chosen thus: the z axis is taken along the axis of symmetry, the y axis passes through the atoms of the diatomic molecule and the x axis is orientated perpendicular to the plane of the system of three atoms.

the intersection of the terms A' and A'' along a curve is replaced by a quasi-intersection, and along any path going through the line of intersection of the terms A and B, the potential energy curve will have a form qualitatively analogous to the curves represented in Fig. 8.

Nonadiabatic transitions between the actual adiabatic terms are caused by the motion of the nuclei in the plane and the rotation of the entire system as a whole. In accordance with this, the operator of nonadiabaticity $\hat{C}$ is

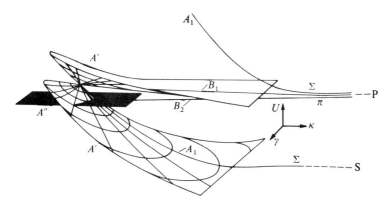

FIG. 15. Qualitative dependence of the adiabatic terms of the M + X system on R and γ close to the configuration of an isosceles triangle. Weak polarizing interaction.

represented in the form

$$\hat{C} = -i\hbar\frac{\partial}{\partial t} = -i\hbar\left(\dot{R}\frac{\partial}{\partial R}+\dot{\gamma}\frac{\partial}{\partial \gamma}\right)-(\omega_1 J_1+\omega_2 J_2+\omega_3 J_3), \qquad (19.2)$$

where the ω_i are the instantaneous angular velocities of rotation of the triatomic system about the principal axes of inertia†; J_i are the projection operators of components of the total angular momentum of the electrons.

In a calculation of the nonadiabatic transition probability between the cones, the rotation of the system as a whole (the second bracket in (19.2)) can be neglected, since the motion of the atoms in the plane (the first bracket in (19.2)) makes the chief contribution. The corresponding dynamical interaction can be reduced to a static interaction (exchange and Coulomb + spin–orbit) if the functions of symmetry $\phi(A_1)$ and $\phi(B_1)$ are considered as the zero-order basis. In this case the transition probability is given by formula (16.19), which contains the parameters of the cone near the vertex, the matrix element of the spin–orbit coupling, and the two components of velocity corresponding to motion along R and γ.

The transitions between the real adiabatic terms (terms of the Hamiltonian H'_e) correlated with the terms A (the lower cone) and B (the approximately horizontal term) of the Hamiltonian H_e, which disregards the spin–orbit coupling, are caused by the motion in the plane and the rotation of the system. In a basis of the functions of the zero-order approximation $\phi(A')$ and $\phi(A'')$, the first interaction leads to a consideration of the spin–orbit coupling, and the second to the calculation of the Coriolis interaction of the orbital angular momentum of the electron with the rotation of the nuclear frame. In accordance with this, by the J_i in the second term in brackets in (19.2) should be understood the projection of the orbital (and not total) angular momentum L_i.

At each small zone near the curve of quasi-intersection of the terms A and B, a local system of coordinates can be introduced, in a similar manner to their introduction in the calculation of vibrational transition probabilities (see §9). Then the two-dimensional problem reduces approximately to the one-dimensional problem, where the transition probability can be calculated by the formulae in §15. An example of a specific calculation of this type is presented in a subsequent paragraph.

The total transition probability can be obtained as a sum of contributions from 'cone–cone' and 'cone–plane' transitions if the distance from the vertex of the cone to the curve of intersection is sufficiently large for the interference of the probability amplitudes to be neglected.

In the neighbourhood of the linear configuration of AX_2, two adiabatic terms of the same symmetry have the form of cones; however, the plane of the

---

† One of the principal axes is always directed along the x axis as a result of the existence of a plane of symmetry in the triatomic system.

§ 19   TRANSLATIONAL ENERGY IN MOLECULAR COLLISIONS   153

third term (an antisymmetric state arising from the degenerate Π-state) goes through the vertex of the cones and touches each of them along a generatrix. In this connection, the interference between the cone–cone and cone–plane transitions must not be neglected. Moreover, the nonadiabatic coupling between the symmetric and antisymmetric states may be very strong, since the average rotational frequencies about the longitudinal axis of the almost linear complex AX_2 are large (as a result of the smallness of the corresponding moment of inertia).

Under thermal collision conditions, the trajectories in the neighbourhood of the vertex of the cone which corresponds to the least energy make the chief contribution to the nonadiabatic transition probabilities. In the present state of quantum-mechanical molecular calculations it is difficult to estimate the difference between the energies corresponding to the vertices of the cones for the linear and triangular configurations of AX_2. Nevertheless, simple qualitative considerations indicate that a very probable situation is such that the energy at the point of intersection of the surfaces for the triangular configuration is less than the energy of the vertex of the cone in the linear distribution of the nuclei. The mean 'cone–cone' transition probability for such a case was calculated in the paper [387]. We will not dwell on the results of the calculation here as the region of applicability of the classical variant of the linear model used in the derivation of (16.19) turns out to be very small. As was noted in § 15, the linear one-dimensional model proves applicable to actual processes because it enables the introduction of a certain classical trajectory (15.28), effectively representing the wave motion along the two electronic terms. For the two-dimensional linear model, it is clearly not possible to introduce such a trajectory. Therefore the condition for the applicability of the classical description of the motion of the nuclei near the vertex of the cone, which is equivalent to the condition of a small change in the velocity v in the region of strong coupling Δl, is formulated in the form†

$$\left(\frac{\Delta U_\infty}{\hbar \alpha v}\right)^{\frac{1}{2}} \frac{kT}{\Delta U_\infty} \gg 1, \qquad (19.3)$$

where ΔU_∞ is the asymptotic difference between the terms $M^*(^2P)$ and $M(^2S)$.

By comparison with (15.12), the left-hand side of (19.3) contains a small factor $kT/\Delta U_\infty$ which strongly limits the region where condition (19.3) is fulfilled.

We will now discuss the case where the vibrational state of X_2 is altered in a nonadiabatic transition.

If the adiabatic potentials acting on the nuclei X differ appreciably in the two electronic states under consideration in the process (19.1), the conversion

† Here, in an estimate of the parameters in (16.19), it is assumed that $F_1 \approx F_2, v_1 \approx v_2, A = 0$.

of part of the energy ΔU_∞ into the vibrational energy of X_2 is very probable. It is convenient to consider this process too as the motion of the representative point over the potential energy surfaces. The possibility of considering it thus is connected with the fact that at temperatures not exceeding 1000 K, molecular collisions, as noted above, are adiabatic not only with regard to the electronic states but also with regard to the vibrations of X_2. The approximate distribution of the adiabatic potential energy surfaces is obtained by the gradual displacement of the terms represented in Fig. 15 by values $\hbar\omega$ upwards (in a more exact investigation it is necessary to take it into account that the distances between the vibrational terms of the excited electronic state differ somewhat from the distances between the vibrational terms of the ground state). In this way many pairs of double cones are obtained, intersecting one another and corresponding to different vibrational quantum numbers. The matrix elements of the interaction between them differ from zero, generally speaking; however, they tend to zero for the symmetric configuration C_{2v} if one of the states corresponds to some 'upper' cone and the other corresponds to a 'lower' cone. The interaction between terms of the same symmetry, differing in vibrational quantum numbers, is small, since these terms do not intersect and the separation between them is large.

To calculate the change in the vibrational energy of X_2 in the nonadiabatic transition (19.1), factors depending on the vibrational coordinates of X_2 should be introduced into the nonadiabatic wavefunction. They must be related to those linear combinations of $\phi(A)$ and $\phi(B)$ which give a stationary solution on leaving the vertex of the cone along a definite trajectory. However, instead of two equations describing a transition without change in n, a system of equations is obtained for the transitions $n \to n'$. The basic difference between this system of equations and (12.3) is that the parameter $\mathfrak{M}$ in them is multiplied by the overlap integral of the vibrational functions $S_{nn'} = \int \chi_n^*(A)\chi_{n'}(B)\,d\tau$ and the times t_1 and t_2 differ for the various pairs of interacting terms.

In the case when we can neglect the interference of the probability amplitudes from the different regions of nonadiabaticity, which correspond to the convergence of surfaces with different n, it is possible to represent the reaction (19.1) as the result of a whole succession of nonadiabatic transitions on intersection by the trajectory of the γ axis, above which lie a row of cones. The apertures of these cones change from comparatively large (if $\Delta n = 0$) to very small (for $\Delta n \neq 0$ and $S_{nn'} \ll 1$). For an increase in Δn, the integral $S_{nn'}$, as a rule, decreases and the probability increases since, as is not difficult to see, after averaging with respect to $|t_1 - t_2|$, the probability P proves to be inversely proportional to $S_{nn'}$. As P approaches unity, the necessity arises for considering repeated transits and, finally, for sufficiently small values of $S_{nn'}$, it is possible to use perturbation theory. In this case the transition probability again becomes small, since it is proportional to the square of the matrix element of the interaction and consequently to $(S_{nn'})^2$.

Thus, if the adiabatic potentials acting on the nucleus X_2 differ appreciably in different electronic states then, along any classical trajectory describing the relative motion of M and X_2 in the plane, regions will exist where the probability of a nonadiabatic transition will be of order unity. Moreover, the initial quantum number n changes by $\Delta n = n' - n$ in such a way that $P(\Delta n = 0)/S_{nn'} \approx 1$.

The condition that the potential should change appreciably is in order that $\langle P \rangle$, determined by (16.19), should at least satisfy the relation $\langle P \rangle \ll S_{n,n+1}$. If this inequality is not satisfied then the contribution of transitions with $\Delta n = 1$ will be negligibly small. As regards the possibility of neglecting the interference of the probability amplitudes, the condition $\omega/\alpha v \gg 1$ favours this. However, without more detailed analysis it cannot be affirmed that, for values of the ratio $\omega/\alpha v$ characteristic of actual processes of the type (19.1), the interference does not have an effect on the total probability.

To conclude this section we note the following. Calculations completed recently of the electronically-excited terms of a diatomic system, in which one of the partners has a closed electronic shell, show that the excited terms can be characterized sufficiently by a deep minimum, even in the case when the term of the ground state corresponds to repulsion [158]. It is very probable that an analogous situation arises in the interaction of an electronically-excited atom with a diatomic molecule having a closed electron shell. At least the qualitative considerations of the role of ionic states, made by Laidler [318] and Mori [379], point to such a possibility. In such a situation, the vertex of the double cone and also the line of intersection of terms of different symmetry may lie below the initial level $E(^2P + {}^1\Sigma_g^+)$. In this case the electronic deactivation of M* takes place in two consecutive stages. In the first stage the intermediate complex M*X_2 is formed and, in the second, the intramolecular nonadiabatic transition in the MX_2 system occurs, leading to the decomposition of the complex. The first stage contains a number of steps: 'adhesion', or capture, of the partners, intramolecular redistribution of energy which, in particular, may cause the reverse decomposition of the complex M*X_2, and finally the exchange of energy between the excited complex and the molecules of the heat reservoir. These problems are considered in the theory of unimolecular reactions (see Chapter 4). If the lifetime τ of the complex is less than the time between consecutive collisions then the last step is generally not considered, and for the rate constants of a process of type (19.1) the expression

$$\kappa = \kappa_c \frac{\kappa_{\text{nonad}}}{\kappa_{\text{dis}} + \kappa_{\text{nonad}}} \qquad (19.4)$$

is obtained, where κ_c is the capture rate of the partners; κ_{dis} is the rate of spontaneous dissociation; κ_{nonad} is the intramolecular nonadiabatic transition rate.

From (19.4) it is evident that for $\kappa_{nonad} \gg \kappa_{dis}$ the limiting stage of deactivation does not generally include a nonadiabatic transition from one electronic state to another. In expression (19.4), to estimate κ_{nonad} we may enlist the considerations given above. However, it should be taken into account that, in a stable complex $(MX_2)^*$, the region of nonadiabaticity may be intersected by the representative point many times during the lifetime of the complex. Unfortunately, to estimate the dissociation rate κ_{dis} within the limits of, for example, the statistical theory (see Chapter 5), knowledge of the binding energy of the complex is necessary, about which information is very limited.

The existing experimental data, which chiefly relate to the quenching of the resonance fluorescence of alkali metal atoms and mercury, do not permit an unambiguous interpretation of the deactivation mechanism to be made within the framework of one of the simple theoretical models considered above. The large deactivation cross-sections (close to gas-kinetic) of Na* in collision with N_2 [285], [391] and also the effectiveness of the reverse process [281], [282], [500] show that, in the process

$$Na^*(^2P) + N_2(^1\Sigma_g^+, n = 0) \rightleftarrows Na(^2S) + N_2(^1\Sigma_g^+, n > 0), \qquad (19.5)$$

vibrationally-excited molecules arise with large probability. This, in turn, implies that the overlap integrals $S_{nn'}$ are sufficiently large even for quantum numbers n and n' which differ markedly. Such a situation is possible with ionic terms (Na^+, N_2^-) taking part (see Fig. 16), although it is not clear to what degree a consideration of this intermediate state may lead to the formation of a complex with a long lifetime. It is clear, however, that, if the ionic term is important, then resonance effects (the conversion of the electronic energy of Na* into the vibrational energy of N_2) cannot be pronounced. As regards the

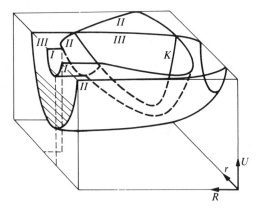

FIG. 16. Qualitative dependence of the adiabatic terms of the $M + X_2$ system on R and r for the linear configuration. The interaction of the ionic state II with the ground state III and excited covalent states I. K is the curve of intersection of the terms II and III.

dependence of the quenching cross-section on the velocity of the partners, it turns out to be negative: in the reaction involving the quenching of the fluorescence of potassium by nitrogen, the cross-section is reduced from 20.2×10^{-16} to 9.6×10^{-16} for an increase in velocity from 9.1×10^4 cm s^{-1} to 14.9×10^4 cm sec^{-1} [234]. From what has been said, it is clear that, for the construction of a theory, the experimental results for processes of type (19.1) are very important; some similar reactions are discussed in Callear's review [186].

Transitions between the fine-structure components

Transitions between the fine-structure components of atoms, induced by the interaction with a diatomic molecule, are the simplest example of a process in which the conversion of the electronic excitation energy of one partner into the vibrational energy of the other occurs with a comparatively small additional emission of energy, in the form of the relative translational energy of the colliding molecules. The change in the electronic energy $\Delta\varepsilon$ in similar reactions is small in the scale of atomic excitation energies. At the same time $\Delta\varepsilon$ may be sufficiently large, in the sense that the collision has an almost adiabatic character ($\Delta\varepsilon/\hbar a v \gg 1$), and the cross-sections for the conversion of electronic energy into kinetic must be very small. In these conditions, the almost-resonant exchange of energy accompanied by the change of vibrational state of the molecule acquires a special importance. We will consider here the mechanism of such a quasi-resonant process, carrying the comparison with the results of § 18 concerning transitions between the fine-structure components of atoms in atomic collisions, as far as possible. Firstly, we will consider the qualitative behaviour of the adiabatic terms.

The correlation diagrams for the dependence of the adiabatic electronic terms on the relative position of the nuclei of the triatomic system, in the case under consideration, can be constructed fairly simply, since we restrict ourselves to a one-electron configuration, in this way avoiding the introduction of many additional undetermined parameters. Later on, we will discuss reactions of the type

$$A(^2P_j) + X_2(^1\Sigma_g^+, n) \rightarrow A(^2P_{j'}) + X_2(^1\Sigma_g^+, n \pm 1) \tag{19.6}$$

$$A(^3P_j) + X_2(^1\Sigma_g^+, n) \rightarrow A(^3P_{j'}) + X_2(^1\Sigma_g^+, n \pm 1), \tag{19.7}$$

where X_2 is a diatomic molecule in the completely symmetric electronic ground state; A is an atom in the doublet or triplet P-state (in a ground or excited state).

In the first stage of the calculation, we will ignore the quantum number n and will build electronic terms of the system $A(^{2S+1}P_j) + X_2(^1\Sigma_g^+)$ assuming the internuclear distance r_{XX} fixed. We will start from atomic wavefunctions of the total angular momentum j, in a representation in which the spin–orbit coupling operator ALS of the atoms is diagonal.†

By the laws of vector addition, these functions are expressed in terms of the coordinate wavefunctions with quantum numbers L, M_L and spin functions with quantum numbers S, S_z. The matrix elements of the interaction energy depend only on the quantum numbers of the coordinate parts, i.e. are expressed ultimately by exchange integrals between the three functions constructed from the wavefunctions of the electronic shell of atom A (P-state) and the wavefunctions of the closed shell of the X_2 molecule ($^1\Sigma_g^+$-state).

For the linear configuration of the nuclei, two parameters remain, the energies E_σ and E_π of the molecular terms Σ and Π, calculated without taking account of the spin–orbit coupling. For the triangular configuration, the interaction energy is expressed in terms of three parameters, E_z, E_y, and E_x, corresponding to the three components of the degenerate P-state of the atom. For the linear configuration of the nuclei (symmetry $C_{\infty v}$) the terms are classified by the quantum numbers Ω. As R decreases and the intermolecular interaction increases, the spin–orbit coupling in the atom is partially broken due to the polarization of the atomic functions (by polarization here is meant the intermolecular interaction due ultimately to the Coulomb interaction of the electrons and nuclei; this is sometimes referred to as the interaction of the electron momentum with the molecular axis). When the splitting $E_\sigma - E_\pi$ becomes larger than A, the spin–orbit coupling can be calculated by first-order perturbation theory.

The correlation diagram is constructed on the basis of the two limiting cases and the rules for the nonintersection of terms. In spite of this, two forms of diagram are possible, depending on the sign of the difference $E_\sigma - E_\pi$. If exchange forces make the chief contribution then $E_\sigma > E_\pi$; for polarization forces, $E_\sigma < E_\pi$. Since the relative distribution of the terms, and not their absolute positions, is of chief interest in considering nonadiabatic transitions, the value of one of the parameters (E_σ or E_π) can be chosen arbitrarily. In particular, we will assume that $E_\pi = 0$ in the first case and $E_\sigma = 0$ in the second. The correlation diagrams, constructed in this way are given in Figs. 17 and 18. These diagrams naturally agree with the diagrams of any pair of atoms $^{2S+1}P_j + {}^1S_0$. For example, for the pair $Hg^*(^3P_j) + Ar(^1S_0)$ for an

---

† The spin–orbit coupling operator can be represented in such a form for Russell–Sanders-type coupling of the angular momentum in the atom A, i.e. when the fine-structure splitting is much smaller than the excitation energy of the nearest atomic term of the same symmetry. For a mercury atom, for example, the ratio $[E(^3P_2) - E(^3P_1)]/[E(^1P_1) - E(^3P_1)]$ is approximately equal to 0·33 and, of course, cannot be considered sufficiently small. In this connection, the interval rule for the states ^3P_j is not fulfilled and the ratio $[E(^3P_2) - E(^3P_1)]/[E(^3P_1) - E(^3P_0)]$ equals 2·6 instead of the theoretical value 2 expected for simple Russell–Sanders coupling. Nevertheless, later on, we will neglect the errors brought in by the deviation from LS-coupling and will take the form of the spin–orbit coupling as indicated above.

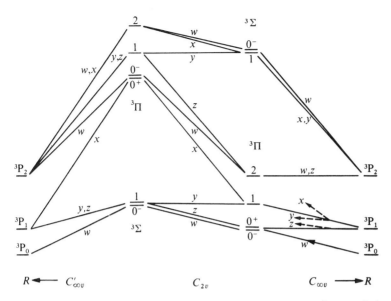

FIG. 17. Correlation diagram of the electronic terms of the system $A(^3P_j)+X_2(^1\Sigma_g^+)$.

interatomic separation $R > 3.5$ Å the actual order of the terms is that given in Fig. 17 [227].

We will now consider the configuration of the isosceles triangle (symmetry C_{2v}). For simplicity, we will assume that $E_x = 0$, so that the adiabatic terms

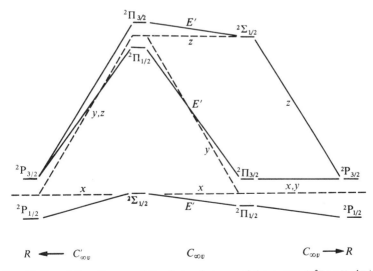

FIG. 18. Correlation diagram of the electronic terms of the system $A(^2P_j)+X_2(^1\Sigma_g^+)$.

will only depend on E_z and E_y. The energy-level diagram for the configuration C_{2v} can be obtained for a fixed value of E_z by gradually increasing E_y. For $E_y = 0$, the triangular system may be formally attributed the symmetry $C_{\infty v}$ (because $E_y = E_x$), so that its terms will coincide with the terms on the right-hand side of Figs. 17 and 18. For $E_y = E_z$ the triangular system is again formally assigned the symmetry $C_{\infty v}$, however, now the axis of symmetry is directed along the x axis (this case is marked by the fact that the symmetry groups to the right and left are designated differently: $C_{\infty v}$ and $C'_{\infty v}$). Therefore the terms of the system will coincide with the terms shown on the left-hand side of the diagram. In the centre of the diagram $E_z \neq E_y \neq E_x = 0$, and the correlation diagram is obtained by applying the rules for the non-intersection of the terms to two limiting cases: namely the terms of the system $C_{\infty v}$ and $C'_{\infty v}$.

In Figs. 17 and 18, correlation diagrams are constructed which give the cross-sections of the systems of adiabatic terms along the following paths:

1. $0 < E_z \leqslant E_z^0$, $E_x = E_y = 0$ (E_z changes; formal symmetry $C_{\infty v}$).
2. $E_z = E_z^0$, $0 < E_y < E_z^0$, $E_x = 0$ (E_y changes; symmetry C_{2v}).
3. $0 < E_z = E_y \leqslant E_z^0$, $E_x = 0$ (E_y and E_z change; formal symmetry $C'_{\infty v}$),

where E_z^0 is some energy parameter limited by the condition $E_z^0 > A$.

The terms of the system $A(^2P) + X_2(^1\Sigma_g^+)$, without taking the spin–orbit coupling into account, are shown by dashes in Fig. 18; for the symmetry C_{2v} they can be classified according to the type of symmetry of x, y, or z (the type of symmetry is indicated in the brackets). Taking the spin–orbit coupling into account, the electronic terms are split; however, for systems with half-integer spin they remain doubly degenerate. In the linear configuration, this degeneracy corresponds to two possible signs for the component of Ω along the symmetry axis. In the non-linear configuration, the degeneracy corresponds to two functions which form the basis of a two-dimensional irreducible representation E' of the two-valued representation of the group C_{2v} [37].

For systems with integral spin the degeneracy of the terms in the C_{2v} configuration is completely removed, and the states can be classified according to the type of symmetry of x, y, z, ω of the single-valued representation of the group C_{2v} (see Table 4). In Fig. 17 the dashes show a portion of the terms which arise when E_y becomes equal to E_z until E_z becomes larger than A. Other sections of a system of potential energy surfaces can be constructed in an analogous way.

On deformation of the symmetric triangular configuration of the system of three atoms, the only symmetry operation which remains is a reflection in the plane going through the stationary nuclei (symmetry group C_s). In accordance with this, the electronic terms can be classified as symmetric (A') and antisymmetric (A'') with respect to reflection, where these terms correlate uniquely with the terms (x, y, z, w) of the symmetric configuration.

As an illustration of the general considerations, we will consider the process of energy exchange in the quenching of the resonant fluorescence of excited mercury atoms Hg*($^3P_1$). Experimental investigation of the quenching indicates a number of possible mechanisms, the simplest of which consists of an almost-resonant conversion of the electronic energy of the atom into the vibrational energy of the molecule [31], [184], [185], [359], [476]:

$$\text{Hg*}(^3P_1) + X_2(^1\Sigma_g^+, n = 0) \rightarrow \text{Hg*}(^3P_0) + X_2(^1\Sigma_g^+, n = 1) - \Delta\varepsilon. \quad (19.8)$$

In considering process (19.8) we will neglect the interaction of the set of terms arising from the excited atomic states $^3P_0$, $^3P_1$, $^3P_2$, with the term correlating with the ground state of the partners $\text{Hg}(^1S_0) + X_2(^1\Sigma_g^+)$. The validity of this assumption can only be based on a detailed analysis of the terms of the ground and excited states, and will not be discussed here. It is clear, however, that if it is confirmed that the mechanism (19.8) is experimentally preferred by comparison with the other possible mechanism

$$\text{Hg*}(^3P_1) + X_2(^1\Sigma_g^+, n = 0) \rightarrow \text{Hg}(^1S_0) + X_2(^1\Sigma_g^+, n') + \Delta\varepsilon, \quad (19.9)$$

then, in calculating the reaction rate of (19.8), the ground-state term can be neglected. In the calculation of the terms and the construction of a qualitative diagram of the type represented in Fig. 17, it is necessary to take into account the exchange interaction of Hg* and X_2, the quadrupole–quadrupole interaction, and the Van der Waals interaction.

As was shown in Chapter 2 for a molecule of the N_2 type, the Massey parameter $\omega/\alpha v$ appreciably exceeds unity for vibrational transitions. Therefore as the zero-order approximation, it is expedient to start from the electronic–vibrational adiabatic functions, considering transitions between states as the result of an interaction of the relative motion of the partners with electronic motion in the Hg* atom and with the vibrations of the N_2 molecule. Such considerations preserve the maximum analogy with the theories of purely vibrational (see § 9) and purely electronic transitions but the interatomic distance r_{N-N} is excluded from the number of independent adiabatic degrees of freedom. These electronic–vibrational terms of the zero-order approximation can be obtained by the successive superposition of correlation diagrams as represented in Fig. 17, each of which is shifted by the amount $\hbar\omega$ relative to the preceding one.

For the case Hg* + N_2 under consideration, the displacement $\hbar\omega$ constitutes approximately $\frac{4}{3}$ of the value of the splitting $E(^3P_1) - E(^3P_0)$. In such a superposition of correlation diagrams, there occurs the intersection of terms characterized by different vibrational quantum numbers n and electronic quantum numbers x, y, z, w. If two terms of the same symmetry were to intersect then this would imply a quasi-intersection in reality, since in a simple superposition of diagrams the interaction which induces an adiabatic change in the potential of the N_2 molecule under the effect of the field of the

Hg* atom is not taken into account. The splitting of the terms at the point of quasi-intersection equal to twice the off-diagonal matrix element of the interaction would be proportional to the small parameter $(\alpha x)_{01}$, however, the equivalent situation does not arise for the system under consideration with the energy parameters adopted $\hbar\omega \sim \frac{4}{3}A$.†

If two terms of different symmetry intersect, then a nonadiabatic interaction is induced between them by that part of the operator $i\hbar(\partial/\partial t)$ which corresponds to the motion of the plane or the symmetry axis relative to which the states under consideration are classified. In the given case, the terms $(w, {}^3P_0, n = 1)$ and $(x, {}^3P_1, n = 0)$ may be of this type, the latter being shown in Fig. 17 by dashes. As seen from Table 4, these two states may be coupled by the rotation of the whole system about the x axis (the character of the product xw agrees with the character of the rotation R_x). On deformation of the configuration of an isosceles triangle, the terms x and w refer to one and the same type of symmetry A'', moreover, if the deformation is small, then the nonadiabatic coupling between them, due to the rotation R_x, may be calculated as for the symmetric configuration. For a small deformation, there appears a non-vanishing matrix element of the electronic–vibrational Hamiltonian between functions of the symmetry of x and w, so that the two correct adiabatic functions $\phi_1(A'')$ and $\phi_2(A'')$ must be written in the form of a linear combination of the functions $\phi_1(w, {}^3P_0, n = 1)$ and $\phi_2(x, {}^3P_1, n = 0)$. Since the coefficients of these combinations depend on the angle of deflection γ' of the X_2Hg system from the isosceles triangle configuration, the motion of the atoms in the plane, analogous to the deformation vibrations of $Hg-X_2$, induces nonadiabatic transitions between the terms $\phi_1(A'')$ and $\phi_2(A'')$, leading ultimately to the electronic–vibrational transition (19.7).

It is precisely the nonadiabatic coupling induced by the motion of the atoms in the plane which makes the chief contribution to the transition probability. In a basis of the functions $\phi(w, {}^3P_0, n = 1)$ and $\phi(x, {}^3P_1, n = 0)$ the nonadiabatic coupling is replaced by a static interaction, where the matrix elements of the total Hamiltonian have the form

$$H_{12} \sim (\alpha x)_{01}(E_x - E_y)\gamma'$$
$$H_{11} - H_{22} \sim \Delta F(R - R_p). \tag{19.10}$$

Returning to § 16 it is not difficult to verify that the relations (19.10) indicate the intersection of the adiabatic terms at a point corresponding to the vertex

† In a subsequent paragraph, a case is considered for which, in a certain sense, it is possible to speak of the quasi-intersection of terms of like symmetry with respect to reflection in the plane of the three atoms. These functions can be written in the form $\phi_s\alpha$ and $\phi_a\beta$, where ϕ_s and ϕ_a are symmetric and antisymmetric space functions. Taking account of the transformation laws of the spin functions under reflection, both total wavefunctions $\phi_s\alpha$ and $\phi_a\beta$ are multiplied by the same factor i on reflection. In accordance with the above-mentioned assertion, the off-diagonal matrix element between these functions for the vibrational $1 \to 0$ transition will be of the order of $(\alpha x)_{10}A$.

of a double cone. A particular feature of this cone is that, in the direction of the γ' coordinate, the aperture of the cone is very large and the splitting of the terms increases very slowly for an increase in γ' and is directly connected with the presence of the small parameter $(\alpha x)_{01}$ in the expression for H_{12}. This, in turn, indicates that the effective region of nonadiabatic coupling in terms of γ' can be comparatively large and thus the approximation of $R - R_p$ and γ' by linear functions of time (cf. (16.16)) proves inadequate. Therefore Bykhovskiĭ and Nikitin [6] proposed that the trajectory of the relative motion of the nuclei in the region of nonadiabatic coupling might be approximated more generally by the expressions

$$\gamma' = \gamma \sin(\bar{\omega}t + \psi)$$
$$R - R_p = \frac{Ft^2}{2\mu} - \frac{\mu v^2}{2F}, \qquad (19.11)$$

from which the linear approximation is obtained, as a particular case, for small frequencies and large amplitudes of the deformation vibrations and sufficiently large energies of the relative translational motion.

The solution of the nonadiabatic problem, within the limits of perturbation theory, for the two energy-level Hamiltonian with the trajectory (19.11) leads to an expression analogous to (15.36), but differing by additional factors due to the contribution of the deformational degrees of freedom. These terms take into account the possibility of a conversion of the energy of the deformation vibrations into the energy of the translational motion of Hg^*-X_2 and the vibrational energy of X_2 and also the statistical weight of the configurations preferred for the nonadiabatic transition under consideration (i.e. for the configuration close to the configuration of an isosceles triangle). Referring the reader to paper [6] for details of the calculation, we will only note here the important difference between the transitions (19.1) and (19.7). If, in the reaction (19.1), the vibrational state of X_2 does not change then the aperture of the double cone is comparatively small and the region of strong nonadiabatic coupling is comparatively small. Therefore the transition $^2P, n \to {}^2S, n'$ for which the double cone proves to be sufficiently sloping makes the chief contribution to the total rate of process (19.1). In process (19.7) such a situation is secured automatically by the presence of the small parameter $(\alpha x)_{01}$ in H_{12}, hence the mean transition probability $\langle P \rangle$ may be of order unity.

A specific estimate of the constants $\kappa(T)$ for different processes of type (19.7) is hampered by the scarcity of information about one of the most important parameters—the activation energy E_a. Existing methods of calculating the adiabatic terms of the triatomic system are too unreliable for the value of E_a found to be used in an evaluation of the exponential term. The difficulty in calculating E_a is further aggravated since the probable value of the activation energy corresponds to the comparable contribution of the attractive polarization potential and the repulsive exchange potential, i.e. corresponds to the region where the variational method of calculation is already very inexact and all the terms in a series of multipole interactions make contributions of the same order.

The results of Matland [359], studying the temperature dependence of the reaction rate of (19.7), indicate that the activation energy E_a is close to the value $\Delta\varepsilon = \hbar\omega - A = 563 \text{ cm}^{-1}$. For such a small activation energy, the tunnel correction and a correction for the excitation of deformation vibrations can be neglected. Then the mean cross-section $\langle\sigma\rangle = \kappa/v$ will be equal to the product of quantities of the order of the gas-kinetic cross-section for Hg*–N$_2$ (~ 1–3×10^{-15} cm^2) times a probability factor $\langle P \rangle$ of the type (15.36). A rough estimate of $\langle P \rangle$ can be made from formula (15.36), omitting the exponential factor corresponding to the tunnel correction. In place of the square of the off-diagonal matrix element a^2, the mean value $\langle H_{12} \rangle^2$ from (19.10) should be substituted.

Taking it into account that the mean value of the square of the amplitude of the angular vibrations is determined by the relation $\langle (\gamma')^2 \rangle \sim kT/E_y$ and estimating ΔF by the formula $\Delta F \sim \Delta\varepsilon\alpha$, we find

$$\langle P \rangle \sim (\alpha x)_{01}^2 \frac{(E_y - E_x)^2 kT}{\bar{v}\hbar\Delta\varepsilon\alpha E_y}. \tag{19.12}$$

This estimate can be compared with expression (15.23), which gives the transition probability between terms of different symmetry and possesses the same sort of temperature dependence. To this end, we separate out the Massey parameter in (19.12), assuming $E_x \ll E_y$ and introducing the factor $Z_{10,\text{vib}} \sim (\alpha x)_{01}^2$:

$$\langle P \rangle \sim Z_{10,\text{vib}} \frac{E_y\mu}{\hbar^2\alpha^2} \cdot \frac{\hbar\alpha\bar{v}}{\Delta\varepsilon}. \tag{19.13}$$

In contrast to (15.23) the expression (19.13), besides the Massey parameter, contains another factor $Z_{10,\text{vib}}(E_y\mu/\hbar^2\alpha^2) \sim E_y/\hbar\omega$ which exceeds unity. Therefore although, under the condition $T \approx 300$ K and $\Delta\varepsilon \approx 500$ cm^{-1}, the ratio $\hbar\alpha\bar{v}/\Delta\varepsilon$ turns out to be smaller than unity, the probability $\langle P \rangle$ may be of order unity. Of course, if it should prove that $\langle P \rangle \geqslant 1$, then it is necessary to go beyond the limits of perturbation theory and calculate more exactly.

The physical difference between the electronic and electronic–vibrational nonadiabatic transitions, in the region of quasi-intersection, is that the probability of the former is proportional to the small parameter $(m_e/M)^{\frac{1}{2}}$, and the probability of the latter does not contain this parameter. Therefore the probability of vibrational or electronic–vibrational transitions induced by the translational or rotational motion of the partners will be small only in two cases; either the adiabatic terms are sufficiently separated ($\Delta U/\hbar \alpha \bar{v} \gg 1$) in all regions of configuration space accessible to the motion of the atoms (as occurs, for example, in the vibrational excitation of molecules in non-degenerate electronic states, see §9) or, for two quasi-intersecting terms, there is an additional small parameter securing a relatively small splitting between them (such as the spin–orbit coupling parameter in the vibrational excitation of molecules in degenerate electronic states, see § 20).

The probability $\langle P \rangle$ in (19.13) depends on the Massey parameter far more weakly than does the probability of the single-quantum vibrational transition induced by the collision of a molecule $X_2(^1\Sigma_g^+)$ with an atom of an inert gas. In the given case, this is ultimately due to the fact that one of the partners in the collision possesses a degenerate electronic state, and the other possesses a spectral structure which allows the intersection to occur of the electronic–vibrational terms arising on removal of the degeneracy under the effect of the intermolecular interaction. The ignoring of these features of the system under consideration prevents us from obtaining the correct order of magnitude for the cross-section and its dependence on the resonance defect. If, for example, the degeneracy of the electronic states of Hg* is neglected and the reaction (19.7) is considered within the framework of the theory of vibrational excitations set out in Chapter 2, then the transition probability proves to be sharply dependent on the temperature and the resonance defect. Going from a molecule of N_2 ($\Delta\varepsilon = 560 \text{ cm}^{-1}$) to CO ($\Delta\varepsilon = 400 \text{ cm}^{-1}$) would have to be accompanied by a ten-fold increase in the cross-section, and the absolute values of the cross-sections would have to be much smaller than gas-kinetic values [212]. If, on the other hand, the vibrational excitation probability is neglected, then the transition cross-section between the fine-structure components of mercury also proves to be extremely small. This is clear from a comparison with the analogous process of the transition between the states $^2P_{\frac{1}{2}} \to {}^2P_{\frac{3}{2}}$, in alkali metal atoms in atomic collisions (§ 19).

In the theory presented above, the interaction of the systems of terms correlating with the states $^3P_j + {}^1\Sigma_g^+$ with the ground term of AX_2 was neglected. For Hg+N_2 collisions this is probably correct. For Hg+CO collisions, it is not, since the experimental results indicate the great effectiveness of CO in inducing the transition $^3P_j \to {}^1S_0$ [290]–[292], [476]. Unfortunately, it is very difficult to evaluate the theoretical nonadiabatic transition probability for $^3P_j \to {}^1S_0$, because it is necessary to have information available about the adiabatic terms corresponding to the different configurations of the mercury atom. It is only possible to assert that, if the electronic state of the X_2 partner does not change in such a transition, then the terms correlating with the $^3P_0$ and $^1S_0$ states of the Hg atom correspond to different symmetry with respect to the plane of the three atoms and are only mixed under the effect of the rotation of this plane. On the other hand, among the three states of the term $^3P_1$ is one whose symmetry coincides with the symmetry of the ground term. Therefore the nonadiabatic transition $^3P_1 \rightarrow {}^1S_0$ will be caused, in the main, by the motion of the atoms in the plane of the system, and the cross-section for deactivation of the state $^3P_0$ of mercury in collisions with diatomic molecules must be appreciably smaller than the cross-section for deactivation of the state $^3P_1$. Experimental data confirm this deduction of the theory [186].

The processes under consideration above clearly illustrate all the complexity of the problems of calculating cross-sections of electronic–vibrational transitions even for the lowest states. Therefore, it is not likely that some form of general theory can be constructed for similar types of processes, like those existing for purely vibrational transitions (see § 9) or for purely electronic transitions (see §§ 15–18).

20. Vibrational transitions between the lowest levels of a diatomic molecule in a degenerate electronic state

The theory of the vibrational excitation of molecules, presented in § 9, is only valid for molecules with closed electronic shells since it is assumed that, at the moment of collision, there is no change in the electronic configuration of the colliding molecules. Such calculations give comparatively satisfactory results for many molecules. However, there exist a large number of cases where the experimental excitation cross-sections exceed the theoretical by several orders of magnitude. The collisions of electron-excited molecules and radicals belong, above all, to this category. It is known that such processes are not only characterized by large vibrational excitation cross-sections but lead comparatively easily to chemical reactions [31]. Attempts to explain the large cross-sections by the strong attraction of the colliding molecules, in certain cases, do not lead to success since it is necessary to assume very great depth for the potential well.

In this section, it is shown that the simultaneous change of the vibrational and electronic states of the colliding molecules with incompletely filled or excited electronic shells can lead to cross-sections considerably in excess of the adiabatic excitation cross-sections. A physical picture of a non-elastic collision is the following. Even if one of the colliding partners has an open shell then, on collision, electronic transitions are possible between the states which arise from the initial state, itself degenerate, on increasing the intermolecular distance without limit. Under the condition that the electronic transition frequency be equal to the vibrational frequency of the nuclei, a kind of resonance arises which causes comparatively large cross-sections for the excitation of vibrations [384].

If the adiabatic functions are used as the zero-order approximation, disregarding the spin–orbit coupling, then, in the semiclassical approximation, the electronic–vibrational wave function of the molecule–atom system can be written in the form

$$\Phi_{lm} = \phi_l(\rho, r, R)\chi_{lm}(r, R)\exp\left[-\frac{i}{\hbar}\int^t U_{lm}(R, \gamma)\,dt\right], \qquad (20.1)$$

where ϕ_l is the electronic wavefunction; χ_{lm} is the vibrational wavefunction of the nuclei; ρ is the set of electronic coordinates; r, R, γ the set of nuclear coordinates, referring to the vibrational and translational motion; l, m are the electronic and vibrational quantum numbers.

The perturbation leading to the nonadiabatic transitions is given by the terms not taken into account in the construction of the function in (20.1):

$$V = \hat{C} + V_{\text{s.o.}}, \qquad (20.2)$$

where $\hat{C} = -i\hbar(\partial/\partial t)$ is the operator which takes account of the nonadiabatic coupling between different electronic–vibrational states Φ_{lm} and is presumed to act on the pre-exponential in the function (20.1); $V_{\text{s.o.}}$ is the spin–orbit coupling.

Since the derivative $\partial/\partial t$ acts on the product of the functions $\phi_l\chi_{lm}$ two terms arise, $\phi_l(\partial\chi_{lm}/\partial t)$ and $(\partial\phi_l/\partial t)\chi_{lm}$. The first term leads to vibrational transitions without change of electronic state, and to the transition probability found in §9. The second term and also the spin–orbit coupling may lead to a change of electronic state, in the first approximation. However, the probability of such transitions may be appreciable only when the above-mentioned resonance occurs. Since all future calculations are conducted within the limits of perturbation theory, the total vibrational transition probability is obtained by summing the contributions from these two processes.

Weak spin–orbit coupling

For definiteness, we will assume that the molecule is in one of the states $^2\Pi_\Omega$ ($\Omega = \frac{1}{2}, \frac{3}{2}$), and that the atom possesses a closed electronic shell. Then the most general process of vibrational deactivation of a single vibrational quantum can be presented in the forms

$$XY(^2\Pi_\Omega, n = 1) + A \rightarrow XY(^2\Pi_\Omega, n = 0) + A. \qquad (20.3)$$

However, this process does not inevitably take place in one stage. For example, in the collisions of XY with A, transitions can occur between the fine-structure components of the electronic term either without a change in the vibrational quantum number n or with a simultaneous change in Ω and n (Fig. 19). In the first case, the energy of the molecule changes by the value A (where A in the constant in the spin–orbit coupling with XY) and, in the second case, by $(\hbar\omega - A)$. Since the existing theory of adiabatic vibrational relaxation, disregarding the electronic–vibrational interaction, leads to the

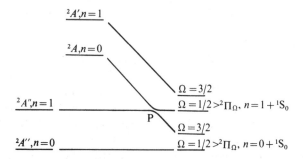

FIG. 19. Correlation diagram of the electronic–vibrational terms of the system $A(^1S_0)+XY(^2\Pi_\Omega)$.

strong dependence of the vibrational deactivation probability on the energy transfer ΔE, it might be deduced that the spin–orbit coupling facilitates deactivation at the expense of a simple reduction in the value of A. In addition, the thermal relaxation rate in the stepwise process

$$(^2\Pi_\Omega, n = 1) \xrightarrow{\kappa_1} (^2\Pi_{\Omega'}, n = 1) \xrightarrow{\kappa_2} (^2\Pi_{\Omega''}, n = 0) \qquad (20.4)$$

will be determined by the slowest stage and, for sufficiently small A when clearly $\kappa_1 \gg \kappa_2$, will not depend on the transition probability between the fine-structure components. However, for $A \ll \hbar\omega$ the effect of the spin–orbit coupling cannot be reduced to the above-mentioned energy effect; the spin–orbit coupling causes nonadiabatic transitions between the $^2\Pi$-state terms split under the effect of the interaction, thus leading to an electronic–vibrational transition.

The adiabatic electronic functions ϕ_l are classified according to their symmetry with respect to a reflection in the plane of the three atoms. In the given case, we are interested in the symmetric functions ϕ_s (representation A') and the antisymmetric functions ϕ_a (respresentation A'') arising on the removal of the degeneracy of the Π-state functions of the free XY molecule under the effect of the intermolecular interaction. For nonadiabatic transitions, those parts of the trajectory of the relative motion of XY and A are important which pass near the line of intersection of the two adiabatic surfaces of potential energy.†

$$W_s(R, \gamma) = W_a(R, \gamma) + \hbar\omega, \qquad (20.5)$$

where the $W_i(R, \gamma)$ are defined as the adiabatic terms of the states Φ_{lm}:

$$W_i(R, \gamma) = \langle \phi_i \chi_{i0} | H | \phi_i \chi_{i0} \rangle, \qquad i = \text{s, a}. \qquad (20.6)$$

The dashed curve S' (see Fig. 4) corresponds to the line of intersection of the potential energy surfaces. At every point of this curve, a system of

† Here, for definiteness, it is assumed that $W_s > W_a$, which in fact occurs in the exchange interaction of the molecules with inert gas atoms.

coordinates x', y', z' is defined whose z' axis is directed along the normal to S'. The motion of the representative point along the y' axis does not cause nonadiabatic transitions since, in this direction, the separation between terms of the zero-order approximation does not change (we neglect here the weak dependence of $V_{\text{s.o.}}$ on R and on γ). Motion along the z' axis causes transitions between the symmetric and antisymmetric terms owing to the spin–orbit coupling. Any motion in the $z'y'$ plane also causes transitions between these terms, owing to the Coriolis interaction, expressed by the operator $-i\hbar(\partial/\partial t)$ in the adopted approximation. Since ϕ_s and ϕ_a differ only in their dependence on the azimuthal angle of the unpaired electron and all corrections due to the intermolecular interaction are neglected in the operator $V_{\text{s.o.}}$, the final expression for V takes the form

$$V = \omega_n l_n + A l_n \mathbf{n} \cdot \mathbf{s}, \tag{20.7}$$

where ω_n is the component of the instantaneous angular velocity of rotation of the plane of AXY in the direction of the axis n of the XY molecule; l_n is the operator of the component of the angular momentum of the electron along this axis; $V_{\text{s.o.}}$ is written down in the same approximation in which the spin–orbit coupling operator is expressed for the free diatomic molecules.[†]

For the given trajectory $R(t)$, $\gamma(t)$, the transition probability P_{10} will depend on the parameters ω_n and $v_{z'}$, ascribed to each point of the line S' or critical surface $S^\star$, produced by the rotation of S' about $\mathbf{n}$.

The averaging over trajectories is carried out in accordance with the approximation of the transition state method, which gives for the rate constants of the process $a \to s$

$$\kappa_{10}^{\text{as}} = \frac{1}{F_{XY}F_A} \int P_{10}^{\text{as}}(\omega_n, v_{z'}) \exp(-\beta H^\star) \frac{d\Gamma^\star}{dt}, \tag{20.8}$$

where the derivative with respect to time of the element of phase space is taken along the reaction coordinate z', and the Hamiltonian (7.15) in which R and γ are connected by relation (20.5) may be used as $H^\star$.

The time dependence of the coordinate z' near the line S' in the Landau–Zener approximation is given by the formula

$$z' = v_z t, \tag{20.9}$$

where v is a constant characterizing the trajectory.

The square of the matrix element coupling the states $(a, 1)$ and $(s, 0)$ equals:

$$|V|^2 = S_{01}^2 \left(\hbar^2 \omega_n^2 + \frac{A^2}{4} \right), \quad S_{01} = \langle \chi_{s0} | \chi_{a1} \rangle, \tag{20.10}$$

[†] For free diatomic molecules, it leads to a splitting of the $^2\Pi$-term into components $\Omega = \tfrac{1}{2}$ and $\Omega = \tfrac{3}{2}$, with a separation between terms equal to $\Delta U_{\frac{1}{2},\frac{3}{2}} = A$.

where the term linear in ω_n is disregarded, since it tends to zero on averaging, in accordance with (20.8).

In first-order perturbation theory, the probability is given by the Landau–Zener formula

$$P_{10}^{as}(\omega_n, v_{z'}) = \frac{4\pi|V|^2}{\Delta F_{as}\hbar v_{z'}}, \quad \Delta F_{as} = \left|\frac{\partial}{\partial z'}(W_s - W_a)\right|. \quad (20.11)$$

If the higher terms in the expansion (20.9) had been taken into account then, instead of (20.11), an expression of the type (15.31) would have been obtained, containing Airy functions. In addition, possibly, it would have been proved necessary to take into account, in the Hamiltonian (7.15), the effect of the cross terms $P_\psi P_\gamma$ corresponding to the kinematic coupling of the plane of rotation of XY–A as a whole with the relative motion. The evaluation of (20.8) with the probability $P_{10}^{as}(\omega_n, v_{z'})$ from (20.11), is carried out comparatively simply. The factor $1/v_{z'}$, in (20.11), cancels with $v_{z'}$, appearing in $d\Gamma^*/dt$, after calculation of the current along the reaction coordinate z'. There remains the element of the critical surface dS^* from the element of volume in configuration space. Integration over the angular coordinates and momenta leads to the replacement of ω_n^2 by the mean value $\langle\omega_n^2\rangle$, whereupon the partition functions in the numerator and denominator cancel. In this way we find

$$\kappa_{10}^{as} = \int S_{01}^2(R,\gamma) \frac{2\pi(\hbar^2\langle\omega_n^2\rangle + A^2/4)}{\Delta F_{sa}(R,\gamma)\hbar} \exp[-\beta W_s(R,\gamma)]\, dS^*, \quad (20.12)$$

where R and γ correspond to points on the critical surface S^*. Since S^* represents a surface of revolution, we can write $dS^* = 2\pi R(\gamma)\sin\gamma\, d\gamma'$, at the same time substituting the equation of the line S' from the equality (20.5) into (20.12)

$$\kappa_{10}^{as} = \int S_{01}^2 \frac{2\pi(\hbar^2\langle\omega_n^2\rangle + A^2/4)}{\Delta F_{sa}(R,\gamma)\hbar} \cdot 2\pi R(\gamma)\exp[-\beta W_s(R,\gamma)]\sin\gamma\, d\gamma'. \quad (20.13)$$

To calculate $\langle\omega_n^2(\gamma)\rangle$ we will go over to the system of coordinates introduced in Fig. 3, in which the rotational Hamiltonian of the three particles takes an especially simple form:

$$H_{AXY,rot} = \left[\frac{\cos^2\psi}{2Mr_e^2} + \frac{\cos^2(\psi+\gamma)}{2\mu R^2}\right]\frac{J^2 - P_\psi^2}{\sin^2\gamma}. \quad (20.14)$$

For the instantaneous rotational velocity with regard to the axis of the XY molecule, we obtain the expression

$$\omega_n = -[\omega_y\cos(\psi+\gamma) + \omega_z\sin(\psi+\gamma)]. \quad (20.15)$$

The inertia tensor is non-diagonal in the variables under consideration, therefore the components of angular velocity ω_y and ω_z can be expressed

§ 20   TRANSLATIONAL ENERGY IN MOLECULAR COLLISIONS   171

in the following way in terms of the components of J_y of the total angular momentum **J** along the y axis:

$$\omega_y = (\mathbf{I}^{-1})_{yy} J_y,$$
$$\omega_z = (\mathbf{I}^{-1})_{zy} J_y, \qquad (20.16)$$

where $J_y^2 = J^2 - P_\psi^2$.

Calculating the tensor $(\mathbf{I}^{-1})_{ik}$ from (7.12) and substituting (20.16) into (20.15), we will represent the quantity we seek in the form

$$\langle \omega_n^2(\gamma) \rangle = \frac{\int \omega_n^2(\psi, \gamma, J, P_\psi) \exp[-\beta H_{\text{AXY,rot}}(\psi, \gamma, J, P_\psi)] \, d\psi \, dJ_z \, dJ}{\int \exp[-\beta H_{\text{AXY,rot}}(\psi, \gamma, J, P_\psi)] \, d\psi \, dJ_z \, dJ}, \quad (20.17a)$$

where the integration extends over the following regions

$$0 \leqslant \psi \leqslant 2\pi,$$
$$-J \leqslant J_z \leqslant J,$$
$$P_\psi \leqslant J < \infty. \qquad (20.17b)$$

Calculation gives

$$\langle \omega_n^2(\gamma) \rangle = \frac{kT}{\sin^2 \gamma} \left(\frac{\cos^2 \gamma}{Mr_e^2} + \frac{1}{\mu R^2} \right). \qquad (20.18)$$

Thus for the mean value of the squared matrix element of the interaction, we find

$$\langle |V|^2 \rangle = S_{01}^2(\gamma, R) \left[\hbar^2 k T \left(\frac{\cotan^2 \gamma}{Mr_e^2} + \frac{1}{\mu R^2 \sin^2 \gamma} \right) + \frac{A^2}{4} \right]. \qquad (20.19)$$

The two limiting cases, in which the second or the first terms in the round brackets can be neglected, correspond respectively to the rotation of the plane of the three atoms either about the stationary collision axis (vector **R**) or about the stationary molecular axis (vector **n**). A specific calculation by formula (20.13) requires a knowledge of the equation $R = R(\gamma)$ for the curve S'. However, several qualitative deductions can be made on the basis of the general expression (20.12). First of all, it is clear that the splitting between the terms ϕ_s and ϕ_a increases as R decreases (for given γ) and decreases as γ decreases (for R = constant). Since, for sufficiently small R, the derivative $\partial E_0/\partial R < 0$, it can be deduced that as $\gamma \to 0$, the energy W_s in (20.12) increases. For molecules between which the chemical forces of attraction are absent, it is natural to assume that W_s possesses a minimum at some value $\gamma = \gamma_0$. In this case, near $\gamma = \gamma_0$, we can put $W_s(\gamma) = W(\gamma_0) + W_{\gamma\gamma}(\gamma_0) \cdot (\gamma - \gamma_0)^2/2$ and calculate (20.13), expanding the integrand in powers of $(\gamma - \gamma_0)$. As regards the overlap integral S_{10}, it can be found by considering the difference $W_s - W_a$ as a perturbation in an expansion of the adiabatic vibrational functions

χ_{s0} and χ_{a1} in terms of the functions χ_n of the free molecule:

$$S_{01} = (x_{01})\frac{\partial}{\partial r}\frac{(W_s - W_a)}{\hbar\omega}\bigg|_{W_s = W_a + \hbar\omega}. \tag{20.20}$$

Denoting the ratio on the right-hand side by the value α'', characterizing the rate of change of the splitting of the adiabatic electronic terms with change of internuclear distance in XY, S_{01}^2 can be written in a form which formally agrees with the expression for Z_{10} (see § 9):

$$S_{10}^2 = |x_{01}|^2 (\alpha'')^2. \tag{20.21}$$

In essence, the difference between $Z_{10,\text{vib}}$ and S_{10}^2 is that $Z_{10,\text{vib}}$ is defined in relation to one electronic state and S_{10}^2 to two. Therefore, in particular, for a calculation of S_{10}, it is necessary to consider the non-linear configuration of nuclei, where the degeneracy between the states is removed; as regards $Z_{10,\text{vib}}$ an evaluation of its order of magnitude can be carried out even for the linear (one-dimensional) model of a collision.

Assuming $\gamma_0 = \pi/2$ for simplicity, we obtain the following expression for κ_{10}^{as}:

$$\kappa_{10}^{\text{as}} = \pi(R_0^\star)^2 \left(\frac{8kT}{\pi\mu}\right)^{\frac{1}{2}} S_{10}^2 \left[\frac{\hbar^2 kT}{M r_e^2 \beta W_{\gamma\gamma}} + \frac{\hbar^2 kT}{\mu(R_0^\star)^2} + \frac{A^2}{4}\right] \left(\frac{\pi}{2\beta W_{\gamma\gamma}}\right)^{\frac{1}{2}} \exp(-\beta E_0), \tag{20.22}$$

where it is assumed that $R_0^\star = R(\gamma_0 = \pi/2)$, $E_0 = E(\gamma_0 = \pi/2)$.

The transition probability can be obtained from (20.22) by the formula

$$\langle P_{10}^{\text{as}} \rangle = \frac{\kappa_{10}^{\text{as}}}{\pi R_0^2 (8kT/\pi\mu)^{\frac{1}{2}}}, \tag{20.23}$$

where R_0 is the gas-kinetic collision cross-section for XY and A; $(8kT/\pi\mu)^{\frac{1}{2}}$ is the mean velocity of the relative motion.

The total rate constant for the vibrational transition $n = 1 \to n = 0$ in process (20.4) is obtained by summation of the current of representative points across the curves S and S' (Fig. 4):

$$\kappa_{10} = \kappa_{10}(S) + \kappa_{10}(S'). \tag{20.24}$$

For a rough estimate of the contribution of the two possible mechanisms to κ_{10} at low and high temperatures, it can be assumed that E_0 is of the order $\hbar\omega$ or exceeds $\hbar\omega$ insignificantly (by a factor of two or three). Then it is clear that, at low temperatures, when $\kappa_{10}(S')$ is given by the law of Arrhenius and $\kappa_{10}(S)$ by the Landau–Teller formula, the preferred mechanism of vibration deactivation will consist of a nonadiabatic transition in the region of the quasi-intersection of terms. It is evident from this though that, in the

§ 20   TRANSLATIONAL ENERGY IN MOLECULAR COLLISIONS   173

region of validity of the Landau–Teller law, the relation

$$\frac{3}{2}\left(\frac{\theta'}{T}\right)^{\frac{1}{3}} \gg \frac{\theta}{T} \qquad (20.25)$$

must be satisfied, whence $\kappa_{10}(S) \ll \kappa_{10}(S')$.

For an explanation of the relative effectiveness of these processes at high temperature, it is necessary to estimate the pre-exponential factor in the expressions for $\kappa_{10}(S)$ and $\kappa_{10}(S')$. From (20.23) we find

$$P_{10}(S) \to Z_{10,\text{vib}} \frac{8\mu k T}{\hbar^2 \alpha^2}. \qquad (20.26)$$

In a calculation of the high temperature limit of $\kappa_{10}(S')$ it should be taken into account that the pre-exponential factor $f(\gamma)$, which formally increases as $\gamma \to 0$, is always cut off by the factor $\exp[-\beta W_s(\gamma)]$, which rapidly decreases as $\gamma \to 0$. Moreover, in the calculation of the transition probability in those configurations which correspond to very small angles γ, it is necessary to introduce a phase factor into the expression for P_{10}^{as}, due to the interference in a double transit through the critical surface (see § 17). This factor leads to a cut-off in the integrand for angles of order $\gamma \approx \lambda/R_0$, where λ is the de Broglie wavelength for the relative motion. Completely neglecting the exponential under the integral sign in (20.13), keeping only the maximum term due to the orbit–rotation interaction, and integrating between the limits $\tilde{\gamma} \leqslant \gamma \leqslant \pi - \tilde{\gamma}$, we obtain

$$P_{10}(S') \approx \frac{2\pi S_{10}^2}{\hbar^2 \omega \alpha \bar{v}} \frac{\hbar^2 k T}{M r_e^2} \ln \tilde{\gamma}. \qquad (20.27)$$

Comparing (20.26) and (20.27), it can be seen that the ratio $P_{10}(S')/P_{10}(S)$ is of order $(\Omega/\omega)\tilde{\gamma} \ln \tilde{\gamma}$, where Ω is the frequency of the first rotational transition in the free XY molecule and $\tilde{\gamma} \ll 1$. Hence it follows that, at a sufficiently high temperature, the vibrational deactivation taking place without a change in the electronic state will be more effective, i.e. the relation $P_{10}(S) \gg P_{10}(S')$ will be fulfilled. The effectiveness of the two mechanisms is comparable at some temperature T_0, whose magnitude depends on the parameters of the intermolecular interaction.

A specific example of competition between the two deactivation mechanisms is considered in a subsequent section in the example of the vibrational relaxation of nitric oxide.

As regards the applications of the theory under consideration to the collision of two molecules, it should be borne in mind that one of the assumptions of the theory was the possibility, in the adiabatic approximation, of classifying the electronic states into symmetric and antisymmetric with respect to the plane of the three atoms. For a system of four atoms (two diatomic molecules), there is, generally speaking, no plane of symmetry and therefore the intermolecular interaction must be taken into account explicitly in the contruction of the adiabatic functions.

Strong spin–orbit coupling

We will now consider qualitatively the most effective vibrational relaxation mechanism in molecules for which the fine-structure splitting is of the order of a vibrational quantum. In this case, the vibrational transitions within each electronic state (in the single-quantum approximation under consideration) and electronic transitions between the fine-structure components (resonance defect $\Delta E = A \approx \hbar\omega$) without change in the vibrational quantum number, followed by electronic–vibrational transitions (also single-quantum) with very small resonance defect $E = A - \hbar\omega$, are competing processes. Transitions of such a type between the states $^2\Pi_{\frac{1}{2}}$ and $^2\Pi_{\frac{3}{2}}$ are represented schematically by the following diagram:

$$^2\Pi_{\frac{3}{2}}, \quad n = 0 \leftrightarrow 1 \leftrightarrow 2$$
$$\updownarrow \quad \updownarrow \quad \updownarrow$$
$$^2\Pi_{\frac{1}{2}}, \quad n = 0 \leftrightarrow 1 \leftrightarrow 2. \tag{20.28}$$

It is of interest to compare the orders of magnitude of the two small transition probabilities $P_{10}(\frac{1}{2} \to \frac{1}{2})$ and $P_{00}(\frac{1}{2} \to \frac{3}{2})$. Since, in a quasi-stationary stepwise process, the rate under observation is determined by the rate of the slowest stage, transitions of type $(\frac{3}{2}, n = 0) \to (\frac{1}{2}, n = 1)$ taking place with large probability, as was shown above, are not generally considered.

To evaluate the transition probability $\langle P_{00}(\frac{1}{2} \to \frac{3}{2}) \rangle$ below, we will completely neglect the contribution of the non-adiabatic coupling of the electronic momentum with the rotation of the plane of symmetry. Furthermore, the transition probability between the fine-structure components in molecular collisions can be estimated on the basis of the formulae of §18. This evaluation is exactly analogous to the calculation carried out in connection with the computation of the transition probabilities between the fine-structure components in atomic collisions (see §19). The only difference is that, for a molecule, the coupling of the electronic momentum with the axis is large and, in the current case, the three-level system, considered in §19, reduces to a two-level system. In connection with this, we must take $\cos\theta = 0$ instead of $\cos\theta = -\frac{1}{3}$.

An estimate of the mean transition probability by the saddle-point method gives

$$\langle P_{00}(\tfrac{1}{2} \to \tfrac{3}{2}) \rangle \sim \exp\left[-3\left(\frac{\pi^2 A^2 \mu}{8\hbar^2 \alpha^2 kT}\right)^{\frac{1}{3}}\right], \tag{20.29}$$

the pre-exponential factor in the expression being of order unity. In the same notation, the probability of a single quantum transition within each electronic state equals

$$\langle P_{10}(\tfrac{1}{2} \to \tfrac{1}{2}) \rangle \sim \exp\left[-3\left(\frac{\pi^2 \omega^2 \mu}{2\alpha^2 kT}\right)^{\frac{1}{3}}\right]. \qquad (20.30)$$

The essential difference between (20.29) and (20.30) is that, for $A \approx \hbar\omega$, the exponent of the exponential in the latter is larger than in the former by a factor of $4^{\frac{1}{3}}$. Since the exponent of the exponential is large, it is clear that the vibrational excitation rate, which includes the simultaneous electronic–vibrational transitions, will be larger than the vibrational excitation rate without change in the electronic state. The condition for the approximate equality of the exponents of the exponentials in (20.29) and (20.30) determines that energy of the fine-structure splitting of the terms for which the components $\Omega = \tfrac{1}{2}$ and $\Omega = \tfrac{3}{2}$ may be considered as independent adiabatic electronic states in the calculation of the vibrational excitation probability.

21. Vibrational relaxation of nitric oxide

The vibrational relaxation of a nitrogen mixture displays sharp anomalies, which are particularly clearly revealed by a comparison of the values of $\langle P_{10} \rangle$ for NO and for N_2 and O_2. The difference in value between $\langle P_{10}(NO) \rangle$ and $\langle P_{10}(N_2) \rangle$ at low temperatures ($T \approx 500$ K) amounts to two to three orders of magnitude and cannot be explained from the point of view of the theory given in §9, which does not take nonadiabatic effects into account.

Before going on to the evaluation of P_{10} in accordance with the theory given in the previous section, it is necessary to examine the qualitative behaviour of the electronic terms of a system of two molecules of NO. To construct the correlation diagram, we will first consider a symmetric collinear arrangement of the two molecules of NO. Since the unpaired electron occupies a π-orbital, as the molecules converge, the following states will arise: $^3\Sigma_g^-$, $^1\Delta_g$, $^1\Sigma_g^+$, $^3\Delta_u$, $^3\Sigma_u^+$, $^1\Sigma_u^-$. The order of their spacing can be established in the following way. All the even states must lie below the odd states because the corresponding wavefunction corresponds to the formation of a chemical bond (of course, the actual formation of a bond between the two radicals still depends on the interaction of the shell of electrons). The order of the spacing of the states with like parity is determined by Hund's rule.

For an arbitrary relative orientation of the molecules, the orbitally degenerate states are split. Among all the terms arising, a pair $^1\Sigma_g^+$ and $^3\Sigma_g^-$ can be found for which the interaction between terms does not disappear, even in the most unfavourable collinear configuration. In fact since, in the operator $\hat{C}$ for the excitation of vibrations, only the differentiation of the electronic function with respect to r is important (where r is the internuclear distance), the selection rule for the matrix elements of the nonadiabatic coupling of the electrons and nuclei runs: $\Delta\Lambda = 0$, $g \leftrightarrow g$ and $u \leftrightarrow u$, $\Delta S = 0$. Then, as a result of the perturbation (20.2), only terms Σ^+ and Σ^- of different multiplicity and the same parity can interact, and it turns out that, with regard for the selection rules indicated above, only the spin-orbit coupling operator in fact appears as the perturbation. Thus the rate constant for a vibrational transition can be calculated approximately on the basis of (20.13), disregarding the contribution of the interaction of the electronic momentum with the rotation and neglecting the dependence on γ of the integrand [62], [382]. In addition, an additional factor g must be introduced into the expression for κ_{10}, taking account of the average over the initial electronic states and the sum over the final states. For the collinear configuration, $g = \frac{3}{16}$. For other configurations, g may exceed this value, since transitions between other pairs of terms prove to be allowed. However, in this case it is necessary to take into account the angular dependence of the integrand in (20.13), which appreciably reduces the contribution of such nonadiabatic transitions. From (20.13) we find

$$\kappa_{10} = 4\pi R_0^2 \frac{\pi S_{10}^2 A^2}{2\hbar^2 \omega \alpha} \exp(-\beta E_0), \tag{21.1}$$

$$\langle P_{10} \rangle = \frac{\kappa_{10}}{\pi R_0^2 \bar{v}}.$$

The chief difficulty in the evaluation of $\langle P_{10} \rangle$ is the calculation of E_0, since it is impossible to calculate the terms of such a complex system as 2NO to the desired accuracy. However, a very rough value for E_0 can be estimated in the following way. The splitting ΔW between the $^1\Sigma_g^+$ and $^3\Sigma_g^-$ terms of the 2NO system is determined just by the exchange interaction of the two π-electrons. This interaction decays exponentially for weak overlap of the electron clouds so that, approximately, we can write

$$\Delta W(R) = \Delta E^* \exp[-\alpha(R - r_e)], \tag{21.2}$$

where r_e is the equilibrium internuclear distance in the NO molecule; ΔE^*, the quantity characterizing the splitting for strong overlap of the wavefunctions, must be of the order 1–2 eV (the corresponding splitting in the O_2 molecule, which also contains two π-electrons in the outer orbital, equals 1·6 eV).

On the other hand, the intermolecular interaction energy can be approximated by a potential of the same form,

$$W(R) = C \exp(-\alpha R), \tag{21.3}$$

where the constant C can be determined from the gas-kinetic radius R_0 by the condition $W(R_0) \sim kT_0$, where T_0 is the temperature at which the value

R_0 is determined. Combining (21.2) and (21.3) we find

$$E_a \sim W(R_0) \sim \frac{\hbar\omega k T_0}{\Delta E^*} \exp(\alpha R_0 - \alpha r_e). \tag{21.4}$$

Assuming here that $\alpha = 2 \times 10^8$ cm^{-1}, $\omega = 2000$ cm^{-1}, $T = 300$ K, $\Delta E^* = 1-2$ eV, $r_e = 1 \cdot 1$ Å and $R_0 = 3$ Å (the mean Van der Waals radius for all six electronic states is equal to 3·6 Å and the equilibrium distance in dimers of $(NO)_2$, in the liquid phase, equals 2·4 Å), we find $E_a/k \approx 1000-2000$ K. In accordance with (21.1) such a value of the energy leads to a weak positive temperature-dependence of the deactivation probability for $T \approx 1000-2000$ K. This conclusion is roughly in line with the experimental results in Robben's paper [444], in which the deactivation probability was found to increase by just a factor of 4 for an increase in temperature from 500 K to 1000 K. This fact, the weak temperature-dependence of $\langle P_{10} \rangle$, already indicates the inapplicability to the case under consideration of the Landau–Teller formula, in accordance with which the experimental results in paper [444] were analysed. We note that a temperature dependence of the Landau–Teller type, valid for the vibrational deactivation of molecules in non-degenerate states with a correct choice of the radius of action, must lead to a hundred-fold change in the cross-section over the same range of temperature.

To evaluate $\langle P_{10} \rangle$ numerically, we will assume that $S_{10}^2 \approx 10^{-2}$, $A = 100-200$ cm^{-1} (the spin–orbit splitting in an NO molecule amounts to 120 cm^{-1}, and in the O_2^+ ion is 200 cm^{-1}), $\omega = 2000$ cm^{-1}, $M \sim \mu/2$, $T = 500$ K. For the collinear configuration the value of g equals $\frac{3}{16}$, and for other configurations may exceed this value. Substituting these values into (21.1) we obtain $\langle P_{10} \rangle \approx 10^{-4}$. The experimental value of the quantity $\langle P_{10} \rangle$ at the same temperature [444] is 7×10^{-4}. A calculation of the cross-section, carried out under the assumption of adiabaticity, would lead to the value 10^{-8}.

To determine the temperature dependence of the deactivation probability $\langle P_{10} \rangle$ over a wide range of temperatures, the contributions of two possible processes must be taken into account: the electronic–vibrational transitions of the type just considered and the vibrational transitions without change in the electronic state of the colliding partners. Such a comparison was made by Wray [566], investigating the vibrational relaxation of NO in shock waves over a wide temperature range. In Fig. 20, taken from his work [566], the contributions of the two possible mechanisms to the vibrational relaxation of NO are shown by dashed curves and the continuous and dotted curves show various experimental results: curve 1 is a calculation in accordance with the theory of Schwartz, Slawsky, and Herzfeld [468] (see §9); curve 2 is a calculation according to a formula obtained by Nikitin [62] (see §20); curve 3 is the theoretical total contribution of processes with and

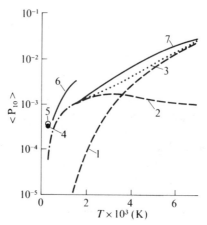

Fig. 20. Vibrational relaxation of nitric oxide: comparison of theory with experiment (Wray [566]).

without change in the electronic state of the partners. The experimental values of P_{10} are obtained by ultrasonic dispersion methods (point 4 is from Bauer and co-workers [143]), by kinetic spectroscopy (point 5 is from Basco and co-workers [138]), and by the investigation of relaxation in shock waves (curve 6 is from Robben [444], curve 7 from Wray [566]).

Fig. 20 illustrates the accuracy of the experimental determination of $\langle P_{10} \rangle$ by different methods. From the data cited, it is clear that combined electronic–vibrational transitions make the principal contributions to relaxation at $T < 3500$ K.

To conclude this section we will point out one difference between the stepwise vibrational relaxation of molecules in a degenerate electronic state and the relaxation of molecules in a closed electronic state. For the former, the expression for the transition probability does not contain a small factor of the Landau–Teller exponential type. This implies that anharmonicity affects the transition probability much more weakly than in the case of molecules in a non-degenerate state, and the linear dependence of the transition probability $P_{n,n+1}$ on n must be satisfied over an appreciably wider range of quantum numbers n.

Thus, the kinetics of vibrational relaxation of molecules of the NO type must conform to the laws governing the relaxation of oscillators with anharmonicity, considered in the paper [148]. In a study of the relaxation of molecules in a non-degenerate electronic state, the appreciable contribution of anharmonicity must certainly be taken into account, which leads to the appearance of a factor γ^n. Finally, the large cross-sections of the vibrational transitions in molecules in a degenerate electronic state signify that competition between vibrational–translational and quasi-resonant vibrational energy exchange becomes likely.

4

UNIMOLECULAR REACTIONS

22. The thermal decomposition and isomerization of molecules as unimolecular reactions

THE term unimolecular reactions is usually understood to refer to chemical changes connected with the structural change of only one molecule or ion. The decomposition and isomerization of polyatomic molecules are typical examples of such reactions. In accordance with this definition, the decomposition of diatomic molecules may also be considered as a unimolecular reaction, although it is often referred to as a bimolecular reaction. This is connected, on the one hand, with the fact that the activation process as a result of binary collisions is of special importance in decomposition reactions of diatomic molecules in comparison with the decomposition of polyatomic molecules. The theory of the dissociation of diatomic molecules is therefore considered separately from the general theory of unimolecular reactions (Chapter 7). On the other hand, decomposition reactions of diatomic molecules are also attributed to bimolecular processes, because, kinetically, they are second-order reactions. However, since, under certain conditions, reactions of polyatomic molecules too can take place in accordance with a second-order law, such a formal separation of the decomposition reactions of diatomic molecules is not justified. Thus unimolecular reactions (in accordance with the above definition) can be characterized as kinetically first or second order. Furthermore, for molecules containing more than two atoms, there exists a transitional pressure range in which it is not generally possible to speak of a definite order of reaction.

Unimolecular processes only proceed at an appreciable rate when the reacting molecule possesses an internal energy greater than a certain threshold value E_0 corresponding to the energy of activation. We will call such molecules *active*. Active molecules arise in the process of chemical reaction as a result of inelastic molecular collisions or other chemical reactions, or on electron bombardment, etc. It is necessary to bear in mind, however, that the concept of a sharp energy threshold was formed before tunnel-effect reactions were discovered. For these reactions, it is not possible to distinguish a sharp boundary dividing molecules able to participate in a unimolecular transformation, with some probability, from non-active molecules. In many cases, however, the corrections due to tunnel transmissions are not large, so that it is expedient to retain the definition given above. Cases where the tunnel transmission is significant will be considered separately.

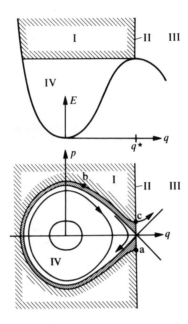

Fig. 21. Phase plane of a system with one degree of freedom.

We will work with the idea of the *activated* molecule, or transition complex, introduced in Chapter 1, the criterion for the reaction to take place being formulated in the form of a requirement that the representative point cross the critical surface S^* in the configuration space of the system under consideration.

The concept of active and activated molecules is illustrated in Fig. 21. Here the phase plane of a system with one degree of freedom is shown in terms of the variables p, q and E, q. The representative points filling the potential well IV correspond to the stable molecule. Region I corresponds to active molecules, line II to the critical surface, and region III to the reaction products. The closed curves in region IV represent the motion of isolated stable molecules, and the trajectories abc represent the motion of active molecules.

Nowadays, the theory of unimolecular reactions has, to some extent, the nature of a model theory. It is limited by the fact that neither the multidimensional surfaces of potential energy nor the exchange of energy between the vibrationally excited polyatomic molecules is sufficiently well known.

If it is a question of the excitation of the lowest vibrational levels, existing experimental data and theory indicate the comparatively small probability of conversion of translational energy into vibrational. With increase in the vibrational quantum number, however, the probability of such a conversion increases and involves an increase in the amplitude of the vibrations and a reduction in the frequency. These effects appear in all molecules, including diatomic molecules, for which there exists a more-or-less well-developed theory of the vibrational–translational energy exchange for the lowest vibrational states (see Chapter 2). In polyatomic molecules, moreover, the anharmonicity of the vibrations plays an essential role, leading to an interaction of the normal vibrations. If, now, it is taken into consideration that the probability of a vibrational–translational transformation of energy depends essentially on the magnitude of the energy being transferred, and for a complex molecule this, in turn, is determined basically by the intensity of the interaction of the normal vibrations, then the important role of the anharmonicity in the intermolecular energy exchange process becomes evident. It is precisely this fact which causes great difficulty in attempts to calculate theoretically the mean energy transfer in collisions of vibrationally excited polyatomic molecules.

The most direct experimental method of investigating the energy transfer in collisions of polyatomic molecules consists of the measurement of the quantum yield of fluorescence and the rates of competing processes on chemical activation as a function of pressure [420]. The values for ΔE^* found in this way, characterizing the mean energy lost by the excited molecule in a single collision, vary over a wide range from tenths to tens of kcal mol^{-1}, and depend essentially on the initial vibrational energy [307], [420]. Since, at present, it is not possible to make a convincing estimate of the cross-sections of similar processes, in regard to the activation mechanism in unimolecular reactions, we have to make various assumptions, whose validity may be checked only by comparison of theory with experiment. Such a comparison, however, includes many additional factors, making the possibility of a direct interpretation of the experiments in terms of the parameters of the theory difficult.

In this connection, it is expedient to consider different models of activation without giving preference to any one of them. In this chapter, in particular, the mechanism of strongly-activating collisions is studied. The mechanism of stepwise excitation is discussed in connection with the diffusion theory of reactions. As regards the intermolecular redistribution of energy, certain general questions are considered in this chapter, but the statistical description of decomposition and isomerization is set forth in connection with the general statistical theory of reactions in the following chapter.

23. Dependence of the rate constant on the pressure. The mechanism of strongly-activating collisions

The kinetic equation

The basic problem in the theory of unimolecular reactions is the determination of the non-equilibrium distribution function of the reacting molecules.

First, we will assume that classical mechanics is correct for the description of the processes of intramolecular transformation of energy, decomposition (and for simplicity in future, decomposition is understood to mean any type of unimolecular reaction), and activation. In this case, the state of the molecule is characterized by a set of momenta p_i and coordinates q_i, and the distribution function of the molecule AB depends on these and also on the time t. The change in the distribution function $f(p_i, q_i, t)$ in time at every point of phase space is determined by the three processes referred to above:

$$\frac{\partial f}{\partial t} = \left(\frac{\partial f}{\partial t}\right)_{\text{int}} + \left(\frac{\partial f}{\partial t}\right)_{\text{diss}} + \left(\frac{\partial f}{\partial t}\right)_{\text{rel}}, \qquad (23.1)$$

where the subscripts int, rel, diss refer respectively to the processes of intramolecular and intermolecular energy exchange (intermolecular relaxation) and dissociation.

The first term on the right-hand side of (23.1) is completely determined by the Hamiltonian $H(p_i, q_i)$ of the molecule under consideration, since $(\partial f/\partial t)_{\text{int}}$ satisfies Liouville's theorem:

$$\left(\frac{\partial f}{\partial t}\right)_{\text{int}} = \sum_i \left(\frac{\partial f}{\partial q_i}\frac{\partial H}{\partial p_i} - \frac{\partial f}{\partial p_i}\frac{\partial H}{\partial q_i}\right). \qquad (23.2)$$

The second term, the rate of decomposition, differs from zero only when, as a result of the collision of AB + C, the representative point of the molecule AB is carried over into the part of phase space corresponding to the decay products A + B. For polyatomic molecules this process can be neglected, since the representative points are mainly sufficiently far away from the critical surfaces. This results from the fact that the mean lifetime of the active molecule, which characterizes the rate of progress of the representative point towards the critical surface, turns out to be appreciably larger than the collision time τ_0 (see below). Therefore the term $(\partial f/\partial t)_{\text{diss}}$ in (23.1) can be omitted, taking account of the dissociation as a boundary condition imposed on the distribution function on the critical surface S^*. On the critical surface, i.e. for $q_r = q_r^*$, the distribution function must tend to zero for that direction of the current which corresponds to the passage of the representative point from the region of A + B into the region of AB. This boundary condition, which supplements eqn (23.1), must be taken into account, of course, in solving (23.1).

The total rate of decomposition is determined by the total current through the critical surface

$$K_{\text{diss}} = \int f^* \dot{q}_r \frac{dp_i}{2\pi\hbar} \prod_{i \neq r} \frac{dp_i\, dq_i}{2\pi\hbar}, \qquad (23.3)$$

where f^* denotes the distribution function of the activated molecule, which is obtained on substituting the critical value of the reaction coordinate q_r^* into the distribution function of the active molecule $f(p_i, q_i)$.

If the phase coordinates p_i and q_i on the left-hand side of (23.1) refer to the active molecule, then the first and second terms on the right also refer to the active molecule AB*. However, the third term, due to the intermolecular interaction on collision, describes transitions between states of the stable and unstable molecules and thus connects the distribution functions $f(p_i, q_i)$, generally speaking, at all points of the phase space of the molecule AB. This prevents a closed equation being obtained for $f(p_i^*, q_i^*)$.

Henceforth we will assume that collisions between the molecules AB can be neglected and that the gas is characterized by the equilibrium Maxwell distribution function for the velocities of the molecules AB and C and by a Boltzmann distribution function for the internal degrees of freedom of C.

In these conditions, the activation (and deactivation) of the AB molecules is completely characterized by the elementary transition rates $Z_0 P(p_i, q_i \to p_i', q_i')$ induced by the collisions of AB and C. Here, as before, the rate of an elementary transition is written down in the form of a product of the gas-kinetic number of collisions Z_0 of the molecules AB and C and the transition probability $P(p_i, q_i \to p_i', q_i')$. Such a representation is very provisional since Z_0 is not exactly determined. Nevertheless, since the probability is normalized with respect to a single collision, this assumption does not affect the final result.

The change in the distribution function as a result of collision can now be represented in the form

$$\left[\frac{\partial f}{\partial t}(p_i, q_i)\right]_{\text{rel}} = Z_0 \int [P(p_i', q_i' \to p_i, q_i) f(p_i', q_i') - P(p_i, q_i \to p_i', q_i') f(p_i, q_i)] \, d\Gamma, \tag{23.4}$$

where Z_0 is the number of collisions of the molecule AB with the C molecules and is directly proportional to the partial pressure of the gas C.

Here the integral is taken over the whole phase volume of AB. The first term under the integral sign describes the transitions, under the effect of collision, from all states (p_i', q_i') into the state (p_i, q_i); the second term describes the transitions from the state (p_i, q_i) into all other states (p_i', q_i'). If the aggregate (p_i, q_i) corresponds to an active molecule, then the first term describes activation and the second describes deactivation.†

The function $P(p_i, q_i \to p_i', q_i')$ possesses two properties which result from its physical meaning. First, the total transition probability from any state (p_i, q_i) does not depend on (p_i, q_i), i.e.

$$\int P(p_i, q_i \to p_i', q_i') \, d\Gamma' = 1, \tag{23.5}$$

† This is not completely exact because, as a result of collision, the active molecule does not inevitably go over into a stable state. If, however, it is taken into account that, in thermal conditions, transitions 'downwards' (i.e. with a reduction of energy) are preferred to transitions 'upwards', such a classification of terms in (23.4) may be retained.

where the integration over $d\Gamma'$ extends over the phase volume of the stable and active molecule. Secondly, the transition rates in the direct and reverse directions $(p_i, q_i \to p'_i, q'_i)$ and $(p'_i, q'_i \to p_i, q_i)$ must be connected in such a way that, in the state of equilibrium, the elementary currents are exactly compensated (the principle of detailed balance):

$$\frac{P(p_i, q_i \to p'_i, q'_i)}{P(p'_i, q'_i \to p_i, q_i)} = \frac{\exp[-\beta H(p'_i, q'_i)]}{\exp[-\beta H(p_i, q_i)]}. \tag{23.6}$$

If it proves possible to describe the processes of activation and deactivation in terms of just the total energy E of the molecules, then the non-equilibrium and equilibrium distribution functions are conveniently presented in a form in which the density of the energy levels $\rho(E)$ of the reacting molecule enters explicitly

$$f(E)\,dE = \frac{1}{F} x(E)\rho(E)\,dE,$$

$$f_0(E)\,dE = \frac{1}{F} \exp(-\beta E)\rho(E)\,dE,$$

$$\rho(E) = \int \delta[E - H(p_i, q_i)]\,d\Gamma,$$

$$F = \int \exp(-\beta E)\rho(E)\,dE. \tag{23.7}$$

In place of the dimensionless transition probability $P(p_i, q_i \to p'_i, q'_i)$, the transition probability density $P(E \to E')$ from the initial level E to the group of levels in the interval dE' close to the final level E' should be introduced:

$$P(E \to E') = \int \frac{P(p_i, q_i \to p'_i, q'_i)\delta[E - H(p_i, q_i)]\delta[E' - H(p'_i, q'_i)]\,d\Gamma\,d\Gamma'}{\rho(E)}. \tag{23.8}$$

Taking it into account that, for the direct $E \to E'$ and reverse $E' \to E$ processes, the intervals dE' and dE are equal, we find, from (23.7),

$$\frac{P(E \to E')}{P(E' \to E)} = \frac{\exp(-\beta E')\rho(E')}{\exp(-\beta E)\rho(E)}. \tag{23.9}$$

In this notation, relation (23.4) is rewritten in the form

$$\left(\frac{\partial f(E)}{\partial t}\right)_{\text{rel}} = Z_0 \int [P(E' \to E)f(E') - P(E - E')f(E)]\,dE'. \tag{23.10}$$

As is evident from (23.1)–(23.4), the equation determining the non-equilibrium distribution function is linear, a fact which substantially simplifies its study.

The independence of the activation rate constant $Z_0 P$ from the required distribution function, is ultimately connected with the fact that we neglect collisions in which two reacting molecules AB take part. If the collisions of the ABs amongst themselves are taken into account, which, of course, must be done in calculating the reaction rates in the pure gas, then the form of the activation term (23.4) is changed. However, cases are possible where the distribution function of AB over the lowest energy levels is close to the equilibrium distribution, and the activation of the molecules is basically due to the collisions of strongly excited molecules of AB with unexcited molecules. Then the latter, to a certain approximation, can be considered as molecules of the equilibrium heat reservoir, so that $(\partial f/\partial t)_{\text{rel}}$ can again be represented in the form (23.4). The only formal difference between these two processes, the decomposition of AB in an atmosphere of C and the decomposition of AB in the pure gas, comes down to the fact that, in the first case, Z_0 is proportional to the concentration of C and, in the second, to the concentration of AB. In each specific case, however, it is necessary to examine the possibility of ascribing the unexcited AB molecules to the heat reservoir.

The general dependence of the rate constant on pressure

To further elucidate the assumptions which are made in the theory of unimolecular reactions, we will mention here the derivation of an expression for the rate constant κ, within the limits of one of the simplest activation mechanisms.

In accordance with the decomposition mechanism indicated above, we will assume that the fraction of the molecules AB decaying from the given state (p_i, q_i) of the active molecule AB* is proportional to the activation rate into this state multiplied by the probability that the decomposition of the molecule takes place during a time $\tau(i)$, in which the molecule does not undergo a deactivating collision. In the derivation of an expression for κ, we will follow the approach proposed by Slater [491], [492] and developed in Thiele's work [518].

The basic assumption in the derivation of κ is that every collision leads to the complete deactivation of an active molecule, independent of its state (p_i, q_i) before collision. In this connection, the effective number of deactivating collisions is introduced, defined by the integral

$$Z^* = \int_{\Gamma - \Gamma^*} Z_0 P(p', q' \leftarrow p^*, q^*) \, d\Gamma', \qquad (23.11)$$

where the integration extends over the phase volume of the stable molecule. (This volume is denoted symbolically by $\Gamma - \Gamma^*$). If the integration is (23.11) were extended over the whole phase volume Γ, then Z^* would agree with Z_0, and consequently would not depend on p^* and q^*. Since the integration in

(23.11) extends over only part of Γ, the requirement that Z^* be independent of (p^*, q^*) imposes a definite restriction on the mechanism of intermolecular energy transfer. The assumption of the existence of such a mechanism is known in the literature as the *strong collision assumption*.[†] In order that condition (23.11) be compatible with the requirement that the rate constant tend to zero in the state of equilibrium, the activation rate (the first term in (23.4)) must be expressed by the following relationship

$$\int Z_0 P(p_i^*, q_i^* \leftarrow p_i', q_i') f(p_i', q_i') \, d\Gamma' = Z^* f_0(p_i^*, q_i^*)$$

$$= \frac{1}{F} Z^* \exp[-\beta H(p_i^*, q_i^*)]. \tag{23.12}$$

The probability that in time τ the molecule does not undergo a single deactivating collision equals

$$P(\tau) = \exp(-Z^*\tau), \tag{23.13}$$

so that for the rate constant of the reaction we can write:

$$\kappa = \frac{1}{F} Z^* \int_\Gamma \exp[-Z\tau(i)] \exp[-\beta H(p_i, q_i)] \, d\Gamma_i, \tag{23.14}$$

where $d\Gamma$ is the element of volume in phase space p_i, q_i, and F is the molecular partition function.

We will now consider a certain trajectory leading to a reaction on the constant-energy surfaces of the active molecule $H(p_i, q_i) = E$. Since, in the interval between collisions, the molecule is a closed mechanical system, motion along this trajectory in the reverse direction (i.e. on changing the sign of the time) also leads to the intersection of the critical surface separating the region of the reaction products from the region of the reactants. For deactivation, only that part of the trajectory exists which corresponds to the unreacted molecule; the integral in (23.14) is taken over precisely these states.

The element of hypersurface $d\Gamma$ can be represented in the form $dt \, dS_\perp$, where dt is an elementary interval of time between two points moving along the chosen trajectory and $dS_\perp$ is an element of the hypersurface with one less dimension than $d\Gamma$, normal to the trajectory. Since the interval dt remains constant along the trajectory and the element of hypervolume $d\Gamma$ is preserved, by Liouville's theorem, the magnitude of $dS_\perp$ also remains unchanged.

[†] Buff and Wilson [176] investigated the case when Z^* depends on p_i^* and q_i^*, but retained the basic assumption of the theory about the complete deactivation of the active molecule in each collision. Taking account of such a dependence somewhat complicates the formulae of this section. We will also note Wilson's work [558], in which the mechanism of strong collisions was introduced into the analysis of the exchange energy in photochemical reactions.

Therefore $d\Gamma = dE \, dt \, dS_\perp$ in (23.14) can be replaced by $dE \, dt \, dS_\perp^*$, where $dS_\perp^*$ is an element of surface at the point of intersection of the critical surface by the trajectory. If τ^* denotes the total time during which the representative point moves along the given trajectory in that region of phase space corresponding to the unreacted molecule, then clearly we may assume $\tau(i) = \tau^* - t(i)$. In this relation, $\tau(i)$ depends only on the parameters of the trajectory at the point of its intersection with the critical surface. Integration over t is easily carried out, and we arrive at the following expression:

$$\kappa = \frac{1}{F} \int_{E_0}^{\infty} \langle 1 - \exp(-Z^*\tau^*) \rangle \exp(-\beta E) \, dE, \qquad (23.15)$$

where $\langle \ldots \rangle$ denotes averaging over $S_\perp^*$:

$$\langle \ldots \rangle = \frac{1}{S_\perp^*} \int (\ldots) \, dS_\perp^*. \qquad (23.16)$$

We note that by $dS_\perp^*$ may be understood the element of the critical surface projected onto a plane perpendicular to the trajectory.

From (23.15) and (23.16) we obtain expressions for the rate constant for high and low pressures:

$$\kappa_\infty = \int_{E_0}^{\infty} \frac{S_\perp^*}{F} \exp(-\beta E) \, dE, \qquad (23.17)$$

$$\kappa_0 = Z^* \int_{E_0}^{\infty} \langle \tau^* \rangle \frac{S_\perp^*}{F} \exp(-\beta E) \, dE. \qquad (23.18)$$

The expressions (23.17) and (23.18) are valid on condition that either the mean decomposition rate is small in comparison with the deactivation rate $(Z^* \langle \tau^* \rangle \gg 1)$, or that the mean lifetime is small in comparison with the time between collisions $(Z^* \langle \tau^* \rangle \ll 1)$. From these relations, it is evident that the reaction rate, in the limiting cases of high and low pressures, is determined by very general properties of the molecule—the total magnitude of the 'projected' critical surface and the mean lifetime of the active molecule. For the calculation in the intermediate case, it is necessary to carry out the averaging of the function $(1 - e^{-Z^*\tau^*})$, which, in principle, can be accomplished if the trajectory of the motion is known. Such an averaging can be represented in the form of an averaging over the lifetime τ^*, with the distribution function $h(\tau^*, E)$. As the simplest assumption, it may be assumed that the lifetimes are randomly distributed and that the distribution function h is characterized by some mean value $1/\kappa(E)$:

$$h(\tau^*, E) = \kappa(E) \exp[-\tau^* \kappa(E)]. \qquad (23.19)$$

Substituting (23.19) into (23.15), we obtain the following expression for the rate constant of a unimolecular reaction:

$$\kappa = \frac{1}{F} \int_{E_0}^{\infty} \frac{Z^* S_1^*}{\kappa(E) + Z^*} \exp(-\beta E) \, dE. \qquad (23.20)$$

The derivation of this formula implicitly contains the assumption about the quasi-stationary nature of the decay of AB. In fact, in calculating the activation rate (formula (23.12)) the concentration of AB is regarded as independent of time and is normalized to unity. Therefore (23.20) can be obtained from the solution in the approximation of a quasi-stationary system of kinetic equations, describing the vibrational relaxation within the limits of the mechanism of strong collisions and decomposition under the assumption of a random distribution for the lifetimes. If, in accordance with (23.7), the non-equilibrium distribution function of the molecule AB is denoted by $f(E)$ then, for the three terms corresponding to the three terms in the relation (23.1), we obtain

$$\left[\frac{df(E)}{dt}\right]_{int} = 0, \quad \left[\frac{df(E)}{dt}\right]_{diss} = -\kappa(E) f(E),$$

$$\left[\frac{df(E)}{dt}\right]_{rel} = -Z^*[f(E) - f_0]. \qquad (23.21)$$

Hence we find the kinetic equation

$$\frac{df}{dt} = -\kappa(E) f - Z^*(f - f_0). \qquad (23.22)$$

By comparison with (23.1) this equation is greatly simplified; first, here there is only one quantity characterizing the state of the molecule AB, namely the internal energy E; second, in (23.22) states with different E are not connected and, consequently, may be attributed to the active molecules. This simplification is achieved by introducing the assumption of the random distribution of the lifetime of AB* with a (as yet unknown) rate constant of spontaneous decomposition $\kappa(E)$ and the assumption of the mechanism of strongly-deactivating collisions.

Quantum theory

To extend the above theory to a quantum system, we should first notice that the state of a molecule cannot be described by giving the coordinates q_i and the conjugate momenta p_i simultaneously. The uncertainty relation

$$\Delta p_i \Delta q_i \geqslant \hbar, \qquad (23.23)$$

due ultimately to the fact that, in quantum mechanics, operators p_i and q_i are ascribed to the dynamical variables $\hat{p}_i$ and $\hat{q}_i$, prevents this. For a definite choice of representation (i.e. for a choice of the complete system of basis functions ψ_α) $\hat{p}_i$ and $\hat{q}_i$ are represented by matrices with matrix elements

$$(p_i)_{\alpha\beta} = \langle \psi_\alpha^* | \hat{p}_i | \psi_\beta \rangle, \quad (q_i)_{\alpha\beta} = \langle \psi_\alpha^* | \hat{q}_i | \psi_\beta \rangle. \tag{23.24}$$

The matrix element of any operator $\hat{L} = \hat{L}(\hat{p}, \hat{q})$ is defined analogously.

To calculate the mean value of L the statistical matrix $W_{\alpha\beta}$ is introduced, replacing the classical distribution function [38]. The basic formula for the averaging has the following form

$$\bar{L} = \sum_{\alpha\beta} W_{\alpha\beta} L_{\alpha\beta}, \tag{23.25}$$

where $W_{\alpha\beta}$ satisfies the normalization condition

$$\sum_\alpha W_{\alpha\alpha} = 1. \tag{23.26}$$

For a rarefied gas the equilibrium statistical matrix of the molecule AB has the form

$$W^0_{\alpha\beta} = \frac{[\exp(-\beta\hat{H})]_{\alpha\beta}}{\sum_\gamma [\exp(-\beta\hat{H})]_{\gamma\gamma}}, \tag{23.27}$$

where H is the Hamiltonian operator of the molecule AB.

From relation (23.27), it is evident that there exists a preferred representation, in which $W^0_{\alpha\beta}$ takes a simple form. In fact, if the eigenfunctions of the Hamiltonian $\hat{H}$ are chosen as the ψ_α, then $W_{\alpha\beta}$ becomes a diagonal matrix of the form

$$W^0_{\alpha\alpha} = \frac{\exp(-\beta E_\alpha)}{\sum_\gamma \exp(-\beta E_\gamma)}, \tag{23.28}$$

where E_α is the energy of the stationary state of the molecule AB:

$$\hat{H}\psi_\alpha = E_\alpha \psi_\alpha. \tag{23.29}$$

If the spectrum of the energy eigenvalues is sufficiently dense, then the energy E can be introduced as a variable in relation (23.28), considering W as a continuous function of E. Summing (23.28) over a small interval dE, we obtain

$$W^0(E)\,dE = \sum_{E \leq E_\alpha \leq E + dE} W^0_{\alpha\alpha} = \frac{\exp(-\beta E)\rho(E)\,dE}{\sum_\gamma \exp(-\beta E_\gamma)}, \tag{23.30}$$

where the energy-level density $\rho(E)$ has been introduced. A sufficiently large energy spectrum density implies that many levels fall in the interval of energy $\delta E \ll 1/\beta$. Such a situation is usually realized in vibrationally-excited polyatomic molecules whose level density grows rapidly with an increase in E.

At the same time, for the lower states, making the main contribution to the sum over γ in (23.30), the intervals allowed ΔE_γ may be comparable with $1/\beta$, therefore the introduction of the level density $\rho(E)$ and the subsequent integration over $\mathrm{d}E$ are, generally speaking, inadmissable. It is evident that (23.30) agrees with (23.7) if, in (23.7), by F is understood the quantum partition function.

If the operator $\bar{L}$ is diagonal in the representation ψ_α, then, on the basis of (23.25), we have

$$\bar{L} = \sum_\alpha W_{\alpha\alpha} L_{\alpha\alpha}. \tag{23.31}$$

The tending to zero of the off-diagonal matrix elements $L_{\alpha\beta}$ implies that, in each stationary state, the quantity L has a limiting value equal to the corresponding diagonal element $L_{\alpha\alpha}$. Then (23.31) ought to be interpreted as a simple average of the possible values $L_{\alpha\alpha}$ with weight functions $W_{\alpha\alpha}$. Hence it follows that $W_{\alpha\alpha}$ gives the normalized population of the states α.

In the absence of equilibrium, the statistical matrix $\hat{W}(t)$ differs from $\hat{W}^0$ in two respects; firstly, the diagonal elements of $\hat{W}(t)$ are not expressed by formula (23.28), and second, $\hat{W}(t)$ has off-diagonal terms differing from zero. The meaning of the off-diagonal matrix elements can be ascertained by considering the transition from the quantum case to the classical.

Turning to a system with one degree of freedom (the anharmonic oscillator), we will introduce, as a generalized momentum, the action variable J determined by the relations (2.17) and (2.18).

The distribution function is then transformed in the following way:

$$\frac{f(p, q, t)\,\mathrm{d}p\,\mathrm{d}q}{2\pi\hbar} \to \frac{f(J, \alpha)\,\mathrm{d}J\,\mathrm{d}\alpha}{2\pi\hbar}. \tag{23.32}$$

If the internal energy of the system $E = H(J)$ is brought in as a variable and it is taken into account that the frequency of vibration $\omega(E)$ satisfies the relation

$$\omega(E) = \frac{\partial E}{\partial J}, \tag{23.33}$$

then the distribution function can be represented in the form

$$f\,\mathrm{d}\Gamma = f(E, \alpha, t)\frac{\mathrm{d}\alpha}{2\pi} \cdot \frac{\mathrm{d}E}{\hbar\omega(E)}. \tag{23.34}$$

The equilibrium distribution, expressed just in terms of the total energy and represented in the form (23.30), is obtained on substituting the Boltzmann factor for $f(E, \alpha, t)$ into (23.34) and integrating over all values of the angular coordinates $(0 < \alpha \leq 2\pi)$:

$$f_0(E)\,dE = \exp(-\beta E)\rho(E)\,dE, \qquad \rho(E) = \int_0^{2\pi} \frac{d\alpha}{2\pi\hbar\omega(E)} = \frac{1}{\hbar\omega(E)}. \qquad (23.35)$$

In the quantum statistical matrix $W_{\alpha\beta}(t)$ of the one-dimensional model under consideration, the discrete energy values E and E' may be chosen as the indices of the states α and β. Thus the quantum analogue of (23.34) takes the form:

$$W_{\alpha\beta}(t) = W_{E,E'}(t). \qquad (23.36)$$

In order to trace the transition to the classical limit in $W_{EE'}(t)$ for $E - E' \ll kT$, a Fourier transform with respect to the difference in the quantum numbers $E - E'$ should be performed on (23.36), introducing the angular coordinate α as a new variable. However, even without carrying out the corresponding calculations, the following is evident. If the non-equilibrium classical distribution function does not depend on the angle variable α, then the corresponding quantum statistical matrix is diagonal. In other words, it implies that the relaxation of the system may be described in terms of the population of the states. In particular cases, the possibility of such a description depends on two conditions: to what degree the non-equilibrium distribution function depends on α (it is determined by the particular relaxation mechanism entering the kinetic equation for f) and to what extent the dependence of f and α affects the mean value of various physical quantities $\bar{L}$. The reaction rate, for example, is mainly determined by the population of the states of the active molecule, and even if the non-equilibrium distribution function depends on the angle variables, the function, averaged beforehand over α, can be used in the calculation of $\kappa(E)$ (compare the derivation of the rate constant $\kappa(E)$ for the Slater model in § 27). On the other hand, the relaxation of the vibrational dipole moment of a harmonic oscillator is determined decisively by the dependence of f on α, because this relaxation generally cannot be described in terms of the population of states (in every stationary state of the oscillator, characterized by a definite population size $x(E)$, the mean value of the dipole moment equals zero). This is easy to interpret on the basis of (23.31), if some quantity L in the energy representation contains only off-diagonal matrix elements (as occurs for the coordinates of the harmonic oscillator), then $\bar{L}$ is different from zero, provided the off-diagonal elements of $W_{\alpha\beta}$ are non-vanishing.

We will now consider the derivation of the expression for the rate constant for the quantum system AB, assuming a random distribution for the lifetimes.† As is evident from (23.21) this assumption enables us to avoid a

---

† In the paper [559] a quantum-mechanical formulation of the problem of the calculation of the rate constant is given without this limitation.

discussion of the detailed behaviour of the non-stationary statistical matrix, and allows us to operate with just the diagonal elements which describe the population of the different states.† We will denote the population of the energy state i by f_i and attribute the spontaneous reaction rate κ_i to this state. If P_{ji} is the dimensionless transition probability between levels i and j in a single effective collision, then the balance equation of the particles has the form

$$\frac{df_i}{dt} = \sum_j Z^* P_{ji} f_j - \sum_j Z^* P_{ij} f_i - \kappa_i f_i \qquad (23.37)$$

(in future, we will consider κ_i zero for $E_i < E_0$). The probabilities of the direct and reverse processes P_{ji} and P_{ij} are connected by the principle of detailed balance:

$$P_{ij} g_i \exp(-\beta E_i) = P_{ji} g_j \exp(-\beta E_j). \qquad (23.38)$$

Within the limits of the mechanism of strong collisions, it is assumed that the activation rate, which is given by the first term of eqn (23.4), always equals the activation rate in equilibrium conditions (the assumption of strongly-activating collisions) i.e., that

$$\sum_j Z^* P_{ji} f_j = Z^* f_i^0, \qquad (23.39)$$

where

$$f_i^0 = \frac{1}{F} g_i \exp(-\beta E_i) \qquad (23.40)$$

and Z^* is the effective number of collisions.

From the condition that the Boltzmann distribution be established in the absence of a reaction, we find in the general case,

$$\sum_j Z^* P_{ji} f_i = Z^* f_i. \qquad (23.41)$$

We thus arrive at an equation of the form

$$\frac{df_i}{dt} = -Z^*(f_i - f_i^0) - \kappa_i f_i, \qquad (23.42)$$

---

†Since, in fact, we deal with the non-stationary states characterized by mean lifetimes $\langle \tau^* \rangle = 1/\kappa(E)$, the energy E_i is determined with an uncertainty δE_i, which is itself determined by the uncertainty principle for the energy $\delta E_i \sim \hbar/\langle \tau^* \rangle = \hbar \kappa(E_i)$. In order to neglect the quantum error δE_i in thermal reactions, we shall assume $\delta E_i \ll kT$, i.e. $\hbar \kappa(E)/kT \ll 1$. This relation is usually satisfied for polyatomic molecules. Apart from this, there exists an uncertainty in the energy ΔE_i, due to the finite time $\tau \sim 1/Z^*$ between collisions. This uncertainty is usually much smaller than kT and may be neglected completely.

which can be easily solved in the quasi-stationary approximation. Assuming $df_i/dt = 0$, we find

$$f_i = \frac{Z^* f_i^0}{Z^* + \kappa_i}. \tag{23.43}$$

The rate constant of the reaction κ can clearly be written in the form

$$\kappa = \sum_i \kappa_i f_i = \sum_i \frac{Z^* \kappa_i f_i^0}{Z^* + \kappa_i}. \tag{23.44}$$

This formula is the quantum analogue to relation (23.20).

The transition to the classical expression of type (23.20) is accomplished by the replacement of the sum over the energy levels E_i by an integral over dE with the additional introduction of the level density $\rho(E)$:

$$\kappa = \frac{1}{F} \int_{E_0}^{\infty} \frac{Z^* \kappa(E)}{Z^* + \kappa(E)} \rho(E) \exp(-\beta E) \, dE. \tag{23.45}$$

This expression agrees with (23.20) and explicitly contains the energy-level density of the active molecule $\rho(E)$.

24. The Slater model

The classical theory

One of the simplest models of active and activated molecules, which permits a sufficiently complete study of the problem of the intramolecular redistribution of energy, was proposed by Slater [492]. Within the bounds of this model, it is assumed that the potential energy of the nuclei $U(q_1, \ldots, q_s)$, depending on s internal coordinates of the molecule, can be represented as a quadratic form:

$$U(q_1 \ldots q_s) = \tfrac{1}{2} \sum_{ij} b_{ij} q_i q_j. \tag{24.1}$$

Writing the potential energy in such a form is valid, as is well known, only near the equilibrium position of the nuclei [10], [15]. The displacements q_i, which are measured from the equilibrium distances, must be appreciably smaller than the characteristic increase in the bond length $\bar{q}$ at which dissociation sets in. Nevertheless Slater assumes that the quadratic approximation (24.1) is also valid for large values of the q_i (namely $q_i \sim \bar{q}$), since the theory claims to describe the dissociation of a molecule. Although the basic assumption of the theory (approximation (24.1)) cannot be considered valid, it is nevertheless of great interest because it is the only one permitting a study of the dynamics of a system of many atoms. In the theory, it is also assumed that the kinetic energy T has the form

$$T = \tfrac{1}{2}\sum_{ij} a_{ij}\dot{q}_i\dot{q}_j, \tag{24.2}$$

where the coefficients a_{ij} do not depend on q. The interaction of the vibrations with the rotation of the molecule is completely neglected. However, in view of the cruder assumption (24.1), this is permissible. After a linear transformation of the coordinates

$$q_i = \sum_k \alpha'_{ik} Q'_k, \tag{24.3}$$

the vibrational Hamiltonian $H = T+U$ can be reduced to diagonal form, i.e. the cross terms in the kinetic and potential energies can be eliminated from H [10], [15].

$$H(\dot{q}_i, q_i) = \sum_{ij}\left(\frac{a_{ij}}{2}\dot{q}_i\dot{q}_j + \frac{b_{ij}}{2}q_iq_j\right)$$
$$\to H(\dot{Q}'_\lambda, Q'_\lambda) = \sum_\lambda \left[\frac{m_\lambda}{2}(\dot{Q}'_\lambda)^2 + \frac{m_\lambda\omega_\lambda^2}{2}(Q'_\lambda)^2\right] = \sum_\lambda H_\lambda. \tag{24.4}$$

It is precisely this step, of dividing the total Hamiltonian into a sum of Hamiltonians, which substantially simplifies the problem of studying the dynamics of the system. At the same time, the separation of the variables implies that the energy of each normal vibration ε_μ is conserved, so that, in the harmonic model of a molecule, intramolecular energy exchange between the normal vibrations is completely forbidden. Energy exchange between the different degrees of freedom q_i remains the only possibility, since each of them is a superposition of normal oscillations.

In future, following Slater [492], it is convenient to change the scale, going over to new coordinates

$$Q_\lambda = \left(\frac{m_\lambda\omega_\lambda^2}{2}\right)^{\tfrac{1}{2}} Q'_\lambda. \tag{24.5}$$

At the same time this succeeds in getting rid of the coefficients m_λ, which have the meaning of a reduced mass and which do not enter the final expression for the rate constants.

In the Q_λ coordinates, the Hamiltonian takes on an especially simple form

$$H(\dot{Q}_\lambda, Q_\lambda) = \sum_\lambda H_\lambda = \sum_\lambda \left(\frac{\dot{Q}_\lambda^2}{\omega_\lambda^2} + Q_\lambda^2\right), \tag{24.6}$$

where the Q_λ are expressed in terms of the energy ε_λ, the frequency ω_λ, and the phase ψ_λ of the normal oscillation in the following way:

$$Q_\lambda(t) = \sqrt{(\varepsilon_\lambda)} \cos(\omega_\lambda t + \psi_\lambda). \tag{24.7}$$

Formulae (24.3) and (24.7) describe the time variation of any internal coordinate of the molecule.

To calculate the rate constant of the intramolecular transformation, it is necessary to determine the condition for the completion of a reaction. This condition is formulated in the Slater theory in the form of a requirement that the reaction coordinate q_r intersect the critical surface:

$$q_r = q_r^\star. \tag{24.8}$$

The significance of the internal coordinate q_r is not made specific—in particular, the extension of some bond of the molecule or the increase in the valence angle may appear as such a quantity. The coordinate q_r is expressed in terms of the normal coordinates Q_λ by the relation

$$q_r = \sum_\lambda \alpha_\lambda Q_\lambda(t) = \sum_\lambda \alpha_\lambda \sqrt{(\varepsilon_\lambda)} \cos(\omega_\lambda t + \psi_\lambda), \tag{24.9}$$

where in the coefficients α_λ, which differ from the $\alpha'_{r\lambda}$ by the transformation factor (24.5), the index has been omitted for simplicity.

From (24.9), it is evident that the Slater theory takes explicit account of the interaction of the reaction coordinate with all the degrees of freedom of the molecule.

We will now determine the lowest energy E_0 at which the reaction is possible. (In future, for definiteness, we will have dissociation in mind). Since $\min T(\dot{Q}_\lambda) = 0$, we have

$$E_0 = \min U(Q_\lambda), \tag{24.10}$$

where the minimum in the potential energy must be found from the additional condition

$$q_r^\star = \sum \alpha_\lambda Q_\lambda. \tag{24.11}$$

A simple calculation gives

$$E_0 = \frac{(q_r^\star)^2}{\alpha^2}, \qquad \alpha^2 = \sum_\lambda \alpha_\lambda^2. \tag{24.12}$$

This relation expresses the activation energy in terms of the properties of the reaction coordinate and the critical surface.

On the basis of the above finding, it is not difficult to calculate the dissociation rate constant κ_∞ at high pressures. In fact the high-pressure condition implies that, in practice, the reaction does not infringe the equilibrium distribution of the active molecule as regards the internal degrees of freedom, therefore the formula of the transition state method can be used to calculate

$$\kappa_\infty = \frac{kT}{2\pi h} \cdot \frac{F^\star}{F} \exp\left(-\frac{E_0}{kT}\right). \tag{24.13}$$

where $F^\star$ denotes the partition function of the activated molecule with the Hamiltonian

$$H^\star = \tfrac{1}{2}\sum_{i,j\neq r} a_{ij}\dot{q}_i\dot{q}_j + \tfrac{1}{2}\sum_{i,j\neq r} b_{ij}q_iq_j + \sum_{i\neq r} b_{ir}q_iq_r^\star. \tag{24.14}$$

Taking it into account that the vibrational partition function equals the product of the partition functions of the normal oscillations, we have

$$F = \prod_{i=1}^{s} \frac{kT}{\hbar\omega_i}, \qquad F^\star = \prod_{i=1}^{s-1} \frac{kT}{\hbar\omega_i^\star}, \tag{24.15}$$

where $\omega_i^\star$ are the normal frequencies of the activated molecule, which are found from the solution to the vibrational problem for a system with the Hamiltonian $H^\star$ (formula (24.14)).

From (24.15) and (24.13) we find

$$\kappa_\infty = v^\star \exp(-\beta E_0), \qquad v^\star = \frac{v_1 v_2 \ldots v_s}{v_1^\star v_2^\star \ldots v_{s-1}^\star}. \tag{24.16}$$

On the basis of the theory of small vibrations, it can be shown that each frequency $v_i^\star$ lies in the interval $v_i \leqslant v_i^\star \leqslant v_{i+1}$, if the $v_i^\star$ and v_i are arranged in increasing order [492]. Hence it follows that the frequency factor in the expression of κ_∞ is smaller than the maximum and larger than the minimum frequency of the decomposing molecule. Furthermore, $v^\star$ is expressed in terms of the v_i and the amplitude factors α_i by the expression

$$(v^\star)^2 = \frac{\sum \alpha_i^2 v_i^2}{\sum_i \alpha_i^2} = \sum \mu_i^2 v_i^2, \tag{24.17}$$

where the reduced amplitude factors $\mu_i = \alpha_i/\alpha$ on the right are convenient for future use.

Formulae (24.16) and (24.17) solve the problem of the calculation of κ_∞ in terms of the parameters of the reacting molecule at high pressures. To calculate the rate constant at arbitrary pressures, the distribution function $h = h(\tau^*, E)$ should be determined. First, however, we will consider the simpler problem of the calculation of the mean lifetime of the active molecule $\langle\tau^*\rangle$, which is a basic characteristic of any distribution $h(\tau^*, E)$. In accordance with (23.18) the mean lifetime $\langle\tau^*\rangle$ determines the rate of decay κ_0 at low pressures. On the other hand, κ_0 may be calculated by formula (23.45) in the limit $Z^* \ll \kappa(E)$.

In the Slater model, a molecule for which the reaction coordinate can reach the critical value $q_r^\star$ in any amount of time is considered active. If the normal frequencies are incommensurable, then it is evident (from (24.9)) that, for any initial phases ψ_k a time $t = t^*$ exists when all the factors $\cos(\omega_k t^* + \psi_k)$ will be near to the maximum value, i.e. unity. Hence we find the condition determining the boundary of the phase volume of the active molecule

$$q_r(t^*) = q_r^*,$$

i.e.

$$\sqrt{(E_0)} = \sum_i \mu_i \sqrt{(\varepsilon_i)}. \quad (24.18)$$

If the frequencies are commensurable, then such a time t^* may or may not exist. In particular, if two frequencies ω_l and ω_{l+1} coincide (a degenerate vibration) then the corresponding terms in the sum (24.9) can never be at maxima simultaneously since, due to the linear independence of the coordinates Q_l and Q_{l+1}, the phases ψ_l and ψ_{l+1} must be different. Having chosen Q_l and Q_{l+1} in such a way that ψ_l and ψ_{l+1} differ by $\pi/2$, the two terms may be added to give just one:

$$\sqrt{\varepsilon_l}\left[\alpha_l \cos(\omega_l t + \psi_l) + \alpha_{l+1} \cos\left(\omega_{l+1} t + \psi_l + \frac{\pi}{2}\right)\right]$$
$$= \sqrt{(\varepsilon_l)}(\alpha_l^2 + \alpha_{l+1}^2)^{\frac{1}{2}} \cos(\omega_l t + \psi_l). \quad (24.19)$$

Whereupon (24.9) takes the form of a sum over non-degenerate states with a single reduced effective amplitude factor $\bar{\mu}_l = (\mu_l^2 + \mu_{l+1}^2)^{\frac{1}{2}}$ for the two degenerate vibrations; i.e. the number of effective degrees of freedom is reduced by one. Physically, this is connected with the fact that the phases of the two vibrations remain strongly correlated at all times, therefore the contribution of the normal coordinates Q_l and Q_{l+1} to the motion of the reaction coordinate cannot be estimated without taking the phase relations into account.

On the basis of (23.45), in the limit of low pressures, we obtain

$$\kappa_0 = Z \int \exp(-\beta \sum_i \varepsilon_i) \prod_i \beta \, d\varepsilon_i, \quad (24.20)$$

where the integral is taken over the region $\sum \mu_i \sqrt{(\varepsilon_i)} \geq \sqrt{E_0}$. This integral, introduced and calculated by Slater [492], equals

$$\kappa_0 = Z(4\pi\beta E_0)^{s-\frac{1}{2}} \mu_1 \mu_2 \ldots \mu_s \exp(-\beta E_0), \quad (24.21)$$

where, for the validity of this formula, the inequality $\mu_i \gg (4\pi\beta E_0)^{-\frac{1}{2}}$ must be satisfied. It is not difficult to see that, if the contributions of all the normal vibrations to the reaction coordinate are of the same order then, by virtue of the normalization factors μ_i, we have $\mu_i \sim s^{-\frac{1}{2}}$ for $s \gg 1$. Thus the condition quoted above can be rewritten in the form

$$\beta E_0 \gg s. \quad (24.22)$$

The inequality (24.22) is satisfied for the majority of thermal decomposition reactions. It will be repeatedly employed later on to simplify results. If some of the amplitudes of the factors μ_j are very small, so that the condition $\mu_j \ll (4\pi\beta E_0)^{-\frac{1}{2}}$ is fulfilled, then, in general, they should be disregarded in (24.21), reducing the number of oscillators s correspondingly. Furthermore, for each pair of degenerate vibrations, the quantity $\mu_l \mu_{l+1}$ is replaced by $(\mu_l^2 + \mu_{l+1}^2)^{\frac{1}{2}}$, and, in addition, the number of oscillators s is reduced by one. The expression found can now be used to determine the mean lifetime $\langle \tau^* \rangle$ of the active molecule. On the one hand, for the Slater model κ_∞ and κ_0 are given by formulae (24.16) and (24.21); on the other hand, the general expressions (23.17) and (23.18), in which the two unknown functions $S_\perp^*(E)$ and $\langle \tau^* \rangle$ enter, are valid for κ_∞ and κ_0. Thus we obtain two integral equations, which are easily solved by the method of the Laplace transform [8]. The resulting solution is [73], [385]:

$$\frac{S_\perp^*}{F} = v^* \frac{(E-E_0)^{s-1}}{(s-1)!} \beta^s, \qquad (24.23a)$$

$$\langle \tau^*(E) \rangle = (v^*)^{-1} \left(\frac{E-E_0}{4\pi E_0} \right)^{(1-s)/2} \frac{(s-1)!}{(s-1)/2)!} \mu_1 \mu_2 \cdots \mu_s. \qquad (24.23b)$$

The inverse lifetime $1/\langle \tau^* \rangle$ should be identified with the reaction rate of the active molecule $\kappa(E)$, with the additional condition that the molecules in the beginning are distributed with equal probability over the energy surface $H(p_i, q_i) = E$. For an unequal initial probability distribution the rate constant, attributed to the same energy, will be different.

This dependence of the rate on the initial distribution is particularly strong for the harmonic model under consideration here, since exact conservation of the energy of each normal vibration prevents an arbitrary redistribution of the system over the energy surface $H(p_i, q_i) = E$. Therefore the mean lifetime $\langle \tau^*(E) \rangle$ and the rate constant $\kappa(E)$ furnish insufficient information for the derivation of the expression $\kappa(T)$ at any pressure.

Returning to formula (24.16) it can be seen that, to calculate $\kappa(T)$, it is necessary to know the dependence of the spontaneous reaction rate on the energy ε_i of the normal vibrations. This rate can be identified with the number of intersections of the critical surface by the representative point, i.e. with the number of zeros $G(t_0)$ of the function $q_r(t) - q_r^*$ during the time $t_0 \gg 1/\omega_i$, with the additional condition that only the motion of the reaction coordinate in the positive direction is taken into account. Since both directions of motion are equally probable, on average, we have

$$\kappa(\varepsilon_1, \varepsilon_2, \ldots, \varepsilon_s) = \lim_{t_0 \to \infty} \frac{1}{2t_0} G(t_0), \qquad (24.24)$$

where the dependence of $\kappa = \kappa(\varepsilon_1, \ldots, \varepsilon_s)$ is determined by the parametric dependence of $q_r(t)$ on the energy of the normal vibrations (see (24.9)). The function $\kappa = \kappa(\varepsilon_1, \varepsilon_2, \ldots, \varepsilon_s)$ was determined by Slater [492]. Omitting the details of the calculation of $\kappa(\varepsilon_i)$ set out in Slater's monograph [492], we will quote the final result:

$$\kappa(\varepsilon_1,\ldots,\varepsilon_s) = \frac{1}{\left(\dfrac{s-1}{2}\right)!} \left(\frac{\sum_i \mu_i \sqrt{(\varepsilon_i)} - \sqrt{(E_0)}}{2\pi}\right)^{(s-1)/2} \left(\frac{\sum_i \mu_i v_i^2 \sqrt{(\varepsilon_i)}}{\prod_i \mu_i \sqrt{(\varepsilon_i)}}\right)^{\tfrac{1}{2}}. \qquad (24.25)$$

If the total energy of the molecule is fixed, so that $\sum_i \varepsilon_i = E$, then $\kappa(\varepsilon_i)$ reaches a maximum for

$$\varepsilon_i^* = \mu_i^2 E, \qquad (24.26)$$

i.e. for a completely determined distribution of energy over the normal vibrations.

To investigate the dependence of the decomposition rate near threshold, let us assume that $\varepsilon_i = \varepsilon_i^* + \Delta\varepsilon_i$ and expand $\varepsilon_i^{\frac{1}{2}}$ as a series in the parameter $\Delta\varepsilon_i/\varepsilon_i^*$. Taking three terms of the expansion into account and introducing the approximation $(1-x)^m \approx \exp(-mx)$, valid for $x \ll 1$, $m \gg 1$, we find

$$\kappa(\varepsilon_1,\varepsilon_2,\ldots,\varepsilon_s) = \frac{1}{\left(\dfrac{s-1}{2}\right)!} \left(\frac{\Delta E}{4\pi E_0}\right)^{(s-1)/2} (\mu_1 \mu_2 \ldots \mu_s)^{-1}$$

$$\times \exp\left[\frac{-(s-1)\sum_i ((\Delta\varepsilon_i)/\mu_i)^2}{8 E_0 \Delta E}\right], \qquad (24.27)$$

where ΔE is the excess of the total energy over threshold, equal to $\sum_{i=1}^s \Delta\varepsilon_i$; $\Delta\varepsilon_i$ is the variation in the energy of the separate normal vibrations.

From this expression, it is clearly seen that for a given excess ΔE, the rate $\kappa(E)$ rapidly diminishes if the energy is concentrated on a small number of normal vibrations. For the distribution (24.26), the exponent of the exponential equals $(s-1)\Delta E/8 E_0$, and the final factor in (24.27) can be neglected, in general, on condition

$$\frac{E_0}{\Delta E} \gg s. \qquad (24.28)$$

This condition for the energy of excitation of the active molecule is analogous to the condition (24.22) for thermal reactions.

Relation (24.25) is sufficient for calculating the rate constant of a thermal reaction within the limits of the mechanism of strong collisions:

$$\kappa(T) = \int \frac{Z^* \kappa(\varepsilon_1,\varepsilon_2,\ldots,\varepsilon_s)}{Z^* + \kappa(\varepsilon_1,\varepsilon_2,\ldots,\varepsilon_s)} \exp\left(-\beta \sum_i \varepsilon_i\right) \prod_i \beta\, d\varepsilon_i. \qquad (24.29)$$

After a series of transformations (24.29) can be represented in the form of the following expression for the dimensionless rate constant κ/κ_∞:

$$\frac{\kappa}{\kappa_\infty} = \frac{1}{m!} \int_0^\infty \frac{x^m \exp(-x)\,dx}{1+x^m/\theta} = I_m(\theta), \qquad (24.30)$$

where

$$\theta = \left(\frac{Z^*}{v^\star}\right)(4\pi\beta E_0)^m \mu_1 \mu_2 \ldots \mu_s m!, \qquad m = \tfrac{1}{2}(s-1),$$

and the limitation, mentioned in connection with (24.21), is imposed on the amplitude factors μ_i. The function $I_m(\theta)$, determined by the integral on the right-hand side of (24.30), is tabulated in Slater's book [492].

Quantum theory

In the quantum generalization of the harmonic model considered above, it must be borne in mind, first and foremost, that the reaction condition, formulated in the form of the requirement of the intersection of the critical surface $q_r = q_r^*$, is not as well defined in the quantum model as in the classical model. This is connected with the fact that, in the quantum case, there is no relation between the coordinates q_r^* and the threshold energy E_0, since chemical reactions can, in general, take place by means of tunnelling. Hence it follows that, first of all, the quantum theory of chemical reactions must be based on an evaluation of the contributions of tunnel effects to the rate constant. Although a number of papers [17], [18], [150] have recently been devoted to a consideration of this problem in a general form, it is, nevertheless, very difficult to obtain a specific estimate in various reactions, since the potential energy surface must be known.

The indefiniteness of the relation between q_r^* and E_0 can be removed by an additional assumption, for example, that the tunnel effects at levels with $E < E_0$ do not make significant contributions to the rate constant. Precisely such an assumption was made in Kassel's quantum model (§ 25) and in the statistical theory of chemical reactions. On the other hand, it may be assumed that the condition of the intersection of the surface $q_r = q_r^*$ (including tunnelling intersections) determines the completion of a reaction at any energy. Furthermore, of course, the question of the relationship between the implicit allowance for the tunnel transitions contained in this condition and the proper correction for tunnel transmission which must be made for real potential energy surfaces remains open.

In Fig. 22 are shown various types of potentials which represent qualitatively the dependence of the potential energy on the coordinate q_r and which satisfy the conditions of the harmonic approximation. Details of the potential for $q_r > q_r^*$ are important only for the quantum interpretation of the reactions,

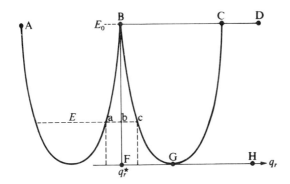

FIG. 22. Possible potentials of a one-dimensional harmonic model.

when, in principle, tunnel corrections may prove to be substantial. The following types of potentials and the processes corresponding to them are presented in Fig. 22:

I. Potential ABCD—dissociation with complete neglect of tunnel effects. Possible only for $E > E_0$.
II. Potential ABFGH—dissociation as a result of above barrier and tunnel transmission across the barrier aBb. Possible for $E > 0$.
III. Potential ABGC—symmetric isomerization on over-barrier and tunnel transmission across the barrier aBc. Possible for $E > 0$.
IV. Potential ABGH—dissociation on above-barrier and tunnel transmission across the barrier aBc. Possible for $E > 0$.

Slater [492] based the quantum-mechanical generalization of his theory on a formal extension of the transition state method to the quantum region. The expression for the reaction rate constant at high pressures is written in the form†

$$\kappa_\infty = \langle \dot{q}_r f(\dot{q}_r, q_r^*) \rangle = \langle \dot{q}_r \rangle P_q(q_r^*), \qquad (24.31)$$

where $\langle \dot{q}_r \rangle$ is the mean velocity of the representative point in the region $q_r > q_r^*$; $P_q(q_r^*)$ is the probability of reaching the critical configuration.

Here, however, the classical distribution function used in the derivation of (24.16) is replaced by the Wigner distribution function. For a single vibration, the Wigner distribution has the form [492]

$$(2\pi\sigma^2\omega_\kappa)^{-1} \exp\left(-\frac{Q_\kappa^2 + \dot{Q}_\kappa^2/\omega_\kappa^2}{2\sigma_\kappa^2}\right) dQ_\kappa \, d\dot{Q}_\kappa, \qquad (24.32)$$

† An analogous approach to the calculation of the reaction rate was formulated by Mayants [46].

where

$$2\sigma_\kappa^2 = \frac{1}{2}\hbar\omega_\kappa \coth\left(\frac{\hbar\omega_\kappa}{2kT}\right).$$

From the property of the Gaussian distributions such as (24.32) it follows that the distribution function for the coordinates q_r and $\dot{q}_r$, which are expressed linearly in terms of Q_κ, will also be Gaussian (as in the classical case) with standard deviations σ and $\dot{\sigma}$ for the coordinate q_r and the velocity $\dot{q}_r$ respectively, where

$$\sigma^2 = \tfrac{1}{2}kT\sum_\kappa \alpha_\kappa^2 y_\kappa \coth y_\kappa,$$

$$\dot{\sigma}^2 = \tfrac{1}{2}kT\sum_\kappa \alpha_\kappa^2 \omega_\kappa^2 y_\kappa \coth y_\kappa, \qquad (24.33)$$

where

$$y_\kappa = \frac{\hbar\omega_\kappa}{2kT}.$$

After averaging (24.31) the following result is obtained:

$$\kappa_\infty = \dot{\sigma}(2\pi\sigma)^{-1}\exp\left[-\frac{(q_r^\star)^2}{2\sigma^2}\right]. \qquad (24.34)$$

The essential difference between this formula and (24.16) is that, primarily, under the condition $\beta\hbar\omega_\kappa \geqslant 1$, the temperature dependence of the rate constant κ_∞ differs sharply from that of Arrhenius. This is connected with the very large contribution of the tunnel transitions to the rate constant κ_∞.

To examine such of the processes enumerated above as correspond to the rate, determined by relation (24.34), we will consider in greater detail the probability of a tunnel transition for the potentials presented in Fig. 22. For simplicity, we will limit ourselves to a one-dimensional model and low temperatures ($\beta\hbar\omega \gg 1$). In this case, the relation (24.34) takes the form

$$\kappa_\infty = \nu \exp\left[-\frac{M\omega(q_r^\star)^2}{\hbar}\right], \qquad (24.35)$$

where the exponential factor agrees, as it must, with the exponential factor of the squared wavefunction $|\psi_0(q_i^\star)|^2$ of the ground state of the harmonic oscillator in the left-hand well (see Fig. 22). (Under the condition $\beta\hbar\omega \gg 1$, the population of the higher states is negligibly small.)

Returning to Fig. 22, we note that, for a potential of type 1, the tunnel transition is in general impossible, therefore the general expression (24.34) is inapplicable in this case. For potentials II and IV, a tunnel transition is possible from the quasi-stationary level E into a state of the continuous energy spectrum. The rate of tunnel decomposition is given by the following

expression:
$$\kappa = \nu\chi, \tag{24.36}$$
where χ is the transmission coefficient under the potential barrier, determined by formula (3.5), for an exponentially small transition probability. Returning to Fig. 22, we find for potentials II and IV:

$$\chi_{\text{II}} = \exp\left\{-\frac{2}{\hbar}\int_a^b \left[2M\left(\frac{M\omega^2}{2}q_r^2 - E\right)\right]^{\frac{1}{2}} dq_r\right\} \approx \exp\left[-\frac{M\omega(q_r^*)^2}{\hbar}\right]$$
$$= |\psi_0(q_r^*)|^2, \tag{24.37}$$

$$\chi_{\text{IV}} = \exp\left\{-\frac{2}{\hbar}\int_a^c \left[2M\left(\frac{M\omega^2}{2}q_r^2 - E\right)\right]^{\frac{1}{2}} dq_r\right\} \approx \exp\left[-\frac{2M\omega(q_r^*)^2}{\hbar}\right]$$
$$= |\psi_0(q_r^*)|^4. \tag{24.38}$$

Comparing (24.38) and (24.37) with (3.5) it is not difficult to see that formula (3.5) corresponds to potential II, but gives a considerably overestimated contribution from tunnel transitions for potential IV.

We will now turn to potential III. In this case, we have a doubly-symmetric potential well of the type realized, for example, in the inversion vibration of ammonia (the coordinate q is proportional to the height of the pyramid and the point $q = q^*$ corresponds to the plane structure). The energy spectrum in such a potential is discrete, and one can only speak of a tunnel transition here in the sense that the delocalized wavefunctions ψ_g and ψ_u which correspond to slightly displaced energies (relative to the level E in one well), belong to stationary states [37]:

$$E_g = E - \frac{\Delta E}{2}, \quad E_u = E + \frac{\Delta E}{2}$$

$$\Delta E = \frac{\hbar^2}{2M}\psi_0(q)\frac{d\psi_0(q)}{dq}\bigg|_{q=q^*}. \tag{24.39}$$

If the wavefunction ψ of the system is initially localized on the left-hand well, then it follows that

$$\psi(t=0) = \psi_0 = \frac{\psi_g + \psi_u}{\sqrt{2}}. \tag{24.40}$$

For $t > 0$ this function will begin to oscillate, since each of the stationary solutions will develop according to its own law, determined by the eigenvalues E_g and E_u.

$$\psi(t) = \frac{[\psi_g \exp(-iE_g t/\hbar) + \psi_u \exp(-iE_u t/\hbar)]}{\sqrt{2}}$$
$$= \frac{\exp(-iEt/\hbar)[\psi_g \exp(i\Delta E t/\hbar) + \psi_u \exp(-i\Delta E t/\hbar)]}{\sqrt{2}}. \tag{24.41}$$

From (24.41) it is evident that in time $\tau = \pi\hbar/\Delta E$ the function $\psi(\tau)$ will have the form

$$\psi(\tau) \sim \frac{(\psi_g - \psi_u)}{\sqrt{2}}, \tag{24.42}$$

i.e. the system will turn out to be localized in the right-hand well.

Substituting the oscillatory function $\psi_0 = (M\omega/\hbar\pi)^{\frac{1}{4}} \exp(-q^2 M\omega/2\hbar)$ into (24.39) and carrying out the differentiation, we obtain

$$\frac{1}{\tau} = v \exp\left[-\frac{(q^*)^2 M\omega}{\hbar}\right]. \tag{24.43}$$

The right-hand part of the expression (24.43) agrees with (24.35). Therefore, if the inverse transition time $1/\tau$ is identified with the rate constant κ_∞, as is sometimes done [340], then formula (24.35) will describe a process of symmetric isomerization. In the general case, however, such an identification is not justified. This is evident from the fact that κ_∞ characterizes an irreversible (in time) reaction process and $1/\tau$ the reversible vibration process. For a systematic investigation of the isomerization process, it is necessary to take into account some form of relaxation mechanism, which will ensure the capture of particles in the right-hand well, after their transition from the left-hand well.†

Thus it is evident that, in a certain sense, formula (24.35) is applicable to the description of isomerization. A process of this type is studied in greater detail in § 37.

What has been said above with reference to transitions in the ground state also immediately carries over to excited states. In this case, due to the many degrees of freedom, the probability of reaching the coordinate q_r^* is expressed by a very complicated formulae [47], however, after averaging over all levels, expression (24.36) is again obtained.

It should be borne in mind that the harmonic approximation is known to be poor for the description of tunnel transitions, since it assumes the existence of a potential barrier whose form may be approximated by two intersecting parabolas. Nevertheless, as a result of the very strong (exponential) dependence of κ_∞ on the critical magnitude of the coordinate q^* the reverse dependence is very weak, so that the interpretation of the experimental values on the basis of relation (24.35) must give reasonable values for q_r^*. Thus, for example, for the *cis-trans*-isomerization of ethylene derivatives, the critical configuration must correspond to a relative rotation of the

† Buff and Wilson [176] considered in a general form such a relaxation process, within the limits of the mechanism of strong collisions. They found that multiple intersections of the critical surface by the representative point in an isomerization reaction could be taken into account by the introduction of a transmission coefficient depending on pressure. In the high-pressure limit, the transmission coefficient tends to unity.

substituted group by $\pi/2$. However, if this value of the critical rotation were to be found, in fact, on the basis of the experimental values of κ_∞, it would not point unambiguously to the satisfactory accuracy of the harmonic model.

Slater showed that, if (24.35) is formally written in the form of an Arrhenius expression with the experimental activation energy [492]

$$E_\mathrm{a} = kT^2 \frac{\mathrm{d}}{\mathrm{d}T} \ln \kappa_\infty, \tag{24.44}$$

then the effective factor before the exponential must be appreciably smaller than the normal values 10^{13}–10^{14} s^{-1}. This deduction is especially interesting in connection with the fact that a number of isomerization reactions actually are characterized by a small frequency factor, a result usually interpreted as due to a nonadiabatic transition via an intermediate triplet state.

Turning finally to potential 1, the expression for κ_∞ is obtained immediately from the quantum formulae of the transition state method with $\eta = 1$:

$$\kappa_\infty = \frac{kT}{2\pi\hbar} \frac{\prod_{i=1}^{s}\left[1-\exp\left(-\frac{\hbar\omega_i}{kT}\right)\right]}{\prod_{i=1}^{s-1}\left[1-\exp\left(-\frac{\hbar\omega_i^\star}{kT}\right)\right]} \exp\left(-\frac{E_\mathrm{a}}{kT}\right). \tag{24.45}$$

In going over to intermediate and low pressures, Slater adopted the general expression for the rate constant in the classical case (23.45), but replaced the classical distribution function by the Wigner distribution. With such a method of approach, it is evident that the rate constant cannot be characterized by the Arrhenius temperature-dependence in the low-pressure region either.

The expression for the reduced rate constant κ/κ_∞ formally preserves the form (24.30), however, the factors μ_i are defined differently

$$\mu_i^2 = \frac{\alpha_i^2 \sigma_i^2}{\sum \alpha_i^2 \sigma_i^2}. \tag{24.46}$$

Since, as shown above, the formulation of the reaction rate in the form of expression (24.34) in many cases overestimates the contribution from tunnel transitions, it is expedient to neglect them completely at low pressures. A calculation of κ_0 can then be carried out on the basis of (23.18), extending the integration to the region $E > E_0$, if the mean lifetime of the active molecule $\langle\tau^*\rangle$ is known.

The simplest method of determining $\langle\tau^*\rangle$ on the basis of quantum theory uses the fact that, at the initial instant, a function ψ_0 is assigned which represents a given localization of energy on the reaction coordinate. The non-stationary Schrödinger equation is solved with this initial condition, and the corresponding wavefunction $\psi(t)$ allows us to trace the energy exchange between the different degrees of freedom of the molecule [70], [74], [386]. The function $\psi(t)$ can be represented in the form of an expansion in terms of the stationary functions $\phi_n(q_1, q_2, \ldots, q_s) \exp(-iE_n t/\hbar)$, as in the expansion for the two states in (24.41). This expansion is linked to the introduction of the uncertainty in the energy δE and, in order that one may speak of the localization of the given energy on some particular bond, it is necessary to demand that δE be small in comparison with the energy parameter of the distribution function $f \sim \exp(-E/kT)$, i.e. $\delta E \ll kT$. The function $\psi(t)$ represents a spreading wavepacket and the matrix element A, determined by the integral

$$A(t) = \int \psi^*(0)\psi(t)\,dq_1\,dq_2\ldots dq_s, \tag{24.47}$$

gives the probability amplitude for finding the initial distribution of energy at time t. As an illustration of this discussion, we will present here the simplest example which allows us to trace the damping out of a packet and the connection between the quantum and classical theories. We will represent the molecule by a system of s harmonic oscillators with frequencies ω'_i, coupled with each other by an interaction of the form $V_{\text{int}} = \sum e_{ik} q_i q_k$. We will assume that the differences $\Delta\omega'_{ik}$ are so small that the stationary wavefunctions $\Phi_{p_i}(Q_i)$ of the normal vibrations may be approximated with sufficient accuracy by a linear combination of vibrational functions $\phi_{n_i}(q_i)$ of oscillators, in which the interaction V_{int} is not taken into account.† Taking no account of V_{int}, the coordinates q_i are normal. After taking V_{int} into account, however, the normal coordinates Q_i are just linear combinations of the coordinates q_i so that, in particular, for the reaction coordinate q_i we have

$$q_i = \sum \mu_i Q_i. \tag{24.48}$$

This transformation also involves the transformations of the functions $\phi(q_i)$ and $\Phi(Q_i)$, in terms of which the function ψ for the whole system of interacting oscillators is expressed. For the solution of stationary problems, both sets ϕ_i and Φ_i are equally convenient, and the difference is just that the total Hamiltonian including the interaction is non-diagonal in the basis ϕ_i but is diagonal in the basis Φ_i. For the solution of a non-stationary problem, the set Φ_i is preferable, because the time dependence of these

---

† This approximation is comparable in accuracy to the approximation used in the discussion of tunnel transitions in a double potential well. The stationary functions ψ^+ and ψ^- are approximated by the sum and difference of the functions ϕ_a and ϕ_b, localized in the left-hand and right-hand wells respectively. The possibility of such an approximation is connected with the fact that the difference between the energies E^+ and E^- is small in comparison with the energy quantum $\hbar\omega$ in a single well.

functions is determined by a simple exponential $\exp(-iE_i t/\hbar)$, where E_i here denotes the energy of a normal vibration of the system taking account of V_{int}.

We will assume that initially the oscillator r (coordinate q_r) is in a state described by the function $\phi_N(q_r)$, and all the remaining oscillators are unexcited, i.e. the functions $\phi_0(q_i)$, $i \neq r$, correspond to them. In subsequent moments, under the influence of the interaction V_{int}, the energy concentrated on the oscillator r will be redistributed, and the condition that the frequencies ω_i' (and also the ω_i) be close enables $A(t)$ to be calculated in an explicit form [70]:

$$A_N(t) = \left[\sum_{\kappa=0}^{s} \mu_\kappa^2 \exp(-i\Delta\omega_\kappa t)\right]^N, \tag{24.49}$$

where μ_κ are amplitude factors, $\Delta\omega_\kappa$ is the difference between ω_κ and the mean frequency $\bar{\omega} = (\sum \mu_\kappa^2 \omega_\kappa^2)^{\frac{1}{2}}$.

Formula (24.49) clearly shows how the spreading of the wavepacket, characterized by the wavefunction $\psi(t)$, proceeds. For $\Delta\omega_\kappa t \gg 1$, there is substantial mutual cancelling of terms in the sum, resulting in a decrease in $A(t)$. The probability of finding the state ϕ_N after a long time is determined by the formula:

$$P_N = \lim_{t_0 \to \infty} \frac{1}{t_0} \int_0^{t_0} |A_N(t)|^2 \, dt, \tag{24.50}$$

which is analogous to the classical formula (24.24). The integration with respect to time in (24.50) leads to an averaging of the interference terms in (24.49) containing the factor $\exp[-i(\Delta\omega_\kappa - \Delta\omega_\kappa')t]$ which vanishes, on averaging, if all the frequencies ω_κ are different. On condition $\mu_\kappa \gg 1/N^{\frac{1}{2}}$ the probability P_N turns out to be equal to

$$P_N = \mu_1 \mu_2 \ldots \mu_s (4\pi N)^{(-s+1)/2} \tag{24.51}$$

If it is now assumed that the state ϕ_N decays with frequency v^*, then for the mean lifetime, we can write

$$\langle \tau_N^* \rangle = \frac{1}{v^* P_N}. \tag{24.52}$$

From formula (24.45) it is not difficult to determine the ratio S_1^*/F in (23.17) for the cases $\hbar\omega \gg kT$. In addition, of course, the integral in (23.17) must be understood as a sum over levels $E_\kappa = \hbar\omega^*\kappa$, $\kappa = N, N+1, \ldots$, but as a consequence of the assumed condition $\delta E \ll kT$, the erosion of the level, determined by the small spread of frequencies ω_κ, can be neglected.

In this way, we obtain $S_1^{\star}/F = v^{\star}$ and, using (3.18) we find [74], [385]

$$\kappa_0 = Z^{\star}\left(\frac{E_N}{4\pi\hbar\omega}\right)^{(s-1)/2} \mu_1\mu_2 \dots \mu_s \exp\left(-\frac{E_N}{kT}\right). \quad (24.53)$$

The energy E_N of level N of the oscillator q_r is introduced into this formula. Comparison of (24.53) with (24.21) shows that the variant of the quantum case considered here differs from the classical only in the replacement of kT by $\hbar\omega$ in the pre-exponential term.

The general formulation of the quantum theory of reactions, even within the limits of the harmonic model, is complicated by the fact that it is necessary to take into account, firstly, the possibility of the violation of the inequality $\delta E \ll kT$, the fulfillment of which allows one to talk approximately of the simultaneous assignment of the total energy of the molecule and its localization on definite bonds: secondly the effect of the reaction (for example, the rupture of the bond q_r) on the intramolecular redistribution of energy.

The spread of the molecular vibrational levels, connected with the quasistationary nature of the state, may affect the 'spreading' of the vibrational packet, if only on account of the removal of some frequencies from resonance. Without dwelling further on these problems, we will refer only to some papers in which simple models of decay [74], [221], [385] and the general formal theory are considered [249], [268], [322], [323], [324].

25. Kassel's model

The classical theory

Kassel's theory is based on the assumptions of strong collisions and on the random, fluctuating nature of the redistribution of energy between the degrees of freedom of the molecule, which is represented by a system of harmonic oscillators in the active and activated states [30], [297]. The latter assumption, in fact, implies an implicit allowance for anharmonicity, ensuring an interaction of the reaction coordinate with the other degrees of freedom, however, the interaction mechanism is not specified further.

The critical surface for this model is the same as for the Slater model, i.e. the reaction is regarded as having been completed on the intersection of the surface $q_r = q_r^{\star}$. Since in the high-pressure region, the rate of the intramolecular redistribution of energy does not, in general, effect the value of the constant, and the equilibrium distribution is sustained at the expense of the intermolecular interaction, the expression for κ_0 can be written down at once, returning to formula (24.16):

$$\kappa_\infty = v^{\star} \exp(-\beta E_0). \quad (25.1)$$

On the other hand, from the physical meaning of the assumption about the random nature of the internal distribution of energy, κ_∞ can be obtained by averaging the rate constant $\kappa(E)$, characterizing the mean lifetime of the active molecule over the equilibrium distribution

$$\kappa_\infty = \frac{1}{F} \int_{E_0}^{\infty} \kappa(E) \exp(-\beta E) \rho(E) \, dE. \tag{25.2}$$

Taking it into consideration that, for a system of s harmonic oscillators, the level density $\rho(E)$, determined by condition (23.7), equals

$$\rho(E) = \frac{E^{s-1}}{(s-1)!} \Big/ \prod_{i=1}^{s} \hbar\omega_i, \tag{25.3}$$

the integral in (25.2) can be inverted by the method of the Laplace transform, after substituting expression (25.3) into the right-hand side. This inversion is single-valued and the validity of the result given below can be verified easily by the direct substitution of $\kappa(E)$ into (25.2):

$$\kappa(E) = v^\star \left(\frac{E - E_0}{E} \right)^{s-1}, \qquad E - E_0 > 0$$

$$\kappa(E) = 0, \qquad E - E_0 = 0. \tag{25.4}$$

Thus the mean lifetime of the active molecule for Kassel's model equals:

$$\langle \tau^* \rangle = \frac{1}{\kappa(E)} = (v^\star)^{-1} \left(\frac{E}{E - E_0} \right)^{s-1}. \tag{25.5}$$

This expression, in which the E in the numerator can be replaced by E_0 with sufficient accuracy, should be compared with formula (24.23b), valid for a strictly harmonic model. In a following section, we will discuss the reason for the difference between them. Here we will only note that neither (25.4) nor (25.5) can be detailed further, as was done, for example, in the expression for $\kappa(E)$ in the Slater model, within whose bounds it was possible to speak of the dependence of the rate constant on the energy of individual normal vibrations (cf. (24.27)). In Kassel's model the rate constant depends only on the total energy $E = \sum_i \varepsilon_i$ of the normal vibrations.

The expression (25.4) permits the interpretation of the rate constant in terms of the probability of fluctuations, in which some energy $E' > E_0$ is concentrated on the oscillator r of the active molecule with energy E. It is evident that, for a fixed value of E', the density of states in the system of the remaining $(s-1)$ oscillators will be proportional to $\rho_{s-1}(E-E')$, so that the probability of a fluctuation with a given value of E' has the form

$$P(E', E) = \frac{\rho_{s-1}(E - E')}{\rho_s(E)}. \tag{25.6}$$

If E' is any value in the range $E_0 \leqslant E' \leqslant E$, then the decomposition rate must be proportional to the sum over all values of E', i.e.

$$\kappa(E) \sim \int_{E_0}^{E} P(E', E) \, dE'. \tag{25.7}$$

The substitution here of P from (25.6) and ρ_s from (25.3) gives a formula of type (25.4). Of course, the frequency factor may be found by such an approach only after a specific critical surface is prescribed and, for the case under consideration, q_r^* is determined by the frequency v^* [239].

The transition to arbitrary pressures is accomplished on the basis of formula (23.45). Moreover, for the reduced rate constant, the expression

$$\frac{\kappa}{\kappa_\infty} = \frac{1}{(s-1)!} \int_0^\infty \frac{x^{s-1} \exp(-x) \, dx}{1 + \frac{v^\star}{Z^*}\left(\frac{x}{x+b}\right)^{s-1}}, \qquad b = \beta E_0 \tag{25.8}$$

is obtained.

The ratio in brackets in the denominator can be expanded in powers of x/b, on fulfillment of the condition $b \gg s$, and limited to the first few terms. In this case we obtain

$$\frac{\kappa}{\kappa_\infty} = \frac{1}{(s-1)!} \int_0^\infty \frac{x^{s-1} \exp(-x) \, dx}{1 + x^{s-1}/\theta'} = I_{s-1}(\theta'), \qquad \theta' = \frac{Z^*}{v^\star} b^{s-1}. \tag{25.9}$$

Thus, in Kassel's theory, when $b \gg s$ (i.e. on condition $E_0 \gg skT$), the reduced rate constant is expressed by the same functions $I_m(\theta)$ as were introduced in Slater's theory. In spite of this, however, the indices and arguments of the functions are different for the two models of the same system:

$$\frac{\kappa}{\kappa_\infty} = I_m(\theta) \begin{cases} m = \dfrac{s-1}{2}, \quad \theta = \left(\dfrac{Z^*}{v^\star}\right)(4\pi\beta E_0)^{(s-1)/2}\left(\dfrac{s-1}{2}\right)! \mu_1 \mu_2 \cdots \mu_s \\ \qquad\qquad\qquad\qquad \text{(Slater's model)} \\[2mm] m = s-1, \quad \theta = \left(\dfrac{Z^*}{v^\star}\right)(\beta E_0)^{s-1}. \\ \qquad\qquad\qquad\qquad \text{(Kassel's model)} \end{cases} \tag{25.10}$$

The physical reason for this difference is discussed in a following section. Here, we will only note that, on the violation of the condition $b \gg s$, the simple relation between the models, established by expression (25.9), breaks down. The range of variation of the parameters k and s, for which this correspondence occurs, is explained in the literature [459].

The expression for κ_0 is obtained from (25.8) and (25.9) in the limit $\theta \ll 1$. Neglecting the unity in the denominator of (25.9) by comparison with x^{s-1}/θ

we obtain

$$\kappa_0 = Z^* \frac{(\beta E_0)^{s-1}}{(s-1)!} \exp(-\beta E_0), \qquad \beta E_0 \gg s. \tag{25.11}$$

Quantum theory

The quantum generalization of Kassel's model is conducted on the assumption of the mechanism of strong collisions. The molecule is represented by a system of s quantum oscillators with the *same* frequencies. This enables the characteristics of the vibrational spectrum necessary to determine the fluctuation probability $P(E', E)$ to be calculated comparatively easily. At the same time, this spectrum—equidistant energy levels with a separation $\hbar\omega$ and a rapidly-growing degree of degeneracy—is very far from the true spectrum of a polyatomic molecule, which is almost continuous for high excitation energies. This defect of the Kassel model is improved within the framework of the general statistical theory of reactions (see Chapter 5).

The multiple degeneracy of level E_n with energy $E_n = \hbar\omega n$ equals[†]

$$g_s(n) = \frac{(n+s-1)!}{(s-1)!n!}. \tag{25.12}$$

The number of states of the activated molecule W^* equals

$$W^* \sim \sum_{n'=n_0}^{n} g_{s-1}(n') = g_s(n-n_0). \tag{25.13}$$

The fluctuation probability of interest to us, obtained from this, equals

$$P(n, n_0) = \frac{W_s^*}{g_s(n)} = \frac{g_s(n-n_0)}{g_s(n)}, \tag{25.14}$$

and the rate constant is given by the expression

$$\kappa(n) = v^* \frac{g_s(n-n_0)}{g_s(n)} = v^* \frac{(n-n_0+s-1)!n!}{(n+s-1)!(n-n_0)!}. \tag{25.15}$$

The frequency factor in this expression can be found from the condition that the quantum formula for $\kappa(n)$ should go over into the classical for $n \to \infty$ (i.e. for $\hbar\omega \to 0$ and $E_n = $ constant), where v^* here coincides with v, as the system of oscillators is considered to have equal frequencies.

[†] We note that the vibrational energy E_n is determined without regard for the zero-order vibration.

Returning now to the general formula (23.44), we will find $\kappa(T)$. The quantum equilibrium distribution function

$$f_n^0 = \frac{\exp(-n\hbar\omega\beta)g_s(n)}{F}, \qquad F = (1-e^{-\hbar\omega\beta})^{-s} \qquad (25.16)$$

should be substituted into (23.44).

In the high-pressure limit, we obtain the formula

$$\kappa_\infty = \frac{\sum_{n=n_0}^{\infty} \kappa(n)g_s(n)\exp(-\beta\hbar\omega n)}{F} = v\exp(-\beta E_0), \qquad E_0 = \hbar\omega n_0. \qquad (25.17)$$

For the reduced rate constant we have

$$\frac{\kappa}{\kappa_\infty} = [1-\exp(-\hbar\omega\beta)]^s \sum_{n=0}^{\infty} \frac{g_s(n)\exp(-\hbar\omega\beta_n)}{1+\frac{v}{Z^*}\frac{g_s(n)}{g_s(n+n_0)}} \qquad (25.18)$$

whence, for low pressures, we obtain

$$\kappa_0 = \frac{Z^* g_s(n_0)(1-e^{-\hbar\omega\beta})^{-1}\exp(-\beta E_0)}{F_s}$$

$$= Z^* \left(\frac{E_0}{\hbar\omega}\right)^{s-1} \frac{(1-e^{-\hbar\omega\beta})^{s-1}}{(s-1)!}\exp(-\beta E_0). \qquad (25.19)$$

Of course, (25.18) and (25.19) go over into the classical formulae (25.8) and (25.11) for $kT \gg \hbar\omega$. As noted above, (25.11) is calculated with a relative error $s/\beta E_0$. The relative error in (25.19) connected with the approximate calculation of the sum (25.17) is of the order $\exp(-\hbar\omega\beta)s\hbar\omega/E_0$.† Therefore it can be shown that, in the limit $\hbar\omega\beta \gg 1$, the result (25.19) is exact, regardless of the dependence on the number of degrees of freedom s. However, this deduction is connected with the defect of the model, which does not take account of the different frequencies of the oscillators. Even for a frequency spread as small as desired, an increase in s leads to such a diffuseness in the spectrum that separate bands arising from a strongly-degenerate level of the system of identical oscillators overlap, so that the spectrum becomes continuous. Moreover, the sum (25.17) again turns into an integral, even under the condition $kT \ll \hbar\omega$.

The contribution of different degrees of freedom to the rate constant for the Slater and Kassel models

For the harmonic Slater model, the mean lifetime is given by relations (24.23b) and (24.52). The former refers to the classical case on condition

† The approximation consists of the replacement of $g_s(n+n_0)$ in the denominator by $g_s(n_0)$.

$E - E_0 \gg \hbar\omega s$, i.e. for an energy of excitation sufficiently far from threshold. The latter applies to a calculation of the lifetime of a system of approximately similar harmonic oscillators close to threshold, on condition $E - E_0 \leqslant \hbar\omega$. As already noted above, (24.23) and (24.52) clearly show how the gradual reduction of the interactions of the reaction coordinate with the normal coordinates Q_j, expressed by the decrease in the reduced amplitude factors μ_j, leads to a decrease in the lifetime of the active molecule. It can be shown, in particular, that, due to the symmetry conditions, certain coefficients μ_j are equal to zero. They should then, in general, be omitted from the formula for the lifetime, and the corresponding number of vibrational degrees of freedom s should be reduced. In the classical case such an effective reduction in the vibrational degrees of freedom may be considered as a result of the complete lack of participation of certain of the normal vibrations in the intramolecular energy exchange process.

In the quantum case, the effective reduction in the vibrational degrees of freedom should be interpreted as resulting from the complete lack of participation in the intramolecular energy exchange of the excitational–vibrational configurations for which part of the excitation is concentrated on normal vibrations ineffective in the classical case (by configuration here is to be understood some distribution of the vibrational quantum numbers over the non-interacting quantum oscillators, each of which represents a normal vibration of the molecule). Moreover, it should be taken into account, of course, that, for degenerate vibrations, s in formulae (24.23b) and (24.52) must be reduced, in accordance with the degeneracy, in comparison with the actual number of degrees of freedom.

We will now consider the linear triatomic molecule, where we will assume that the frequencies of all four of its normal vibrations (two valence and one doubly-degenerate deformation) are close to one another. If the activated state is not a state with axial symmetry then both the valence and the deformation vibrations make contributions to the mean lifetime. Since, however, the deformation vibration is doubly degenerate, we should assume $s = 3$ and not $s = 4$. If the activated state corresponds to axial symmetry, then the deformation vibrations do not interact with the reaction coordinate and $s = 3$ is reduced to $s = 2$. This implies that only configurations of the type $\sigma_1^{n_1}\sigma_2^{n_2}\pi^0$ take part in the expansion of the wavefunction of the activated molecule in terms of the wavefunctions of the vibrationally-excited active molecule. Configurations of the type $\sigma_1^{n_1}\sigma_2^{n_2}\pi^{n_3}$ ($n_3 \neq 0$) do not take part in the expansion. Here n_1, n_2, and n_3 denote vibrational quantum numbers of the two valence and one (degenerate) deformation vibrations, and the symbols σ_1, σ_2, and π denote the corresponding vibrational functions for $n = 1$. By analogy with the electronic configurations of linear molecules, the vibrational configuration $\sigma_1^{n_1}\sigma_2^{n_2}\pi^{n_3}$ can be considered as a state arising from the filling up, by phonons, of the 'one-phonon orbits'. The occupation numbers n_i are connected with the total number of vibrational quanta by the relation $n_1 + n_2 + n_3 = n$.

Thus for Slater's model, the effective number of degrees of freedom s_{eff} may be smaller than the actual number and, therefore, in the analysis of experiments s_{eff} may be interpreted in some sense as an empirical parameter satisfying the condition $s_{eff} < s$.

As regards Kassel's model, the interpretation of s as an empirical parameter leads to known difficulties. In fact, for $s_{eff} < s$, it has to be assumed that the anharmonic interaction between groups of certain normal vibrations is much smaller than the anharmonic coupling within these groups. What is more, the interaction of the reaction coordinate with certain normal vibrations (in principle, different from zero even for the harmonic model) must be less than the anharmonic interaction between groups of different normal vibrations. In these arguments, however, the anharmonicity is already considered from the point of view of the dynamics within the molecular energy exchange whereas, within the bounds of Kassel's theory, only the role of an interaction, which ensures the statistical nature of the energy redistribution but whose magnitude is not specified, is assigned to the anharmonicity.

In the formal interpretation of s_{eff} as a semi-empirical parameter it has sometimes turned out, in fact, that the effective number of degrees of freedom is only a fraction of the total number of vibrational degrees of freedom [31], [449].

However, this is connected with the incorrect application of classical theory to the interpretation of experiments, for which the condition of applicability of this theory ($\hbar\omega \ll kT$ for thermal decomposition and $E - E_0 \gg \hbar\omega s$ for the decay of ions in mass spectrometric conditions) is not fulfilled. Staying consistently within the bounds of Kassel's model, it is impossible to separate out definite vibrationally-excited configurations which may have been excluded from consideration of the intramolecular energy exchange, since such an exclusion would contradict the essence of the model itself. However, if the activated state possesses definite symmetry then the total wavefunctions of the vibrationally-excited molecule, which take account of the anharmonic coupling between oscillators but correspond to different types of symmetry, cannot be mixed, as a result of intramolecular interactions. Formally, this effect may appear as a reduction in the effective number of degrees of freedom in comparison with the true number.

For clarification of what has been said, we turn to the above-mentioned model of a linear triatomic molecule. If the activated state possesses axial symmetry then the total wavefunction of the active and activated molecules will correspond to a definite value of the component of the vibrational momentum along the axis of symmetry, so that the vibrational terms are classified by quantum numbers Σ, Π, ... This implies that, in the most general form, the wavefunction of the activated molecule must differ in the functions of all the configurations of the harmonic approximation; however, from every configuration, multiply degenerate in the general case, only those terms will take part in this expansion which correspond to the same component of the vibrational momentum. For the lifetime $\langle \tau^* \rangle$ of the lowest vibrational state of the active molecule $\langle \tau^* \rangle = (v^\star) n_0^2/4$ is obtained [386], where $n_0 = E_0/\hbar\omega$. The Slater theory for the same model gives $\langle \tau^* \rangle = (v^\star)^{-1}\sqrt{(\pi n_0)}$. If it is assumed that the energy can be redistributed in a statistical manner between all the states of the active molecule, and not just between states with the same momentum, then $\langle \tau^* \rangle = (v^\star)^{-1} n_0^3/3$ is obtained. Hence it is evident that the absence of interaction between terms of different symmetry can be reduced formally to a decrease in the effective number of oscillators in comparison with the actual number; however, in spite of this, it is not possible to point to a definite type of vibration which would not interact with the reaction coordinate.†

We note finally that the assumption of a statistical distribution of the energy, according to the terms of different symmetry, implicitly assumes the existence of a sufficiently strong vibrational–rotational interaction, since the vibrational momentum loses its meaning as a quantum number only under this condition. The analogous problem of separating out the inactive degrees of freedom in the classical case is discussed in the paper [495].

A calculation of the effect of anharmonicity on the rate of intramolecular redistribution of energy as a rule requires a study of the equations of motion of a closed system, such as a molecule, in the interval of time between collisions. However, to a certain approximation, all the effects of anharmonicity can be reduced to the random exchange of quanta between the harmonic oscillators representing the molecule. In connection with this, an additional parameter, the mean exchange time, is introduced into the theory.

Then the probability that n quanta will be concentrated on a single (active) oscillator, which represents the dissociating bond, will be determined not by expression (25.14) but by the solution of a kinetic equation, describing the intramolecular redistribution of energy and the spontaneous decomposition (or isomerization) of the molecule. In this case, the fraction of molecules which are active P^n will be some complicated function of time. Then, by analogy with (23.12), we can write

$$\kappa = Z^* \sum_{n=n_0}^{\infty} \int \left(-\frac{dP^n}{dt} \right) \exp(-Z^* t) \, dt, \qquad (25.20)$$

† For example, all the degenerate configurations $\sigma_1^{n_1}\sigma_2^{n_2}\pi^{n_3}$ (for even n_3) contribute to the total vibrational function of the Σ-state in the harmonic approximation. In particular, in the CO_2 molecule, the first and second vibrational terms of symmetry Σ_g^+, whose splitting is due to the anharmonicity, are represented by a superposition of the configurations $\sigma_g^1\sigma_u^0\pi_u^0$ and $\sigma_g^0\sigma_u^0\pi_u^2$.

where the integral under the summation sign gives the probability of spontaneous decomposition, averaged over the distribution of lifetimes of the active molecule, and the sum over n gives the rate of equilibrium activation.

Wilson [556] and Brauner and Wilson [169] studied a simple model of the intramolecular redistribution of energy, assuming that the interaction of the active oscillator with the remaining oscillators of the molecule is proportional to its vibrational energy. In this case, for the fraction of active molecules P_m^n, whose total energy E_n equals $\hbar\omega n$ of which $E_m = \hbar\omega m$ is concentrated in the active oscillator, the following kinetic equations are obtained [169]:

$$\frac{dP_m^n}{dt} = \gamma m(n-m-1)P_{m-1}^n + \gamma(m+1)(n+s-m-2)P_{m+1}^n$$
$$- \gamma[(m+1)(n-m) + m(n+s-m-1)]P_m^n - v^\star U(m-n_0)P_m^n. \quad (25.21)$$

The first three terms on the right-hand side of this system of n equations describe transitions between adjacent levels $m \to m+1$ of the active oscillator, induced by the interaction with the other oscillators of the molecule; the parameter γ characterizes the rate of these transitions. The last term takes account of spontaneous decomposition which takes place from all levels $m > n_0$ of an active oscillator with frequency $v^\star$; the function $U(m-n_0)$ equals unity for $m \geq n_0$ and zero for $m < n_0$. In the limiting case $n, s \to \infty$, $\gamma \to 0$, $v^\star/\gamma \gg 1$, $n\gamma = \text{constant}$, the system (25.21) goes over into a system of equations of the type (38.6), describing the relaxation and dissociation of a 'truncated' quantum oscillator in a heat reservoir. This is quite natural, since a sufficiently complex molecule in certain cases can be considered as a heat reservoir with regard to the degrees of freedom describing a chemical reaction. In contrast to (38.6), however, in (25.21) it is taken into account that the excitation of the active oscillator leads to a 'cooling' of the reservoir (the quadratic dependence of the transition probability $m \to m \pm 1$ on m, which effectively disappears for $n \gg m$ is connected precisely with this).

The function P^n in (25.21) is obtained from P_m^n by averaging over the various initial distributions of the active oscillator according to the levels after collision of the reacting molecule with a molecule of the reservoir. The function $P^n(t)$ can be written down, in the general case, in the form of a sum of terms decreasing exponentially with time, so that, after the integration of (25.20), an expression is obtained for κ analogous to (23.44) provided the partial rates κ_i are assumed dependent on $Z^\star$. The existence of such a dependence, however, in fact implies that the κ_i cannot be interpreted as rate constants of spontaneous molecular decomposition. Only in the limiting case of very rapid intramolecular redistribution of energy, when $\gamma \gg v^\star$, are the expressions (25.15) and (25.18) obtained for κ_i and κ. In spite of this, however, a violation of the initial assumption of the small effect of anharmonicity on the structure of the vibrational spectrum of the molecule is possible. At present, it is not clear which physical model of the kinetics of the *statistical* redistribution of energy is described by the system of equations (25.21). It is interesting to note that (25.20) describes a case, in some sense the opposite of the quantum Slater model, for which the redistribution of energy occurs as the result of *dynamical* beats in the system of oscillators.

For the model under consideration, Brauner and Wilson [169] found that the dimensionless rate constant κ/κ_∞ in the intermediate pressure range was less than the rate constant for Kassel's model, and a reduction in the rate of intramolecular energy exchange led to a widening of the transition region.

26. The effect of anharmonicity on the reaction rate

From formula (25.10), it is not difficult to see that, for an appropriate choice of the effective number of collisions Z^* for $b \gg s$, the expressions for the reduced rate constant κ/κ_∞ agree in the theories of Kassel and Slater, if the number of vibrational degrees of freedom of the activated molecule in Slater's theory is taken twice as large as the number of degrees of freedom of the activated molecules in Kassel's theory. The physical reason for the difference between these theories is clear. Weak anharmonicity, taken into account implicitly in Kassel's theory, substantially increases the phase volume of the active molecules and, at the same time, leads to an obliteration of the difference in effectiveness of the different degrees of freedom. In particular, this implies that if, in the interpretation of the experimental data, Kassel's formula is used with an effective number of oscillators less than the actual number of vibrational degrees of freedom, then it has tacitly been assumed that the interaction of the reaction coordinate with the ineffective degrees of freedom is less than the interaction between the normal coordinates due to anharmonicity. From this the difficulties encountered in the attempt to separate the active and inactive degrees of freedom can be perceived. The mean lifetime of the active molecule $\langle \tau^* \rangle$ in Slater's theory (expression (24.23b)) is less than in Kassel's theory (formula (25.5)). This is in agreement with the physical meaning of $\langle \tau^* \rangle$, as the lifetime of the unstable complex AB* (an active molecule) formed by the bimolecular association of the products of A + B. In the absence of anharmonicity, the representative point goes through the region corresponding to the active molecule along a trajectory of purely harmonic motion. For the small anharmonicity implicitly assumed in Kassel's theory, the representative point, apart from motion along these trajectories, in addition performs slow transitions between them, falling for some time into that portion of phase space from which there is no exit into the region of reaction products along trajectories of harmonic motion.

Very rough limits of applicability of the strictly-harmonic models may be estimated in the following way. Anharmonicity affects the exchange of energy between normal vibrations most strongly when the overtones (or fundamental tone) of one type of vibration are close to the overtones of another type. In the case of exact resonance (of which the Fermi resonance in triatomic molecules is an example) the frequency of the energy exchange can be identified with the displacement δv in the frequency of the vibrations of the nuclei, caused by the anharmonic terms in the potential. Since the constants of anharmonicity in polyatomic molecules are, as a rule, of the same order as in diatomic molecules, the magnitude of δv can be estimated, for example, from the Morse oscillator model. For the Morse oscillator, the dependence of the frequency v on the vibrational energy E has the form

$$v = v_0 \left(\frac{D-E}{D}\right)^{\frac{1}{2}},$$

where v_0 is the frequency of small vibrations; D is the dissociation energy. For a change of frequency δv in comparison with v_0 we obtain

$$\delta v = v_0 - v \approx \tfrac{1}{2} v_0 \frac{E}{D}. \tag{26.1}$$

In order that the harmonic model apply to a reacting molecule, it is necessary that the mean lifetime of the active molecule, calculated in the harmonic approximation, be appreciably smaller than the characteristic time τ_{anh} of the energy exchange between the normal vibrations, i.e.

$$\langle \tau^* \rangle \ll \tau_{anh}. \tag{26.2}$$

If $1/\delta v$ from (26.1) is substituted in place of τ_{anh} and as E, the mean energy of the active molecule falling on a single bond $\bar{E} \sim E_0/s$ is introduced, then we obtain a sufficient condition for the applicability of the harmonic model, since $1/\delta v$ gives a lower bound to τ_{anh}.

Although such an estimate for τ_{anh} leads to an unnecessarily strict condition for the applicability of the harmonic model, it nevertheless reproduces an important qualitative effect correctly; as the number of degrees of freedom increases, so the harmonic model becomes less applicable to describe the *dynamics* of intramolecular energy exchange over the *long* intervals of time important in the calculation of the rate of decomposition (or isomerization). On the other hand, with an increase in the number of degrees of freedom, the mean value of the energy of the active molecule falling on a single degree of freedom, and in consequence the anharmonic corrections to the energy levels are reduced, so that the harmonic model becomes more applicable to the description of the *statistical* redistribution of energy over the different degrees of freedom. It is evident, however, that the statistical description is suitable only in those cases when a sufficiently long time elapses between the act of excitation of the molecule AB and its decomposition. For processes in which comparatively short time intervals are significant, the harmonic approximation may be completely satisfactory.

The separation of processes into two types, characterized by long and short time intervals, can be illustrated by the example of the recombination of $A + B$. During the time of collision of $A + B$ ($\sim 10^{-13}$ s) there takes place the conversion of the relative energy of $A + B$ into the vibrational energy of AB. This energy, concentrated initially in the newly-forming bond, gradually goes over into the other degrees of freedom. In quantum mechanics such a process is described by the spreading of a wavepacket during a time of the order of the inverse of the difference between the frequencies of the normal vibrations $1/|\omega_i - \omega_\kappa|$ (qualitatively, this is clear from (24.49)). After this the second stage begins, when the mean energy of each normal vibration does not change and the concentration of a large portion of the energy in a given degree of freedom can be described by the mean time $\langle \tau^* \rangle$ for the return of the system to the initial state. Although the dependence of $\langle \tau^* \rangle$ on the magnitude of the anharmonic interaction should be calculated in accordance with the dynamical or statistical model (the Slater or Kassel model), the rate of the first stage can usually be found in the harmonic approximation. And what is more, since, during the first stage, the energy does not have time to be redistributed over the whole molecule, the actual number of degrees of freedom does not play an essential role. In this sense, the problem of the probability of the formation of the excited molecule AB* in the collision of A with a polyatomic fragment of B is close to the problem of the attachment of A to the surface of a solid body. Referring the reader to the original papers on this problem [39], [43], [44], [363], we will note only one simple model, permitting an exact solution.

An atom A is incident on an endless chain B–B ... , where the interaction of A with the outer atom B is described by the potential of a truncated harmonic oscillator. If the frequency of vibration in the well of the truncated oscillator is greater than the maximum frequency of the normal vibrations of the chain, then part of the energy liberated on recombination remains localized in the bond formed (there arises a so-called localized vibration). The subsequent dissipation of the localized energy takes place only as a result of the anharmonic interaction of the localized vibration with the normal vibrations of the lattice.

As an illustration of this discussion, we will cite the results of a calculation of the temporal change of the reaction coordinate q_r of a system of an atom + an infinite one-dimensional chain of identical atoms. The force constants b_{ij} of the paired harmonic potential between neighbouring atoms of the chain are assumed all equal to b, but the force constant b_0 of the interaction A–B of the potential ABCD (see Fig. 22) differs from b. For simplicity, we will assume that the mass of atom A is the same as the mass of B. An exact

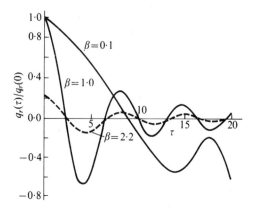

Fig. 23. Time dependence of the reaction coordinate for the process of 'attachment' of an atom to a linear harmonic chain (Mazhuga [44]).

solution of the problem enables us to obtain the time dependence of the distance $q_r = R_{AB}$ for given initial values of $\dot{q}_r(0)$ and $q_r(0)$ on the assumption that the atoms of the chain are stationary up to the collision. If the depth of the well appreciably exceeds the energy of atom A (as, in fact, occurs in the recombination of a radical), then the initial energy of A can be neglected and we may assume $\dot{q}_i(0) = 0$. The dependence of $q_r(t)$, determined under these assumptions, is shown in Fig. 23 for different values of the parameter $\beta = b_0/b$. The dimensionless time $\tau = 2(b/M)^{\frac{1}{2}}t$ is plotted along the abscissa and the ratio of the instantaneous value of the reaction coordinate $q_r(\tau)$ to the initial value $q_r(0)$, which coincides with the critical value q_r^*, is plotted along the ordinate. The continuous curve $\beta = 1$ illustrates the general nature of the damping-out of the vibrations of the bond AB as a result of the harmonic interaction with the chain. On reducing the coupling constant ($\beta < 1$), the rate of damping the vibrations $q_r(\tau)$ is reduced, and the asymptotic value of the amplitude of vibrations q_r tends to zero (curve $\beta = 0.1$). In a more detailed study of the problem of the vibrations, it becomes clear that the tending of the amplitude to zero (or to a small value E_0/s for finite s) is connected with the fact that, in the vibrational spectrum of an isolated chain, there exist frequencies exactly coinciding with the frequency of the isolated bond AB. This resonance is responsible for the normal vibrations of the system (chain + atom) being completely delocalized. This, in turn, implies that the amplitude factors μ_k are approximately of the same order, and from the normalization condition ($\Sigma_k \mu_k^2 = 1$), it is clear that $\mu_k \sim s^{-\frac{1}{2}}$.

On increasing the coupling constant ($\beta > 1$), the frequency of the isolated bond AB finally turns out to be higher than the maximum frequency of the isolated chain. In this case, the damping rate of the vibration $q_r(\tau)$ remains approximately the same as for $\beta = 1$; however, the asymptotic value of the amplitude $q_r^\infty(\tau)$ proves to be finite. The contribution δq_r of the normal vibrations of the chain to the amplitude of the reaction coordinate $q_r(\tau) = q_r^\infty(\tau) + \delta q_r$ for $\beta = 2 \cdot 2$ is shown by the dashed curve in Fig. 23. It is clear that only the contribution δq_r tends to zero. The non-zero value of $q_r^\infty(\tau)$ is connected with the fact that the amplitude factors differ sharply in magnitude; only one of them will be of order unity and all the remaining ones will decrease very rapidly with increase in t. To these two qualitatively different types of vibration in Slater's theory, there correspond different conditions satisfied by the amplitude factors μ_k. If all the μ_k satisfy the condition mentioned in connection with (24.21), then the vibrations should be considered delocalized, and the mean time for the coordinate q_r to reach a certain value $q_r' < q_r^*$ (or energy $E' < E_0$) can be calculated by formula (24.23b). If only one normal vibration satisfies the condition $\mu_i \gg 1/4\pi\beta E'$ for some E', then the total number of degrees of freedom s, introduced in (24.23b), must be reduced to $s-1$. Then the value obtained for the mean time will be of order $(v^\star)^{-1}$; this just implies that the energy E' is localized on the coordinate q_r, which is very close to the coordinate of a single vibration.

We will now consider the effect of anharmonicity on the intramolecular energy exchange rate. Numerical analysis of the dynamics of the vibrations of anharmonic chains shows that the rate of redistribution of the energy between the normal vibrations depends decisively on the structure of the vibrational spectrum of the same chain in the absence of anharmonicity [228], [229]. In particular in the papers [27], [228], [229], [283], the role of resonance of the overtones (Fermi-type resonance) on the redistribution of energy is considered, and a qualitative interpretation is given of the results of a calculation, first carried out by Fermi and coworkers [225], for a linear anharmonic chain.

There are at present several papers which investigate the effect of anharmonicity on the dynamical behaviour of atoms in molecules. Some of them are devoted to the solution of the vibrational problem. In particular, Thiele and Wilson [517] studied the model of two dynamically-coupled Morse oscillators. They found that when the energy of the symmetric vibrations exceeds the threshold value $0 \cdot 6 \, D$, the small antisymmetric vibrations do not appear stable: their amplitude starts to grow rapidly at the expense of a transfer of energy from the symmetric vibrations to the antisymmetric. Another model, studied in the work of Tredgold [529], enables the effect of anharmonicity on the energy exchange between two oscillators with the same frequencies to be traced analytically.

The investigation of the effect of anharmonicity on the spontaneous decomposition rate, of great interest in the theory of unimolecular reactions, is carried out in a series of papers [180], [181], [279], [519]. The determination of the distribution function $h = h(\tau^*)$ of the active molecules in terms of lifetimes must be the principal result of such a study. Below, we will consider a model of the N$_2$O molecule, founded on the results of a calculation by Bunker [181]. We will return to this molecule in connection with nonadiabatic reactions and a calculation of the thermal decomposition rate.

To simplify the problem we will assume that the atoms of the triatomic molecule ABC move only in a fixed plane. To specify the relative positions of the atoms three coordinates are sufficient, either the internuclear distances $r_{AB} = r_1$, $r_{BC} = r_2$, and the angle $\alpha = <\mathrm{ABC}$, or the distances between the nuclei A and B, the distance R between C and the centre of mass of A and B, and the angle γ between the vectors $\mathbf{r}$ and $\mathbf{R}$. In addition, there is one other coordinate ψ describing the rotation of the whole system in the plane, but it is clear that the potential does not depend on ψ. The kinetic energy, expressed in terms of the coordinates r, R, γ, and ψ and their conjugate momenta, has the form (7.15), where it should be assumed that $P_\psi = J$ (restriction to plane motion) and

$$M = \frac{m_A m_B}{m_A + m_B} \quad \text{and} \quad \mu = \frac{m_C(m_A + m_B)}{(m_A + m_B + m_C)}.$$

In studying the problem of small vibrations, r_1, r_2, and α should be taken equal to their equilibrium values $r_{1,\mathrm{eq}}$, $r_{2,\mathrm{eq}}$, and α_{eq}, and P_ψ be taken equal to zero.

A reasonable approximation to the potential energy can be obtained if $U(r_1, r_2, \alpha)$ is represented in the form of a sum of two Morse potentials and a harmonic function in $(\alpha - \alpha_{\mathrm{eq}})$:

$$U = U_1 + U_2 + U_\alpha = D_1\{1 - \exp[-\gamma_1(r_1 - r_{1,\mathrm{eq}})]\}^2$$
$$+ D_2\{1 - \exp[-\gamma_2(r_2 - r_{2,\mathrm{eq}})]\}^2 + \frac{k_\alpha}{2}(\alpha - \alpha_{\mathrm{eq}})^2. \tag{26.3}$$

The parameters of this potential D_i, γ_i, k_α, $r_{i,\mathrm{eq}}$ are calculated to conform to values of the energies of rupture of the bonds AB, BC, the frequencies of the valency and deformation vibrations of the molecule, and the equilibrium configuration parameters.

The function $h = h(\tau^*)$ is found by means of a statistical analysis of the trajectories, accomplished on computers with random prescribed initial conditions compatible with a fixed specified total energy. For each trajectory, the time τ is determined from the initial moment of motion up to the intersection of the critical surface, which in terms of the coordinates r_1, r_2, α is defined by the condition $r_2 = r_2^*$ or $U(r_1^*) = E_0$. Next, the number of molecules characterized by the given value of τ is found.

The parameters of the model in its application to the N$_2$O molecule are shown in the first row of Table 5.

Instead of the parameters γ_1, γ_2, and k_α, the frequencies of normal vibrations agreeing with the frequencies of the nitrous oxide molecule, which

TABLE 5
Parameters of models describing the unimolecular decomposition of a triatomic molecule

Kinematic constants						Dynamical constants				
m_A m_B m_C	$r_{1,eq}$ (Å)	$r_{2,eq}$ (Å)	χ_{eq}	D_1 (kcal mol^{-1})	D_2 (kcal mol^{-1})	ω_1(cm^{-1})	ω_2(cm^{-1})	ω_3(cm^{-1})	E_0 (kcal mol^{-1})
14 14 16	1·15	1·23	π	114	84·7	2228	1266	591	60
1 24 24	1·15	1·23	π	114	84·7	5086	1377	1028	60

unambiguously determine γ_i and k_α, are indicated here. The fact that the dissociation energy E_0 differs from the energy of rupture of the bond is due to the nonadiabatic nature of dissociation. This problem is discussed in a following section.

Within the limits of Kassel's theory, the decomposition rate constant of such a model is found from formula (25.4), in which we take $s = 3$†. As regards the frequency factor v^*, it may be calculated either by formula (24.16) or by (24.17). For the harmonic model, (24.16) and (24.17) give the same results: $\ln v^*_{harm} = 13 \cdot 78$ (where v is expressed in s^{-1}). However, the first expression can be used for the anharmonic model. In this case the v^*_i are defined as the frequencies of small (harmonic) vibrations of the activated complex whose potential energy surface takes into account the contribution of anharmonicity arising from the great extension of the bond $r_2 = r_2^*$. Such an analysis gives $\ln v^*_{anh} = 13 \cdot 75$.

We will now return to the result of the numerical investigation of the dynamics of the N_2O system.

1. The distribution function $h(\tau^*)$ proved to be exponential, which confirms the assumption about the random nature of the distribution of energy over the normal vibrations.
2. The rotation does not contribute to the spontaneous decomposition rate, in the sense that varying the initial momentum P_ψ does not change $\kappa(E)$.
3. The rate constant $\kappa(E)$ can be represented in the form:

$$\kappa(E) = v_{eff} \left(\frac{E - E_0}{E} \right)^{s_{eff} - 1},$$

with the parameter $s_{eff} = 3 \cdot 19 \pm 0 \cdot 32$ and the frequency factor $\ln v_{eff} = 13 \cdot 78 \pm 0 \cdot 24$.

---

† The actual number of vibrational degrees of freedom of a linear triatomic molecule is equal to 4 (two valence + one doubly-degenerate deformation vibrations); however, for the planar model under consideration, only the deformation vibrations lying in a fixed plane are allowed.

From these deductions, it is evident firstly that, within the limits of error of the calculation, the effective number of degrees of freedom agrees with the actual number $s = 3$. In this sense, the dynamical calculation confirms the assumption of the statistical theory. Second, the effective frequency factor is close to v_{harm} and to v_{anh}. Such close agreement is undoubtedly due to the comparatively small contribution of the anharmonicity to v_{eff} and to the level density $\rho(E)$. In the given model, anharmonicity is comparatively small, since the dissociation proceeds, in fact, until the vibrational energy becomes comparable to D_2. If the condition $E_0 \sim D_2$ were demanded for the same model, then v_{harm} and v_{anh} would prove to be four times larger than v_{eff}; however, the actual reduction of v_{eff} in comparison with v^* is compensated to a substantial degree by the increase in statistical weight of the activated molecule, so that Kassel's model again proves to be applicable, though with less accuracy.

Returning to Table 5, it can be seen that the overtone of the vibration ω_3 is close to the fundamental tone ω_2, but the overtone of ω_2 is close to the fundamental tone ω_1. This impels us to assume that the statistical redistribution of the energy is basically due to the effect of Fermi resonance. In fact, investigation of the model in which the fundamental frequencies are markedly different (the second line in Table 5) but the anharmonicity is approximately the same as for the N_2O molecule shows [181] that the distribution function $h(\tau^*)$ is not exponential. This fact may explain the indication that the distribution of the lifetimes is determined not only by the intramolecular statistical redistribution of energy but also by the kinetics of this distribution. If, for example, the distribution function is approximated by the sum of two exponentials with very different exponents then, for such molecules, the dependence of the rate constant on the pressure must be step-shaped; on increasing the pressure, at first, degrees of freedom with a long redistribution time become ineffective, and the curve $\kappa = \kappa(Z^*)$ reaches the first asymptote. On this asymptote, the parameters are such that the intermolecular energy exchange rate is higher than the intramolecular exchange rate for one subsystem of degrees of freedom but lower for the other.

For an even greater increase in Z^*, the intermolecular exchange rate begins to exceed the intramolecular exchange rate for any group of degrees of freedom, and the function $\kappa = \kappa(Z^*)$ reaches a true asymptote of the first order.

From physical considerations, and also from the general formula (23.14), it follows that, in the distribution function $h(\tau^*)$ of the active molecules for the lifetimes, as the pressure increases shorter and shorter times begin to play a main role. In connection with this, the question arises of the lower bound of τ_{eq} for which the concept of the statistical nature of the redistribution of the intramolecular energy still remains valid. This lower limit is determined not only by the properties of the active molecule as an isolated system, but by the intermolecular energy exchange mechanism. If, as a result of collision, the representative point of the molecule fills up phase space on a constant energy surface more-or-less uniformly, then this aids the rapid estabishment of the statistical equilibrium distribution of energy to an appreciable degree. If mainly separated portions of phase space are filled up on activation, then the time for the establishment of the statistical distribution may be rather large. In the latter case, it can be shown that, on increase in pressure, the rate constant of a unimolecular reaction at first increases and then, going through a maximum, decreases, tending asymptotically as $Z^* \to \infty$ to zero. A decrease in the rate, in the limit of high pressures, does not take place within the bounds of the mechanism of strong collisions for which, properly, expressions (23.20) and (23.45) are also valid.

On excitation, as a result of strong collisions the active molecules emerge as close as one pleases to the critical surface, and therefore the increasing (with increase in Z^*) deactivation rate of the molecules near the critical surface is completely compensated by the growing rate of appearance of molecules in this region of phase space. If, as a result of collisions, the active molecules emerge at some distance l from the critical surface, whose passage requires a time $\tau_{\min}$, which is independent of Z^*, then the increase in the deactivation rate obstructs the progress of the active molecule towards the critical surface, and the reaction rate constant falls. A similar situation arises on the impulsive excitation of vibrations in the nuclei, when, after collision, the nucleus of the molecule does not change its position but receives a comparatively large impulse.

The one-dimensional model considered by Baetzold and Wilson [130], [175] enables an analytical expression to be obtained for the decomposition rate constant, which displays a non-monotonic dependence on Z. The effect of the initial conditions on the distribution function $h(\tau^*)$ for polyatomic systems can be followed with the aid of computers, on the basis of an investigation of the dynamics of the intramolecular redistribution of the energy. Baetzold and Wilson [131], having considered harmonic and anharmonic models of triatomic and tetratomic systems, obtained a qualitative estimate for $\tau_{\min}$. For a system simulating the vibrations of a CO_2 molecule by a Morse potential for the CO bond it was found that the distribution function $h(\tau^*)$ for impulsive excitation agrees, in practise, with the distribution function for random excitation (the uniform filling-up of the hypersurfaces of total energy) after 15 periods of vibration of the nuclei. For the harmonic model the distribution functions preserved the 'memory' of the mechanism of excitation for a long time.

Information about the distribution function $h(\tau^*)$ for short lifetimes can be obtained experimentally by studying the reaction rates at high pressures. In addition, of course, it is desirable to have active molecules with a very non-uniform filling-up of phase space, since the deviation from the statistical redistribution is easier to detect in precisely this situation.

The butyl radicals, formed as a result of the chemical activation process

$$H + CH_3CH = CHCH_3 \rightarrow CH_3CH_2\dot{C}HCH_3^* \qquad (26.4)$$

and studied by Placzek, Rabinovitch, and Dorer [403], serve as an example of such a system.

The radicals, possessing energies of about 40 kcal mol^{-1}, which exceeds the energy of rupture of the C–C bond by 7 kcal mol^{-1}, can either decompose

$$CH_3CH_2\dot{C}HCH_3^* \rightarrow CH_3 + CH_3\dot{C}H = CH_2 \qquad (26.5)$$

or be stabilized on collision with molecules of the heat reservoir

$$CH_3CH_2\dot{C}HCH_3^* \rightarrow CH_3CH_2\dot{C}HCH_3. \qquad (26.6)$$

At a chemical activation rate proportional to the concentration of molecules in the reservoir (which is exactly achieved in the experiment under description), the observed decay rate in the high-pressure limit must not depend on Z if the distribution function $h(\tau^*)$ is to remain exponential for short lifetimes. Such a constancy was, in fact, observed for the above-mentioned reaction, for a variation in the pressure of the molecular hydrogen forming the heat reservoir from 0·036 atm to 115 atm [402]. In the high-pressure region, the time between successive collisions amounted to about 3×10^{-13} s, therefore, it can be affirmed that, at least for lifetimes τ^* exceeding 10^{-13} s in the active radical $CH_3CH_2\dot{C}HCH_3^*$, statistical equilibrium, characterized by an exponential distribution function for the lifetimes, is established at the expense of intramolecular redistribution of energy.

The large role of the anharmonicity found in these model calculations does not, however, imply that we cannot meet cases where Slater's theory would be applicable. As an example, we cite nonadiabatic reactions in electronically-excited states, where the value of $\langle \tau^* \rangle$ can be shown to be small in connection with the comparatively small degree of vibrational excitation, so that (26.2) will be satisfied.

The discussion presented above clearly points to the necessity of creating a theory which takes into account the effect of anharmonicity on the dynamics of the intramolecular motion. In view of its absence, Slater's theory remains the only theory simulating the detailed mechanism of a reaction.

The effect of anharmonicity on the decomposition rate is decreased on going to high pressures, where the relative role of the intramolecular energy exchange mechanism becomes less important. Therefore the frequency factors v calculated within the framework of Slater's theory, characterizing the rate of unimolecular transformations at high pressures, change comparatively little on account of anharmonicity [243], [495].

In order to demonstrate the dependence of the pre-exponential factor on the choice of reaction coordinate, we present a calculation of the frequency factor for the thermal isomerization of cyclopropane and propylene (Table 6).

TABLE 6

Dependence of the frequency factor of the thermal isomerization rate constant of cyclopropane on the choice of the reaction coordinate

Reaction coordinates	$v \times 10^{-14}$ (s^{-1})	Symmetry factor
Distance between chemically non-bonded atoms H···C	4·50	12
Extension of C=C bond	0·92	3
Difference between distances H···C and C=C	4·20	12
Total change of distance on extension of C=C bond and torsional oscillations of CH$_2$	1·68	3

27. Nonadiabatic reactions

The classical and quantum theories of unimolecular reactions considered above refer to adiabatic reactions, where, to a sufficient approximation, it is possible to assume the existence of a definite potential, whose properties determine the motion (classical or quantized) of the representative points in phase space. However, there is always a whole set of such potentials or potential energy surfaces, motion of the nuclei along which corresponds to one of the possible adiabatic processes. The motion of the nuclei induces transitions between these surfaces, and the transitions take place mainly in a region where two surfaces approach one another or intersect. If the reaction path includes a transition between different potential energy surfaces at some point, then the reaction is called nonadiabatic. In accordance with this definition, all reactions which might, in principle, proceed by an infinitely-slow change of the coordinates of the nuclei should be referred to as adiabatic. In fact, however, in speaking of transitions between various electronic terms, very often terms based not only on the adiabatic approximation but also on a number of other approximations are considered. With such an imprecise definition of the concept of adiabatic electronic terms, it can turn out that transitions between them are caused not only by the dynamical interaction of the electrons and nuclei but also by static interactions of terms, which have not been taken into account in the calculation.

We will now consider, for example, the rupture of the N—O bond in the N$_2$O molecule. The qualitative path of the potential curves is shown in Fig. 24. As a result of dissociation, the transition goes from the $^1\Sigma$-state of the

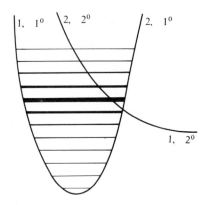

FIG. 24. Adiabatic terms (curves 1 and 2) and quasi-stationary vibrational states for a weak interaction between terms of the zero-order approximation (curves 1^0 and 2^0).

linear N_2O molecule into the $^1\Sigma$- and ^3P-states of the N_2 molecule and the O atom, where the change of multiplicity of the terms occurs close to the point of intersection of the terms of the stable $^1\Sigma$-state (term 1^0) and the unstable $^3\Pi$-state (term 2^0) of the N_2O molecule. This process is considered to be nonadiabatic, since, as a result of the reaction, the multiplicity is changed and consequently also the electronic quantum number of the molecule. However, if the spin–orbit coupling is taken into account then the electronic terms cannot be characterized by a definite value of the spin. In the case under consideration, the terms can be characterized by a definite value of the spin, to any degree of accuracy, only far from the point of quasi-intersection of the $^1\Sigma$ and $^3\Pi$ curves.

It is precisely in this case, when the spin–orbit coupling is small, that it can be considered as the cause of the change in the multiplicity of the term during the transition from the initial state of the system (equilibrium internuclear distances in $N_2O(^1\Sigma)$) to the final state (a free $O(^3P)$ atom and an $N_2(^1\Sigma)$ molecule).

For large spin–orbit coupling, the adiabatic electronic terms must be calculated taking it into account, of course, even in the zero-order approximation, as was done, for example, in the investigation of transitions between the fine-structure components of atoms (see §§ 19 and 20). For the N_2O molecule under consideration, the spin–orbit coupling could also have been included in the calculation of the adiabatic terms in the zero-order approximation. Then, the component of the total momentum Ω (spin and orbital) along the axis of the molecule would have been chosen as an electronic quantum number and the decomposition of N_2O could have been considered as a typical adiabatic process, taking place without change of electronic state and characterized by the quantum number $\Omega = 0$. This confusion in the definition of nonadiabatic transitions and reactions, now established in the literature, should always be borne in mind. (See the footnote on p. 193 in Kodrat'ev's book [31] in connection with this). In future, we will limit consideration to small spin–orbit coupling, and by adiabatic terms will understand those electronic states which are characterized by a definite value of the total electronic spin. In such cases, both the spin–orbit coupling and the dynamical interaction of the electrons with the nuclei must be considered as causing nonadiabatic transitions.

Usually the vibrational energy of an active molecule in decomposition reactions is so large that, from the point of view of the energy, transitions into excited electronic states of the molecule are quite possible. The corresponding potential energy surfaces can be classified into three types with regard to the reactions considered. First, it may turn out that an adiabatic motion along such a surface does not lead to a reaction in general. The AB(l) molecule in such an electronic state must be considered as active, nevertheless, since the probability of its decomposition is determined not only by the intramolecular energy exchange but also by the probability of a nonadiabatic transition to another surface l', leading to adiabatic decomposition. Secondly, the surface l may correspond to an adiabatic reaction path. If the lifetime of the molecule in this electronic state appreciably exceeds the characteristic time of vibration of the nuclei, then such a molecule must be considered as active. If the lifetime of the molecule in an excited electronic state is of the order of the period of vibration of the nuclei, then the corresponding potential energy surface may be classified as of the third type, and the molecule should be considered as activated.

The terms $^3\Pi$ of the isoelectronic molecules CO_2 and N_2O serve as examples of the last two types of surface. For CO_2 the electronic excited state $^3\Pi$ apparently corresponds to a sufficiently deep potential well [341], so that the lifetime of the molecule can be comparatively large. For N_2O, this state is characterized by a wavefunction, strongly antibonding with regard to the N–O bond, and thus dissociation of the N–O bond during the transition of the nitrous oxide molecule into this state must take place during a time of the order of 10^{-13}–10^{-14} s.

Nonadiabatic transitions may occur, of course, when the vibrational energies E are lower than the threshold value E_0. In this case, they represent part of the activation process of the molecule. If the rate of such transitions is higher than the rate of the vibrational excitation of the reacting molecules on collision then the transitions cannot limit the excitation rate and thus have practically no effect on the reaction rate. If, though, the nonadiabatic transition rate is lower than the vibrational excitation rate, then it is very possible that just the former will determine the reaction rate. The recent experiments on the decay of CO_2 [167], [367] and CS_2 [237] behind a shock wave [167], in which the effective dissociation energy found proved to be smaller than the true energy were interpreted from this point of view.†

The basic difficulty of the theory of nonadiabatic transitions in complex molecules is that the nonadiabatic transition has to be studied in a multi-dimensional space, and also that the effect of intermolecular collisions on the intramolecular transitions (transitions of the type occurring in the forced predissociation of a diatomic molecule) must be taken into account. Qualitative discussions of these prolems, in connection with the dissociation of diatomic molecules, have been held by Rice [440] and Nikitin [65]. The theory developed of nonadiabatic transitions in diatomic molecules can be applied to calculate the transition probability only in those cases where the complex motion of the representative point in the region of greatest nonadiabatic coupling can be reduced to one-dimensional motion.

The basic defect of the one-dimensional model is that it does not explicitly take into account the interaction of the reaction coordinate with the other degrees of freedom of the active molecule. This interaction can be traced only in sufficiently simple models. In particular, for a model of a strictly harmonic molecular vibrations, a nonadiabatic transition of the predissociation type along one of the bonds of the molecule, in fact, implies predissociation for all the normal vibrations. This, in turn, leads to the appearance of many closely-lying points of intersection (on the energy axis) [74], [386]. Nevertheless, it may be assumed that the one-dimensional model will give the correct order of magnitude of the nonadiabatic transition probability in a number of cases. Therefore we will consider in more detail certain results which can be obtained for the one-dimensional model.

Returning, for definiteness, to a process of the predissociation type (see Fig. 24) we will assume that the interaction between the terms in the zero-order approximation (intersecting terms) is so small that the probability of a transition from the state 1^0 into the state 2^0 may be calculated by perturbation theory. The possibility of the decomposition of the molecules from vibrational states lying higher than the asymptotic value of the term 2^0 becomes apparent in the broadening of the vibrational energy levels, which is taken into account by attributing to the state n the complex energy $E_n - i\Gamma_n/2$. For the one-dimensional model Γ_n is expressed in terms of the nonadiabatic transition probability by the following form [37]:

$$\Gamma_n = \hbar v_n P^0_{12}(E_n), \tag{27.1}$$

where v_n is the frequency of vibration in the state n; P^0_{12} is the transition probability during a single vibration (i.e. during a double transit of the region of intersection of the terms).

† It should be borne in mind that an underestimate of the dissociation energy is obtained in this case, if the rate constant is represented by an Arrhenius form with a temperature-independent pre-exponential factor.

In the presence of intersecting terms, the linear approximation is usually sufficient, therefore P_{12} can be calculated by the formulae of § 15. Here it is important to obtain just an estimate of the transition probability for typical unimolecular processes.

First, we will assume that the effect of the tunnel transitions can be neglected. Then the second factor in (15.36) can be taken equal to unity and, in the first factor, for the intersection of terms represented in Fig. 24, we will take $\Delta F \sim F \sim E_0 \alpha$, where E_0 is a quantity of the order of the binding energy in the molecule; $1/\alpha$ is the radius of action of the exchange forces $((0\cdot3-0\cdot5) \times 10^{-8}$ cm). The effective vibrational temperature T_{vib} for non-equilibrium processes may differ from the temperature of the translational motion of the molecules; however, for thermal reactions, it may be assumed that, approximately, $T_{\text{vib}} \approx T$. For allowed spin–orbit couplings causing transitions between those molecular states whose wavefunctions differ substantially close to the nuclei of the atoms, the magnitude of a must be of the same order as the spin–orbit coupling constant in the free atoms. A transition of just such a type occurs in CO_2, where the change of symmetry in the total wavefunction ($^1\Sigma \to {}^3\Pi$) occurs as the result of a change of symmetry in the hybridized atomic orbitals. For this case $a \approx 100-200$ cm^{-1}. Substituting these values in (15.36) and assuming $T \approx 1000$ K, we find that the nonadiabatic transition probability is of the order $10^{-1}-10^{-2}$, without taking account of tunnel corrections. If a spin–orbit transition, allowed by general selection rules, connects states whose wavefunctions differ only in the weights of the atomic orbitals entering into them, and the atomic orbitals themselves are not changed by the transition, then the electronic matrix element of the spin–orbit coupling may be appreciably smaller than the corresponding atomic constant. Thus, for transitions between the lower states of molecules with π-electrons (molecules of the benzene type), a turns out to be about 1 cm^{-1}. By comparison with the preceding example, this indicates a reduction in the transition probability of four orders of magnitude. The value obtained $P \approx 10^{-6}$ is usually employed to evaluate the probability of intercombinational non-radiative transitions in aromatic molecules [296]. A value of the same order of magnitude is adopted for the singlet–triplet transition probability in *cis-trans*-isomerization reactions [16], [340], which were mentioned in connection with the discussion of the Slater theory. Of course, the spin–orbit coupling in such molecules is very markedly changed by the replacement of light atoms by heavy atoms. This should be borne in mind in an evaluation of transition probabilities, in the general case.

We note finally that, for nonadiabatic transitions induced by the dynamical interaction between the nuclei and the electrons, the matrix element must depend on the velocity of the nuclei and increase with its increase. Thus for the above-mentioned transition between the Σ- and Π-terms of the linear molecule as a result of the interaction of the rotation of the nuclei with the electronic momentum, the matrix element of the interaction a is approximately equal to the mean energy of a rotational quantum.

We will now consider the tunnel corrections which are described by the second factor in (15.36). For atoms of mean atomic weight $\mu \approx 10$ and with the same values of the parameters $\Delta F, F, E_0, \alpha$, we find that the exponent of the exponential is greater than unity when $E_0/kT > 10$. Since for thermal decomposition reactions, in many cases, this inequality is indeed satisfied, the result obtained indicates that tunnel transitions play a role even when the energy barriers are comparatively wide.

The contribution of tunnel transitions can be particularly large when the energy barriers are narrow, when terms diverge very slowly (i.e. $\Delta F \ll F$) close to the points of intersection. Just such an almost-parallel path of the terms is very probable in polyatomic molecules, where the density of the electronically-excited states may be high. In this case, it might be possible to characterize the probability of non-radiative tunnel transitions, from excited states to the ground state, by certain parameters which describe the mean change in the form of the potential energy surfaces on electronic excitation (for example, the change in the mean curvature at the minimum or the mean square deviation of the minimum of the ground state or excited terms). Such an approach to the description of nonadiabatic transitions is set out in paper [280]. We finally note that the existing experimental data on the effect of isotopic substitution on the intensity of phosphorescence in complex molecules indicate the important role of tunnel effects in nonadiabatic transitions [446].

We will now turn to the case when the interaction V_{12} cannot be considered small. Then, as the zero-order approximation, it is expedient to chose adiabatic terms U_1 and U_2 and functions ϕ_1 and ϕ_2, considering the operator $\hat{C}$

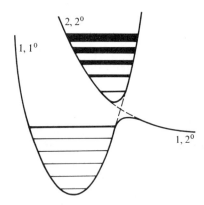

FIG. 25. Adiabatic terms (curves 1 and 2) and quasi-stationary vibrational states for a strong interaction between terms of the zero-order approximation (curves 1^0 and 2^0).

(see Chapter 1) as a perturbation giving rise to the transition. The path of the terms is shown in Fig. 25. The fundamental difference between this situation and the previous one (see Fig. 24) is that the quasi-stationary energy levels now refer not to the non-interacting terms 1^0 and 2^0 but to the adiabatic terms 1 and 2, in which the interaction V_{12} is taken into account in the adiabatic approximation. As a result of the nonadiabatic coupling of the terms 1 and 2, the vibrational states in the upper well will decay so that a width $\Gamma_n = \hbar v_n P_{12}(E_n)$ can be attributed to them. Since there is sense in characterizing the energy level E_n by the width Γ_n, provided it is small in comparison with the separation between neighbouring levels $E_n - E_{n-1}$, it is evident that, for the lower levels shown in Fig. 25, the condition $P_n \ll 1$ must be satisfied. On the basis of the general theory given in § 15, it is to be expected that this condition will be satisfied by the exponential smallness of P_n. The nonadiabatic coupling will increase as n increases and, in the end, the description of a quasi-stationary state in terms of widths will become inapplicable. In this limit, the transition must be made to the case considered earlier (see Fig. 24), since the increase in the kinetic energy leads to a reduction of the Massey parameter in the transition region, i.e. to the effective reduction of the interaction V_{12}.

The existence of quasi-stationary states in the upper potential well is exhibited in the emergence of resonances in the reflection coefficient Q for the motion of a particle along the lower terms. Investigation of this problem for a model of linear terms (Fig. 26) was carried out by Ovchinnikova [83], who gave the following expressions for the nonadiabatic transmission and

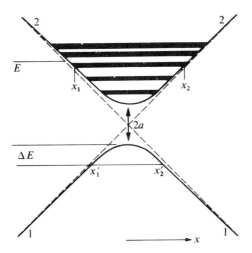

FIG. 26. Adiabatic terms and quasi-stationary vibrational states for the linear model.

reflection coefficients for a particle whose energy E exceeds the energy of the upper well's minimum.†

$$P = \frac{1+\cos J}{1+\lambda+\cos J}, \tag{27.2a}$$

$$Q = \frac{\lambda}{1+\lambda+\cos J}, \tag{27.2b}$$

where

$$\lambda = \frac{\exp\left(-\frac{4\pi a^2}{\Delta Fvh}\right)}{2\left[1-\exp\left(-\frac{2\pi a^2}{\Delta Fhv}\right)\right]} \tag{27.3a}$$

$$J(E) = \frac{2}{\hbar}\int_{x_1}^{x_2} \{2\mu[E-U_2(x)]\}^{\frac{1}{2}}\,dx. \tag{27.3b}$$

It is evident that $Q(E)$ reaches a maximum value $Q_{\max} = 1$ for $\cos J = -1$, i.e. for $J = 2\pi(n+\frac{1}{2})$. Since this condition determines the position of the quasi-classical energy level in the upper well in the adiabatic approximation, it is clear that the total reflection of particles is determined by its resonant capture in this well. The form of the resonance for small λ is found from (27.2b) by expanding $J(E)$ in a series in the neighbourhood of the level E_n:

$$Q = \frac{\Gamma_n^2/4}{(E-E_n)^2+\Gamma_n^2/4}, \tag{27.4}$$

where the level E_n and the width Γ_n are determined by the relations

$$J(E_n) = \frac{2}{\hbar}\int_{x_1}^{x_2}(2\mu)^{\frac{1}{2}}[E-U_2(x)]^{\frac{1}{2}}\,dx = 2\pi(n+\tfrac{1}{2}), \tag{27.5a}$$

$$\Gamma_n = \hbar v_n 2\exp\left(-\frac{2\pi a^2}{\Delta Fhv_n}\right), \qquad v_n = \left(\frac{E_n}{2\mu}\right)^{\frac{1}{2}}. \tag{27.5b}$$

As expected, the width of the level (or the decay probability) is described by the Landau–Zener formula, where Γ_n increases with increase in n. A description of the state in terms of the width becomes inapplicable when the quadratic expansion of $\cos J$ in (27.2b), which leads to (27.4), proves to be insufficiently exact. In such conditions it is necessary to revert to the general formulae (27.2). In particular, if the resonance structure of P and Q is not of interest, i.e. P and Q are averaged over an energy interval ΔE including

---

† Strictly speaking, the energy E must be so large that the vibrations in the upper well are quasi-classical.

several of the vibrational levels in the well, we arrive at the following expressions:

$$\bar{P} = \frac{1 - \exp(-2\pi\bar{\delta})}{1 - \tfrac{1}{2}\exp(-2\pi\bar{\delta})}, \qquad (27.6a)$$

$$\bar{Q} = \frac{\tfrac{1}{2}\exp(-2\pi\bar{\delta})}{1 - \tfrac{1}{2}\exp(-2\pi\bar{\delta})}, \qquad (27.6b)$$

$$\bar{\delta} = \frac{a^2}{\Delta F \hbar \bar{v}},$$

where $\bar{v}$ is some mean velocity corresponding to the energy $\bar{E}$ in the interval ΔE.

The mean value $\bar{P}$ agrees with the value which can be found by a simple summation of the currents, calculated with a probability P_{12} for 'up-down' transitions. Taking into account the multiple reflections of particles from the turning points at $x_1 = x$ snd $x = x_2$ and the nonadiabatic transitions with probability P_{12} at the point $x = 0$, we find

$$\bar{P} = (1-P_{12}) + P_{12}^2(1-P_{12})\sum_{n=0}^{\infty}(1-P_{12})^{2n} = \frac{1-P_{12}}{1-\tfrac{1}{2}P_{12}}. \qquad (27.7)$$

Substituting the Landau–Zener expression (15.18) for P_{12} in here, we obtain (27.6a). It is interesting to compare relation (27.7) with the formula for the transmission coefficients χ given in the book [126] (see [126], Chapter 16). If the probability ρ of the points reflection prior to reaching the activated state and the probability ρ of reflection on transition out of the activated state is introduced then, after a summation analogous to (27.7), χ may be written in the form

$$\chi = \frac{1-\rho}{1+\rho}, \qquad (27.8)$$

which differs from (27.7).

From (27.6) it is evident that, for $E \to \infty$, the transmission coefficient is reduced but the reflection coefficient increases. Such a dependence is directly contrary to what would be expected for the usual potential barrier. This result may be of interest in bimolecular chemical reactions, where the potential hump responsible for the activation energy is always accompanied by an upper potential well hanging over it.

The usual features of reflection and transmission can only be obtained for comparatively small energies and large adiabatic splitting, when the nonadiabatic coupling is unimportant.

In particular, if the peak of the potential hump is classically inaccessible, P differs from zero only as a result of the tunnel effect. For a single electronic term and an exponentially small transmission, P is given by expression (3.5) where the integral is taken over the classically-inaccessible region ab (see Fig. 1). A calculation of the effect of the nonadiabatic coupling with the upper term shows that the exponential factor in the expression for P remains unchanged, however, a factor B appears in front of the exponential, differing from unity and depending on the parameter δ:

$$P = B(\delta) \exp\left[-2 \int_{x_{1'}}^{x_2'} \frac{|p(x)|}{\hbar} dx\right]. \tag{27.9}$$

For $\delta \ll 1$ and $\delta \gg 1$ the following expressions are valid:

$$B(\delta) = \begin{cases} 2\pi\delta = \dfrac{\pi}{4} b|\varepsilon|^{-\frac{1}{2}}, & \delta \ll 1 \tag{27.10a} \\ 1, & \delta = \dfrac{b}{8|\varepsilon|^{\frac{1}{2}}} \gg 1 \tag{27.10b} \end{cases}$$

It is evident that, for $\delta \ll 1$, the nonadiabatic coupling is important and reduces the transmission coefficient in comparison with the coefficient for the adiabatic tunnel transmission (for which $B = 1$ always). Expression (27.10) enables the contribution of nonadiabatic effects to the tunnel-transmission coefficient to be estimated. As is to be expected, this contribution is small if in the region where the terms approach the condition for the adiabatic approximation $\Delta E \Delta l / \hbar v \gg 1$ is satisfied. Taking into account that, in the case under consideration, $\Delta E = 2a$ and $\Delta l \approx a/\Delta F$, it is easily seen that the condition $\Delta F \Delta l / \hbar v \gg 1$ is equivalent to the condition $\delta \gg 1$, for which $B = 1$. Thus expressions (3.5), (3.6), (27.2), and (27.10) enable the construction of a very general picture of the transmission of particles over (or under) a potential barrier with regard for the possible nonadiabatic effects.

It should be noted that a description of nonadiabatic intramolecular processes, in terms of the one-dimensional model considered, requires care. The diatomic molecule is the only object for which such a model is always valid, since the conservation of angular momentum, in fact, reduces the three-dimensional problem to the one-dimensional. For a polyatomic system, nonadiabatic processes must be considered jointly with the dynamical motion in adabatic regions. The process of electronic deactivation of M* in collision with X_2 with the formation of the complex MX_2^* (see § 19) may serve as an illustration. In point of fact, this complex in no way differs from an active molecule capable of spontaneous decomposition.

5

THE STATISTICAL THEORY OF REACTIONS

28. The basic assumptions of the statistical theory

THE basic idea of the statistical or so-called quasi-equilibrium theory is that a molecule capable of spontaneous chemical transformation (i.e. an active molecule, in accordance with the definition introduced in Chapter 4) is considered, during the time between collisions, as a closed system whose states are distributed with equal probability over hypersurfaces of constant total energy $H(p_i, q_i) = E$. Thus, for this system, the validity of the micro-canonical distribution is assumed.

It is evident that the statistical theory applies to the description of a reaction only when the reaction proceeds via an intermediate complex with a long lifetime τ^*. As regards an estimate of the lower bound of τ^* for which the basic assumption of the statistical theory is valid, that remains an open question. One of the main requirements imposed on the system is that the non-linear coupling between different degrees of freedom must ensure a sufficiently intensive energy exchange between them, although, in the calculation of the features of the energy spectrum, an approximation permitting a separation of the variables is often used. The assumption of the equiprobable distribution is invalid, for example, if the separate parts of the phase volume of the active molecule are divided by a sufficiently high potential barrier or if there are special selection rules (for example, on the type of symmetry) for the interaction of the different degrees of freedom. On the basis of comparison of theory with experiment and of the deductions of the statistical theory with the results of averaging the dynamical calculations for a number of model systems, it can be affirmed that such a complex is actually formed in uni-molecular decomposition and isomerization and also in ion–molecular reactions. However, even in those reactions for molecules arising as a result of some chemical process or on electron or ion collisions in conditions of a markedly non-equilibrium initial distribution, the appearance of correlations between the lifetime of the active molecule τ^* and the initial state in which it proves to be after excitation is possible.

The existence of other integrals of motion (apart from the total energy) hinders the uniform 'spreading' of the distribution function $f(p_i, q_i)$ over the surface $H = E$, which can lead to a substantial reduction in the lifetime τ^* of the active molecule in comparison with the lifetime obtained for an equiprobable distribution. As an example, the harmonic model of Slater can be cited, for which such a situation actually occurs as a result of the conservation of the energy of each normal vibration.

Another example is a polarization complex (see § 31), for which the law of conservation of the total angular momentum essentially restricts the dimension of the 'area' accessible to the system on the constant total-energy hypersurface.

In connection with the assumption about the 'spreading' of the distribution function in the statistical theory, a separation of the degrees of freedom into active, inactive, and adiabatic is frequently introduced. To clarify this separation, we will turn to the Hamiltonian H of an isolated polyatomic molecule. Denoting by q_i the internal coordinates describing the vibrations of the nuclei, the internal rotation, and the rotation of the molecule as a whole, we will represent H in the form

$$H = \sum_i H_i(q_i) + V(q_1, \ldots, q_{3N-3}), \qquad (28.1)$$

where the $H_i(q_i)$ are the Hamiltonians of the separate degrees of freedom and $V(q_1, \ldots, q_{3N-3})$ is the interaction between them.

If the interaction V is neglected, then the total energy E of the system, which is an integral of the motion, can be written in the form of a sum of energy contributions ε_λ, each of which is conserved as a result of the separation of the Hamiltonian into a sum of independent terms

$$E = \sum \varepsilon_\lambda. \qquad (28.2)$$

If now the reaction condition is formulated in the form of a requirement that the representative point intersect a certain critical surface, it is possible to show that some of the coordinates (for example, $q_{s+r_i}, \ldots, q_{3N-3}$) do not, in general, come into the equation of this surface. Since in the approximation (28.2) the energy corresponding to the motion with respect to these coordinates is conserved, these coordinates generally make no contribution to the motion of the representative point along the trajectories leading to a reaction, therefore they are referred to as inactive degrees of freedom. Those of the normal vibrations, in Slater's model, for which the amplitude factor μ_i equals zero or is very small ($\mu_i \ll 1/4\pi(E-E_0)$) are examples of inactive degrees of freedom, as is also the rotation of the molecule as a whole.

If the interaction $V(q_1, \ldots, q_{3N-3})$ is taken into account then the energy E cannot be represented by the form (28.2).

However, it can be shown that the molecular system possesses, apart from the total energy, certain other integrals of motion, which we will denote by J. Taking advantage of this, we will eliminate the corresponding variables from (28.1), and write the effective Hamiltonian as

$$H_J(q_1, \ldots, q_s) = \sum H_{J,i}(q_i) + V_J(q_1, \ldots, q_s). \qquad (28.3)$$

Now the intramolecular exchange of energy is limited to the system of s degrees of freedom remaining from the $s+r$ degrees of freedom after the elimination of a certain set of coordinates connected by the integrals of motion J. These coordinates are called adiabatic, since the integrals of motion corresponding to them do not change during the time between successive collisions. An example of an adiabatic degree of freedom is the rotation of the molecules under which the angular momentum of the nuclei is conserved. This special case is considered in § 31. Another example is provided by the electronic degrees of freedom. In the form (28.1) it has been assumed that the electronic motion can be separated out from the vibrations and rotations of the nuclei, so that, in fact, to the Hamiltonian H in (28.1) must be added the index l of the electronic state, playing the role of an integral of motion of the system. If, however, nonadiabatic transitions between electronic terms are probable under motion of the nuclei, then the electronic degrees of freedom lose the meaning of being adiabatic and the decomposition (or isomerization) of the molecule via various electronic states must be considered as reactions in which the electronically active degrees of freedom take part. In particular, in the original theory of Rosenstock, Wallenstein, Wahrhaftig, and Eyring [447] it was assumed that the redistribution of the energy took place as a result of fast non-radiative nonadiabatic transitions between the excited electronic states of the reacting molecule and the electronic ground state. These transitions must lead to preferential population of the ground state, as a result of its large vibrational statistical weight.

At present, information about the electronically-excited states of complex molecules is so imprecise that it is difficult to express general *a priori* views as to the role of the excited states in the decomposition and isomerization reactions. Only for thermal reactions where the decay channels into many excited states turn out to be closed (see § 33) can more-or-less definite assertions be made.

In future, we will limit ourselves to an account of a theory describing the reactions only in a single electronic state.

Disregarding the interaction $V_J(q_1, \ldots, q_s)$ in (28.1), the total energy of the molecule can be written in the form

$$E(J) = \sum \varepsilon_\lambda(J), \tag{28.4}$$

so that now the index J is added to each energy level. With the interaction $V_J(q_1, \ldots, q_s)$ in, the total energy of the molecule cannot be written in the form of a sum of independent contributions $\varepsilon_\lambda(J)$. Nevertheless, the most widespread variant of the statistical theory, based on the harmonic model of an active molecule, assumes that for a calculation of the level density the approximation (28.2) is satisfactory. The role of the interaction is reduced simply to ensuring an equiprobable redistribution of energy over the normal vibrations.

Separation of the degrees of freedom into active and adiabatic may be justified in many cases on the basis of more-or-less general considerations (conservation laws, the large distances between the electronic terms, etc.). The separation into active and inactive can be carried out only as a result of a concrete estimate of the interaction V. Since the statistical theory relinquishes precisely this step in the calculation, any separation of the inactive degrees of freedom is of a semi-empirical nature.

To construct a correlation diagram of the energy levels of the active and activated molecules, we first of all note that, for given J, the minimum value of the total energy of a molecule differs from zero. Introducing the notation

$$\min E(J) = E_j \tag{28.5}$$

it is possible to write approximately

$$E = E^* + E_j, \tag{28.6}$$

where E^* is measured from zero. It often turns out that E^* depends on J considerably more weakly than does E_j; in this case, E_j may be identified approximately with the energy of the active degrees of freedom of the molecule AB*. For the activated molecule AB⋆ an analogous approximate relation

$$E - E_a = E^\star + E_j^\star \tag{28.7}$$

is valid.

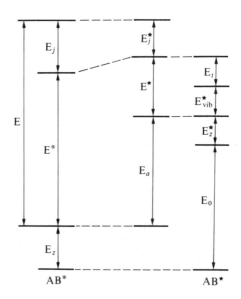

FIG. 27. Energy diagram of the active AB* and activated AB⋆ molecules.

The difference between E_j and E_j^* is caused by the fact that E_j basically depends on the equilibrium separation of the atoms in the molecule and E_j^* on the separations in the activated complex. The latter themselves depend on J, so that the function E_j is rather complicated. This problem is considered in § 31 for the case where J is the total angular momentum of the molecule. A correlation diagram of the energy levels for the system AB* → AB★ is given in Fig. 27.

The statistical theory of Landau

The statistical theory, first proposed by Landau [321], considers the decomposition of a molecule as the result of a fluctuation in the redistribution of the internal energy and the concentration of a fraction E_0 of it in one of the bonds of the molecule. Considering the molecule, during the time between collisions, as a closed system, Landau used the thermodynamical theory of fluctuations to determine the probability P for the formation of an activated molecule. This probability can be written in the form

$$P \sim \exp\left\{\frac{[S^\star(E, E_a) - S^*(E)]}{k}\right\}, \qquad (28.8)$$

where $S^\star(E, E_a)$ is the entropy of the activated molecule and $S^*(E)$ is the entropy of the active molecule.

If the entropy contribution of the reaction coordinate is neglected, then, for a sufficiently large number of internal degrees of freedom, the entropy of the activated molecule can be written in the form $S^\star = S^*(E - E_a)$. The rate constant for spontaneous unimolecular transformation is then represented by the form

$$\kappa(E) = A \exp\left[\frac{S^*(E - E_a) - S^*(E)}{k}\right], \qquad (28.9)$$

where A is a frequency factor characterizing the frequency of intramolecular motion.

The simultaneous consideration of the deactivation and decomposition processes leads to the conclusion that, at high pressures, chiefly those molecules decompose whose energy exceeds the activation energy by an amount equal to the mean energy of a molecule; and the rate constant can be written in the form

$$\kappa_\infty = A \exp\left(-\frac{E_a}{kT}\right). \qquad (28.10)$$

At low pressures, those molecules whose total energy is close to the activation energy make the main contribution to the decomposition rate and the rate constant equals

$$\kappa_0 = Z^* \exp\left[\frac{S(E_a)}{k}\right] \exp\left(-\frac{E_a - \bar{E}}{kT}\right), \tag{28.11}$$

where Z^* is the number of collisions per unit time leading to the deactivation of the molecules, and $\bar{E}$ is the mean vibrational energy per molecule of the molecules that react per unit time.

It is not difficult to trace these deductions within the framework of the more detailed statistical theory set out below and also within the bounds of Kassel's simple theory.

In comparison with the theories of Kassel and Landau, the general statistical theory, developed recently by a number of authors [345], [346], [438], [448], [450], permits a number of refinements: an explicit evaluation of the frequency factor A, on the basis of the properties of the molecule and the activated complex, can be carried out and a calculation made of the density of the vibrational spectrum, based on a more realistic model of a system of harmonic oscillators with different frequencies and with account taken of the effect of anharmonicity and internal rotation on the density of the vibrational levels.

Calculation of the rate constant of the reaction of an active molecule according to the general statistical theory

The starting point for the calculation of the reaction rate constant of an active molecule is the assumption that, in the active molecule, the total energy is statistically redistributed over the active degrees of freedom and that the passage of the representative points of the molecule through a certain critical surface separating the region of the initial molecule from the region of the reaction products in configuration space corresponds to the completion of the reaction.

In accordance with the basic assumptions of the statistical theory, we will assume that the number of representative points dN of the active molecule is proportional to an element of phase volume $d\Gamma$. The rate constant of the reaction $\kappa(E)$ is defined to be the normalized current dN/dt across the critical surface $q_r = q_r^*$. The general formula for $\kappa(E)$ is analogous to formula (2.8) of the transition state method, in which the canonical Boltzmann distribution is replaced by the microcanonical. In accordance with this, we obtain

$$\kappa(E) = \frac{\int \delta(E - H^*) \, d\Gamma^* \, dE_t}{2\pi \hbar \int \delta(E - H) \, d\Gamma}. \tag{28.12}$$

The integral in the denominator, by definition, is the density $\rho(E)$ of the energy states of the active molecule.† Assuming that in the activated molecule the motion with respect to q_r is separated from other forms of motion, we will represent H^* in the numerator in the form of (2.11). Part of the integral in the numerator, which is taken over $d\Gamma^*$ with due regard for the displacement by E_a of the origin from which the energy is measured, is expressed in terms of the energy density of the states of the activated molecule $\rho^*(E-E_t-E_a)$ with an arbitrary fixed value of the kinetic energy E_t along the reaction coordinate. Thus (28.12) takes the form

$$\kappa(E) = \frac{\int \rho^*(E-E_a-E_t)\,dE_t}{2\pi\hbar\rho(E)}, \qquad (28.13)$$

where the integration over dE_t, taking into account the reaction of the active molecules for different rates of intersection of the critical surface by the representative points for which $\rho^*(E-E_a-E_t)$ differs from zero, is extended over the whole region $E_t > 0$. Since the functions $\rho(E)$ and $\rho^*(E-E_a-E_t)$ are defined basically by the characteristics of the spectrum of the active and activated molecules, it is clear that the frequency factor is also determined by them. From this deduction, it is evident that the theory considered does not take account of the nonadiabatic and tunnel transitions. Moreover, very important limitations are imposed by the assumption that it is possible to separate a single coordinate, since such a separation of variables must be carried out in a region whose dimensions would not satisfy the condition for quasi-classical motion along the coordinate q_r.

For a calculation of the reaction rate, it is convenient to introduce the phase volume of the active and activated molecules $W(E)$ and $W^*(E)$ respectively, which are connected with $\rho(E)$ and $\rho^*(E)$ by the relations

$$\rho = dW/dE, \qquad \rho^* = dW^*/dE, \qquad (28.14)$$

then expression (28.13) can be written in the form

$$\kappa(E) = \frac{W^*(E-E_a)}{2\pi\hbar\rho(E)}. \qquad (28.15)$$

The calculation of the functions $W^*(E)$ and $\rho(E)$ is the main problem in the statistical theory.

---

† In this chapter the level density ρ and number of states W always refer to molecules capable of a spontaneous reaction, i.e. to active molecules, therefore later on we will drop the sign *. This does not lead to misunderstandings because the energy E will always be measured from the ground state level E_z of the molecule under consideration.

29. Harmonic model of an active molecule

Calculation of the phase volume by the combinatorial method

The idea behind the combinatorial method of calculating the phase volume of a system of harmonic oscillators with different frequencies is to separate the system into groups of oscillators whose frequencies, to some degree of approximation, can be considered identical. These groups may, in fact, correspond to the amalgamation of various oscillators; however, they can also correspond to oscillators representing a real system only in the sense of its energy. We will assume that there are s oscillators

$$a, b, c, \ldots, r, s. \tag{29.1}$$

If n_i oscillators possess frequencies close to v_i, where i takes values from 1 to l (l is the number of groups of oscillators), then it is possible to carry out the following partition

$$|a, b, c, | d \ldots \vdots \ldots | \ldots, r, s|. \tag{29.2}$$
$$\quad n_1 \quad\quad n_2 \quad n_i \quad n_l$$

The total energy $E\{k_i\}$ of the whole system is equal to the sum of the energies of the separate groups, each of which comprises k_i vibrational quanta. Thus, for the total energy $E\{k_i\}$ and the corresponding statistical weight $g\{k_i\}$ the following relations can be written down

$$E\{k_i\} = \sum_i \hbar\omega_i k_i$$

$$g\{k_i\} = \prod_{i=1}^{l} \binom{n_i + k_i - 1}{k_i}, \tag{29.3}$$

where $\binom{n_i + k_i - 1}{k_i}$ is the number of combinations of the k_i from $(n_i + k_i - 1)$ elements. Each of the factors in this product is the statistical weight of the system of degenerate oscillators.

The function $W[E\{k_i\}]$ changes very irregularly: it has a stepwise form, increasing by unity each time the energy passes through one of the allowed values of the energy spectrum of the system under consideration (if the corresponding energy state is degenerate, then $W(E)$ is increased by a value equal to the multiplicity of the degeneracy). Therefore $W(E)$ can be represented by the form

$$W(E) = W_0(E) + \Delta W(E), \tag{29.4}$$

where $W_0(E)$ is a smoothly-varying monotonic function; and $\Delta W(E)$ is a function having the form of an irregular 'saw' and possessing the property that its mean value over a small energy interval is equal to zero.

Since this type of averaging almost always occurs, in some form or other, in the approximations of the statistical theory, the function $W_0(E)$ is of fundamental interest.†

To obtain a smooth function $\rho_0(E)$, k_i can be considered as a continuous variable and (29.3) can be integrated over all k_i with regard for the conservation of the total energy:

$$\rho_0(E) = \int g\{k_i\} \delta[E - E\{k_i\}] \, dk_1 \ldots dk_l. \tag{29.5}$$

Schlag and Sandsmark [460] showed how an integral of the type (29.5) and the phase volume corresponding to it $W_0(E)$ should be calculated in practice. In particular, it proved possible in this way to obtain a correction to the so-called semiclassical approximation of Marcus and Rice [345]. For a system of identical oscillators this is obtained as a result of the approximation

$$\binom{n+k-1}{k} \approx \frac{(k+n/2)^{n-1}}{(n-1)!}, \tag{29.6}$$

valid for $k \gg n$. Marcus and Rice [345] generalized this approximation to a system of different oscillators, introducing as the sole quantum correction the total energy of the zero-point vibration of the whole system $E_z = \frac{1}{2} \sum_i \hbar \omega_i$:

$$W_{\text{semiclass}}(E) = \frac{(E+E_z)^s}{s! \prod_i \hbar \omega_i}. \tag{29.7}$$

Taking the following correction terms into account in formula (29.6) and introducing them into (29.7), Schlag and Sandsmark obtained a more precise semiclassical approximation [460]:

$$W_0(E) = W_{\text{semiclass}}(E) \left[1 - \frac{s(s-1) \sum (\hbar \omega_i)^2}{24(E+E_z)^2} + \ldots \right]. \tag{29.8}$$

---

† It should be borne in mind that, as a result of the quasi-stationary nature of the quantum state of an activated molecule, the energy levels are broadened and, as Rice showed [438], for adiabatic decomposition reactions this broadening is of precisely the order of magnitude necessary to smooth out $W(E)$. Thus, for this reason, $W(E)$ in (29.4) should be considered as a continuous function of E. In accordance with this, by $\rho(E)$ should be understood the density of vibrational levels averaged over an interval of energy δE, which is larger than the separation between discrete levels of the activated molecule but small in comparison with the broadening of the spectrum of the distribution of the active molecule over energies $\overline{\Delta E}$.

However, cases can be encountered where such a smooth distribution for $W(E)$ and $\rho(E)$ cannot be introduced. It is clear that the discrete structure of the vibrational spectrum can appear only on condition that the width Γ of level E be small in comparison with the separations between the neighbouring levels ΔE, which must in turn be larger than $\overline{\Delta E}$ (for thermal reactions, it is evident that $\overline{\Delta E} \approx kT$). Such relations between the magnitudes of Γ, $\overline{\Delta E}$, and ΔE are perfectly possible for nonadiabatic reactions [74], [386].

On the basis of the correction in the brackets, the region of applicability of the semiclassical approximation can be estimated for each specific case.

Lin and Eyring [334] noted that an integral of the type (29.5) can be calculated by the saddle-point method for a variable z if, for each oscillator, a generating function $g(v_i, z)$ is introduced whose expansion coefficients in a series in z agree with the multiplicity of the degenerate energy level E_i, and then the total generating function $\Phi(z) = 1/z \prod_i g_i(v_i, z)$ is formed.

We will now examine another possible partition of system (29.1). Following Vestal, Wahrhaftig, and Johnston [535], we will divide up the whole of phase volume $W(E)$ into partial volumes $U_p(E)$, each of which corresponds to the excitation of only p oscillators. For $U_3(E)$, for example, the following (shown by bold letters) oscillators from the total system (29.1) take part in the system of three oscillators:

$$\mathbf{a, b, c}, d \ldots r, s$$
$$\mathbf{a, b}, c, \mathbf{d} \ldots r, s$$
$$\ldots \ldots \ldots \ldots$$
$$a, b, c, d \ldots \mathbf{r, s}, \quad (29.9)$$

so that $U_3(E)$ represents on average the phase volume of three excited oscillators.

If all the oscillators were identical, then $W(E_k)$ would have the form

$$W(E_k) = \sum_p \binom{s}{p} U_p(E_k), \quad U_p(E_k) = \frac{k!}{p!(k-p)!}, \quad E_k = \hbar\omega k. \quad (29.10)$$

This expression, exact for a system of degenerate oscillators, can be used for an approximate calculation of the phase volume of the different oscillators if the different groups of p oscillators are characterized by different frequencies ω_p. To this end, we will make the substitution:

$$U_p(E_k) = \frac{k!}{p!(k-p)!} \Rightarrow U_p(E_{k,p}) = \frac{k_p!}{p!(k_p-p)!}, \quad k_p = \frac{E}{\hbar\omega_p}, \quad (29.11)$$

where the vibrational quantum number k_p (now different for different groups) is defined in terms of the total energy E and the mean frequency of the group ω_p.

In paper [535] it is shown that, with sufficient accuracy, it can be assumed that $k_p = \sigma_p k$, where the factor σ_p is defined in terms of the geometric mean frequency $\tilde{\omega}$ of the whole system of s oscillators and the normal frequencies ω_i by the relation

$$\sigma_p = \left[\sum_{\substack{i,j,\ldots,l \\ p}} \left(\frac{\tilde{\omega}}{\omega_i} \cdot \frac{\tilde{\omega}}{\omega_k} \ldots \frac{\tilde{\omega}}{\omega_l} \right) \right]^{1/p}, \quad (29.12)$$

in which the summation goes over all the different groups of oscillators.

For an approximate calculation of W, the coefficients σ_p can be calculated by a simpler method than the direct summation in (29.12), and the function U_p can be approximated by an expression of the form

$$U_p = \frac{1}{p!}\left(k_p \sigma_p - \frac{p-1}{2}\right)^p. \tag{29.13}$$

Thus for $W(E)$ we obtain

$$W(E) = \sum_p \frac{s}{p} \cdot \frac{1}{p!}\left(\frac{\sigma_p E}{\hbar\tilde{\omega}} - \frac{p-1}{2}\right)^p, \tag{29.14}$$

where the summation goes from 0 to s if the expression in brackets turns out to be positive; otherwise, the summation breaks off at the last positive term. With this summation rule, (29.14) is a very good interpolation formula, valid even for small energies, when the approximation of U_p by a function of the form of (29.13) is not justified. This is connected with the fact that the method just considered of dividing the system of oscillators into groups at small energies automatically excludes precisely those contributions to the total phase volume which are taken into account incorrectly in the semi-classical approximation.

As an example, we may cite the results of a calculation of $W(E)$ for a model system of harmonic oscillators which approximately describes the distribution of frequencies in propane: 8 oscillators with a frequency $\tilde{\omega}/3$, 11 with a frequency $\tilde{\omega}$, and 8 with a frequency $3\tilde{\omega}$. As the energy changes from 0 to $15\hbar\tilde{\omega}$ the true value of $W(E)$ changes over a range from 1 to 3×10^{12} and, when calculated by formula (29.14), from 1 to $2·95 \times 10^{12}$. At the same time, a calculation within the framework of Kassel's model leads to values for W differing from the true values by more than an order of magnitude at high energies [535].

In the limiting case, when the number of quanta significantly exceeds the number of oscillators, the expressions in each of the brackets in (29.14) can be expanded in a series in powers of the ratio p/k_p and then summed, retaining the largest terms in each of the series of the expansion. In this way, the semi-classical approximation (29.7) given above is obtained.

As an illustration of the method under consideration, we will now discuss the dependence of the decomposition rate close to threshold on the excess energy $E - E_a$. This question arises, for example, on estimating the lifetimes of vibrationally-excited ions with energies close to the decomposition threshold or in evaluating the lifetime of the complex formed as a result of the bimolecular association of polyatomic molecules or radicals. The classical expression for the lifetime (25.5) is valid, on condition $E - E_a \gg \hbar\omega s$. Although for the majority of cases which are of interest kinetically it is not satisfied, nevertheless, formula (25.5) is often used without foundation, to calculate $\kappa(E)$. The quantum-mechanical calculation gives a considerably shorter lifetime than the classical calculation. Thus, Wolfsberg [562] showed that even in the threshold region (where the excess of the energy E over the activation energy E_a is less than 0·1 eV) the lifetime of the $C_3H_8^+$ ion relative to the decomposition into $C_3H_7^+$ and H is 10^{-10} s, though the classical calculation gives the values 3–4 orders of magnitude longer. This behaviour of the rate close to threshold is illustrated in Fig. 28, where the abscissa is the excess energy in units of the mean vibrational quantum, and the ordinate is the decomposition rate constant. The calculation is performed for the model system mentioned above of 27 oscillators with frequency $\tilde{v} = 3 \times 10^{13}$ s^{-1}. The solid curve is the quantum calculation; the dashed curves are calculations using the classical formula (25.5) with effective numbers of oscillators in the active molecule $s = 27, 14,$ and 7.

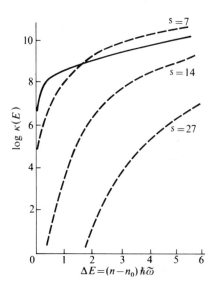

FIG. 28. The dependence of the decomposition rate of a C_3H_8 molecule on the energy E above the threshold E_a (solid curve). (Vestal, Wahrhaftig, and Johnston [535].)

From Fig. 28, it is clear that, in an attempt to interpret the solid curve classically, it would have to be assumed that the effective number of oscillators s_{eff} taking part in the redistribution of energy was less than the actual number. The results of experimental work studying the mass spectra of ions lead independently to the same conclusion.

Calculation of the phase volume by the method of the Laplace transform

One of the convenient methods of calculating the phase volume $W(E)$ is to invert the formula for the partition function F of the system under consideration:

$$F(\beta) = \int_0^\infty \rho(E)\exp(-\beta E)\,dE = \int_0^\infty W(E)\exp(-\beta E)\beta\,dE. \quad (29.15)$$

This is not difficult to do, if it is noted that the integrand in (29.15) contains an exponential which can be used to construct the δ-function:

$$\delta(E-E') = \frac{1}{2\pi}\int_{-i\infty}^{i\infty} \exp[i\lambda(E-E')]\,d\lambda. \quad (29.16)$$

We will multiply (29.15) by $\exp(\beta E')$ and integrate in the complex plane of the variable β along a line parallel to the imaginary axis. The direction of the path of integration depends on the fact that the integral representation of the δ-function contains an oscillatory exponential. In the theory of the Laplace transform it can be shown that the path of integration must be taken to the right of all the singularities of the function F/β [8]. In this way, we finally obtain the formula

$$W(E) = \frac{1}{2\pi i}\int_{\sigma-i\infty}^{\sigma+i\infty} \frac{F(\beta)}{\beta}\exp(\beta E)\,d\beta, \quad (29.17)$$

the basis for a strict deduction of which can be found, for example, in the book [8].

It is very important that the integral (29.17) can be calculated by the theory of residues, and that the contributions from the different singularities in the integrand enable the function to be presented in a form convenient for application.

Thiele [521] and Haarhof [247], [248] showed, by the example of the system of harmonic oscillators, that the function $W_0(E)$ can be obtained from the general formula (29.17) if, of all the residues of the integrand, only the residue at the point $\beta = 0$ is taken into account. This result may be understood if it is taken into consideration that residues at points far from the real axis will contribute oscillatory factors due to the presence of the exponential $\exp(\beta E)$ in the integrand of (29.17). For a system of s harmonic oscillators, the partition function has the form

$$F^s = \prod_{i=1}^s [1-\exp(-\beta\hbar\omega_i)]^{-1}, \quad (29.18)$$

where the function $F^s(\beta)$ has a pole of order $s+1$ at the origin. A calculation of the contribution from this pole enables $W_0(E)$ to be represented in the form of a finite sum:

$$W_0(E) = \frac{(E+E_z)^s}{s!\prod_i \hbar\omega_i} \sum_{i=0}^s D_i^s(\omega_1,\omega_2,\ldots,\omega_s)\binom{s}{i}(E+E_z)^{-i}, \quad (29.19)$$

where the summation only goes over the even indices i, so that the number of terms in the sum equals half the number of oscillators s. The coefficients

D_i^s depend on the frequencies of the oscillators and can be found by an expansion of F^s in a series in powers of β.

The first terms of the series are

$$D_0^s = 1$$
$$D_2^s = -\frac{1}{3}\sum_i \left(\frac{\hbar\omega_i}{2}\right)^2. \tag{29.20}$$

It is not difficult to see that, taking account of these two terms, expression (29.19) is equivalent to expression (29.8) and the expansion parameter is proportional to the square of the ratio $E_z/(E+E_z)$. A calculation of further terms in (29.19), for which a general expression [521] is known, enables $W_0(E)$ to be calculated for any value of the ratio $E_z/(E+E_z)$.

It is important to notice that, although (29.19) has the form of an expansion in terms of the parameter $E_z/(E+E_z)$, which has the meaning of a quantum correction to the classical expression (i.e. vanishes as $\hbar \to 0$), nevertheless, (29.19) cannot immediately be obtained by the inversion of the Laplace transform applied termwise to the expansion of the partition function in a series in powers of $\beta\hbar\omega_i$. In fact, the quantum partition function of any system at sufficiently high temperatures can be written in the form

$$F(\beta) = F_{\text{class}}(\beta)(1 + a_1\beta + \ldots) \tag{29.21}$$

where $F_{\text{class}}(\beta)$, the classical partition function for the models under consideration, equals

$$F_{\text{class}}(\beta) = C\beta^{-n}. \tag{29.22}$$

Applying the inverse Laplace transform to (29.21) we find

$$W(E) = C\left[\frac{E^n}{\Gamma(k+1)} + \frac{a_1 E^{n-1}}{\Gamma(k)} + \ldots\right]. \tag{29.23}$$

Expression (29.23) represents an infinite series in terms of the parameter E_z/E, where E_z is a certain effective zero-point energy for the system. The transformation from (29.23) to (29.19) is very complicated: firstly, in (29.23) it is necessary to discard terms connected with the fluctuating part of the phase volume ΔW, and, second, it is necessary to pass from an expansion in E_z/E to an expansion in $E_z/(E+E_z)$, which leads to a recombination of the infinite series (29.23) into the finite form (29.19).

How this recombination is carried out may be understood qualitatively from the example of the system of s oscillators. For this system, the parameter a_1 equals E_z, so that the expression (29.23) takes the form

$$W^s(E) = \frac{E^s}{s!\prod_i \hbar\omega_i}\left(1 + \frac{sE_z}{E} + \ldots\right). \tag{29.24}$$

To an accuracy of terms of first order in E_z/E this expression can also be written in the form

$$W^s(E) = \frac{(E+E_z)^s}{s! \prod_i \hbar\omega_i}. \qquad (29.25)$$

But now the region of applicability of (29.25) is considerably extended (of course, this can be seen only on the basis of the evaluation contained in (29.8)): in particular, $W_{\text{semiclass}}$ differs from $W_0(E)$ being a factor of 2 smaller, if the condition $E \geq \sqrt{(s)}E_z/2$ is satisfied [471].

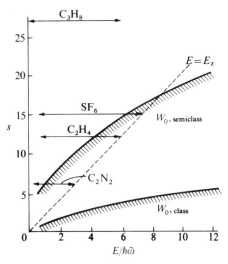

Fig. 29. Range of applicability of the classical and semiclassical approximations for the calculation of the phase space of a system of harmonic oscillators.

In Fig. 29 the regions of applicability of the classical and semiclassical approximations are shown. The boundaries of the regions correspond approximately to those values of E and s at which the corresponding approximation gives an error of about 100 per cent. For comparison, the range of variation of the energy is also given here, lying beyond the limits of applicability of the semiclassical approximation, for which the decomposition rate of $C_3H_8^+$ was calculated proportional to the phase volume of the activated molecule $(C_3H_8^+)^\star$ with 26 effective oscillators (see Fig. 28). For a fixed value of the energy E and an increase in the number of degrees of freedom, the semiclassical expression finally becomes inapplicable. However, if the spread of frequencies $\langle\omega_i^2\rangle - \langle\omega\rangle^2$ is sufficiently small, then the system under consideration can be approximated by a group of s degenerate oscillators, for which a closed expression for $W(E)$ is known, valid for all values of E and s. This enabled Haarhof [248] to deduce an approximate formula, which makes it possible to go beyond the range of the semiclassical approximation.

Computation of the phase volume of a system of harmonic oscillators

In cases where the criterion of applicability is not satisfied by any of the analytical expressions for the phase volume, the quantity $W(E)$ can be calculated by direct summation of the number of states of the system for energies less than the given value E. It is convenient to present the results of the calculation in the form of a semi-empirical formula, analogous to the formula of the semiclassical approximation

$$W(E) = \frac{(E + aE_z)^s}{s! \prod_i \hbar \omega_i}. \qquad (29.26)$$

The dimensionless function $a = a(E)$ was determined by Whitten and Rabinovitch [543], for a large number of molecules, by comparing the computer calculations with calculations using formula (29.26). The results of this work enable the selection of a suitable model without much difficulty, and the evaluation of the value of W for a very wide class of molecules. Of course, the direct summation of the states for complex molecules is only possible with the help of high-speed computers.

It is convenient to represent the monotonic part $W_0(E)$ of the function $W(E)$ also in the form of (29.26), where it proves to be possible to approximate the corresponding parameters a_0 by the expression

$$a_0 = 1 - (s-1) \frac{\sum \omega_i^2}{(\sum \omega_i)^2} f_0\left(\frac{E}{E_z}\right), \qquad (29.27)$$

where $f_0(E/E_z)$ is a function which is almost independent of the structure of the vibrational spectrum and can therefore be used to calculate W_0 for any molecule.

The fluctuating character of W at small energies is revealed in that the quantities a and f in the exact formula, (29.26) differ from a_0 and f_0 in the corresponding formula for the monotonic part W_0. In Fig. 30, taken from the paper [543], the function $f_0(E/E_z)$ is shown by the solid curve, and the dots are separate values of $f_0(E/E_z)$ for certain molecules. It is evident that, for an increase in the ratio E/E_z, the fluctuating part of f dies out. The range of the variable $0.1 < E/E_z < 1$ is also shown on Fig. 29, to illustrate the connection between the exact calculation and the semiclassical approximation.

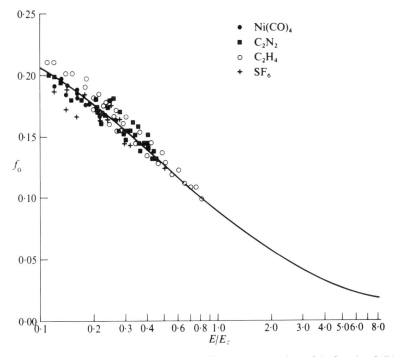

FIG. 30. Plot of the function $f_0(E/E_z)$ (solid curve). The points are values of the function $f_0(E/E_z)$ for various molecules.

At sufficiently high energies, expression (29.19) can be used to calculate the phase volume, giving W_0 in the form of a finite sum. Such a programme was carried out by Schlag, Sandsmark, and Valance [463], who determined exact values of W_0 for a number of molecules. Comparison of W and W_0 enables the contribution of the fluctuating part ΔW to be evaluated. It turns out that, for cyclopropane, for example, the fluctuating part of the level density $\rho(E)$ is reduced from 9 per cent to 2 per cent for a change in the energy E from 10 kcal mol^{-1} to 20 kcal mol^{-1}. Since for small energies a direct calculation of the number of states is carried out fairly rapidly, but for large energies the fluctuating part introduces a small contribution, these two methods—the computer summation of W at small energies and the calculation of W_0 by formula (29.19) at large energies—successfully complement one another.

30. An anharmonic model of an active molecule

The methods of calculating the phase volume and level density, described in the previous section, are based on the model of a molecule as a system of harmonic oscillators. Within the framework of this approximation, the effect of anharmonicity on the quantities $W(E)$ and $\rho(E)$ is completely neglected, although it is precisely this which is ultimately responsible for the statistical distribution of the representative points over the accessible phase volume. If the theory is to lay claim to a qualitative description of the decomposition or isomerization of a molecule, then it must take into account the possibility of a transition from one type of motion in the active molecule into another type in the activated molecule (for example, the transition from free internal rotations relative to the C–C bond in strongly vibrationally-excited ions and molecules of saturated hydrocarbons into torsional vibrations relative to the same bond in the activated molecule). At the same time, for a specific calculation of the energy level density, the possibility of separating (even though approximately) the vibrations and the internal rotations has to be assumed. Physically, it is evident that a transition from one type of motion to another is connected with a very large anharmonic interaction between the different degrees of freedom. Therefore, at present it is still difficult to express a definite opinion as to the compatibility of the assumption about the separation of the different degrees of freedom and the taking into account of the large anharmonicity, in a calculation of the level density.

As regards the rotation of the molecule as a whole, the corresponding degrees of freedom are usually considered adiabatic. The rotation of the active molecule leads to two effects which can be indicated qualitatively. Firstly, the centrifugal forces can change (usually decrease) the activation energy. This change is analogous to the corresponding effect for the diatomic molecule (see § 41) and can be calculated for polyatomic molecules by the approximation of separating the internal motion from the rotation as a whole. Second, the rotation of the molecule dynamically interacts with the internal motion. This interaction can either affect the energy-level density of the active molecule very weakly, leading only to an interaction between the vibrational terms of different symmetry and a speeding up of the energy exchange between the different degrees of freedom, or it may change the energy spectrum of the active molecule substantially. In the latter case, the effect of rotation on the reaction rate is not adiabatic and therefore can be taken into account only by explicitly introducing the corresponding degrees of freedom as active.

Accounting for the anharmonicity

The effect of anharmonicity on the distribution of the vibrational levels of a complex molecule is two-fold: it increases the level density at high excitation energies in comparison with the harmonic oscillator model by reducing the size of the vibrational quanta and, furthermore, it imposes restrictions on the allowed values of the excitation energies of a single oscillator.

The limitations imposed by the finite energy capacity of the oscillators are comparatively unimportant. Thus, for cyclopropane at $E \approx 200 \text{ kcal mol}^{-1}$ the effect of the finite energy capacity changes the phase volume by no more than 1 per cent [534]. An appreciable difference is obtained only when the total energy turns out to be of the order of the sum of the energies of all the bonds in the molecule. Here, however, the statistical description of the reaction becomes invalid, due to the very short lifetime of the excited molecule. In thermal conditions, when the proportion of molecules with high energies is usually very small, the corresponding corrections can be neglected.

The effect of anharmonicity on the level density is more important. For an approximate estimate of this effect, we will use formulae (29.3) and (29.5), but now to the factors of the type $(n_i+k_i-1)!/k_i!(n_i-1)!$ we will associate the energy not of a system of harmonic oscillators $E_{i,\text{harm}} = \hbar\omega_i k_i$ but the energy $E_{i,\text{anh}}$, which is obtained when each of the groups k_i of oscillators has the same energy

$$E_{i,\text{anh}} = \hbar\omega_i k_i \left[1 - x_i \left(\frac{2k_i+1+n_i}{n_i+1}\right)\right], \qquad (30.1)$$

where x_i is the coefficient of anharmonicity.

It is evident that this approximation only takes account of the anharmonicity on average, and its accuracy can be ascertained by comparing this approximation with the results of a direct numerical summation of the number of states. Such a comparison shows [462] that the approximation (30.1) underestimates the true anharmonic correction by 2–10 per cent, an accuracy quite acceptable in kinetic calculations. From general considerations, it is clear that, as the number of degrees of freedom increases, the anharmonicity has less effect on the density of energy levels since, on average, a smaller proportion of the excitation energy falls on each bond.

As an example illustrating this discussion, we quote the result of a calculation by Schlag, Sandsmark, and Valance [462]. The ratios of the phase volumes of the anharmonic and harmonic models are given below, showing the energy dependence for CD_4 and C_3H_6 molecules, whose vibrational spectra are approximated by the spectra of four and five different groups of oscillators respectively:

Energy (kcal mol^{-1})	20	50	100	150
Ratio of phase volume of anharmonic and harmonic models for molecules				
CD_4	1·16	1·36	3·08	5·19
C_3H_6	1·18	1·39	1·85	2·46

A simple analytical expression for the correction factor B, which should be introduced into the level density ρ of the harmonic system to take account of the anharmonicity, can be obtained for a system of s non-interacting

Morse oscillators. In this case, the coefficient of anharmonicity x is expressed in terms of the frequency ω and energy of dissociation D, and the factor B takes the form [248]

$$B = \left(\frac{E+E_{z,\text{anh}}}{E+E_{z,\text{harm}}}\right)^{s-1} \exp\left[\frac{E}{2D} + \frac{E^2}{(s+1)D^2} + \dots\right], \quad (30.2)$$

where $E_{z,\text{anh}}$ and $E_{z,\text{harm}}$ are the energies of the zero-point vibrations of the anharmonic and harmonic systems.

This expression can be employed to calculate the effect of anharmonicity on the thermal reaction rate. Taking it into account that, for a Morse oscillator

$$E_{z,\text{harm}} - E_{z,\text{anh}} = (\hbar\omega_0/8D_0)E_{z,\text{harm}},$$

expression (30.2) (on condition $s \gg 1$) can be written in the form

$$B = \exp\left(\frac{E_{z,\text{harm}}^2}{4DE} + \frac{E}{2D} + \dots\right). \quad (30.3)$$

Since the states lying close to the energy threshold make the main contribution to the reaction rate in thermal conditions, we can assume that $E \approx D$ in (30.3).

Thus the anharmonic correction turns out to be small. More exact calculations leave this conclusion unchanged [465], [555]. For molecules with a small number of atoms, the anharmonic correction can be substantial, for example, for a triatomic molecule, taking the anharmonicity into account leads to an increase of several orders of magnitude in the calculated thermal decomposition rate [438].

Accounting for the rotational degrees of freedom

As noted above, the contribution of the active rotational degrees of freedom to the total phase volume is usually calculated in the approximation of a small interaction between the vibrations and rotations. In this approximation, the statistical weights g_{rot} and g_{vib} of the rotational and vibrational degrees of freedom are independent so that, for the energy level density, we can write

$$\rho(E) = \frac{1}{\Delta E} \sum g_{\text{vib}}(E_{\text{vib}}) g_{\text{rot}}(E_{\text{rot}}), \quad E < E_{\text{vib}} + E_{\text{rot}} < E + \Delta E, \quad (30.4)$$

where ΔE is some small energy interval.

The statistical weight of the rotational states g_{rot} equals the product of the statistical weights of the independent rotations:

$$g_{\text{rot}}(E_{\text{rot}}) = \prod_m g_{m,\text{rot}}(E_{m,\text{rot}}), \quad g_m = \begin{cases} 2j_m & \text{for a two-dimensional rotator} \\ 2 & \text{for a one-dimensional rotator.} \end{cases}$$

$$(30.5)$$

Since, in the majority of cases of practical interest, the rotation can be regarded as quasi-classical,† the summation in (30.4) over the rotational quantum numbers is replaced by an integration with regard for the connection between the rotational energies $E_{m,\text{rot}}$ and the angular momentum j_m:

$$E_{m,\text{rot}} = \hbar^2 j_m^2 / 2 I_m. \tag{30.6}$$

In addition, the following expression is obtained for ρ

$$\rho(E) = \sum g_{\text{vib}}(E_{\text{vib}}) \int \prod_m (2j_m)^{d_m - 1} \, dj$$

$$= \left(\frac{2}{\hbar^2}\right)^{r/2} \left\{ \frac{\prod_m I_m^{d_m/2} \Gamma(d_m/2)}{\Gamma(r/2)} \right\} \sum_{E_{\text{vib}} \leq E} g_{\text{vib}}(E_{\text{vib}})(E - E_{\text{vib}})^{(r/2) - 1}, \tag{30.7}$$

where r, the total number of degrees of freedom for the internal rotation, equals $\sum_m d_m$.

If the moments of inertia I_m are expressed in terms of the corresponding partition functions then (30.7) takes the form

$$\rho(E) = \frac{F_{\text{rot}}}{(kT)^{r/2} \Gamma(r/2)} \sum_{E_{\text{vib}} \leq E} g_{\text{vib}}(E_{\text{vib}})(E - E_{\text{vib}})^{r/2 - 1}, \tag{30.8}$$

where F_{rot} is the total partition function for the internal rotation.

Although the temperature clearly comes into this expression, in fact, the density $\rho(E)$ does not depend on T, since the factor $(kT)^{r/2}$ is implicitly contained in F_{rot}.

The expression for $W(E)$, corresponding to (30.7), has the form

$$W(E) = \frac{F_{\text{rot}}}{(kT)^{r/2} \Gamma(r/2 + 1)} \sum_{E_{\text{vib}} \leq E} g_{\text{vib}}(E_{\text{vib}})(E - E_{\text{vib}})^{r/2}. \tag{30.9}$$

If the semiclassical approximation with a correction factor a_0 is used for g_{vib}, then (30.9) is somewhat simplified:

$$W(E) = \frac{F_{\text{rot}}(E + a_0 E_z)^{s + r/2}}{(kT)^{r/2} \Gamma(s + r/2 + 1) \prod_i \hbar \omega_i}, \tag{30.10}$$

† If the quantization of the rotation is important then, for the rotational phase volume, an approximation analogous to the semiclassical approximation (29.7) can be used. Moreover, it turns out that the replacement of a summation by an integration can be taken into account, formally, by the introduction of an effective zero-point energy, although the actual zero-point energy of a rotator equals zero [326]. For molecules of the propane type, the classical approximation for the rotational statistical weight introduces an error no greater than 2 per cent for energies equal to 100 cm^{-1} [544].

where

$$a_0 = 1 - \lambda f_0\left(\frac{E}{E_z}\right); \quad \lambda = \frac{(s-1)(s+r/2)\sum \omega_i^2}{s\left(\sum_i \omega_i\right)^2},$$

where f_0 is the universal function given in Fig. 30.

For a propane molecule, for example, a calculation of W using (30.10) differs from a calculation using (30.9) by approximately 3 per cent, even at the comparatively low energy $E \approx 1500 \, \text{cm}^{-1}$ [544].

If there is no free internal rotation in the active molecule, then it should be taken that $r = 0$ in the formulae quoted above. In the limit $r \to 0$, (30.10) simply goes over into (29.26).

The final expression for the rate constant $\kappa(E)$ is obtained from (30.7) and (30.9):

$$\kappa(E) = \frac{W^\star_{\text{vib,rot}}(E - E_a)}{2\pi \hbar \rho_{\text{vib,rot}}(E)}. \tag{30.11}$$

Here the phase volume of the activated molecule $W^\star$ is expressed by formula (30.9), except that the energy E is changed to $E - E_a$ and the number of rotational degrees of freedom $r^\star$ and vibrational degrees of freedom $s^\star$ may differ from r and s respectively.

31. Statistical theory with regard for the conservation of angular momentum

The statistical theory considered up to now assumes that the size of the phase volume accessible to the active and activated molecules is limited only by the condition that the energy be constant (vibrational energy if the rotational degrees of freedom are considered inactive, and vibrational + rotational if the rotation is considered active). In reality, for a closed system, such as an excited molecule, there exists yet another obvious integral of motion—the total angular momentum of the system, whose conservation imposes an additional restriction on the size of the phase volume of the active and activated molecules. This restriction is especially important for 'loose' activated molecules, where internal rotation is possible. Furthermore, the conservation of angular momentum is important when the active molecule arises as a result of excitation by electron, or photon, impact. This is connected with the fact that, due to the electron, or photon, impact, as a rule a large amount of electronic–vibrational excitation energy is transferred to the molecule, but the rotational momentum of the molecule is very little changed on collision.

In these conditions, it can be shown that the equiprobable redistribution of energy between different degrees of freedom is impossible by virtue of the law of conservation of the angular momentum. In this section, following the papers [75], [76], [331], [332], [398], [399] we will consider the effect of the law of conservation of angular momentum on the decomposition rate of an active molecule.

Derivation of the formula for the rate constant

In future, for definiteness, we will consider the decomposition reaction of an active molecule AB* into an atom A and a polyatomic fragment B

$$AB^* \to A + B. \qquad (31.1)$$

On the basis of a discussion completely analogous to that presented in connection with the derivation of formula (28.12) for the decomposition rate constant $\kappa(E, J)$ of an active molecule with energy E and total angular momentum J, we have:

$$\kappa(E, J) = \frac{\int \delta(E - H^\star)\delta(J - M^\star)\,d\Gamma^\star\,dE_t/2\pi\hbar}{\int \delta(E - H)\delta(J - M)\,d\Gamma}. \qquad (31.2)$$

In comparison with (28.12) an additional δ-function is introduced into the integrand, securing the law of conservation of the angular momentum of the disintegrating fragments; here M denotes the total angular momentum of the system expressed as a function of the generalized coordinates and momenta. In order to take the conservation law into account in an explicit form, it is convenient to go over to those variables which act directly as arguments of the δ-function, as was done previously, in fact, in the derivation of (28.13).

We note that, as the rate constant does not depend on the component of the angular momentum along an axis fixed in space, the argument in the second δ-function in the numerator and denominator of (31.2) is the difference between the moduli $J - M$.

To calculate the integral in the denominator of (31.2) we will assume that, in the Hamiltonian of the active molecule, vibrations and rotations are separated. It is then convenient to introduce, as variables of integration, the vibrational energy of the molecule E_{vib} and the three generalized coordinates and momenta characterizing the rotation, the modulus of the total angular momentum M, the component M_z of the vector $\mathbf{M}$ along a fixed z axis, and the generalized momentum $\mathbf{K}$ characterizing the orientation of $\mathbf{M}$ in relation to the system of coordinates connected with the molecule AB and coordinates conjugate to them.

In the quantum-mechanical treatment of rotation, the ratios $M/\hbar$ and $K/\hbar$ are quantum numbers usually employed to classify the energy levels of an asymmetric top.† In the particular case of a symmetrical top, K

---

† Everywhere in future, rotation will be considered as quasi-classical so that, for all angular momenta l, the condition $l/\hbar \gg 1$ will be assumed satisfied. To the same approximation, the number of different components, equal to $2l + 1$, will be changed to $2l$.

acquires the meaning of the component of the vector **M** along the axis of symmetry of the molecule.

With such a choice of variables, the expressions for H and $d\Gamma$ take the forms

$$H = E_{\text{vib}} + E_{\text{rot}}(M, K),$$

$$d\Gamma = \rho_{\text{vib}}(E_{\text{vib}})\, dE_{\text{vib}}\, dM_z\, dM\, dK/\hbar^3. \tag{31.3}$$

Substitution of (31.3) into (31.2) enables the integration over dE_{vib}, dM_z, and dM to be carried out:

$$\int \delta(E-H)\delta(J-M)\, d\Gamma = \left(\frac{2J}{\hbar}\right)\frac{1}{\hbar}\int \rho_{\text{vib}}[E-E_{\text{rot}}(J,K)]\frac{dK}{\hbar}. \tag{31.4}$$

The integral on the right, by definition, is the density of the vibrational and rotational states of the active molecule, i.e. the density of energy levels with given values E, J per unit interval of angular momentum:

$$\rho(E, J) = \frac{1}{\hbar}\int \rho_{\text{vib}}[E - E_{\text{rot}}(J, K)]\frac{dK}{\hbar}. \tag{31.5}$$

On condition that the rotation is quasi-classical, it is sometimes convenient to reduce this relation to a form in which the energy E_{rot} acts as a variable of integration. We will assume that the functions $E_{\text{rot}} = E_{\text{rot}}(J, K)$ are known from a quantum-mechanical calculation of the energy levels of a top. Considering E_{rot} as a continuous function of the momenta J and K and introducing the energy-level density of rotation $\rho_{\text{rot}}(E_{\text{rot}}, J)$,

$$\rho_{\text{rot}}(E_{\text{rot}}, J) = \frac{1}{\hbar}\left[\frac{\partial E_{\text{rot}}(J, K)}{\partial K}\right]^{-1}, \tag{31.6}$$

we will rewrite (31.5) in the form

$$\rho(E, J) = \frac{1}{\hbar}\int \rho_{\text{vib}}(E - E_{\text{rot}})\rho_{\text{rot}}(E_{\text{rot}}, J)\, dE_{\text{rot}}, \tag{31.7}$$

where the integration over dE_{rot} is extended to all values of the energy for which ρ_{rot} differs from zero.

It is known that the function $E_{\text{rot}}(J, K)$ can be found in an explicit form only for a symmetric top, two of whose three principal moments of inertia are equal. This enables the integral in (31.7) to be calculated in a closed form. For the general case of an asymmetric top, the level density can be calculated by the general formula as an integral over all the variables of phase volume taken over constant energy surfaces with the constancy of the total angular momentum J and its z component J_z as additional restrictions. In particular, we will choose a moving system of coordinates, in which the z axis is directed along one of the principal axes of inertia of the top and the x

axis is along the line of intersection of the plane perpendicular to the z axis and the plane going through the z axis and the vector $\mathbf{J}$. In such coordinates, the rotational Hamiltonian can be represented in the form of a function of the two generalized momenta J and K and a coordinate ψ conjugate to the momentum K. The angle ψ characterizes the rotation of the principal axes of inertia I_{xx} and I_{yy} with respect to the chosen axis Ox: K is the component of $\mathbf{J}$ along the z axis.

The general expression for the density of rotational energy levels takes the form

$$\rho_{rot}(E_{rot}, J) = \frac{1}{\hbar} \int \delta[E_{rot} - H_{rot}(J, K, \psi)] \frac{dK \, d\psi}{2\pi\hbar}. \tag{31.8}$$

If H_{rot} does not depend on ψ, then we again return to expression (31.6).

For the symmetric top, the rotational energy E_{rot} is given by the expression

$$E_{rot}(J, K) = AJ^2 + (C - A)K^2, \tag{31.9}$$

where A is expressed in terms of the moments of inertia I_{xx} or I_{yy} and C in terms of the differing moment of inertia I_{zz}:

$$A = 1/2I_{xx} = 1/2I_{yy}, \qquad C = 1/2I_{zz}. \tag{31.10}$$

The restriction $J \geqslant K$, imposed on the momenta J and K, is connected with the physical meaning of these parameters. Taking this restriction into account, it is not difficult to find the following expression for the level density:

$$\rho_{rot}(E_{rot}, J) = \frac{\sqrt{2}}{\hbar^2 |C - A|^{\frac{1}{2}} |E_{rot} - AJ^2|^{\frac{1}{2}}} \begin{cases} AJ^2 \leqslant E_{rot} \leqslant CJ^2 \\ \text{for } C > A \text{ (prolate top)} \\ CJ^2 \leqslant E_{rot} \leqslant AJ^2 \\ \text{for } C < A \text{ (oblate top)} \end{cases} \tag{31.11}$$

It is not difficult to verify that $\rho_{rot}(E_{rot}, J)$ is normalized over the total number of states differing in the components K per unit interval of the variable J:

$$\int \rho_{rot}(E_{rot}, J) \, dE_{rot} = \frac{(2J/\hbar)}{\hbar}. \tag{31.12}$$

The expressions (31.11) are simplified in two particular cases. For $C = A$ (spherical top) the range of variation for the allowed values E_{rot} contracts to the point at which ρ_{rot} tends to infinity. From the normalization condition (31.12) it is clear that, in this case, ρ_{rot} is represented by the form

$$\rho_{rot}(E_{rot}, J) = \frac{2J}{\hbar^2} \delta(E_{rot} - AJ^2). \tag{31.13}$$

The other limiting case corresponds to a rotator, one of whose moments of inertia vanishes ($I_{zz} = 0, C = \infty$). In this case, relation (31.11) formally gives an infinite range of variation for $E_{\rm rot}$. However, this result is obtained only in the approximation of considering K as a continuous variable. If it is taken into account that K takes a discrete set of values $K = 0, \pm\hbar, \pm 2\hbar$ ..., then, in the case under consideration, it should be restricted to only one value $K = 0$, so that $E_{\rm rot}$ for fixed J takes a single value. Thus we find

$$\rho_{\rm rot}(E_{\rm rot}, J) = \frac{1}{\hbar}\delta(E_{\rm rot} - AJ^2). \tag{31.14}$$

Comparing (31.14) with (31.13) we see that the difference between them is determined only by factor $(2J/\hbar)$, which comes from the different statistical weight of the energy levels of the rotator and the symmetrical top.

Thus, in all subsequent formulae, if convenient, the integration over dK can be replaced by an integration over $dE_{\rm rot}$ on the basis of the following relation:

$$dK/\hbar \to \rho_{\rm rot}(E_{\rm rot}, J)\, dE_{\rm rot}. \tag{31.15}$$

In order to carry out an analogous calculation of the numerator of (31.2), it is expedient to go over to those variables which act directly as arguments of the δ-function.

We will choose as the critical surface the sphere of radius $R^\star$ which separates the molecule AB from the decomposition products A and B in configuration space. We will consider $R^\star$ to be so large that the interaction of A and B can be neglected, so that the total Hamiltonian and total angular momentum can be represented in the form

$$H^\star = H_A + H_B + E_{\rm trans},$$
$$\mathbf{M}^\star = \mathbf{j} + \mathbf{l}, \tag{31.16}$$

where $\mathbf{l}$ is the relative angular momentum of A and B, $\mathbf{j}$ is the intrinsic angular momentum of B, $E_{\rm trans}$ is identified with E_t (as a result of the critical surface chosen).

By analogy with (31.3) we will write $d\Gamma^\star$ in the form:

$$d\Gamma^\star = d\Gamma_B\, d\Gamma^\star_{\rm A-B, rot}$$

$$d\Gamma_B = \rho_{B,\rm vib}(E_{B,\rm vib})\, dE_{B,\rm vib}\frac{dj_z\, dj\, dk}{\hbar^3} \tag{31.17}$$

$$d\Gamma^\star_{\rm A-B, rot} = \frac{dl_z\, dl}{\hbar^2}$$

where the meaning of the angular momenta $j_z, j,$ and k is analogous in meaning to the quantities $M_z, M,$ and K introduced above.

Since the Hamiltonian H^* does not depend on l, l_z, and j_z the integration over them can be carried out, taking the factor $\delta(J - M^*)$ into account. Here, some difficulty arises in that $\mathbf{M}^*$ is a function of the vectors $\mathbf{J}$, $\mathbf{l}$. It is easy to avoid this difficulty, if the well-known addition rule for momenta is employed: for two momenta $\mathbf{a}$ and $\mathbf{b}$, with fixed moduli a and b respectively, the modulus of the momentum $\mathbf{c}$, the vector sum of $\mathbf{a}$ and $\mathbf{b}$, varies between the limits $|a-b| \leqslant c \leqslant |a+b|$.

Thus the integral taken over dl_z and dj_z, for fixed values of l, j, and M^*, gives the density of states of the modulus of the vector $\mathbf{M}^*$, with regard for the degeneracy with respect to directions in space:

$$\int \frac{dl_z\, dj_z}{\hbar^2} = \left(\frac{2M^*}{\hbar}\right) \frac{dM^*}{\hbar}. \tag{31.18}$$

The assumption about the separation of the vibrations and rotations in H_B and the substitution of (31.16) and (31.17) (with regard for (31.18)) into the numerator of (31.2) enables the following formula to be obtained, in which the integration over $dE_{B,vib}$ and dM^* has been carried out:

$$\int \delta(E - H^*)\delta(J - M)\, d\Gamma^* = (2J/\hbar) \int \rho_B(E - E_a - E_{trans}, j) \Phi(J, j, E_{trans})\, dj/\hbar. \tag{31.19}$$

Here, the density of the vibrational–rotational states of the molecule,

$$\rho_B(E, j) = 1/\hbar \int \rho_{B,vib}[(E - E_{rot}(j, k))]\, dk/\hbar, \tag{31.20}$$

and the number of states characterizing different values of l for which decomposition is possible,

$$\Phi(J, j, E_{trans}) = \int dl/\hbar, \tag{31.21}$$

have been introduced.

The integral on the right-hand side of (31.21) depends on J and j, due to the restrictions imposed by the law of conservation of total angular momentum, and depends on E_{trans}, due to the condition of intersection of the critical surface by the representative point. The question of the specific calculation of Φ is considered below in connection with the model of a polarization complex. Here we will only consider the general formula for the decomposition rate constant. Substituting (31.19) and (31.5) into (31.2) we obtain

$$\kappa(E, J) = \frac{\int \rho_B(E - E_a - E_{trans}, j)\Phi(J, j, E_{trans})(dj/\hbar)(dE_{trans}/2\pi\hbar)}{\rho_{AB}(E, J)}. \tag{31.22}$$

The double integration in the numerator takes into account the decomposition for all allowed relative energies of A and B and for all allowed angular momenta of the molecule B.

We will now trace the conditions under which (31.22) goes over into formula (28.13), in which the law of conservation of the total angular momentum has not been taken into account.

It is natural to assume that such a transition takes place in the limit of small J, when the active molecule has small rotational energy. Therefore, in the calculation of $\rho_{AB}(E, J)$, the contribution of $E_{rot}(J, K)$ to E may be neglected and, in the calculation of Φ, the condition $j \gg J$ may be used, enabling $\Phi(J, j, E_{trans})$ to be identified with the number of states of the fragments A and B which differ by momentum l, which lies in the range $j - J \leqslant l \leqslant j + J$. In this way, we find

$$\rho_{AB}(E, J) = (2J/\hbar^2)\rho_{AB,vib}(E)$$
$$\Phi(J, j, E_{trans}) = 2J/\hbar. \qquad (31.23)$$

After the substitution of (31.23) into (31.22) and cancellation of the common factor $2J/\hbar^2$, it is not difficult to see that (31.22) agrees with (28.13) if the level density of the activated complex is related to the density of the vibrational–rotational states of the fragment B in the following way:

$$\rho^\star_{AB}(E) = \int \rho_{B,vib}[E - E_{B,rot}(j, k)]\frac{dj\, dk}{\hbar^2}. \qquad (31.24)$$

Such a relation is completely in keeping with the physical nature of the problem; within the limits of the model under consideration, the number of degrees of freedom of the activated complex is greater by two than the number of degrees of freedom of the polyatomic fragment B, and (31.24) simply expresses the fact that the level density of the system increases with the addition of the two degrees of freedom. In the given case these are the rotational degrees of freedom; within the framework of another model of the activated complex they might be the degrees of freedom of a deformation vibration of the diatomic bond.

Model of a polarization complex

We will consider a simple model for which the function $\Phi(J, j, E_{trans})$ can be found easily. We will assume that the interaction of the fragments A and B is described by a central potential $W(R)$ in the range of the variable R, important in determining the decomposition rate. It is evident that, for $R \to 0$, this assumption is violated since, in the region of the close approach of A and B, the directed nature of the chemical bonds is at variance with a central potential $W(R)$. At sufficiently large R, however, when the interaction is determined by the polarization forces, such an approximation may be satisfactory.

It is convenient in future discussions to introduce the idea of a reaction channel as a set of trajectories corresponding to a given set of quantities J, j, E_{trans}, and l in the phase space of the system $A + B$.

If, in the motion along a trajectory of a given channel, it is possible to go out of the region of phase space of the active molecule AB^* into the region of the final molecules A and B, then such a channel is called open. Otherwise, that is, if the transition is impossible, the channel is called closed.

In accordance with this, the function $\Phi(J, j, E_{\text{trans}})$ may be said to be simply the number of open decay channels of the active molecule AB^*, differing only in the values of the momentum l. The condition that a channel be open imposes an additional restriction on the region of integration over dl in (31.21).

For the model under consideration these restrictions are formulated as the requirement that the radial kinetic energy of the fragment be greater than zero,

$$\frac{\mu \dot{R}^2}{2} = E_{\text{trans}} - W(R) - \frac{l^2}{2\mu R^2} \geq 0, \qquad (31.25\text{a})$$

$$E_{\text{trans}} = E - E_{\text{a}} - E_{\text{B,vib}} - E_{\text{B,rot}}. \qquad (31.25\text{b})$$

This relation must be satisfied by any value of R, including the greatest 'bottleneck' when the effective potential energy $W_{\text{eff}} = W + l^2/2\mu R^2$ is a maximum. The condition for a maximum,

$$\frac{\partial}{\partial R}\left[W(R) + \frac{l^2}{2\mu R^2}\right] = 0, \qquad (31.26)$$

together with relation (31.25a), where $\dot{R} = 0$ is assumed, determines the value of the momentum l^* which separates the regions of open and closed channels. Joint solution of (31.25) and (31.26) also gives the critical value of the reaction coordinate R^* which determines the configuration of the activated complex. For the potential $W(R) = -Q(R_0/R)^n$ the momentum l^* equals [331]:

$$l^*(E_{\text{trans}}) = (2\mu E_{\text{trans}} R_0^2)^{\frac{1}{2}} \left(\frac{nQ}{2E_{\text{trans}}}\right)^{1/n} \left(\frac{n}{n-2}\right)^{\frac{1}{2}-1/n}. \qquad (31.27)$$

Here, for convenience for future analysis, the gas-kinetic radius R_0 of the pair $A + B$ is introduced and also the energy of attraction Q, which would be obtained by a formal extrapolation of the potential from R to R_0 (in reality, near R_0 the potential of the attraction W is appreciably changed due to the contribution of the repulsive force. Assuming $Q = ae^2/2R_0^4$, $n = 4$, in (31.27) we find for the ion–molecule polarization potential:

$$l^\star(E_{\text{trans}}) = (8\alpha e^2 \mu^2 E_{\text{trans}})^{\frac{1}{4}}. \tag{31.28}$$

Since the gas-kinetic radius is determined by the condition of approximate equality of the long-range and short-range forces, the condition for the validity of the assumed approximation for W is

$$\frac{l^\star}{\mu v} \gg R_0. \tag{31.29}$$

In accordance with (31.25b), E_{trans} depends on j for fixed values of E and $E_{\text{B,vib}}$. The substitution of $E_{\text{trans}}(j)$ from (31.25b) into (31.27) gives the function $l^\star = l^\star(j)$. The corresponding curve separates the regions of open and closed channels on the j, l coordinate plane (see Fig. 31). The former region is enclosed between the coordinate axes and the hatched curve 2.

In these same coordinates, it is now necessary to demonstrate the possible range of variation of j and l, for which the vector sum of $\mathbf{j}$ and $\mathbf{l}$ would give the total momentum $\mathbf{J}$. In Fig. 31 is shown the range of variation of the momenta j and l contributing to the sum J (region 1). The overlap region of these two parts of the plane also gives the range of variation of the momenta j and l for which the decay channels are open and the law of conservation of the total

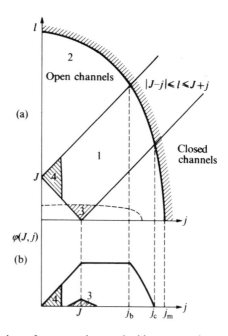

FIG. 31. Phase plane of a system characterized by two angular momenta j and l.

momentum is satisfied. In this representation, the integration over j and l has already been carried out, so that $\Phi(J, j, E_{\text{trans}})$ simply equals the integral $\int dl$ over region 1. A graph of this function is shown in Fig. 31(b), which also demonstrates a correlation between the shape of $\Phi(J, j)$ and some characteristic values of j ($j = J, j_b, j_c$). From Fig. 31(a), it is evident that, depending on the relationship between the momenta j_m, l_m, and J (where j_m and l_m are the maximum values which can be achieved with conservation of total energy), two limiting cases are possible.

1. On condition $l_m \ll J \leqslant j_m$, the allowed zone of the phase plane corresponds to region 3. This takes place, for example, in the dissociation of a hydrocarbon molecule RH as a result of electron impact. In this case, as is clear from the diagram, Φ is approximately represented by a triangular function:

$$\Phi(J, j, E_{\text{trans}}) = l_m(E_{\text{trans}}) - |J - j|, \quad l_m(E_{\text{trans}}) = l^*(E_{\text{trans}} + E_{\text{rot}}). \quad (31.30)$$

Neglecting the rotational energy E_{rot} in comparison with the vibrational, the integration over dj and dk can be carried out explicitly, and gives

$$\rho(E, J) = \left(\frac{2J}{\hbar^2}\right) \rho_{\text{vib}}(E)$$

$$\int \Phi(J, j, E_{\text{trans}}) \frac{dk}{\hbar} \cdot \frac{dj}{\hbar} = \frac{2J}{\hbar^2} \cdot \frac{l_m^2(E_{\text{trans}})}{\hbar^2} \approx \frac{2J[l^*(E_{\text{trans}})]^2}{\hbar^3}$$

$$\int \rho_{B,\text{vib}}(E - E_a - E_{B,\text{rot}} - E_{\text{trans}}) \Phi(J, j, E_{\text{trans}}) \frac{dk \, dj \, dE_{\text{trans}}}{2\hbar^3 \pi}$$

$$= \frac{2J}{\hbar^2} \int \rho_{B,\text{vib}}(E - E_a - E_{\text{trans}})[l^*(E_{\text{trans}})]^2 \frac{dE_{\text{trans}}}{2\pi\hbar}. \quad (31.31)$$

Substituting (31.31) into (31.2) we find for the polarization interaction for $n = 4$:

$$\kappa(E) = \frac{1}{2\pi\hbar\rho_{AB,\text{vib}}(E)} \int \rho_{B,\text{vib}}(E - E_a - E_{\text{trans}}) \frac{4\mu}{\hbar^2} \left(\frac{e^2\alpha}{2}\right)^{\frac{1}{2}} (E_{\text{trans}})^{\frac{1}{2}} dE_{\text{trans}}. \quad (31.32)$$

This expression can be obtained on the basis of statistical theory, without taking account of the conservation of angular momentum, when the activated molecule is represented by the representative point at the peak of the centrifugal barrier [305].

We will now consider the limits of applicability of the simplest variant of the statistical theory, not taking account of the conservation of total angular momentum. It is not difficult to ascertain that the inequalities [75], [76]

$$l_m \ll J \ll j_m \qquad R_0 \ll R^* \quad (31.33)$$

take the following form

$$\frac{\mu^2}{m_B} \ll \frac{E_{rot}^2(J)}{E_{trans}Q} \ll \frac{\bar{E}_{trans}}{Q} \ll 1, \tag{31.34}$$

where μ is the reduced mass of A and B; m_B is the mass of B; $E_{rot}(J) \approx J^2/2m_B R_0^2$; and the mean energy $\bar{E}_{trans}$ is substituted for E_{trans}.

We will now examine to what extent condition (31.33) is satisfied in actual cases of dissociation of ions. Typical values of the mean energy $\bar{E}_{trans}$, found experimentally, vary over the range 0·1–0·2 eV [305]. For the separation of an H atom from the hydrocarbon ions it should be assumed that $\mu = 1$, $m_B \approx 50$, $R_0 \approx 4$ Å, $\alpha = 1$ a.u., $Q \approx 2$ eV. Then it is not difficult to see that (31.34) is satisfied for values of the rotational energy E_{rot} which correspond to a temperature exceeding 500 K. For the detachment of heavier fragments from an excited ion, the range of variation of $\bar{E}_{trans}$, determined by relation (31.34), shrinks considerably.

2. If the relation $j_m \ll J \ll l_m$ is satisfied by the components l_m, J, and j_m, then the allowed zone of phase space corresponds to region 4. In this case l varies close to J, and we have

$$\Phi(J, j, E_{trans}) = 2j, \tag{31.35}$$

whence we obtain

$$\kappa(E, J) = \frac{\int \rho_{B,vib}(E - E_a - E_{rot} - E_{trans})\rho_{B,rot}(E_{B,rot}, j) 2j \, dj \, dE_{trans} \, dE_{rot}}{2\pi\hbar^3 (2J/\hbar)\rho_{AB,vib}(E)}, \tag{31.36}$$

where the integration over dj, dE_{rot}, and dE_{trans} goes over all open channels. In particular, for a diatomic fragment B for which rotation about the axis of symmetry is absent, expression (31.14) should be substituted for ρ_{rot}. For this case we obtain

$$\kappa(E, J) = \frac{\int \rho_{B,vib}(E - E_a - E_{trans} - Aj^2) 2j \, dj \, dE_{trans}}{2\pi\hbar^3 \rho_{AB,vib}(E) 2J/\hbar}. \tag{31.37}$$

From (31.37) it is evident that the rate constant does not, in general, depend on the characteristics of the potential $W(R)$. This means that the activated molecule cannot be represented by a representative point at the peak of the centrifugal hump. The limiting stage of decomposition results not in the passage of the system over the potential hump but in the rotational excitation of B. This excitation is accompanied by an increase in the intrinsic momentum **j** of the molecule B which, being added vectorially to the total momentum **J**, extends the range of allowed values of l and increases the number of open channels. It is evident that, for processes of this type, the law of conservation of total momentum is very important.

We note finally that the condition $j_m \ll J \ll l_m$ runs

$$\frac{\bar{E}_{trans}}{Q} \ll 1, \quad \frac{\bar{E}_{trans}}{Q} \ll \frac{E_{rot}^2(J)}{\bar{E}_{trans}Q} \ll \frac{\mu^2}{M^2}. \tag{31.38}$$

Dissociation of molecules under electron and photon impact

The statistical theory of decomposition was used by a number of authors [192], [271]–[273] to calculate the distribution function over the rotational states of molecule B, on dissociation. Dissociation takes place as the result of electronic excitation of the molecule into an unstable state due to collisions with electrons or photons. The complete process is described by the scheme

$$AB(E, J) + \begin{Bmatrix} e \\ h\nu \end{Bmatrix} \rightarrow AB^\star(E', J') + \{e\} \rightarrow A(j_A) + B(j_B), \tag{31.39}$$

where J' is very close to J, so that the photon (or electron) changes its initial angular momentum very little.

The authors, mentioned above, assumed that the distribution functions $f^A(j_A)$ and $f^B(j_B)$ are proportional to that part of the phase volume $\Delta\Gamma(j)$ of the system $AB^\star$ from which there is an exit into the region of the reaction products A and B, with account taken of the laws of conservation of energy and momentum. Since $f(j)$ is proportional to the total number of molecules, the assumption $f^A(j_A) \approx \Delta\Gamma^A(j_A)$ implies that the decomposition rate by channel j_A is considered to be proportional to the portion of the phase volume $\Delta\Gamma^A$ and not to the current $d\Gamma^A/dt$ flowing out of this part.† This assumption may be interpreted in the sense that, in the excited state, the energy does not have time to be redistributed, but the excited molecules already arise with a distribution over phase volume, which corresponds to filling in this volume uniformly. It is evident that $\Delta\Gamma(E', J', j_B)$ can be obtained from (31.5) by integrating the current over the time along the reaction coordinate. Assuming that $dt = dR/v$, we find from (31.5):

$$\Delta\Gamma(E', J', j_B) = \int \frac{d\Gamma}{dt} dt \approx \int \rho^\star(E' - E_0 - E_{trans} - H_{B,rot})$$

$$\times dk\Phi(J, j_B, E_{trans}) \frac{dR}{v} dE_{trans}. \tag{31.40}$$

We will now examine what form this expression takes for the decomposition of a triatomic molecule, if the excitation of vibrations can be neglected. Moreover, for simplicity, we will assume the validity of the inequality $l_m \gg j_m$.

† In the original theory, Light [331] also proceeded from this assumption. However, it was subsequently abandoned [398] (see also [514]).

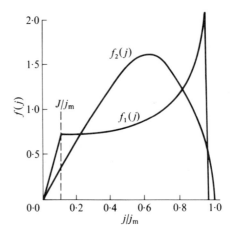

FIG. 32. Normalized distribution functions of the rotational states of diatomic fragments: $f_1(j)$ is the distribution function of the triatomic molecule (H_2O) on dissociation; $f_2(j)$ is the distribution function of a tetratomic molecule (H_2O_2) on dissociation into two similar fragments. In both cases, the total angular momentum of the system $J = 0.1\, j_m$.

A calculation carried out under this assumption gives

$$f_1^B(j_B) = \Phi(J, j_B, E' - E_0)\left[1 - \frac{j_B^2}{2I_B(E' - E_0)}\right]^{-\frac{1}{2}}. \quad (31.41)$$

The function f_1^B, normalized to unity, is shown in Fig. 32. Just such a type of distribution over the rotational states was observed experimentally in the investigation of the spectra of the OH radical arising from the decomposition of H_2O as a result of electron, and photon impact [271], [272]. For this case, the condition $l_m > j_m$ is hardly satisfied, therefore the more exactly calculated distribution turns out to be somewhat smoother in the region of large j in comparison with the distribution (31.41).† It is not difficult to see that the conservation of the total angular momentum appreciably affects the form of the distribution: the simple statistical theory, which does not take the conservation laws into account properly, describes the distribution only for $j < J$.

We will now consider the case when both decomposition products possess finite moments of inertia. It is then necessary to consider three momenta: the two momenta $\mathbf{j}_A$ and $\mathbf{j}_B$, corresponding to the internal rotation of the products, and the relative momentum $\mathbf{l}$. On redistribution of the energy, the sum of $\mathbf{j}_A$, $\mathbf{j}_B$, and $\mathbf{l}$ must remain constant, equal to the total momentum $\mathbf{J}$. The function $\Phi(J, j_A, j_B, E_{\text{trans}})$ can be found by successively considering the addition of $\mathbf{j}_A$

† The limiting effect of l_m on the distribution function $f^B(j_B)$ is considered in the paper [192].

and **l** to form the vector **κ**, and then **κ** and **j**$_B$ to form the vector **J**. An analytic expression for $\phi(J, j_A, j_B, E_{trans})$ on condition $j_A, j_B, J \ll l_m$ has the form [272]

$$\Phi(J, j_A, j_B, E_{trans}) = \begin{cases} 4j_A j_B & \text{when } j_A + j_B \leqslant J \\ -(J-j_B)^2 + 2(J+j_B) - j_A^2 & \text{when } |j_A - j_B| \leqslant J \leqslant j_A + j_B \\ 4Jj_A & \text{when } j_B \geqslant J + j_A \\ 4Jj_B & \text{when } j_A \geqslant J + j_B. \end{cases} \quad (31.42)$$

For comparison we will quote the expression for $\Phi(J, j_B, E_{trans})$, valid on condition $j_B, J \leqslant l_m$:

$$\Phi(J, j_B, E_{trans}) = \begin{cases} 2j_B & \text{when } j_B \leqslant J \\ 2J & \text{when } j_B \geqslant J. \end{cases} \quad (31.43)$$

The distribution function over the states of a single fragment, analogous to f_1^B, equals

$$f_2^B(j_B) = \int \Phi(J, j_A, j_B, E_{trans}) \, dj_A \left[1 - \frac{j_A^2}{2I_A(E'-E_0)} - \frac{j_B^2}{2I_B(E'-E_0)} \right]^{-\frac{1}{2}}. \quad (31.44)$$

The calculation of this function by another method for $I_A = I_B$ is carried out in the paper [272] with a view to interpreting the distribution over the rotational states of the OH radicals arising as a result of the dissociation of H_2O_2 on electron impact. The diagram of the normalized function $f_2^B(j_B)$ taken from the paper [272] is shown in Fig. 32.

As is evident from (31.42) the simple statistical model is applicable on condition $j_A + j_B \leqslant J$. This follows from the first equality since, only on this condition is $\Phi(J, j_A, j_B, E_{trans})$ equal to the statistical weight of the two independent rotations $4j_A j_B$. As a concrete example for which this condition is satisfied, we may cite the vibrationally-excited ethane molecule, which in the activated state can be represented by two independent three-dimensional rotators (CH_3 groups).

Although application of statistical theory to the calculation of the distribution functions in reactions of the type considered above is not sufficiently well founded, nevertheless, both in these cases and in many others, the statistical theory furnishes an important limiting case, which should always be borne in mind in the interpretation of experiments.

32. The isotope effect

One of the interesting consequences of the quantum statistical theory of unimolecular reactions is the prediction of various isotope effects due to the dependence of the rate constant on the isotopic composition of the molecule.

One of the reasons for the dependence of the rate constant on the isotopic composition of the molecule is that the frequency factor of the reaction rate constant at high pressures depends on the mass of the atoms. This dependence is very complicated, generally speaking, since the reaction coordinate is related to the effective motion of all the atoms of the molecule. However, for the classical harmonic model, it can be shown that if the reaction coordinate corresponds to the rupture of the bond between atoms A and B then the ratio of the frequency factors v'/v for the isotopic substitutions $A \to A'$ and $B \to B'$ is $(\mu_{AB}/\mu_{A'B'})^{\frac{1}{2}}$. Thus a reduction in the mass of atom A or B leads to an increase in the pre-exponential factor. On taking account of the quantization of the vibrations and its effect on the rate of decomposition under equilibrium conditions, it is found that, in addition to the isotope effect in the frequency factor, the effective energy of dissociation also changes. Its change is connected with the dependence of the energy of the zero-point vibrations of the molecule on the mass of the atoms. Since this energy increases on a reduction in the atomic mass, the effective energy of dissociation, equal to the difference between the energies of the initial molecule and the reaction products, usually decreases, which leads to an increase in the decomposition rate for molecules with heavy isotopes. These effects are called *normal primary* isotope effects (see § 5).

Under non-equilibrium conditions, changes in the masses of the atoms reveal themselves not only by a change in the activation energy and the frequency factor but also by a change in the energy-level density of the active molecule. Consider, for example, the decomposition of molecules with an initial vibrational distribution of a δ-function form. On the basis of (28.13), the rate constant can be written in the form (for simplicity we will neglect the partition functions of the adiabatic degrees of freedom here and assume the absence of internal rotations):

$$\kappa(E) = \frac{\sum_{E^*_{\text{vib}} \leq \Delta E} \rho(E^*_{\text{vib}})}{2\pi\hbar \rho^*(E_0 + \Delta E)}, \qquad \Delta E = E - E_0, \tag{32.1}$$

where ΔE is the excess of energy above the threshold.

In the limit as $\Delta E \to 0$, the number of possible states of the activated molecule equals one, so that the ratio of the rate constants for light and heavy molecules is inversely proportional to the density of the vibrational levels close to threshold. Since for heavy molecules the density of vibrational levels is greater than for light molecules, the rate of decomposition will be lower for a heavy molecule (*normal secondary* isotope effect).

As an illustration of the normal secondary isotope effect, we cite the results of an experimental investigation of the isomerization of substituted cyclopropane molecules. The reaction consists of the opening of the cyclopropane ring and its conversion into the propylene structure. In order to separate out just the secondary effect to the greatest extent, it is necessary to study molecules in which, if possible, isotopic substitution does not change the nature of the bonds in the molecules taking part directly in the isomerization reaction. In other words, the reaction coordinate must be isolated as much as possible from the coordinate describing the motion of the substituted atom.

Dorer, Rabinovitch, and Placzek [213], [214] investigated the isomerization of molecules of the type $CH_3(CH_2)_nC_3H_5$ and $CF_3(CH_2)_nC_3H_5$. The initial energy of these molecules is approximately 110 kcal mol^{-1}, whereas the activation energy of the isomerization changes, for different molecules, over a small range close to 63–65 kcal mol^{-1}. The experimental results, and also the results of the theoretical calculation of the isomerization rate constant within the limits of the statistical model, are shown in Table 7. The model of the activated molecule was chosen so that it would give the correct values for the frequency factors v in thermal reactions, which for different molecules vary around a mean value of approximately 7×10^{14} s^{-1}. Of especial note is the correspondence between the calculated and experimental values of the rate constants (and not just the ratios) for the isomerization of propylcyclopropanes, for which the difference between the activation energy ΔE_a of the fluorosubstituted and ordinary molecules is less than 1 kcal mol^{-1}. Even for other pairs, for which the difference in activation energies is greater, the small variation in the value ΔE_a does not destroy the satisfactory agreement between theory and experiment which is a very serious test of the applicability of the statistical theory.

For thermal decomposition at low pressures (see § 33), which is also characterized by strong violation of the equilibrium distribution, the increase in the density of the vibrational levels close to threshold, on condition $\hbar\omega_i \gg kT$, increases the rate of decomposition, thus leading to the *inverse secondary* isotope effect.

TABLE 7

The ratio of the rate constants for the non-equilibrium isomerization of isotopically substituted molecules of cyclopropane [213], [214]

Molecule	$(\kappa_H/\kappa_F)_{theor}$	$(\kappa_H/\kappa_F)_{exp}$
$CH_3-C_3H_5$ $CF_3-C_3H_5$	10	8·1
$CH_3-CH_2-C_3H_5$ $CF_3-CH_2-C_3H_5$	4·2–5·6	4·5
$CH_3-CH_2-CH_2-C_3H_5$ $CF_3-CH_2-CH_2-C_3H_5$	5·7–4·2	4·8

Although the experimental isolation of any one type of isotope effect is very difficult, these effects have nevertheless been detected in various non-equilibrium systems (see review [420]).

33. The application of the statistical theory to thermal reactions

If the active molecule arises in thermal conditions, as a result of intermolecular energy exchange, then the statistical theory of unimolecular reactions enables the rate constant of the thermal process to be calculated. In spite of this, however, an additional assumption must be introduced into the statistical theory, regarding the mechanism of activation and deactivation in collisions. At present, we do not have the information available which would enable the construction of a detailed theory of activation. It is necessary therefore to consider various models, of which the model of strong collisions (see § 23) and the model of stepwise excitation (activation) (see § 35) are two limiting cases.

In the high-pressure limit (first-order reaction), when the intermolecular energy exchange rate becomes higher than the intramolecular exchange rate, both models give identical results, since any activation mechanism in these conditions ensures the equilibrium distribution of the active molecules over the degrees of freedom in the region of phase space which makes the main contribution to the reaction rate. In the low-pressure region (second-order reaction) both models give the same results, provided the number of effective collisions responsible for activation is regarded as an empirical parameter in both theories. The difference between the two models which cannot be eliminated by any choice of the arbitrary parameters occurs only in the intermediate pressure region, where the order of the reaction changes from the first to the second.

In this section, we will consider the expression for the rate constant for thermal reactions in the high- and low-pressure region and will quote the general formula for the rate constant in the intermediate region, obtained within the limits of strong collision theory [32].

Reactions of the first order

In the high-pressure region, the rate constant κ_∞ is determined by the general formulae of the transition state theory (see Chapter 1):

$$\kappa_\infty = g^\star \chi \frac{kT}{2\pi\hbar} \cdot \frac{F^\star_{AB}}{F_{AB}} \exp\left(-\frac{E_a}{kT}\right). \tag{33.1}$$

It is convenient, in future, to represent the partition functions $F^\star_{AB}$ and F_{AB} of the activated and unexcited molecule AB in the form of a product of the partition functions F_1 and F_2 of the adiabatic and active degrees of freedom:

$$\kappa_\infty = g^* \chi \frac{kT}{2\pi\hbar} \cdot \frac{F^*_{1,AB}}{F_{1,AB}} \cdot \frac{F^*_{2,AB}}{F_{2,AB}} \exp\left(-\frac{E_a}{kT}\right). \tag{33.2}$$

In this expression the active and adiabatic degrees of freedom enter with equal importance. This is understandable, since frequent collisions in fact obliterate any difference between them, leading to transitions between states in both the system of active degrees of freedom and in the system of adiabatic degrees of freedom. If the molecule is represented by a system of harmonic oscillators and the transmission coefficient is close to unity then, from (33.2), it is not too difficult to see that, at high temperatures ($kT \gg \hbar\omega$), the pre-exponential factor is of the order of the frequency of the molecular vibrations. Moreover, from the theory of small oscillations the factor $(kT/2\pi\hbar) \cdot (F^*_{AB}/F_{AB})$ in (33.2) can be shown always to lie between the lowest and highest frequencies of the molecule. Nevertheless, in principle, the general treatment does not restrict the upper value of the pre-exponential factor. However, since the frequency factor for most of the unimolecular reactions studied experimentally proves to be of the order of the frequency of the vibrations, cases of reactions where the pre-exponential factor exceeds a value $\sim 10^{14}\, \text{s}^{-1}$ are often considered in the literature, to be reactions with an anomalously large pre-exponential factor.

In connection with these remarks, with reference to the oscillatory model of the molecule AB, it can be affirmed that the high pre-exponential factors are connected with the manifestation of large anharmonicity in the intramolecular vibrations. Anharmonicity leads to the loosening of the intramolecular bonds and to the formation of a weakly bound activated molecule. This effect is often referred to as a high entropy contribution of the activated molecule to κ_∞ [432], [494], [502], [503]. The simplest and, from the quantitative point of view, the most important manifestation of anharmonicity on dissociation appears in the transformation of certain types of vibrations of the active molecule into the free rotation of the fragments of the activated molecule.

We will estimate the order of magnitude of the contribution made by the transformation of a single vibrational degree of freedom of the molecule AB into a rotational degree of freedom of the activated molecule. Under classical conditions, with $kT \gg \hbar\omega_i$, we have

$$F_{i,\text{vib}} = \frac{kT}{\hbar\omega_i}, \qquad F_{i,\text{rot}} = \frac{(2\pi I kT)^{\frac{1}{2}}}{2\pi\hbar}. \tag{33.3}$$

Evaluating the frequency ω on the basis of the Morse potential, we obtain $F_{\text{rot}}/F_{\text{vib}} \approx (D/kT)^{\frac{1}{2}}$, where D is an energy of the order of a chemical bond's energy.

As an example for which the frequency factor proves to be anomalously large, we cite the thermal decomposition of ethane into two CH_3 radicals. A calculation carried out by Setser and Rabinovitch [472] represents the C_2H_6 molecule in the activated state by a system of 12 oscillators (two non-interacting CH_3 radicals) and 5 active rotators (the free internal rotation of the CH_3 group around the C–C bond and 4 degrees of freedom, corresponding to a rotation of the plane of the H_3 atoms of each CH_3 group). The C–C bond is the reaction coordinate. The calculation gives a value $\approx 10^{17}\,s^{-1}$ at 600°C for the pre-exponential factor. This value agrees with the known value, which can be obtained from the experimental value of the bimolecular recombination constant, with the help of the connection between the dissociation and recombination constants, in terms of the equilibrium constant.

Formula (33.1) for the rate constant κ_∞ is exact if by F^* is understood the total partition function of the activated molecule. All the difficulty of calculating the rate constant is transferred to formulating the problem of the correct determination of the critical surface. The exact calculation of χ is equivalent, in fact, to the solution of the many-body problem, which means giving up the simplifying assumptions regarding the possibility of introducing a reaction coordinate. However, since we assume that it can be introduced for the calculation of χ, the expressions of §§ 3, 15–17, and 27 can be used.

Calculating the partition function in (33.1) and representing the decomposing molecule by a system of s oscillators and r rotators, and the activated molecule by a system of s^* oscillators and r^* rotators, we obtain the following formulae for the temperature dependence of the pre-exponential:

$$A(T) \sim \chi(T)T^{(r^*-r)/2+1} \quad \text{for } \hbar\omega_i \gg kT, \tag{33.4a}$$

$$A(T) \sim \chi(T)T^{(r-r^*)/2} \quad \text{for } \hbar\omega_i \ll kT, \tag{33.4b}$$

where, usually, $r^* - r > 0$. Thus, excluding the transmission coefficient $\chi(T)$, A is characterized by a positive temperature-dependence at low temperatures and a negative temperature-dependence (or no dependence for $r^* = r$) at high temperatures.

Reactions of the second order

The thermal reaction rate at low pressures can be calculated by the general formulae of Chapter 4. However, here, it should also be taken into account that in the presence of adiabatic degrees of freedom, the reaction condition cannot be formulated in the form of an energy relation $E > E_a$, since the conservation of the quantum numbers of these degrees of freedom during the time between successive collisions imposes definite restrictions on the intramolecular redistribution of energy. In contrast to the approach taken in § 23, we will characterize the molecule AB by the total energy E and the

§ 33   THE STATISTICAL THEORY OF REACTIONS   277

set of quantum numbers J of the adiabatic degrees of freedom. Introducing the density of states of the active molecule in the area $dE\, dJ$, we will write the rate constant κ in the form

$$\kappa_0 = \frac{1}{F} Z^* \int \rho(E, J) \exp(-\beta E) \beta\, dE\, dJ. \tag{33.5}$$

This expression corresponds to the assumption of the mechanism of strong collisions; at low pressures the decomposition rate equals the equilibrium activation rate. In order to find the limits of integration in (33.5) over dE and dJ, we should go over from the parameters which characterize the active molecule to the parameters of the activated molecule. To this end, we will represent $\rho(E, J)$ in the form

$$\rho(E, J) = \rho_2(E^*), \tag{33.6}$$

and express the energy of the active degrees of freedom E^* in terms of the energy of the activated state. From (33.5), we find

$$\kappa_0 = \frac{1}{F} Z^* \int \rho_2(E_a + E^* + E_j^* - E_j) \exp(-\beta E_a - \beta E^* - \beta E_j^*)\, dE^*\, dJ. \tag{33.7}$$

Assuming now that, under a variation of E_j^*, ρ_2 changes more slowly than $\exp(-\beta E_j^*)$, the integration can be carried out over dJ, replacing $E_j^* - E_j$ by the mean value $\langle E_j^* - E_j \rangle$, calculated with the distribution function of the activated complex. The integral over dJ gives the partition function of the adiabatic degrees of freedom of the activated state, so that (33.7) takes the form

$$\kappa_0 = Z^* \frac{F_1^*}{F} \left[\int_0^\infty \rho_2(E_a + E^* + \langle E_j^* - E_j \rangle) \exp(-\beta E^*)\, dE^* \right] \exp(-\beta E_a), \tag{33.8a}$$

$$\kappa_0 \approx Z^* \frac{F_1^*}{F_1 F_2} \rho_2(E_0 + \langle E_j^* - E_j \rangle) kT \exp\left(-\frac{E_a}{kT}\right). \tag{33.8b}$$

Expression (33.8b) is obtained on the assumption that $\rho_2(E)$ changes very much slower than the exponential $\exp(-\beta E)$ near E_0. It would be possible to take ρ_2 out of the integral (33.8a) and replace E^* by the mean value kT; however, this correction is usually small. From this, it is also clear that it makes sense to leave the mean value $\langle E_j^* - E_j \rangle$ in (33.8b) only when it appreciably exceeds kT. As a rule, the mean value $\langle E_j^* - E_j \rangle$ is less than zero. For example for r adiabatic rotators, its value equals

$$\langle E_j^* - E_j \rangle = \sum_{n=1}^{r} kT(1 - I_n^*/I_n). \tag{33.9}$$

Therefore the level density $\rho_2(E_a + \langle E_j^* - E_j \rangle)$ turns out to be somewhat too low in comparison with the analogous quantity calculated without taking account of the effect of the adiabatic degrees of freedom. At the same time, the ratio F_1^*/F_1 exceeds unity, so that the total contribution of the adiabatic degrees of freedom leads to an increase in the reaction rate. Physically, this is not difficult to understand: although the adiabatic degrees of freedom hinder the statistical redistribution of energy within the molecule, at the same time, their energy contribution leads to an effective reduction in the activation energy. An analogous effect also occurs for diatomic molecules, where the rotation reduces the dissociation energy owing to the centrifugal extension of the bond (see § 41). In the limiting case, it can be shown that the energy of the adiabatic degrees of freedom does not change in the transition from the active to the activated molecule. Under these conditions $F_1^*/F_1 = 1$ and $\langle E_j^* - E_j \rangle = 0$, i.e. the adiabatic degrees of freedom become inactive, so that they do not in general contribute to κ_0.

In considering the temperature dependence of the pre-exponential in the expression (33.8b), we will neglect the adiabatic degrees of freedom, representing the molecule by a system of harmonic oscillators and rigid rotators. The additional problem (in relation to the statistical theory) of the temperature dependence of the effective number of collisions Z^* arises here. Within the framework of the assumptions of strong collision theory, Z^* is usually identified with the gas-kinetic number of collisions Z_0:

$$Z^* = Z_0 \sim T^{\frac{1}{2}}. \qquad (33.10)$$

It is possible, however, that, in each collision, energy is transferred which, on average, is small in comparison with kT. Then the expression for the rate constant κ_0, written formally in the form of (33.8b), will contain as a factor some effective number of collisions Z^*, which may depend on the temperature in a very complicated way. Some information about Z^* can be obtained from definite activation models. In particular, Buff and Wilson [174] assumed that single-quantum transitions occurred on collision of AB with molecules of the reservoir. The probability they obtained is given by a formula which is a generalization of the Landau–Teller formula for a system of s identical harmonic oscillators:

$$P_{n+1, n} = (n+s) P_{10}. \qquad (33.11)$$

The solution of the system of equations describing stepwise activation and decomposition, whose spontaneous rate, by assumption, is given by formula (25.15), leads to the following expression for κ_0:

$$\kappa_0 = Z_0 P_{10}[1 - \exp(-\hbar\omega\beta)] N \frac{(N+s-1)!}{(s-1)! N!} \exp[-N(\hbar\omega\beta)][1 - \exp(-\hbar\omega\beta)]^{s-1}.$$

$$(33.12)$$

§ 33 THE STATISTICAL THEORY OF REACTIONS

For the model $Z^* = Z_0 P_{10}(1 - e^{-\hbar\omega\beta})$, where $Z^* \ll Z_0$ in absolute magnitude and increases rapidly with increase in temperature.

If the conditions of classical statistics are satisfied and the mean energy exchange turns out to be much less than kT, then diffusion theory (see Chapter 6) gives the following expression for Z^*:

$$Z^* = Z_0 \frac{\langle \Delta E^2 \rangle}{(kT)^2}. \tag{33.13}$$

It is natural to assume that, in the activation of polyatomic molecules, combinational transitions occur with high probability when different normal vibrations are excited and deactivated simultaneously. The value of the energy transfer here is not restricted by any conditions. The only criterion which can be used in the evaluation of $\langle \Delta E^2 \rangle$ is probably the Massey criterion; the energy transferred with maximum probability satisfying the condition [55]:

$$\frac{\Delta E^\star}{\hbar \alpha \bar{v}} \approx 1, \tag{33.14}$$

where $\bar{v}$ is the mean velocity of the colliding molecules; $1/\alpha$ the range of the intermolecular potential.

On the basis of (33.14), assuming $\langle \Delta E^2 \rangle \sim (\Delta E^\star)^2$, we find

$$\frac{\langle \Delta E^2 \rangle}{(kT)^2} \sim \frac{1}{T}. \tag{33.15}$$

In accordance with this estimate, the stepwise excitation mechanism leads to the appearance of an additional factor T^{-1} in comparison with the mechanism of strong collisions.

From (33.8b), it is evident that the dependence of the pre-exponential on the temperature is determined, apart from the factor $Z^*(T)$, by the factor $F_2(T)$. In this way we find

$$A(T) \sim Z^*(T) T^{-r/2+1} \prod_{i=1}^{s} \left[1 - \exp\left(-\frac{\hbar\omega_i}{kT}\right) \right], \tag{33.16}$$

where r is the number of active rotational degrees of freedom and s is the number of normal vibrations of the molecule AB.

If the semiclassical approximation (30.10) is used to calculate the level density, then the expression for κ_0 (33.8) takes the form

$$\kappa_0 = Z^* \frac{\left(\frac{E_a + E_z}{kT}\right)^{s+r/2-1}}{\Gamma(s+r/2)} \cdot \frac{F_{2,\text{class}}}{F_2} \exp\left(-\frac{E_a}{kT}\right), \tag{33.17}$$

where $F_{2,\text{class}}$ and F_2 are the classical and quantum partition functions of the active degrees of freedom respectively.

At high temperatures, when $F_{2,\text{class}}/F_2 \to 1$ and with $r = 0$, (33.17) goes over, of course, into Kassel's classical formula (25.11). At low temperatures ($\hbar\omega_i > kT$), however, the pre-exponential (33.17), equal to

$$A = Z^* \frac{(E_a + E_z)^{s-1} kT}{(s-1)! \prod_{i=1}^{s} \hbar\omega_i}, \qquad (33.18)$$

differs from the corresponding factor in Kassel's quantized formula, which is obtained from (25.19) within the limits of the semiclassical approximation

$$A_{\text{Kass}} = \frac{Z^* \left(\dfrac{E_a + E_z}{\hbar\omega}\right)^{s-1}}{(s-1)!}. \qquad (33.19)$$

This difference is explained by the fact that, in Kassel's model, the discrete form of the spectrum for any energy of excitation is retained whereas, within the framework of the statistical theory, it is assumed that close to the level $E = E_a$ the spectrum is, in practice, continuous. In this connection, the calculations carried out in the work [402], showing how a small spread in the frequencies of the normal vibrations affects the structure of the vibrational spectrum at high excitation energies, are very instructive.

The intermediate pressure region

In the transitional region, where the rates of intramolecular and intermolecular exchange are comparable, the expression for $\kappa(T)$ is given by formula (23.45), valid for the mechanism of strong collisions. Substituting the explicit expression for $\kappa(E)$ into (23.45), we obtain

$$\kappa(T) = \frac{kT}{2\pi\hbar} \frac{F_1^*}{F_1 F_2} \exp\left(-\frac{E_a}{kT}\right) \int_{E^*=0}^{\infty} \frac{g^* W_2^*(E^*) \exp(-E^*/kT) \, dE^*/kT}{1 + \dfrac{g^* W_2^*(E^*)}{2\pi\hbar\rho_2(E) Z^*}} \qquad (33.20)$$

where, for simplicity, no account has been taken of the adiabatic degrees of freedom† and it has been assumed that $\chi = 1$.

For the mechanism of stepwise excitation, attempts to write down an explicit expression for $\kappa(T)$ in the transition region, in the general case, have proved unsuccessful, since the non-equilibrium distribution function must be determined from the solution of a system of coupled equations or from the Fokker–Planck equation (see Chapter 6). Some results, however, can be obtained for the comparatively simple models considered in § 39.

† The effect of the adiabatic degrees of freedom, which is relatively small for polyatomic molecules, is taken into account in the work of Marcus [352].

The thermal decomposition of nitrous oxide

As an illustration of the application of the theory of unimolecular decomposition, we will consider the thermal decomposition reaction of nitrous oxide. The rate constant of this reaction has been experimentally determined in the high, intermediate, and low pressure regions [226], [286], [393], [433]. The reaction proceeds nonadiabatically, and was first considered from this point of view by Stearn and Eyring [501]. Using the experimental decomposition rate constant at high pressures and the activation energy known at the time, $E_a = 52$ kcal mol^{-1}, the authors found that the probability of a nonadiabatic transition between the terms $^1\Sigma^+$ and $^3\Pi$ must be of the order 10^{-4}. Later experiments [286] gave the value 60 kcal mol^{-1} for the activation energy. In this case, as Gill and Laidler [241] showed, the nonadiabatic transition probability should be 10^{-1}–10^{-2}. The experimental value of the frequency factor of the rate constant κ_∞ at $T = 888$ K is 8×10^{11} s^{-1}, and may reasonably be interpreted within the limits of the statistical theory on the basis of formula (33.1).

Gill and Laidler [241] applied the classical variant of formula (33.20) to the calculation of the rate constant at intermediate pressures, although the condition of applicability of this variant ($\hbar\omega_i \ll kT$) is not satisfied at the temperatures of the experiment (characteristic temperatures $\hbar\omega_i/k$ for the valence vibrations of N$_2$O are 1700 K and 3000 K). Nikitin [57] showed that the quantum statistical theory of decomposition can explain the experimental value of the decomposition rate at low pressures, which is obtained by extrapolation of the experimental data to low pressures.† For this, it is necessary to take it into account that the activation rate, on collision, may be appreciably smaller than the gas-kinetic number of collisions, due to the comparatively large value of the vibrational quantum (400–500 cm^{-1}) close to the threshold of the reaction. The low activation rate implies, in fact, that the interpretation of the decomposition rate at intermediate pressures must be based on the mechanism of stepwise excitation and not on the mechanism of strong collisions. Moreover, in calculating the probability of vibrational transitions in collisions, not only the possibility of the conversion of the translational energy into vibrational should be taken into account but, as Osipov [92] noted, also the resonant exchange of the vibrational energy between two molecules of N$_2$O. Finally, Wieder and Marcus [551] calculated the rate of decomposition of N$_2$O within the framework of the statistical theory of unimolecular reactions by formula (33.20). In addition, it was shown that the quasi-equilibrium theory cannot explain the dependence of the decomposition rate constant on pressure. Since this theory gives very satisfactory results for other reactions, it might be anticipated that the

---

† As regards the basis of the statistical approximation for the description of the redistribution of the vibrational energy in N$_2$O, it rests on the model calculations of Bunker [181], carried out within the limits of the classical approximation.

discrepancy is connected with the fact that the decomposition of N_2O, under consideration, does not satisfy the assumptions of the mechanism of strong collisions (retaining the equilibrium activation rate in clearly non-equilibrium conditions).

For the adiabatic decomposition of molecules, the mechanisms of stepwise excitation and strong collisions may not differ appreciably, since the adiabatic rupture of the bond always leads to a substantial reduction in the magnitude of the vibrational quanta of the active molecule in comparison with the vibrational quanta of the vibrationally-unexcited molecule. It is precisely the small magnitude of the vibrational quanta of the decomposing molecule close to the dissociation threshold which determines the comparatively weak dependence of the transition probability on the number of vibrational levels, and this, in turn, prevents a sharp distinction being drawn between the two types of activation.

Furthermore, in calculating the pressure dependence of the rate constant, it is necessary to take into account that the mean transition probability $\langle P \rangle$ between the electronic states and, in particular, the nonadiabatic tunnel corrections make contributions to the temperature dependence of the rate constant of a nonadiabatic reaction. This leads to an additional negative temperature-dependence of the pre-exponential which, for the decomposition of N_2O, can be expressed formally as a lowering of the activation energy by ~ 3 kcal mol^{-1} ($T = 1500$–2000 K). [393].

6

DIFFUSION THEORY OF REACTIONS

34. Diffusion in phase space

THE diffusion theory of chemical reactions, first formulated by Kramers [312], regards a chemical reaction as a process of diffusion of the representative point in phase space from the region corresponding to the reactants into the region corresponding to the reaction products. The basic assumption of the theory is that, in each act of interaction between the system under consideration and the environment or, as we will say in future, between the representative point and the heat reservoir (heat bath), there is very little change in the distribution function $f(p_i, q_i)$. The assumption of a small change in the distribution function implies that many random acts of interaction occur during a characteristic time of change τ of the distribution function. It is convenient to represent each of them by the collision of a particle, representing the representative point of the system, with the particles of the heat reservoir. Such a representation is very arbitrary, since phase space (p_i, q_i) does not contain the coordinates of the heat reservoir and therefore cannot serve to describe the motion of the molecules of the heat reservoir. Nevertheless, it obviously characterizes the interaction between the system and the heat reservoir: during the time between collisions, the representative point moves on the trajectory $p_i = p_i(t)$, $q_i = q_i(t)$, which is obtained from the solution of the dynamical problem of a closed system with the Hamiltonian $H = H(p_i, q_i)$; then, as a result of collision, the system changes trajectory so that, after collision, the representative point moves along another trajectory $p_i = p_i'(t)$, $q_i = q_i'(t)$. It is evident that the idea of separate collisions is valid only when the duration of each collision τ_0, i.e. the time during which the system changes trajectory, is appreciably smaller than the time between successive collisions τ_1. Hence it follows that, in the general case, we have three characteristic times τ, τ_0, and τ_1, which must satisfy the conditions

$$\tau \gg \tau_1, \tag{34.1a}$$

$$\tau_1 \gg \tau_0. \tag{34.1b}$$

The second condition ($\tau_1 \gg \tau_0$) characterizes the behaviour of the heat reservoir. A reservoir consisting of a rarefied gas satisfies this condition. Since we will consider the application of diffusion theory to gas kinetics, the inequality (34.1b) will always be considered as satisfied.

The fulfilment of the first condition depends on what is understood by the quantity τ. Therefore, for each specific case it will be necessary to inquire into the meaning of the characteristic time τ satisfying condition (34.1a) and, by making use of the smallness of the parameter τ_1/τ, to obtain the basic equations of diffusion theory. It is precisely the smallness of the parameter τ_1/τ which implies that the distribution function $f(p_i', q_i')$ at the point p_i', q_i', after the collision, depends mainly on the distribution function $f(p_i, q_i)$ before the collision, in the closed neighbourhood of the point p_i, q_i, consequently, $f(p_i', q_i')$ can be expressed in terms of $f(p_i, q_i)$ and the partial derivatives $\partial f/\partial p_i$ and $\partial f/\partial q_i$.

Below, following Kramers [124], [312], we will assume that the collision time τ_0 is so small that the change in coordinates of the representative point in each act of interaction can be neglected. We then arrive at a more distinct picture of the migration of the representative point in phase space: during the time between successive collisions, the point moves on a trajectory which is determined by the equations of motion of the isolated system; at the time of collision, which is considered instantaneous, the representative point does not change its coordinates but undergoes a change of momentum which is so small that is may be written in terms of the derivatives of the distribution function with respect to the momentum.

In the deduction of the diffusion equation for the distribution function, for the sake of simplicity, we will limit ourselves to the one-dimensional case. The basic problem consists of determining those of its terms which describe the interaction of the representative point with the heat reservoir. In accordance with what has been said above, the increase (decrease) in the number of particles per unit of phase volume moving along a trajectory of the isolated system is due to just the reduction (increase) in the diffusion current j_p, directed along the phase coordinate p. Therefore we can write

$$\frac{df}{dt} = -\frac{\partial}{\partial p} j_p, \qquad (34.2)$$

where d/dt denotes the derivative with respect to time in a system of coordinates moving together with the point along the undistorted trajectory.

The diffusion current j_p must be proportional, firstly, to the difference between the distribution functions at two neighbouring points, i.e., in the chosen approximation, must contain the derivative $\partial f/\partial p$; moreover, it must also contain, of course, the distribution function f. Hence we obtain

$$j_p = A(p, q) f(p, q) + B(p, q) \frac{\partial f}{\partial p}, \qquad (34.3)$$

where A and B are certain functions, as yet unknown.

Second, the current must tend to zero if the distribution over the momenta has the Maxwellian form $f(p, q) \sim \exp(-p^2/2mkT)$. From these

requirements we find the connection between the functions A and B.†
As a result, the relation (34.3) is conveniently rewritten in the form

$$j_p = -z\left(pf + \frac{m}{\beta}\cdot\frac{\partial f}{\partial p}\right), \quad \beta = \frac{1}{kT}. \tag{34.4}$$

Here z is some function whose dimension (inverse time) enables it to be ascribed the sense of an effective number of collisions per unit time between the diffusing particles (representative points) and the particles of the reservoir. The form of this function, which is often called the coefficient of friction, depends on the interaction mechanism between the representative point and the heat reservoir, and may be determined in an explicit form on the basis of some specific model or other. Substituting (34.4) into (34.2) and writing df/dt in the form of partial derivatives, we obtain the Fokker–Planck diffusion equation:

$$\frac{df}{dt} \equiv \frac{\partial f}{\partial t} + \frac{\partial f}{\partial q}\dot{q} + \frac{\partial f}{\partial p}\dot{p} = \frac{\partial}{\partial p}z\left(p + \frac{m}{\beta}\frac{\partial}{\partial p}\right)f. \tag{34.5}$$

It is convenient to transform the given equation into another form, using the equations of motion

$$\dot{q} = \frac{p}{m}, \quad \dot{p} = -\frac{\partial U}{\partial q}. \tag{34.6}$$

This enables (34.5) to be written in just the variables p and q, and introduces the potential $U = U(q)$, in whose field the unperturbed motion of the point takes place, into this equation in an explicit form:

$$\frac{\partial f}{\partial t} + \frac{\partial f}{\partial q}\dot{q} - \frac{\partial f}{\partial p}\cdot\frac{\partial U}{\partial q} = \frac{\partial}{\partial p}(zpf) + \frac{\partial}{\partial p}\left(D\frac{\partial f}{\partial p}\right), \tag{34.7}$$

where the diffusion coefficient D is introduced

$$D = \frac{zm}{\beta}. \tag{34.8}$$

The relation (34.8) in diffusion theory reflects the principle of detailed balance, which is formulated in the form of the relation (23.6), in the general case.

We will now clarify the meaning of the diffusion coefficient D. To this end, following the method employed in paper [105], we will substitute into

---

† In such a derivation of the diffusion equation, in general, it is not made clear why the interaction between the heat reservoir and the system under consideration finally leads to a state of equilibrium. A consistent theory of Brownian motion must not include the condition of the relaxation of the distribution function f to the equilibrium distribution function f_0 as an assumption, but, on the contrary, must prove the inevitability of such a relaxation, on the basis of a a study of the general properties of the interaction of the system with the reservoir. As a simple example of a similar investigation, we cite the literature [107], [524], [525].

(34.7) the distribution function which describes a single particle with momentum p_0:

$$f(p, q) = \phi(q)\delta(p - p_0). \tag{34.9}$$

We then multiply eqn (34.7) by $(p - p_0)^2$ and integrate over dp. On the left-hand side of (34.7) we obtain the mean square of the momentum transmitted by the representative point per unit time due to the interaction with the heat reservoir:

$$\int (p - p_0)^2 \frac{df}{dt} \, dp = \frac{d}{dt} \langle \Delta p^2 \rangle. \tag{34.10}$$

On the right-hand side, after a double integration by parts and using the properties of the δ-function, the coefficient D is separated out, so that we have finally

$$\frac{D(p_0)}{2} = \frac{d}{dt} \langle \Delta p^2 \rangle. \tag{34.11}$$

The arguments given in the derivation of (34.11) enable the order of magnitude of the time scale, for which a description of the distribution function by the diffusion equation is valid, to be clarified. The time derivative in (34.11) should be understood as the limit:

$$\frac{d}{dt} \langle \Delta p^2 \rangle = \lim \frac{\langle [p(\Delta t) - p(0)]^2 \rangle}{\Delta t}, \tag{34.12}$$

in which Δt, on the one hand, is appreciably smaller than the characteristic time τ_1 and, on the other hand, exceeds by far the collision time τ_0. Under the condition (34.1b) it is possible to select Δt so that

$$\tau_1 \gg \Delta t \gg \tau_0. \tag{34.13}$$

In this case, the difference $\langle [p(\Delta t) - p(0)]^2 \rangle$ is clearly equal to the mean square of the momentum $\langle \Delta p^2 \rangle$, transmitted in a single collision, multiplied by the probability of the single collision during the time Δt.† The latter quantity equals $\Delta t / \tau_1$, so that (34.11) takes the form

$$D(p) = \tfrac{1}{2} Z_0 \langle \Delta p^2(p) \rangle, \quad Z_0 = \frac{1}{\tau_1}. \tag{34.14}$$

The quantity Z_0, introduced here, has the meaning of a gas-kinetic number of collisions. It characterizes the *frequency* of the interaction between the system under consideration and the heat reservoir. On the other hand,

---

† More strictly, it should be noted that $\langle [p(\Delta t) - p(0)]^2 \rangle = \langle \Delta p^2 / \tau_2 \rangle \Delta t$ where the averaging $\langle \cdots \rangle$ goes over all parameters of the collision (for example, over the velocities). In future estimates, however, it can be assumed that $\langle \Delta p^2 / \tau_1 \rangle \sim \langle \Delta p^2 \rangle / \tau_1$.

the effective number of collisions z characterizes the *intensity* of this interaction. On the basis of (34.14) we obtain the following relation:

$$z = Z_0 \frac{\langle \Delta p^2 \rangle}{2kTm}. \tag{34.15}$$

From eqn (34.5) it is evident that the characteristic time of change of the non-equilibrium distribution function has the order of magnitude of the only parameter with the dimensions of time, i.e. $\tau \sim 1/z$. Hence it follows that the basic assumption under which the diffusion equation is valid, namely the smallness of the change in the distribution function during the time between successive collisions, can be represented in the form

$$z \ll Z_0, \quad \langle \Delta p^2 \rangle \ll kTm. \tag{34.16}$$

This relation illustrates the meaning of the inequality (34.1). For a one-dimensional problem, the relation (34.16) is satisfied on condition that the mass m of the diffusing particles appreciably exceeds the mass μ of the particles of the reservoir. Then the particles of the reservoir, on collision, transfer momentum $\langle \Delta p^2 \rangle \sim \mu kT$, and the inequality (34.16) is satisfied as a consequence of the condition $\mu/m \ll 1$. In the three-dimensional case the same condition, $\mu/m \ll 1$, secures the smallness of the transferred momentum and, for the model of rigid spheres for example, the number of effective collisions equals [125]:

$$z = Z_0 \left(\frac{4}{3} \cdot \frac{\mu}{m} \right), \quad Z_0 = \pi a^2 \left(\frac{8kT}{\pi \mu} \right)^{\frac{1}{2}}. \tag{34.17}$$

(Here and in future, the number of collisions is normalized to unit concentration of the molecules of the reservoir.) Expression (34.17) is applicable to microscopic diffusing particles of radius a. In the derivation of (34.17) the assumption is made that the particles of the reservoir which collide with the diffusing particles can be described by the equilibrium distribution function. For this to be so, it is necessary that the free-path length l of the particles in the reservoir be much greater than the radius a of the diffusing particles; only under this condition can each subsequent collision be really considered as a collision with new particles coming from the reservoir.

For a macroscopic Brownian particle, the opposite condition $l \ll a$ is satisfied, under which, conversely, the particle of the reservoir remains close to a diffusing particle for a long time. This implies that, in a calculation of the coefficient of friction, the violation of the equilibrium distribution of the particles of the reservoir close to the moving Brownian particle should certainly be taken into account.

In the theory of transport phenomena in gases [125], it is shown that the coefficient of friction z_B of a Brownian particle can be expressed in terms of the viscosity η by the relation:

$$z_B = \frac{6\pi a\eta}{m}, \qquad (34.18)$$

in which η only depends on the properties of the particles in the reservoir. By introducing different coefficients of friction for the microscopic and macroscopic particles, the applicability of the diffusion equation to the latter case is maintained, since, strictly speaking, the distribution function of the particles of the reservoir at distances of the order of l away from the Brownian particle is not the equilibrium distribution function. Comparison of (34.17) and (34.18) also shows that, in a disturbance of the equilibrium in the reservoir, the coefficient of friction depends on those characteristics of the particles of the reservoir which determine the rate of relaxation processes in the reservoir itself. An analogous situation, as we will see below, occurs in the dissociation of molecules; the macroscopic rate of dissociation is determined by the rate of the relaxation process for the molecules in bound states, which appear as a kind of reservoir in relation to the dissociated states.

Before going over to the solution of the specific problem of calculating the rates of relaxation and chemical reactions, it should be noted that the solution of the diffusion equation (34.7) is equivalent to the solution of the classical equations of motion for the representative point, in which a random force $F(t)$ is introduced to represent the effect of the heat reservoir.† These equations have the form

$$\dot{q} = \frac{p}{m}, \qquad \dot{p} = -\frac{\partial U}{\partial q} - zp + \frac{1}{m}F(t). \qquad (34.19)$$

Since there is only one independent parameter, the coefficient of friction z or the diffusion coefficient D in eqn (34.7), it is clear that the statistical properties of $F(t)$ must depend on this parameter. It can be shown [124] that eqns (34.7) and (34.19) are equivalent if the transferred momentum δp, connected with the force by the relation

$$\delta p = \int_t^{t+\Delta t} \frac{1}{m} F(t)\,dt, \qquad (34.20)$$

complies with the distribution function

$$W(\delta p)\,d(\delta p) = \left(\frac{4\pi mz\Delta t}{\beta}\right)^{\frac{1}{2}} \exp\left[-\frac{\beta(\delta p)^2}{4mz\Delta t}\right] d(\delta p). \qquad (34.21)$$

---

† The physical significance of the separation of the interaction between the system and the reservoir into two parts, the random force $F(t)$ and the dissipative term zp, is considered in the literature [524], [525] and in the book [107]. In addition, the reason is examined for the irreversible relaxation of the system under the influence of the reservoir (see note on p. 285).

We will now consider, in more detail, the solution of the Fokker–Planck equations, with a constant coefficient of friction, for particles in a potential $U(q)$. This solution enables the basic features of relaxation and dissociation of an oscillator in a heat reservoir to be examined. To be precise, we will have in mind the phase plane shown in Fig. 21.

We will assume that the temperature of the reservoir is sufficiently low so that the relation $\beta E_0 \gg 1$ is satisfied. In this case, it is possible to limit consideration to just the lower part of the potential curve, for which the parabolic approximation $U(q) = m\omega^2 q^2/2$ can be taken. Eqn (34.5) takes the form:

$$\frac{\partial f}{\partial t} + \dot{q}\frac{\partial f}{\partial q} - \omega^2 q \frac{\partial f}{\partial \dot{q}} = z\frac{\partial}{\partial \dot{q}}\left(\dot{q}f + \frac{1}{m\beta}\frac{\partial f}{\partial \dot{q}}\right), \qquad \dot{q} = \frac{p}{m}. \qquad (34.22)$$

For the model described by this equation, the condition for shortness of collision time τ_0 reads

$$\omega \tau_0 \ll 1. \qquad (34.23)$$

To investigate the general case, it is sufficient to find the fundamental solution $W(p, q; p', q', t)$ satisfying the initial conditions

$$W(p, q; p', q', t)|_{t=0} = \delta(p-p')\delta(q-q'). \qquad (34.24)$$

Then the non-equilibrium distribution function $f(p, q, t)$ for an arbitrary initial distribution $f^0(p, q)$ has the form

$$f(p, q, t) = \int W(p, q; p', q', t) f^0(p', q')\, dp'\, dq'. \qquad (34.25)$$

The solution W has a fairly complicated form, therefore it is convenient to follow the change in time of the mean values of some physical quantities during the relaxation of the non-equilibrium distribution function. Of interest, in particular, are the mean values of the coordinate $\langle q(t) \rangle$ and the total energy $\langle E(t) \rangle$, which are expressed in terms of f by the relations:

$$\langle q(t) \rangle = \int q f(p, q, t)\, dp\, dq.$$

$$\langle E(t) \rangle = \int \left(\frac{p^2}{2m} + \frac{m\omega^2}{2}q^2\right) f(p, q, t)\, dp\, dq. \qquad (34.26)$$

Instead of this, in order to extract explicit expressions for $\langle q \rangle$ which depend on the initial values q_0 and p_0, a differential equation can be obtained for $\langle q \rangle$, on the basis of (34.22). Multiplying (34.22) by q and integrating over the whole of phase space, we obtain an equation connecting $d\langle q \rangle/dt$, $\langle q \rangle$, and $\langle p \rangle$. The same procedure, multiplying (34.22) by p, gives the second equation. It is not difficult to verify that the system of equations obtained agrees with (34.19), with the difference that, in the new system, q and p are replaced by $\langle q \rangle$ and $\langle p \rangle$ and, furthermore, it is assumed that $\langle F(t) \rangle = 0$. This is natural because (34.19) and (34.22) are completely equivalent. The elimination of $\langle p \rangle$ from the system of equations obtained gives the following equation for $\langle q \rangle$:

$$\langle \ddot{q} \rangle + z \langle \dot{q} \rangle + \omega^2 \langle q \rangle = 0, \tag{34.27}$$

enabling cases with different initial distributions to be studied easily.

Such a method can be applied to derive the relaxation equations for the mean energy $\langle E \rangle$ of an oscillator. Straightforward calculations lead to the equation [524]

$$\frac{d}{dt} \langle E \rangle = -z \left(\frac{\langle p^2 \rangle}{m} - \frac{1}{\beta} \right). \tag{34.28}$$

Hence it is evident that the rate of change of the mean energy at each moment of time is determined by just the mean kinetic energy at that moment. This is accounted for by the mechanism assumed for the interaction of the oscillator with the heat reservoir—on collision, only a change in the momentum of the oscillator occurs.

To investigate relaxation, it is often convenient to introduce, in place of the phase coordinates p and q, new coordinates, one of which should be an integral of motion for the isolated system. For a system with one degree of freedom, such coordinates are the action J and the angle variable α, defined by relations (2.17) and (2.18). In the new variables, the Hamiltonian of the system does not depend on α and, in particular, for the harmonic oscillator, takes the form $H = J\omega$.

In an investigation of the relaxation distribution function $f(J, \alpha, t)$ written in the new variables, two processes can be distinguished; relaxation of the population, which leads to a Boltzmann distribution for the energy levels, and phase relaxation, which ultimately leads to the system totally 'forgetting' the initial value of the angular coordinate. These two processes are not independent, as is clear from the solution of the relaxation equations for $\langle q \rangle$ and $\langle E \rangle$. Assuming, in particular, that the initial distribution function does not depend on α and taking account of the relations satisfied in connection with this

$$\frac{m\omega^2}{2} \langle q_0^2 \rangle = \frac{\langle p_0^2 \rangle}{2m} = \frac{\langle E_0 \rangle}{2},$$

$$\langle q_0 p_0 \rangle = 0, \tag{34.29}$$

we find [124]:

$$\phi_q(t) = \frac{\langle q(t)q(0)\rangle}{\langle q^2(0)\rangle} = \begin{cases} \exp\left(-\frac{zt}{2}\right)\left(\cos\omega_1 t + \frac{z}{2\omega_1}\sin\omega_1 t\right), & \omega > \frac{z}{2} \\ \exp\left(-\frac{zt}{2}\right)\left(\cosh\gamma_1 t + \frac{z}{2\gamma_1}\sinh\gamma_1 t\right), & \omega < \frac{z}{2}, \end{cases} \quad (34.30)$$

$$\phi_E = \frac{\langle E(t)\rangle - E(\infty)}{E(0) - E(\infty)}$$

$$= \begin{cases} \exp(-zt)\left(1 + \frac{z}{\omega_1}\cos\omega_1 t \sin\omega_1 t + \frac{z^2}{2\omega_1^2}\sin^2\omega_1 t\right), & \omega > \frac{z}{2} \\ \exp(-zt)\left(1 + \frac{z}{\gamma_1}\cosh\gamma_1 t \sinh\gamma_1 t + \frac{z^2}{2\gamma_1^2}\sinh^2\gamma_1 t\right), & \omega < \frac{z}{2}, \end{cases}$$

$$\omega_1 = \left(\omega^2 - \frac{z^2}{4}\right)^{\frac{1}{2}}, \quad \gamma_1 = \left(\frac{z^2}{4} - \omega^2\right)^{\frac{1}{2}}. \quad (34.31)$$

From these expressions it is evident, for example, that, although the initial distribution function depends only on the energy (and not on the phase), the mean energy undergoes relaxation according to a law, in which traces of the dynamical process are clearly seen, whose characteristic feature is the periodic dependence of the quantities on time. This is connected with the fact that the relaxation of the population induces phase relaxation, and the latter leads to the appearance of terms depending on the phase in the non-equilibrium distribution function. Such a situation is illustrated in Fig. 33, where part of the phase plane of the oscillator is shown. The lines of constant action (or energy) are the ellipses, and the phase is equal to the polar angle of the representative point.

We will assume that initially the system is characterized by a δ-function distribution function with respect to the energies

$$f^0(J, \alpha)\, dJ\, d\alpha = \delta(E - E_0)\, dJ\, d\alpha, \quad (34.32)$$

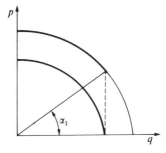

FIG. 33. Transitions of oscillators in the phase plane, induced by the impulsive interaction.

i.e. the representative points are uniformly distributed over one of the ellipses (for example, the inner ellipse). As a result of the interaction with the heat reservoir, the system may pass over to the outer ellipse; however, all values of the polar angle α less than a certain value α_1 prove to be inaccessible, since only those transitions are possible in which the coordinate q does not change. During the time between collisions, the representative points will move over elliptical trajectories and those portions of the trajectory which prove to be unfilled as a result of the first collision remain unfilled later on. However, by the time the next collision occurs, these unfilled regions are at another portion of the trajectory, so that later transitions obliterate the non-uniform distribution of the representative points with respect to the phases, which arises as a result of the first collision. The superposition of processes of this type leads to the joint relaxation of the phase and the population.

We will now trace how, in the limit of $z \ll \omega$, eqn (34.28) is simplified, on whose right-hand side the oscillating function $\langle p^2 \rangle / m$ occurs. For a small interaction between the oscillators and the heat reservoir, there is little change in the energy E during the time of a single oscillation, therefore both parts of the equation can be averaged over a time interval T_0, over which, on the one hand $\langle E \rangle$ can be considered constant and, on the other hand, $\langle p^2 \rangle / m$ can be replaced by the mean kinetic energy of an oscillator over the time of many oscillations. Taking into account that $\langle p^2 \rangle / m = \langle E \rangle$ for the mean energy $\langle E \rangle$, we obtain the closed equation

$$\frac{d}{dt} \langle E \rangle = -z \left(\langle E \rangle - \frac{1}{\beta} \right), \tag{34.33}$$

whose solution leads to a simple exponential decay

$$\phi_E = \frac{E(t) - E(\infty)}{E(0) - E(\infty)} = \exp(-zt). \tag{34.34}$$

This result can be obtained also, of course, from (34.31) in the limit $z \ll \omega$. The derivation given of (34.33) shows that introducing the smoothed-out distribution, averaged over an interval of time T_0 ($1/\omega \ll T_0 \ll 1/z$), successfully reduces the full relaxation problem to the relaxation of just the population.

We will now consider, in more detail, the solution of the Fokker–Planck equation for the transmission of particles across a potential barrier. We will have in mind the situation represented in Fig. 21. The rate constant of the reaction, which is identified with the rate constant of over-barrier penetration, equals the mean current of particles across the surface $q = q_r^\star$:

$$\kappa = \int_{-\infty}^{\infty} \dot{q} f(p, q^\star) \, dp, \tag{34.35}$$

where the distribution function is normalized to one particle in the well on the left of the maximum and is zero at sufficiently large distances to the right of it. For the present, we will assume, purely formally, that the time t_0 for the motion of the representative point along an undisturbed trajectory of the type abc satisfies the condition $zt_0 \gg 1$. Then, it is natural to expect that the calculation of the non-equilibrium distribution function on the critical surface may be carried out taking account of just the region Δq immediately adjoining it, since, on leaving the critical surface to the left, the disturbance of the equilibrium distribution is reduced sufficiently quickly (condition $zt_0 \gg 1$). Close to the critical surface, the potential can be represented approximately by a parabola $U(q) = E_0 - [m(\omega^\star)^2](q-q^\star)^2/2$, and the Fokker–Planck equation takes the form

$$\frac{\partial f}{\partial t} + \dot{q}\frac{\partial f}{\partial q} + (\omega^\star)^2 \Delta q \frac{\partial f}{\partial \dot{q}} = z\frac{\partial}{\partial \dot{q}}\left(\dot{q} + \frac{1}{\beta}\frac{\partial}{\partial \dot{q}}\right)f, \quad \Delta q = q - q^\star. \quad (34.36)$$

Following Kramers [312] we will solve this equation in the quasi-stationary approximation, assuming $\partial f/\partial t = 0$. Using the fact that the coefficients of the derivatives in (34.36) are proportional to $\dot{q}$ and Δq, we will look for a solution in the form

$$f(\dot{q}, q) = \exp\left[-\frac{\beta m \dot{q}^2}{2} + \beta m \frac{(\omega^\star)^2}{2}(\Delta q)^2\right]\Phi(\xi),$$

$$\xi = \dot{q} - \gamma \Delta q, \quad (34.37)$$

where the constant γ must be chosen from the condition that (34.37) satisfy an equation in one variable ξ. Substituting (34.37) into (34.36), we find

$$\frac{d\Phi}{d\xi}(z-\gamma)\xi = \frac{z}{\beta}\frac{d^2\Phi}{d\xi^2}, \quad \gamma = \frac{z}{2} + \sqrt{\left[\frac{z^2}{4} + (\omega^\star)^2\right]}. \quad (34.38)$$

The boundary conditions imposed on $\Phi(\xi)$ follow from the boundary conditions for the function $f(\dot{q}, \Delta q)$, which state that, to the left of the potential barrier ($\Delta q \to -\infty$), f must go over into the equilibrium distribution function and, to the right ($\Delta q \to +\infty$), f must tend to zero:

$$f(\dot{q}, \Delta q) = \begin{cases} \exp[-\beta H(\dot{q}, \Delta q]/F, & \Delta q \to -\infty, \\ 0, & \Delta q \to +\infty, \end{cases} \quad (34.39)$$

where $F = kT/\hbar\omega$ is the partition function of the particles in the potential well, and its value is determined, as usual, by the potential only in the neighbourhood of its minimum.

This normalization of f makes possible the identification of the current through the barrier with the rate constant. In order to join the function

(34.39) onto the solution of (34.37), it is necessary to substitute the Hamiltonian close to the peak of the barrier $H = \beta m \dot{q}^2/2 - \beta m(\omega^*)^2(\Delta q)^2/2$ for $H(\dot{q}, \Delta q)$ in (34.39). Such a substitution is legitimate when the equilibrium distribution to the left of the barrier is established at distances δq small in comparison with q^*, i.e. if z and $1/\omega$ are connected by the relation $z \gg 1/\omega$. We obtain the boundary condition for $\Phi(\xi)$ from (34.39).

$$\Phi(\xi) = \begin{cases} \hbar \omega \beta & \text{for } \xi \to \infty \\ 0 & \text{for } \xi \to -\infty. \end{cases} \quad (34.40)$$

Integrating (34.38) we obtain

$$\Phi(\xi) = \left[\frac{(\gamma-z)\beta}{2\pi z \mu}\right]^{\frac{1}{2}} \int_{-\infty}^{\xi} \exp\left[\frac{-\beta(\gamma-z)y^2}{2z\mu}\right] dy \cdot \hbar \omega \beta. \quad (34.41)$$

Substituting (34.41) into (34.35) we find the following expression for the rate constant:

$$\kappa = \frac{\omega}{2\pi \omega^*} \left\{ \left[\frac{z^2}{4} + (\omega^*)^2\right]^{\frac{1}{2}} - \frac{z}{2} \right\} \exp(-\beta E_0). \quad (34.42)$$

35. Diffusion through energy states

As shown above, the function ϕ_E is close to an exponential for $z \ll \omega$. This implies that relaxation over the energy levels is practically unconnected with the phase relaxation. In view of this, we can attempt to eliminate the angle variable from the Fokker–Planck equation from the very beginning. Transforming eqn (34.5) to an equation in the variables J and α, it takes the form

$$\frac{\partial f}{\partial t} + \omega \frac{\partial f}{\partial \alpha} = z \frac{\partial}{\partial J} J \left(1 + \frac{1}{\beta \omega} \frac{\partial}{\partial J}\right) f + F\left(\frac{\partial^2 f}{\partial \alpha^2}, \frac{\partial^2 f}{\partial J \, \partial \alpha} \cdots\right). \quad (35.1)$$

On the left-hand side, the derivative $\partial f / \partial J$ is absent, since J is an integral of the motion for a free harmonic oscillator. On the right-hand side of the equation, the function F, whose explicit form we will not quote here, depends on the derivative $\partial^2 f / \partial \alpha^2$ and on the mixed derivative $\partial^2 f / \partial J \, \partial \alpha$. It is precisely the presence of this latter term which causes the non-equilibrium distribution function, initially independent of the angle coordinate, to relax towards the equilibrium distribution (also independent of α) via a series of functions dependent on α. However, it may be assumed beforehand that the distribution function does not depend on α at all times. This assumption, enabling the terms $\partial f / \partial \alpha$, $\partial^2 f / \partial \alpha^2$, $\partial^2 f / \partial J \, \partial \alpha$ in (35.1) to be discarded, in fact implies a rejection of the previously-assumed mechanism of diffusion of the representative points and the passing over to another mechanism. In certain model cases, it is possible to show how the structure of the diffusion terms of eqn (35.1) is connected with the type of interaction between the system and the heat reservoir. In particular, Prigojin and coworkers [107] derived the Fokker–Planck equation for an oscillator connected by a weak anharmonic interaction to a system of oscillators in thermal equilibrium. It was shown that, for this case, the function F has the form

$$F = \frac{1}{4\omega J \beta} \frac{\partial^2 f}{\partial \alpha^2}. \qquad (35.2)$$

In this case (35.1) differs from the equation which can be obtained from (34.5), the difference being that, in (35.1) taking account of (35.2), the interaction term of the oscillator with the medium contains the momentum and coordinate to the same degree.

Assuming, in future, an equiprobable initial distribution of the phases, we will consider the equation which is obtained from (35.1) if we take $\partial f / \partial \alpha = 0$.

$$\frac{\partial f}{\partial t} = z \frac{\partial}{\partial J} J \left(1 + \frac{1}{\beta \omega} \frac{\partial}{\partial J} \right) f. \qquad (35.3)$$

This equation is not equivalent to eqn (34.22); however, in the limit of small friction, the solutions to eqns (35.3) and (34.22) will be close to one another.

The law of relaxation of the mean energy can be obtained immediately from (35.3). Introducing the energy $E = \omega J$ as an independent variable, multiplying (35.3) by E, and integrating over dE, we find

$$\frac{d}{dt} \langle E \rangle = \int_0^\infty zE \frac{\partial}{\partial E} E \left(1 + \frac{1}{\beta} \frac{\partial}{\partial E} \right) f(E) \, dE, \quad \langle E \rangle = \int_0^\infty E f(E) \, dE. \qquad (35.4)$$

Integrating the right-hand side twice by parts and taking into consideration the conservation of the normalization of the distribution function, we obtain eqn (34.33) and solution (34.34).

We will now consider another important consequence of the phase independence of the distribution function. In the derivation of eqn (34.5), it was assumed that the collision time τ_0 was vanishingly small in comparison with the characteristic time of change of the configuration of the system $1/\omega$. In eqn (35.3), in general, the coordinate characterizing the configuration does not appear. Therefore, it might be thought to remain true also when the relation $\omega \tau_0 \ll 1$ is violated, provided, in addition, a condition is satisfied which secures the sufficiently-fast uniform distribution of the representative points over the states of constant energy. In Kramer's theory, this condition has the form

$$z \ll \omega. \qquad (35.5)$$

From physical considerations, it is clear that, on the fulfilment of (35.5) in any variant of diffusion theory, including the case when the relation $\omega \tau_0 \ll 1$

is not satisfied, the non-equilibrium distribution function will depend only on the action variables. An analogous situation can occur in systems with many degrees of freedom if there is one separated-out variable for which the disturbance from the equilibrium distribution must be taken into account. In this case, we again return to the diffusion equation for one variable, in which, however, the coefficient of friction acquires another meaning. As an important example, we will mention the statistical model of a unimolecular reaction (see Chapter 4), within whose limits it is assumed that equilibrium is established over all the degrees of freedom, in practice, until the distribution function over the energies changes substantially.

Turning to systems with many degrees of freedom, we will introduce, instead of the multi-dimensional distribution function $f(p_i, q_i, t)$, a distribution function over the energies $x(E)$, defined by the relation

$$x(E, t)\rho(E)\,dE = \int f(p_i, q_i, t)\delta[E - H(p_i, q_i)]\,d\Gamma,$$

$$\rho(E) = \int \delta[E - H(p_i, q_i)]\,d\Gamma. \tag{35.6}$$

The diffusion equation, which describes transitions between states differing in total energy, has a form analogous to (34.2):

$$\frac{\partial}{\partial t}x(E, t)\rho(E) = -\frac{\partial}{\partial E}j_E. \tag{35.7}$$

The current j_E must tend to zero in a state of equilibrium and must contain the first derivative of the distribution function, since it is precisely this which secures the diffusional nature of the current. On the basis of these considerations, we find

$$j_E = -b(E, \beta)\rho(E)\left[x(E) + \frac{1}{\beta}\frac{\partial x(E)}{\partial E}\right], \tag{35.8}$$

where $b(E, \beta)$ is some coefficient of proportionality depending on E and β.

Substituting (35.8) into (35.7) we obtain the equation

$$\frac{\partial x}{\partial t} = \frac{1}{\rho(E)}\cdot\frac{\partial}{\partial E}b(E, \beta)\rho(E)\left(1 + \frac{1}{\beta}\frac{\partial}{\partial E}\right)x. \tag{35.9}$$

The meaning of the coefficient b is made clear as the result of reasoning exactly analogous to the reasoning in the derivation of (34.11)

$$b(E, \beta) = \frac{1}{2}\frac{\partial}{\partial t}\langle\Delta E(E, \beta)^2\rangle, \tag{35.10}$$

i.e. b is equal to the mean square of the energy transferred to the system per unit time.

Finally, the Fokker–Planck equation in energy space takes the form

$$\frac{\partial x}{\partial t} = \frac{1}{\rho(E)} \cdot \frac{\partial}{\partial E}\left[\frac{1}{2}\left(\frac{\partial}{\partial t}\langle \Delta E^2 \rangle\right)\rho(E)\left(x + \frac{1}{\beta}\frac{\partial x}{\partial E}\right)\right]. \qquad (35.11)$$

The function $b(E, \beta)$ like $D(\rho_0)$ in (34.11) is an independent parameter of diffusion theory.

To take account of a chemical reaction, it is now necessary to modify eqn (35.11) in such a way that, besides the diffusion of the representative points over the energy levels describing the relaxation, it also incorporates the reduction in the concentration of representative points caused by chemical reaction. Since it is assumed that the 'spreading' of the system over all the dynamical variables except energy takes place very rapidly, the reaction rate constant $\kappa(E)$ of the active molecule with energy $E \gg E_0$ can be introduced. The term $-\kappa(E)x(E)\rho(E)$, responsible for the reduction in the concentration of the active molecules as a result of the decomposition, should then be added to the diffusion term on the right-hand side of (35.11). The diffusion equation (35.11) then takes the form

$$\frac{\partial x}{\partial t} = \frac{1}{\rho}\frac{\partial}{\partial E}\left[b\rho\left(x + \frac{1}{\beta}\frac{\partial x}{\partial E}\right)\right] - \kappa(E)x. \qquad (35.12)$$

The solution of this equation with an arbitrary initial distribution function completely describes the relaxation of the system and its decomposition in the approximation of diffusion through the energy states. However, no account has been taken of the reverse reaction, which leads to partial repopulation of the levels of the active molecule due to recombination.† Therefore (35.12) is correct in conditions far from the state of chemical equilibrium.

Eqn (35.12) reduces to (35.11) not merely in the absence of a chemical reaction, when $\kappa(E) = 0$. For a very small activation rate $[b(E) \ll \kappa(E)]$, the molecules, having reached a certain threshold level E_0, can be considered as disappearing instantly. Then the region of energy $E > E_0$ need not be considered, in general, the reaction having been replaced by the condition for total absorption $x(E_0) = 0$. Under this condition, the solution to eqn (35.11) can easily be obtained in the quasi-stationary approximation. The continuity condition for the current is written in the form

$$j_E = -b\beta\rho\left(x + \frac{1}{\beta}\frac{\partial x}{\partial E}\right) = \text{constant}. \qquad (35.13)$$

---

† In the literature [172], [300] an attempt was made to take into account the reverse reaction within the framework of Kramer's diffusion theory.

This equation, of the first order with regard to $x(E)$, is easily integrated; to determine the two constants of integration (one of which equals the unknown current j_E and the second of which appears on integration of the equation) there are two conditions: the total absorption of the particles at level E_0

$$x(E_0) = 0 \tag{13.14}$$

and the normalization condition for the distribution function of one particle

$$x(E)|_{E \to 0} \to \frac{1}{F}, \qquad F = \int_0^\infty \exp(-\beta E)\rho(E)\,dE, \tag{35.15}$$

which enables j_E to be identified with the rate constant κ.

It is evident that the second condition only has meaning when the number of active molecules make a vanishingly small contribution to the normalization; hence we obtain an additional condition for the validity of the quasi-stationary approximation: $\beta E_0 \gg 1$.

From (35.13), (35.14), and (35.15) we find the general expression for the decomposition rate constant:

$$\kappa_{\text{diss}} = \left[F(\beta) \int_0^{E_0} \exp(\beta E) \frac{dE}{b(\beta, E)\rho(E)} \right]^{-1}. \tag{35.16}$$

The quasi-stationary approximation considered is valid only on condition that the characteristic time of chemical relaxation is appreciably greater than the time of relaxation over the low energy levels. In the general case, eqn (35.12) can be solved by the method of expanding the required function for $x(E, t)$ in terms of the eigenfunctions $\phi_\mu(E)$ of the operator on the right-hand side of eqn (35.12). The solution has the form

$$x(E, t) = \sum_\mu A(\mu) \exp(-\mu t)\phi_\mu(E), \tag{35.17}$$

where μ is the eigenvalue of the equation

$$\frac{1}{\rho(E)} \frac{\partial}{\partial E}\left[\beta b(E, \beta)\rho(E) \left(\frac{1}{\beta} \frac{\partial \phi_\mu}{\partial E} + \phi_\mu \right) \right] - \kappa(E)\phi_\mu = \mu \phi_\mu(E). \tag{35.18}$$

The functions ϕ_μ must satisfy the boundary conditions which the equilibrium distribution function satisfies:

$$\phi_\mu(E) \to 0, \qquad E \to \infty,$$

$$\frac{d\phi_\mu(E)}{dE} \to \beta, \qquad E \to 0. \tag{35.19}$$

If the first eigenvalue μ_0 is appreciably smaller than all the others then, at times $t \gg 1/\mu_1$, including the main decomposition region, the kinetics will be described by an exponential function, and consequently μ_0 must be identified with κ_{diss}. Thus a calculation of the non-equilibrium decomposition rate reduces to finding the smallest eigenvalue of eqn (35.12). In this connection, it is convenient to make use of the variational principle in an approximate calculation of κ_{diss}. It is not difficult to see that if the minimum of the functional

$$K(y) = \frac{\int \exp(-\beta E)[b(\mathrm{d}y/\mathrm{d}E)^2 + \kappa(E)y^2]\rho(E)\,\mathrm{d}E}{\int \exp(-\beta E)y^2\rho(E)\,\mathrm{d}E} \qquad (35.20)$$

is found, where

$$y(E) = x(E)\exp(\beta E),$$

then the problem reduces to the solution of eqn (35.18). If by $y(E)$ is understood the trial function $\tilde{y}$ with the boundary conditions

$$\tilde{y}\exp(-\beta E) \to 0, \qquad E \to \infty,$$

$$\frac{\mathrm{d}\tilde{y}}{\mathrm{d}E} \to 0, \qquad E \to 0, \qquad (35.21)$$

then, on the basis of the variational principle, the relation

$$\kappa_{\text{diss}} \leqslant K(\tilde{y}) \qquad (35.22)$$

must be satisfied.

These relations are very useful in an investigation of the general dependence of the rate constant on the number of collisions Z and for a qualitative discussion of different variants of the non-equilibrium theory of decomposition.

36. Relaxation and the transmission of particles across a potential barrier

In this section we will consider the kinetics of the relaxation of the mean energy of an oscillator interacting with a heat reservoir and the expression for the rate constant of a reaction represented by the transmission of particles over a potential barrier. On the basis of the theory set forth above, it is convenient to distinguish three limiting cases which depend on the relationship of the three parameters of the problem z, ω^*, and ω.

1. *Large friction.* We will assume that the relations

$$\omega\tau_0 \ll 1 \qquad (31.1\text{a})$$

$$\frac{z}{\omega} \gg 1 \qquad (36.1\text{b})$$

$$\frac{z}{\omega^*} \gg 1 \qquad (36.1\text{c})$$

are satisfied. Condition (36.1a) enables the diffusion over states to be described in the approximation of instantaneous collisions, i.e. it enables eqn (34.22) to be used with a constant coefficient of friction. Turning to (34.31), we employ condition (36.1b) to simplify the expression for $\phi_E(t)$, expanding the exponent of the exponential in a series in terms of the small parameter ω/z:

$$\phi_E(t) = \tfrac{1}{2}\exp(-zt) + \tfrac{1}{2}\exp\left(-\frac{\omega^2}{z^2}zt\right). \tag{36.2}$$

From this, it is clear that the relaxation of the population is characterized by two widely-differing relaxation times $\tau_1 \sim 1/z$ and $\tau_2 \sim z/\omega^2$. To elucidate their meanings, we will return to eqn (34.7). We will find the solution of this equation in the form

$$f(p, q, t) = (2\pi mkT)^{-\tfrac{1}{2}} \exp\left(-\frac{p^2}{2mkT}\right) F\left(\frac{zq - \dot{q}}{z}, t\right). \tag{36.3}$$

Substituting (36.3) into (34.7) and integrating over dp, we obtain the relation

$$\frac{\partial}{\partial t}\int f(p, q, t)\,dp = \int \frac{\partial}{\partial(q - \dot{q}/z)}\left(\frac{kT}{mz}\frac{\partial f}{\partial q} + \frac{1}{z}\frac{\partial U}{\partial q}f\right)\frac{dp}{m}. \tag{36.4}$$

Considering an interval of time far exceeding the time between successive collisions $\tau_1 \approx 1/z$, the distribution function over velocities may be regarded as almost Maxwellian, and the factor F in (36.3) will be changed by a slowly varying function of $\dot{q}$ (this is reflected explicitly in the specific dependence of the argument of F on the number of collisions z). Then, in calculating the integral in (36.4), the dependence of F on $\dot{q}$ can be neglected. Moreover, the derivative with respect to the variable $(q - \dot{q}/z)$ can be replaced by a derivative with respect to the variable q if the potential $U(q)$ changes little over distances of the order of $\delta q \approx \overline{|\dot{q}|}/z \approx (kT/m)^{\tfrac{1}{2}}/z$. For an oscillator, characteristic distances important in vibrational relaxation are of the order of the amplitude of vibrations $\Delta q_0 \approx (kT/m\omega^2)^{\tfrac{1}{2}}$. Therefore the condition for the constancy of $\partial U/\partial q$ can be written in the form $\delta q \ll \Delta q_0$, and it is easy to verify that this condition is in fact equivalent to the condition $\omega \ll z$.

These approximations enable an equation to be obtained pertaining only to configuration space.† Introducing a new function

$$n(q) = \int f(p, q)\,dp, \tag{36.5}$$

---

† Detailed consideration of the transition from the Fokker–Planck equation to the diffusion equation in configuration space (Smolukhovskiĭ's equation) is discussed in articles [172], [563].

which has the meaning of an instantaneous density of particles at the point q in configuration space, we will reduce (36.4) to the form

$$\frac{\partial n}{\partial t} = D\frac{\partial}{\partial q}\left(\frac{\partial n}{\partial q} + \frac{\partial U}{\partial q}\frac{n}{kT}\right) \qquad D = \frac{kT}{mz}. \tag{36.6}$$

This is the usual diffusion equation in an external field $U(q)$. The relaxation rate in configuration space is equal, in order of magnitude, to the diffusion coefficient divided by the square of the characteristic radius of the configuration space accessible to the oscillator, so that we obtain, for the inverse quantity, the relaxation time $\tau_2 \approx \Delta q^2/D \approx z/\omega^2$. Hence it is clear that the slow constituent of the function ϕ_E corresponds to relaxation with respect to the coordinate q. Consequently, the fast constituent should be attributed to relaxation with respect to the momenta.

As regards the transition rate across the potential barrier, for this, in the limit $z/\omega^\star \gg 1$ we find, from (34.42),

$$\kappa = v\frac{\omega^\star}{z}\exp(-\beta E_0), \qquad v = \frac{\omega}{2\pi}. \tag{36.7}$$

For small values of the ratio $\omega^\star/z$, the representative point undergoes many activating and deactivating collisions with the particles of the reservoir, at the time of the transition across the potential barrier, so that, in the region of the barrier, the distribution function over velocities will differ little from the equilibrium distribution function. The total transition rate, proportional to the diffusion coefficient in configuration space, decreases with increase in the number of collisions z, and the factor $\omega^\star/z$ in formula (36.7) can be interpreted as a transmission coefficient, which determines the probability of 'unimpeded' transmission of the point across the barrier.

This effect is, to some extent, analogous to the reduction in the rate of unimolecular reactions on increase in pressure. A reduction in the rate can appear, in theory, for the mechanism of strong collisions (see § 23).

2. *Small friction.* We will consider the solution of eqn (34.22) under the following conditions:

$$\omega\tau_0 \ll 1, \tag{36.8a}$$

$$\frac{z}{\omega} \gg 1, \tag{36.8b}$$

$$\frac{z}{\omega^\star} \ll 1. \tag{36.8c}$$

The first two relations are analogous to (36.1a) and (36.1b); therefore vibrational relaxation occurs under these conditions, just as in the case of large friction. However, an expression is obtained for the rate constant of decomposition κ, which differs appreciably from (36.7) because (36.8c) means that, during the time of transit of the system across the potential barrier, the representative point does not undergo collisions with the particles of the reservoir. Returning to (34.42) we find

$$\kappa = v \exp(-\beta E_0). \tag{36.9}$$

This formula can be obtained within the framework of the transition state method with a transmission coefficient $\chi = 1$. We recall that, in the transition state method, the integration over the velocities corresponding to the reaction coordinate is carried out only for the current in one direction—in the direction of the reaction products. In a more consistent method, integration should be carried out for currents in both directions, as is indicated, in particular, in formula (34.35).

Approximations connected with the transition state method can be understood better if the non-equilibrium distribution function in the velocities is considered in the small-friction approximation. From (34.41), it is not difficult to see that, on reduction of the ratio ω^*/z, the integrand approximates more and more to the δ-function, so that, in the limit $\omega^*/z \to 0$, the function Φ takes a stepped form. The non-equilibrium distribution function on the critical surface has a Maxwell–Boltzmann form for the velocities which are directed towards the reaction products, and for the velocities in the opposite direction it is equal to zero. Just such a distribution function justifies the basic assumption of the transition state theory, namely: 'the activated complex X^* in the reaction $X \to X^* \to Y$, having been formed from X, is always converted into Y and never into X' [118].

The equilibrium distribution for the current in the direction of the reaction products is sustained at the expense of the intensive interaction of the system with the reservoir to the left of the barrier. The condition for a sufficiently-intensive interaction, for the one-dimensional model under consideration, is described in the form of the relation (36.8b). However, for more-or-less realistic models, the frequences ω and ω^* are quantities of the same order. Therefore the inequalities (36.8b) and (36.8c) cannot be satisfied simultaneously. Hence it follows that the approximation of small friction cannot be realized for the simplest system with one degree of freedom—namely, for a diatomic molecule in a heat reservoir.

Physically, the meaning of the condition $z/\omega \gg 1$, used in the derivation of formula (36.9), is that, during the time of the motion of the representative point though the phase volume of the active molecule, many random acts of interaction occur with the particles of the reservoir. For a system with many degrees of freedom, this condition can clearly be represented in the form†

$$z\langle \tau^* \rangle \gg 1. \tag{36.10}$$

---

† Here by $\langle \tau^* \rangle$ should be understood the mean lifetime of the active molecules with the energy which makes the main contribution to the rate of decomposition.

The mean lifetime $\langle \tau^* \rangle$, as was shown above (see Chapter 4), can far exceed the period of vibration $1/\omega$, so that the two conditions (36.8b) and (36.10) are completely consistent for polyatomic molecules. The fulfilment of condition (36.10), within the framework of the mechanism of strong collisions, means that the reaction proceeds in the limit of high pressures. Thus expression (36.9), which gives the reaction rate in the diffusion model of activation, corresponds to κ_∞ (relation (24.16)) for the mechanism of strong collisions.

3. *Diffusion through the energy levels.* We will assume that the effective number of collisions z is much smaller than the rate of spontaneous decomposition $1/\langle \tau^* \rangle$. This implies that the condition $z \ll \omega$ is also satisfied, guaranteeing a comparatively fast phase relaxation, which has a negligibly small effect on the relaxation of the population.

On condition

$$\frac{z}{\omega} \ll 1, \tag{36.11}$$

eqn (35.12) serves as a base for the study of relaxation and decomposition. In this equation, it is convenient, for future use, to express the diffusion coefficient in terms of the gas-kinetic number of collisions Z and the mean square energy transfer per collision $\langle \Delta E^2 \rangle$. Returning to a system with one degree of freedom, we will express the level density $\rho(E)$ in terms of the frequency of the vibrations $\omega(E)$, taking $\rho(E) = 1/\hbar\omega(E)$. Then (35.12) takes the form

$$\frac{\partial x}{\partial t} = \omega(\varepsilon) \frac{\partial}{\partial E} \left[\frac{Z}{2} \cdot \frac{\langle \Delta E^2 \rangle}{\omega(\varepsilon)} \left(x + \frac{1}{\beta} \frac{\partial x}{\partial E} \right) \right]. \tag{36.12}$$

The decomposition rate constant for the model under consideration is given by expression (35.16), where it should be assumed that $b = (Z/2)\langle \Delta E^2 \rangle$, $\rho = 1/\hbar\omega(\varepsilon)$. As regards the nonequilibrium distribution function, it has the form

$$x(E) = \frac{\exp(-\beta E) \int_E^{E_0} \dfrac{\exp(\beta\varepsilon)\omega(\varepsilon)2\, d\varepsilon}{\langle \Delta E^2 \rangle}}{\int_0^{E_0} \exp(\beta\varepsilon) \dfrac{\omega(\varepsilon)2\, d\varepsilon}{\langle \Delta E^2 \rangle}}. \tag{36.13}$$

For certain cases, as a reasonable approximation it may be assumed that, close to the threshold of dissociation, the ratio $\omega/\langle \Delta E^2 \rangle$ depends on E much more weakly than the exponential factor. Then from (35.16) we get

$$\kappa \approx \frac{Z}{2}(\langle \Delta E^2 \rangle \rho(E)\beta)_{E=E_0} \exp(-\beta E_0)/F(\beta) \tag{36.14}$$

and from (36.13), at E close to E_0,

$$x(E) \approx \frac{\exp(-\beta E)}{F(\beta)}[1 - \exp(-\beta E_0 + \beta E)], \qquad E \leqslant E_0. \qquad (36.15)$$

For example, for the model of a truncated harmonic oscillator we have

$$\rho = 1/\hbar\omega_0, \qquad F = kT/\hbar\omega, \qquad \langle \Delta E^2 \rangle = 2E\langle \Delta E_{0,\text{class}} \rangle,$$

and thus from (36.14) we obtain

$$\kappa \approx Z\frac{\langle \Delta E_{0,\text{class}} \rangle}{kT} \cdot \frac{E_0}{kT} \cdot \exp\left(-\frac{E_0}{kT}\right). \qquad (36.16)$$

Region of applicability of the classical diffusion theory

To determine the range of validity of the classical diffusion theory, we will consider, for simplicity, diffusion through energy states. The concept of diffusion can be employed as long as the displacement of the representative point, during a time characterizing the fluctuation of the heat reservoir, remains small in comparison with typical energies at which the distribution function changes appreciably. The random walk over the energy states is characterized by the mean square energy $\langle \Delta E^2 \rangle$ transferred to the system per collision. On the other hand, the function f changes in an energy interval of order kT. Thus a necessary condition for the applicability of the diffusion theory is

$$\langle \Delta E^2 \rangle^{\frac{1}{2}} \ll kT. \qquad (36.17)$$

However, at the minimum value of $\langle \Delta E^2 \rangle$, a condition is imposed under which the quantization of the levels can be neglected:

$$\langle \Delta E^2 \rangle^{\frac{1}{2}} \gg \hbar\omega. \qquad (36.18)$$

Taking joint account of these relations, two different situations can arise. First, it may turn out that the quanta are so small that the inequality $\hbar\omega \ll kT$ is satisfied for all states. Then, in principle, there is an interaction mechanism between the system and the reservoir such that the mean square energy transfer satisfies the condition $\hbar\omega \ll \langle \Delta E^2 \rangle^{\frac{1}{2}} \ll kT$. Just such a case was studied in the preceding section. For sufficiently-strong interactions of the system with the reservoir, the random walk may become so great that condition (36.17) is violated. In this case, a stochastic process can no longer be described by a differential equation, and it is necessary to study an integral equation of the type (23.4).

Secondly, it may turn out that the quanta are comparable with, or greater than, kT, i.e. $\hbar\omega \geqslant kT$. In this case, the classical description of transition processes between different states itself is no longer applicable, and the differential equation must be replaced by a difference equation, in order to

37. Tunnel transitions in a double potential well

The tunnel transmission of a system under a barrier separating two potential minima (see Fig. 22) is the quantum analogue of the transmission of a particle across a potential barrier. It is well known [37] that, for the double symmetric potential well, the spectrum of energy eigenvalues E below the energy of the maximum of the barrier is a set of doublet states, the splitting between which $\hbar\omega$ is, generally speaking, appreciably larger than the splitting $\hbar\Delta\omega$ within each doublet. The splittings ω and $\Delta\omega$ only become comparable in magnitude close to the peak of the barrier. Considering the lower states for which the splitting due to the tunnel effect is appreciably smaller than the frequency of vibrations in each of the potential minima, a situation can be visualized where the interaction with the reservoir will give rise to intensive transitions between each pair of states but will prove to have very little effect on states corresponding to different doublets. A molecule of ammonia is an example for which such a picture is valid. The comparatively-weak electric field arising in the collision of the NH_3 with a molecule of the reservoir completely polarizes the molecule if the energy of the dipole interaction appreciably exceeds a quantum of the inversion vibration ($\Delta v = 2\cdot 4 \times 10^{10}$ s^{-1}). At the same time, the field is still weak enough to change the nature of the vibration of the pyramidal (i.e. polarized) NH_3 molecule appreciably.

Thus, for the description of the relaxation within each doublet state, a model Hamiltonian can be used which only takes into account a pair of close-lying states of the isolated system in a symmetric potential well and also the interaction responsible for transitions between these states.

We will proceed from a quantum equation analogous to the classical equations (34.19). Denoting the wavefunction of the system by Ψ, we will write the non-stationary wave equation

$$i\hbar\frac{\partial \Psi}{\partial t}(q,t) = H_0(q)\Psi(q,t) + V(q,t)\Psi(q,t), \qquad (37.1)$$

where H_0 is the model Hamiltonian of the isolated system; V is the random interaction of the system with the heat reservoir.

For a given realization of the random interaction, the problem of the calculation of the mean value $\bar{L}(t)$ of any operator $\hat{L}$ may be considered solved if the wavefunction or the bilinear form (the density matrix) corresponding to it,

$$\rho(q, q', t) = \Psi^*(q', t)\Psi(q, t), \tag{37.2}$$

is known. The mean value $\bar{L}$ equals

$$\bar{L}(t) = \int \delta(q-q')\hat{L}(q')\rho(q, q', t)\,dq\,dq'. \tag{37.3}$$

For a random process, however, the value of $\bar{L}(t)$ pertaining to some definite realization of the interaction $V(t)$ is of less interest than its mean value $\langle \bar{L}(t) \rangle$ obtained after averaging $\bar{L}(t)$ over all possible realizations of the interaction $V(t)$. Introducing

$$W(q, q', t) = \langle \rho(q, q', t) \rangle. \tag{37.4}$$

in which the symbol $\langle \ldots \rangle$ denotes the average over random processes, we will write (37.3) in the form

$$\langle \bar{L} \rangle = \int \delta(q-q')\hat{L}(q')W(q, q', t)\,dq\,dq'. \tag{37.5}$$

This expression has the form of a trace of the product of two matrices $\delta(q-q')\hat{L}(q')$ and $W(q, q', t)$, in which the coordinates q and q' appear as indices.

In any other representation, formula (37.5) is written in the form

$$\langle \bar{L}(t) \rangle = \sum_{i,\kappa} L_{i,\kappa} W_{\kappa,i}(t). \tag{37.6}$$

Hence it is apparent that $W_{\kappa,i}(t)$ is the statistical matrix introduced in Chapter 4. Its determination forms the main problem of the quantum theory of relaxation.

We will write the Hamiltonian $H_0 + V$ of the two level system under consideration in the form

$$H_0 + V = \frac{\hbar}{2}\begin{pmatrix} \Delta\omega + \omega_z & \omega_x \\ \omega_x & -\Delta\omega - \omega_z \end{pmatrix}, \tag{37.7}$$

where $\hbar\Delta\omega$ denotes the energy difference of the zero-order Hamiltonian H_0; $\hbar\omega_z$ is the difference between the diagonal elements of the perturbation ($\hbar\omega_z = V_{11} - V_{22}$); $\hbar\omega_x/2$ is the non-diagonal element V_{12}.

It is expedient, for later use, to introduce the spin matrices

$$s_z = \frac{1}{2}\begin{bmatrix} 1 & 0 \\ 0 & -1 \end{bmatrix}, \quad s_x = \frac{1}{2}\begin{bmatrix} 0 & 1 \\ 1 & 0 \end{bmatrix}, \quad s_y = \frac{1}{2}\begin{bmatrix} 0 & -i \\ i & 0 \end{bmatrix}, \tag{37.8}$$

whose rules of operation are well known. The Hamiltonian (37.7) can be written, using (37.8), in the form

$$H_0 + V = \hbar\Delta\omega s_z + \hbar\omega_z(t)s_z + \hbar\omega_x(t)s_x. \tag{37.9}$$

The density matrix ρ satisfies the equation which follows from (37.1) and (37.2):

$$i\hbar\frac{\partial\rho}{\partial t} = [H, \rho]. \tag{37.10}$$

Any 2×2 Hermitian matrix, including the density matrix, can be represented in the form of an expansion in terms of the spin matrices s_x, s_y, s_z and the unit matrix $I = \begin{bmatrix} 1 & 0 \\ 0 & 1 \end{bmatrix}$:

$$\rho = \tfrac{1}{2}I + Zs_z + Xs_x + Ys_y, \tag{37.11}$$

where the coefficients Z, X, and Y are certain real functions of time.

From (37.11) it is not difficult to see that the normalization condition $\Sigma\rho_{ii} = 1$ is satisfied as a result of the properties of the spin matrices s_k. Substituting (37.11) into (37.10) we obtain the following system of equations for X, Y, and Z [33]:

$$\dot{X} = -[\Delta\omega - \omega_z(t)]Y,$$
$$\dot{Y} = [\Delta\omega + \omega_z(t)]X - \omega_x(t)Z,$$
$$\dot{Z} = \omega_x(t)Y. \tag{37.12}$$

The result of averaging over different realizations of the random interaction corresponding to (37.4) can be represented in the form

$$W = \tfrac{1}{2}I + \langle Z \rangle s_z + \langle X \rangle s_x + \langle Y \rangle s_y. \tag{37.13}$$

The mean values $\langle X \rangle$, $\langle Y \rangle$, and $\langle Z \rangle$ can be found either by averaging the functions X, Y, and Z over different realizations of the random interaction or by obtaining an equation satisfied by $\langle X \rangle$, $\langle Y \rangle$, and $\langle Z \rangle$. The first way is more direct but much more complicated, since it requires a preliminary solution of the dynamical problem (i.e. the Schrödinger equation). The second way avoids the solution of the dynamical problem but, to make up for it, uses some properties of the random interaction to carry out the correct averaging of (37.12). It is not difficult to see that the direct averaging of (37.12) does not give an equation for $\langle X \rangle$, $\langle Y \rangle$, and $\langle Z \rangle$, since the means of products $\langle \omega_x(t)Y \rangle$, etc., appear on the right-hand side.

In the random interaction models the collisions in the gas, then $V(t)$ represents a series of surges, the duration τ_0 of each of them being small in comparison with the time between them. In this case (37.12) leads to the following system for the mean values [33]:

$$\langle \dot{X} \rangle = -\left(\frac{z-z_1}{2}\right)\langle X \rangle - \Delta\omega \langle Y \rangle,$$

$$\langle \dot{Y} \rangle = \Delta\omega \langle X \rangle - \left(\frac{z+z_1}{2}\right)\langle X \rangle,$$

$$\langle \dot{Z} \rangle = -z_1 \langle Z \rangle,$$

$$z = \frac{1}{\tau_1} \left\langle \frac{1-\cos\Omega\tau_0}{\Omega^2}(2\omega_z^2 + \omega_x^2) \right\rangle,$$

$$z_1 = \frac{1}{\tau_1} \left\langle \frac{1-\cos\Omega\tau_0}{\Omega^2}\omega_x^2 \right\rangle, \qquad (37.14)$$

where τ_0 is the collision time; τ_1 is the time between collisions.

In this system, the equation for $\langle Z \rangle$ is not connected with the equations for $\langle X \rangle$ and $\langle Y \rangle$. It describes the exponential equalization of the population of the two levels.

To investigate the temporal behaviour of $\langle X \rangle$ and $\langle Y \rangle$, it is convenient to compare the mean values of some physical quantities with these functions. It is not difficult to see that the coordinate operator (measured from the centre of the potential hump), in the representation chosen, has the form

$$\hat{q} = q_0 s_x, \qquad (37.15)$$

where q_0 is some constant.

Substituting (37.15) into (37.3) we find that $\langle q(t) \rangle$ is proportional to $\langle X(t) \rangle$. Then on the basis of (37.14) we obtain the following equation for the mean value of the particle's coordinate in the double potential well

$$\langle \ddot{q} \rangle + z \langle \dot{q} \rangle + \left[(\Delta\omega)^2 + \frac{z^2}{4} - \frac{z_1^2}{4} \right] \langle q \rangle = 0. \qquad (37.16)$$

This equation describes the damped harmonic vibrations. However, in contrast to the classical case in (34.27), two parameters of the interaction of the system with the medium (z and z_1) come into it, in accordance with the two different mechanisms of this interaction. If the external perturbation leads only to a shift in the levels, then $\omega_x = 0$ and consequently $z_1 = 0$. If, on the other hand, the total effect of the perturbation is transmitted by the off-diagonal elements $V_{12}(t)$, then $\omega_z = 0$ and $z - z_1 = 0$. Since the latter case is close to the problem of the relaxation of a classical oscillator,† we will consider it in somewhat greater detail.

† If the relaxation of a classical oscillator is considered as a limiting case ($\beta\hbar\omega \gg 1$) of the relaxation of a quantum mechanical oscillator, then the operator of the perturbation $V(t)$ must be written in the form $-xF(t)$. For an interaction of this form, only the off-diagonal matrix elements connecting neighbouring states differ from zero.

Of the two independent solutions of eqn (37.16), we will consider the one which dies away the more slowly. This means that the rate constant of the relaxation κ is identified with the smaller (negative) real part of the root of the characteristic equation corresponding to (37.16). In this way we find

$$\kappa = \text{Re}\left\{\frac{z}{2} - \sqrt{\left[\frac{z^2}{4} - (\Delta\omega)^2\right]}\right\}. \qquad (37.17)$$

As in the classical case, it is convenient to distinguish here the limiting cases of large and small friction.

For small friction ($z/2 \ll \Delta\omega$) the mean value of the coordinate $\langle q(t) \rangle$ undergoes damped harmonic vibrations, where the rate of damping $\kappa = z/2$ increases with increase in pressure.

For $z/2 = \Delta\omega$ the relaxation rate is at a maximum and is equal to the frequency of the tunnel splitting $\Delta\omega$. In these conditions, the tunnel transmission is analogous to the classical transmission across the potential barrier in the region of small friction (formula (36.9)), where the transition rate does not depend on the properties of the interaction between the system and the reservoir. In fact, in the classical case, the kinetics turn out to be of the first order when the heat reservoir sustains the equilibrium distribution over the degrees of freedom of the active molecule (in this sense, the interaction of the system with the reservoir is still sufficiently strong), but does not interact with the system at the time of the representative point's transit across the peak of the barrier (in this sense, the interaction is already sufficiently weak).

In the quantum case, with the condition $z/2 = \Delta\omega$, the system goes under the barrier during a time of the order of the time between successive collisions, so that the reservoir is practically no obstruction to tunnel infiltration. However, in the region of one of the wells (which might be called the configuration space of the active molecule), since the system spends time of the same order of magnitude as the time of transit under the barrier, a kinetic first-order process is, strictly speaking, correct, for the system of two levels under consideration, only at the one point $z = 2\Delta\omega$.

In the region $z \gg 2\Delta\omega$ the rate of relaxation $\kappa = (\Delta\omega)^2/z$ falls with increase in pressure. In Kramer's classical theory the same dependence is obtained for the above-barrier transmission rate for large friction (formula (36.7)). However, for above-barrier transitions the conditions for the approximation of large friction are very rarely satisfied, whereas for the tunnel transitions the fulfillment of the inequality $z \gg 2\Delta\omega$ is very probable. The fact that κ is proportional to the square of the splitting may be interpreted in the following way. The splitting $\Delta\omega$ in the double potential well is expressed in terms of the potential corresponding to the barrier by formula (24.39). On the other hand, the coefficient of tunnel transmission χ out of the left well into the right might have been calculated by formula (24.38) if the energy spectrum in the region of the right-hand minimum were continuous (i.e. if the actual potential of type AaBcGC were replaced by the potential AaBcGH—see Fig. 22). From a comparison of (24.39) and (24.38) it is evident that, with exponential accuracy, $(\Delta\omega)^2 \sim \chi$. It should now be taken into account that, for a large number of collisions ($z \gg 2\Delta\omega$), the energy levels are so broadened that, for a particle undergoing a tunnel transition, the broadened levels in the right-hand well appear as some portion of a continuous spectrum. Therfore it is to be expected that, in the region of large friction, the relaxation rate must be proportional to the transmission coefficient χ of the particle, i.e. to $(\Delta\omega)^2$. The theory given above of relaxation in the presence of tunnel transitions in a double potential well can be checked experimentally, for example, by examining the pressure dependence of the inversion spectrum of molecules of the NH_3 type, since the absorption contour $I(\omega)$ is given by the Fourier transform of the function $\langle q(t) \rangle$; Ben-Reuven's paper is devoted to this question [151].

38. Random walks over the discrete energy levels

The example considered in the preceding section, which enables the transition between the limiting cases of large and small friction to be followed, does not permit a transition to the classical limit, since for $\hbar \to 0$, in general, no relaxation takes place, as the system proves to be isolated in one of the wells. With the aim of studying the transition from the quantized case to the classical, we will consider here a model with many levels, the transitions between which are responsible for vibrational relaxation and chemical reactions.

In § 34 it was noted that, if the internal frequency of the classical system far exceeds the characteristic frequency of collisions z, then the relaxation of the population proceeds independently of the phase relaxation. As analogous arguments are also valid for the quantum-mechanical system, we will consider the relaxation of an oscillator in a heat reservoir on the assumption that the frequency of collisions z 'mixing up' the phase is far higher than the collision frequency corresponding to the relaxation of the population. We will operate, therefore, with just the diagonal elements of the density matrix $\rho_{nn} \equiv x_n$. The system of equations describing the transitions between neighbouring levels has the form

$$\frac{dx_n}{dt} = Z_0 P_{n-1,n} x_{n-1} - Z_0(P_{n,n-1} + P_{n,n+1})x_n + Z_0 P_{n+1,n} x_{n+1}, \quad (38.1)$$

where Z_0 is the number of collisions of the system with molecules of the reservoir; $P_{n,n+1}$ is the probability of a transition between the levels n and $(n + 1)$.

Since the reservoir is in equilibrium the whole time, the transition probabilities satisfy the principle of detailed balance:

$$\frac{P_{n,n+1}}{P_{n+1,n}} = \exp[-\beta(E_{n+1}-E_n)]. \tag{38.2}$$

We will have in mind the energy levels of a system in the potential well represented in Fig. 21. We will assume from the start that one of the upper levels of the system at $n = N$ is not connected with the next level, i.e. that $P_{N,N+1} = 0$. Then (38.2) is a finite system of linear differential equations, whose method of solution is well known. A solution $x_n(t)$ is sought for in the form of an expansion in eigenvectors $l_n(\mu)$ of the matrix of the right-hand side of the system (38.1):

$$x_n(t) = \sum_\mu A_\mu \exp(-\mu t) l_n(\mu). \tag{38.3}$$

The eigenvectors and eigenvalues satisfy the system of equations

$$\sum_n B_{m,n} l_n(\mu) = -\mu l_m(\mu), \tag{38.4}$$

where $B_{m,n}$ is the kinetic matrix composed of the elements $P_{n,n\pm 1}$ of the right-hand side of system (38.1).

The coefficients A_μ in (38.3) are determined by the initial conditions. One of the eigenvalues of the system (38.4) is known beforehand: $\mu_0 = 0$; the eigenvector $l_n(0)$, whose components agree with the equilibrium distribution function, corresponds to it. All the other eigenvalues μ_k, which prove to be positive [374], [452], [483], depend on the specific form of the transition probabilities $P_{n,n\pm 1}$.

If a process is considered in which transitions occur only between the lower levels of the system then the problem simplifies further, since the levels can be considered equidistant (a property of the harmonic oscillator) and the transition probability is linear in the vibrational quantum number (a result of applying first-order perturbation theory to calculate the transition probabilities for the linear interaction of an oscillator with the reservoir). Returning to formula (8.18) we will assume:

$$P_{n,n+1} = (n+1)P_{01}, \qquad P_{n+1,n} = (n+1)P_{10}, \qquad P_{01} = \alpha P_{10}, \qquad \alpha = e^{-\hbar\omega\beta}, \tag{38.5}$$

so that the system of equations (38.1) takes the form:

$$\frac{dx_n}{dt} = Z_0 P_{10} \alpha n x_{n-1} - Z_0 P_{10}[n+\alpha(n+1)]x_n + Z_0 P_{10}(n+1)x_{n+1}. \tag{38.6}$$

This system of equations was first studied by Landau and Teller [320], who showed that, as a result of the simultaneous linearity in the vibrational quantum numbers of both the oscillator energies and the transition probabilities, a closed equation is obtained for the mean energy $\langle E \rangle = \sum_n \hbar\omega n$ from system (38.6).† Multiplying (38.6) by n and summing, we obtain

$$\frac{d}{dt}\langle E \rangle = -Z_0 P_{10}(1-\alpha)(\langle E \rangle - E_0), \qquad E_0 = \frac{\hbar\omega\alpha}{1-\alpha}. \qquad (38.7)$$

This equation agrees in form with the classical equation (34.33), and goes over into it exactly on fulfillment of the condition $\hbar\omega\beta \ll 1$ when $E_0 = kT$.

The non-equilibrium distribution function for different initial conditions was found by Montroll and Shuler [374], [375]. Without quoting the results of the calculation, we will note only one important consequence: the equilibrium distribution function, which at first was a Boltzmann distribution function with temperature T_0, undergoes relaxation to an equilibrium distribution function x_n^0 with the temperature T_1 of the reservoir, through a sequence of Boltzmann distributions whose temperatures depend on time. Thus this system is an example where to speak of a definite vibrational temperature, even in non-equilibrium conditions, has some meaning.

The system of equations (38.1), which can be considered as a single equation in finite differences, leads to a differential equation if the first difference (i.e. a difference of type $x_{n+1} - x_n$) is approximated by the first derivative and the second (i.e. a difference of the type $x_{n+1} - x_{n-1}$) is expressed in terms of the first and second derivatives of the distribution function $x(n)$ with respect to the argument n. Moreover, the argument should be considered as varying continuously, and the condition under which such an approximation is applicable has the form $\hbar\omega\beta \ll 1$. If now the energy $E = \hbar\omega n$ is introduced as a variable, then the differential equation, thus found, agrees with eqn (35.3), which describes the relaxation of the distribution function of a classical oscillator.

We will now study the system (38.1), taking the absorption of the particles which have reached the highest level N into account. This boundary condition represents the decomposition reaction of a diatomic molecule in the initial stage, when recombination can be neglected. Two problems arise here: the first is connected with calculating the decomposition rate, the second with elucidating the effect of decomposition on the vibrational relaxation rate. We will discuss only the former, having ascertained when the mutual effect of decomposition and relaxation can be neglected. The system of equations (38.1) must now be solved with the boundary condition $x_N = 0$. From physical considerations, it is clear that the condition for absorption at the level N will ultimately lead to the total disappearance of the molecules at all levels, therefore the lowest eigenvalue of the matrix B in this case must differ from zero. We will retain the notation μ_0 for it and will assume that the first

---

† The question of the conditions under which the relaxation of the moments of the distribution function obey independent equations is discussed in papers [484], [487].

eigenvalue μ_0 is appreciably smaller than all the others

$$\mu_0 \ll \mu_1, \mu_2 \ldots \quad (38.8)$$

Then at times $t \gg 1/\mu_1$ the decomposition of the molecule will follow an exponential law, and the change in the total concentration of the molecules will be described by the equations

$$\frac{dc}{dt} = -\mu_0 c, \qquad c = \sum_n x_n(t). \quad (38.9)$$

The condition for the absorption of particles at the level N can be formulated in the form of the requirement $P_{N,N+1} \to \infty$, which is imposed on one of the coefficients of the infinite system of equations. It can be seen from this that the problems of relaxation with and without decomposition differ only in the changing of the value of the coefficient $P_{N,N+1}$.

The change in one of the elements of the matrix B can be considered as a perturbation, and the corresponding perturbation theory [40] can be used to examine the correlation between the eigenvalues of the matrix $B_0 (P_{N,N+1} = 0)$ and the eigenvalues of the matrix $B_\infty (P_{N,N+1} = \infty)$. Such an investigation [52] shows that, under the condition (38.8), the eigenvalues μ_k of the matrices B_0 and B_∞ differ but little. This means that the kinetics of vibrational relaxation, for which times of the order $t \approx 1/\mu_1$ are important in both the decomposing and non-decomposing oscillator, will be described by eqn (38.7). Thus we see that, in all relaxation processes under the condition (38.8) two independent stages can be distinguished. The first, rapid stage with characteristic times $\tau_{vib} \approx 1/\mu_1$ corresponds to vibrational relaxation, where the relaxation is accomplished before an appreciable proportion of the molecules has decomposed. The second, slow stage results in the decomposition of the molecules. It is characterized by times $\tau_{diss} \approx 1/\mu_0$, and proceeds mainly after the mean energy has undergone relaxation to the equilibrium value.

A change in the concentration, independent of the mean energy, does not mean, however, that the dissociation of the molecule is characterized by the equilibrium distribution function over the vibrational states for all levels; the condition for the completion of relaxation by the time the dissociation begins means that the distribution function is close to the equilibrium form only for the lower vibrational levels, which are important for the mean vibrational energy. For the higher vibrational states, close to the dissociation boundary, the Boltzmann distribution proves to be substantially disturbed during the whole process of decomposition. The equilibrium distribution is only re-established for them with the approach of the system to chemical equilibrium.

The lowest eigenvalue for the system of equations (38.1), on condition (38.8), was calculated in the papers [53], [112], [129], [152]. The result of the calculation is

$$\mu_0 = \kappa_{\text{diss}} = \left[\sum_{j=0}^{N} \exp(-\beta E_j) \sum_{n=j}^{N+1} \frac{\exp(\beta E_n)}{Z_0 P_{n,n-1}} \right]^{-1}. \quad (38.10)$$

In this expression, the parameter $P_{N+1,N}$ still remains undetermined. If it is assumed that $P_{N+1,N} = 0$ (decomposition forbidden) then, from (38.10), we obtain $\mu_0 = 0$. If, on the contrary, $P_{N+1,N} = \infty$ is taken, then the last term disappears from the sum over n in connection with the fact that the fastest stage does not, in general, affect the kinetics of the subsequent process.

We will now trace how this formula goes over into the classical expression (35.16). To this end, we will assume that the transition probability $P_{n,n-1}$ and the vibrational quanta $\hbar\omega_{n,n-1}$ depend only weakly on the vibrational quantum number, so that we can assume

$$Z_0 \langle \Delta E_n^2 \rangle = Z_0 P_{n,n-1}(\hbar\omega_{n,n-1})^2 + Z_0 P_{n,n+1}(\hbar\omega_{n,n+1})^2 \approx 2P(n) Z_0 (\hbar\omega_n)^2. \quad (38.11)$$

On the other hand, the sum over n is changed into an integral over dE if the condition $\beta\hbar\omega_n^* \ll 1$ is satisfied in the region of Δn near the n^* which makes the main contribution to the sum (38.10):

$$\sum_{n=j}^{N} \frac{\exp(\beta E_n)}{Z_0 P_{n,n-1}} \to \int_{E_j}^{E_0} \frac{\exp(\beta\varepsilon) \, d\varepsilon}{Z_0 P(\varepsilon) \hbar\omega(\varepsilon)}. \quad (38.12)$$

Furthermore, the inner sum over n will be considered independent of the lower limit, which is permissible on condition $\beta E_0 \gg 1$. Then the external sum is simply the partition function $F(\beta)$ and we arrive at the expression

$$\kappa_{\text{diss}} = \left[F(\beta) \int_0^{E_0} \frac{\exp(\beta\varepsilon) \, d\varepsilon}{Z_0 P(\varepsilon) \hbar\omega(\varepsilon)} \right]^{-1} = \left[F(\beta) \int_0^{E_0} \frac{\exp(\beta\varepsilon) 2\hbar\omega(\varepsilon) \, d\varepsilon}{Z_0 \langle \Delta E(\varepsilon) \rangle^2} \right]^{-1}. \quad (38.13)$$

which is identical with (35.16).

We will mention yet another method of calculating the non-equilibrium rate constant, to which papers [304], [547], [550] are devoted. The method is based on a calculation of the mean time $\bar{t}$ necessary for the transition of a molecule from an initial state characterized by some distribution $x_n(0)$ into a dissociated state. The time $\bar{t}$ is determined by the relation

$$\bar{t} = -\int_0^\infty t \frac{dc}{dt} \, dt = \int_0^\infty c(t) \, dt = \int_0^\infty \sum x_n(t) \, dt. \quad (38.14)$$

A formal expression for $\bar{t}$ in terms of the kinetic matrix B has the form

$$\bar{t} = \sum_i (B-I)_{ii}^{-1} x_i(0), \quad (38.15)$$

so that the mathematical problem reduces to the inversion of the matrix B. If all the $x_n(t)$ decay according to an exponential law, then $\bar{t}$ equals $1/\mu_0$. However, the question of the temporal law of decomposition is not raised in a general form, so that $\bar{t}$ can also be calculated when relaxation and dissociation overlap.

Finally, we will pause to consider the quasi-stationary non-equilibrium distribution function $l_n(\mu_0)$, for which the expression [129], [381]

$$l_n(\mu_0) = \exp(-\beta E_n)\left(1 - \frac{\mu_0}{M_n}\right)$$

$$M_n = \sum_{j=0}^{N} \exp(-\beta E_j) \sum_{n=j}^{n+1} \frac{\exp(\beta E_n)}{Z_0 P_{n,n-1}} \quad (38.16)$$

is valid where, as is clear from (38.10), $M_N = \mu_0$. If total absorption does not occur at the highest level, then $l_N(\mu_0)$ does not tend to zero. In particular, for the model of a truncated harmonic oscillator, the non-equilibrium distribution function has the form [52], [54], [482]

$$l_n(\mu_0) = \exp(-\beta\hbar\omega n)\left\{1 - \exp[\beta\hbar\omega(N-n)]\frac{P_{N,N+1}}{P_{N,N+1} + P_{10}(1-\alpha)n}\right\}. \quad (38.17)$$

This distribution corresponds to a rate constant

$$\mu_0 = Z_0 P_{10}(1-\alpha)\frac{P_{N,N+1}}{\frac{P_{N,N+1}}{(1-\alpha)n} - P_{10}}\exp(-\beta\hbar\omega N). \quad (38.18)$$

The non-stationary distribution function $x_n(t)$ for the same model was calculated in the work of Shuler [482], [483].

39. Mechanism of activation and the non-equilibrium distribution function in unimolecular reactions

The mechanism of stepwise activation, an alternative to the mechanism of strong collisions considered in Chapter 4, lies at the basis of the diffusion theory of chemical reactions. The actual activation mechanism is intermediate between these two cases. In this connection the question arises, to what extent an experimental investigation of the dependence of the rate constant of unimolecular reactions on the pressure and on the distribution functions over the energies of the reacting molecules can lead to a clarification of the activation mechanism.

From this point of view, we will consider the simplest variant of a unimolecular reaction for which first- and second-order kinetics are possible.

Taking Kassel's classical model, we will assume the following relations

$$\hbar\omega \ll kT \ll \frac{E_0}{s-1} \qquad (39.1)$$

to be satisfied. The problem consists of finding a quasi-stationary solution of the diffusion equation (35.12) for $\kappa(E) = v[(E-E_0)/E]^{s-1}$. The essential feature of this equation is that, near the reaction threshold, $\kappa(E)$ depends on E much more strongly than the other functions coming into the equation. Therefore, in the equation

$$\frac{\partial}{\partial E}\left[\tfrac{1}{2}Z_0\langle\Delta E^2\rangle\beta\rho(E)\left(1+\frac{1}{\beta}\frac{\partial}{\partial E}\right)x(E)\right] - \kappa(E)\rho(E)x(E) = 0, \qquad (39.2)$$

quantities of the type $(\partial\rho/\partial E)/\beta\rho$ can be regarded as changing comparatively slowly [61]. This enables us to reduce the general equation (39.2) to a simpler model equation which still takes into account all the basic features of (39.2) [61], [79].

Going over to new independent variables $\xi = \beta E$, we will introduce the functions

$$z(\xi) = \tfrac{1}{2}Z_0\langle\Delta E^2(E)\rangle^2\beta^2,$$

$$\Psi = \frac{\mathrm{d}(z\rho)}{\mathrm{d}\xi}\bigg/ z\rho$$

$$\phi_0 = \frac{1-\Psi}{2},$$

$$\phi = \left[\left(\frac{1-\Psi}{2}\right)^2 + \frac{\kappa}{z}\right]^{\tfrac{1}{2}}, \qquad (39.3)$$

in terms of which the non-equilibrium distribution function will be expressed. Expressions for the non-equilibrium distribution functions in the low- and high-pressure limits are given in Table 8. For comparison, formulae obtained on the assumption of the mechanism of strong collisions are quoted in the same table. The subscripts d and s distinguish quantities referring to the diffusion mechanism and the mechanism of strong collisions. As regards the ratio $(x/x_{eq})_d$, it should be noted that the formulae quoted in Table 8 are invalid at the dissociation threshold itself, where the solution of the model equation cannot be represented by elementary functions. From the expressions presented in Table 8, it is evident that, at low pressures, the diffusion mechanism of activation, in general, is not able to maintain an appreciable

TABLE 8

Characteristics of unimolecular reactions for the mechanism of strong collisions and the diffusion mechanism

	Low pressures	High pressures
Reduced rate constant	$\left(\dfrac{\kappa}{\kappa_\infty}\right)_s = \dfrac{Z_0 k T \rho(E_0)}{F} \ll 1$	$\left(\dfrac{\kappa}{\kappa_\infty}\right)_s = 1$
	$\left(\dfrac{\kappa}{\kappa_\infty}\right)_d = \dfrac{\langle \Delta E^2 \rangle}{2(kT)^2}\left(\dfrac{\kappa}{\kappa_\infty}\right)_s \ll 1$	$\left(\dfrac{\kappa}{\kappa_\infty}\right)_d = 1$
	$\left(\dfrac{\kappa}{\kappa_\infty}\right)_d < \left(\dfrac{\kappa}{\kappa_\infty}\right)_s$	
Distribution function above threshold	$\left(\dfrac{x}{x_{eq}}\right)_s = 0$	$\left(\dfrac{x}{x_{eq}}\right)_s \approx \dfrac{Z_0}{Z_0 + v(1 - \xi_0/\xi)^{s-1}}$
	$\left(\dfrac{x}{x_{eq}}\right)_d = 0$	$\left(\dfrac{x}{x_{eq}}\right)_d \approx \exp\left[-\int_{\xi_0}^{\xi}(\phi - \phi_0)d\xi\right]$
Distribution function below threshold	$\left(\dfrac{x}{x_{eq}}\right)_s = 1$	$\left(\dfrac{x}{x_{eq}}\right)_s = 1$
	$\left(\dfrac{x}{x_{eq}}\right)_d = 1 - \exp(\xi - \xi_0)$	$\left(\dfrac{x}{x_{eq}}\right)_d = 1$

concentration of active molecules. At high pressures, when the rate constant reaches the asymptotic equilibrium value, the distribution functions for the energies E exceeding the threshold differ from the equilibrium distribution, the difference being the greater the further the level E is from E_0. Although a disturbance of the equilibrium distribution of this type takes place both with the diffusion mechanism and with the mechanism of strong collisions, from the expressions quoted it is evident that, in the diffusion mechanism, the depletion of the higher levels sets in more rapidly.

We will now consider the thermal decomposition of a non-linear triatomic molecule in more detail. The case is interesting, because a non-linear triatomic molecule is the simplest system with the minimum number of internal degrees of freedom $s = 3$ for which first- and second-order kinetic processes are possible. Moreover, the non-equilibrium distribution function of just such a system (thermal decomposition of NO_2 [325]) has been investigated experimentally. For $s = 3$, the non-equilibrium distribution function can be expressed in terms of parabolic cylindrical functions, and the rate constant in the intermediate pressure region can be found in closed form.

The calculations give [79]

$$\left(\frac{\kappa}{\kappa_\infty}\right)_d = \frac{1}{2}\gamma_d \frac{\Gamma\left(\frac{3}{4}+\frac{\sqrt{\gamma_d}}{16}\right) - \frac{\gamma_d^{\frac{1}{4}}}{4}\Gamma\left(\frac{1}{4}+\frac{\sqrt{\gamma_d}}{16}\right)}{\Gamma\left(\frac{3}{4}+\frac{\sqrt{\gamma_d}}{16}\right) + \frac{\gamma_d^{\frac{1}{4}}}{4}\Gamma\left(\frac{1}{4}+\frac{\sqrt{\gamma_d}}{16}\right)},$$

$$\gamma_d = \frac{z(\xi_0)(\beta E_0)^2}{\nu}. \tag{39.4}$$

The population of the threshold level $E = E_0$, corresponding to the rate (39.4), is

$$\left[\frac{x(E_0)}{\exp(-\beta E_0)}\right]_d = \frac{\frac{\gamma_d^{\frac{1}{4}}}{2}\Gamma\left(\frac{1}{4}+\frac{\sqrt{\gamma_d}}{16}\right)}{\left[\Gamma\left(\frac{3}{4}+\frac{\sqrt{\gamma_d}}{16}\right) + \frac{\gamma_d^{\frac{1}{4}}}{4}\Gamma\left(\frac{1}{4}+\frac{\sqrt{\gamma_d}}{16}\right)\right]}. \tag{39.5}$$

It is interesting to compare this expression with the formulae, obtained on the assumption of strong collisions:

$$\left(\frac{\kappa}{\kappa_\infty}\right)_s = \frac{1}{2}\int_0^\infty \frac{x^2 \exp(-x)}{1+x^2/\gamma_s}dx, \quad \gamma_s = Z_0\frac{(\beta E)^2}{\nu}, \tag{39.6}$$

$$\left[\frac{x(E_0)}{\exp(-\beta E_0)}\right]_s = 1. \tag{39.7}$$

If γ_d is identified with γ_s, then, in the limiting cases of low and high pressures, eqns (39.4) and (39.6) are identical. The real difference between the two activation mechanisms, which it is impossible to eliminate by a semi-empirical choice of the parameter characterizing the 'effectiveness' of the activation, appears only in the transitional region from a second-order to a first-order kinetic process. In Fig. 34, curves showing the dependence of the rate constant and population of the threshold level on the pressure (parameter γ) are given. It is clear that the dependence of the ratio κ/κ_∞ on the activation mechanism is much weaker than the dependence of the non-equilibrium distribution function. For example, after replacing $s = 3$ by $s = 3 \cdot 5$, curve 1a can be displaced into curve 1b almost exactly, i.e. kinetically, the mechanism of strong collisions for $s = 3 \cdot 5$ is practically indistinguishable from the diffusion mechanism for $s = 3$, however curves 2a and 2b remain unchanged.

A quantum-mechanical variant of the two activation mechanisms was discussed in the work of Buff and Wilson [174], mentioned in § 33. Averaging over all the oscillators, the probability of a single-quantum transition is

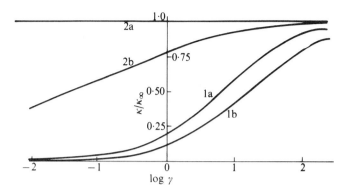

FIG. 34. Pressure dependence of the reduced rate constant of a unimolecular reaction for $s = 3$: 1a—mechanism of strong collisions; 1b—mechanism of stepwise excitation (diffusion mechanism). Curves 2a and 2b show the relative population of the threshold level for these two mechanisms.

given by formula (33.11) and the rate constant in the low-pressure region by (33.12). The latter goes over into the classical expression (23.45), in the limit $\beta\hbar\omega \ll 1$. At arbitrary pressures, the quantum-mechanical expression for the rate constant has a somewhat complicated form, so that we will not quote it here. We will only note that, as in the classical case, the curves describing the dependence of $(\kappa/\kappa_\infty) = f(\gamma)$ for the two collision mechanisms are close to one another, so that in the interpretation of experimental data with semi-empirical parameters s or Z_0 it is hardly possible to give preference to one or other of the mechanisms. Nevertheless, Hung and Wilson [278] showed that the two mechanisms can be distinguished by studying the unimolecular decomposition of a molecule taking place via two channels at the same time.

$$AB \xrightarrow{\kappa_1} A_1 + B_1,$$
$$AB \xrightarrow{\kappa_2} A_2 + B_2, \quad E_{a,1} < E_{a,2}, \qquad (39.8)$$

In this case, the ratios of the rate constants κ_1/κ_2 for the mechanisms of strong collisions and stepwise excitation differ considerably at low pressures.

The increase in the ratio κ_1/κ_2, on a reduction in the number of collisions, is easily interpreted as follows. A reaction with a lower activation energy leads to a depletion of the levels lying above $E_{a,1}$; as a result, the rate of a second reaction with higher activation energy will prove to be lower than the value it would be in the absence of the first reaction. Moreover, it is evident that the more localized the distribution function for the energy transfer, the greater is the disturbance of the equilibrium distribution above the lower reaction threshold.

It might be expected that the actual activation mechanism could be elucidated from an investigation of the pressure dependence of the fluorescence spectrum of molecules participating in photochemical decomposition reactions. The quantum yield of reactions and the intensity of individual vibrational bands are characteristic of photochemical processes. The calculations of Porter and Connelly [408], and also of Wilson and coworkers [557], [558], showed that the quantum yield depends only very weakly on the details of the inelastic molecular collisions. However, the distribution of intensity in the spectrum does depend on the path—single quantum or multiple quantum—by which the vibrational deactivation of the electronically-excited states of the radiating molecule takes place [470]. Although it is rather difficult to establish the experimental conditions necessary for such an investigation, Levitt [325] nevertheless succeeded in observing the depopulation of the vibrational levels in the emission spectrum of NO_2^*, caused by the predissociation of $NO_2^* \rightarrow NO + O$. Levitt found that the energy dependence of the relative disturbance of the equilibrium distribution below the predissociation limit is of the form [325]:

$$\frac{x(E)}{\exp(-\beta E)} \sim 1 - \exp\left(\frac{E - E_0}{kT}\right), \tag{39.9}$$

where E is the energy of the vibrational level; E_0 is the dissociation limit.

This expression agrees with the diffusion mechanism of excitation.

The transition between the two limiting mechanisms of activation can be traced only on the basis of the general kinetic equation (23.10), in which some specific function $P(E \rightarrow E')$ must be substituted. Troe and Wagner [530] considered a number of models for which $P(E \rightarrow E')$ depends only on the difference $E - E'$,† which enables eqn (23.10) to be solved by the method of the Laplace transform. In particular, if the probability of vibrational transitions $E \rightarrow E'$ is approximated by the expression

$$P(E \rightarrow E') = \frac{1 + \Delta E^*/kT}{2 + \Delta E^*/kT} \cdot \frac{1}{\Delta E^*} \begin{cases} \exp\left(-\frac{E'-E}{kT} - \frac{E'-E}{\Delta E^*}\right), & E' > E \\ \exp\left(-\frac{E-E'}{kT}\right), & E' < E \end{cases} \tag{39.10}$$

then the non-equilibrium distribution function in the low-pressure region is given by the expression

$$\frac{x(E)}{\exp(-E/kT)} = 1 - \frac{1}{1 + \Delta E^*/kT} \exp\left(-\frac{E_0 - E}{kT}\right), \tag{39.11}$$

from which (39.9) is obtained in the limit $\Delta E^* \ll kT$.

† In the diffusion description of excitation, this approximation corresponds to a constant coefficient of diffusion.

As regards the effective number of collisions Z^* in the expression for κ_0 (33.8b), for the case under consideration it equals

$$Z^* = Z_0\left(1 + \frac{kT}{\Delta E^*}\right)^{-2}. \tag{39.12}$$

Naturally, formulae (33.10) and (33.13), which are correct for the mechanism of strong collisions and the diffusion method, are obtained in the limiting cases for $\Delta E^* \gg kT$ and $\Delta E^* \ll kT$ respectively.

7

DISSOCIATION OF DIATOMIC MOLECULES AND THE RECOMBINATION OF ATOMS

40. Equilibrium theory of decomposition and recombination

IF a diatomic molecule possesses vibrational energy exceeding the dissociation energy, then it undergoes decomposition during a time of the order of the period of vibrations (the only exceptions are cases of nonadiabatic decomposition, where the probability factor may be appreciably less than unity, see § 27). Since this time is considerably less than the time between successive collisions of the molecule, the decomposition of an active molecule (in the sense of the definition given in § 22) can be considered instantaneous, and the observed rate of decomposition must always be determined by the rate of the thermal excitation of the molecule to the dissociation limit. Early theories of the thermal decomposition of diatomic molecules [434], [435], were based on the assumption that thermal activation proceeded at the equilibrium rate and differed only in the dependence of the probability of dissociation on the vibrational energy of the molecule. Within the limits of this approximation, which enables the transition state method to be employed, the rate constant of the dissociation process

$$AB + C \xrightarrow{\kappa_{diss}} A + B + C \quad (40.1)$$

can be written in the form

$$\kappa_{diss} = \frac{kT}{2\pi\hbar} \cdot \frac{F^\star_{ABC}}{F_{AB}F_C} \exp\left(-\frac{D}{kT}\right), \quad (40.2)$$

where $F^\star_{ABC}$ is the partition function of the transition complex; F_{AB} and F_C are the total partition functions of the partners AB and C (C can be an atom or a molecule); D is the dissociation energy of AB.

The various equilibrium theories differ only in the specific definition of the factor $F^\star_{ABC}$ in (40.2).

Rice [435] and Widom [546] assumed that, in the collisions of AB and C, only those molecules AB* dissociate whose vibrational energy lies in a range kT close to the dissociation threshold.† In this case, $F^\star_{ABC}$ is

† In this chapter, we will understand by AB* molecules capable of decomposition on collision with C, if it is a question of dissociation, or capable of stabilization, if it is a question of recombination. Such a notation in relation to diatomic molecules is formally analogous to the designation of an active polyatomic molecule. In a number of cases, however, AB may be considered as an active molecule within the framework of the definition given in § 22, if account is taken of the possibility of tunnel transmission under the centrifugal barrier.

represented in the form

$$F^\star_{AB-C} = F^\star_{AB} F^\star_{AB-C}. \qquad (40.3)$$

Here $F^\star_{AB}$ is the internal partition function of AB*; $F^\star_{AB-C}$ is the partition function of the activated complex, whose critical surface is determined by the condition $R_{AB-C} = R^\star$, and in which the molecule AB is considered as a system devoid of internal degrees of freedom. Then (40.2) is rewritten in the form

$$\kappa_{diss} = A \exp\left(-\frac{D}{kT}\right) = \left[\pi(R^\star)^2 \left(\frac{8kT}{\pi\mu}\right)^{\frac{1}{2}}\right]$$
$$\times \frac{(F^\star_{AB})_{vib}(F^\star_{AB})_{rot}(F^\star_{AB})_{el}}{(F_{AB})_{vib}(F_{AB})_{rot}(F_{AB})_{el}} \exp\left(-\frac{D}{kT}\right). \qquad (40.4)$$

The first factor in this expression equals the number of 'gas-kinetic' collisions of AB* with C. The second pre-exponential factor (the ratio of the partition functions) is dimensionless and exceeds unity in magnitude, since $(F^\star_{AB}/F_{AB})_{vib}$ equals the number of vibrational levels Δn^* in a band kT close to the dissociation limit; $(F^\star_{AB})_{rot}$ exceeds $(F_{AB})_{rot}$ due to the large moment of inertia of the vibrationally excited molecule, and $(F^\star_{AB}/F_{AB})_{el}$ equals the number of bonding electronic states converging to the dissociation limit. For example, in the thermal decomposition of haloid molecules, the quantities entering the pre-exponential factor have the following values: $(F^\star_{AB}/F_{AB}) \approx 5$, $\Delta n^* \approx 10$, $(F^\star_{AB}/F_{AB})_{rot} \approx 5$, so that A may exceed Z_0 by two orders of magnitude. For a hydrogen molecule, the vibrational level density is low ($\Delta n^* \approx 1$), but the effect of the centrifugal extension is great $[(F^\star_{HH}/F_{HH})_{rot} \approx 10]$, and A may exceed the gas-kinetic number of collisions by an order of magnitude.

Rice's theory formally explained the large values of A for the rate constants of thermal decomposition, which were obtained from the thermodynamical equilibrium constant using the experimental values of the recombination rate constant:

$$\frac{\kappa_{diss}}{\kappa_{rec}} = K_{eq} = \frac{F_A F_B}{F_{AB}} \exp\left(-\frac{D}{kT}\right). \qquad (40.5)$$

From (40.2) and (40.5) we find

$$\kappa_{rec} = \frac{kT}{2\pi\hbar} \cdot \frac{F^\star_{ABC}}{F_A F_B F_C}, \qquad (40.6)$$

where F_A, F_B, and F_C are the partition functions for atoms A, B, and C.

If, instead of $F^\star_{ABC}$, F_A, F_B, and F_C, their explicit expressions are substituted, it is not difficult to ascertain that a weak positive temperature-dependence is obtained for κ_{rec}. However, it was found experimentally [177], [409] that the values of κ_{rec} for many reactions are considerably reduced on a rise in temperature (a so-called negative temperature-dependence). Thus Rice's theory, which may be called the impact equilibrium theory of dissociation, is unable to explain the temperature dependence of κ_{rec}.

In this connection, Rice [436], [439] proposed another mechanism for the dissociation of diatomic molecules, which proceeds via the formation of an intermediate weakly-bound complex between a decomposition product atom and the inert molecule C:

$$AB + C \underset{\kappa_2}{\overset{\kappa_1}{\rightleftarrows}} AB-C, \qquad (40.7a)$$

$$AB-C \overset{\kappa}{\to} A + BC. \qquad (40.7b)$$

It might be expected that the formation of weakly-bound complexes, after dissociation, would somewhat lower the effective dissociation energy. In fact, the rate constant of the total decomposition process corresponding to such a mechanism, can be written in the form

$$\kappa_{diss} = K_{eq}\kappa, \qquad K_{eq} = \kappa_1/\kappa_2. \qquad (40.8)$$

It is, of course, essential that the concentration of the complexes AB–C is an equilibrium concentration and that this equilibrium is not disturbed in the reaction (40.7b), which represents the unimolecular decomposition of the complex AB–C along the bond AB. Assuming the equilibrium rate of the thermal activation, the rate constant κ_{diss} can be calculated in accordance with the theory of unimolecular reactions, where by F and $F^\star$ should be understood the partition functions of the complex AB–C and its activated state respectively. As regards the effective dissociation energy of the bond AB in the complex AB–C, it may differ somewhat from the dissociation energy in the free AB molecule. This difference is apparently very slight [436] and, in future, we will neglect it. To the same approximation, the energy ε_0 of the bond between the molecules AB and C and also between B and C must be considered equal, which makes it possible to take the formation of the complex BC in the reaction (40.7b) into account, even in considering reaction (40.7a). The corresponding calculations give

$$\kappa = v^\star \left(\frac{F_{el}^\star}{F_{el}}\right)\left(\frac{r^\star}{r_e}\right)^2 \exp\left(-\frac{D}{kT}\right). \qquad (40.9)$$

This expression is analogous to formulae (33.1) and (33.2). The factor $(r^\star/r_e)^2$ takes account of the increase in the density of the rotational states in the activated molecule AB–C in comparison with the unexcited molecule AB–C. Here r_e denotes the equilibrium interatomic distance in AB and $r^\star$ is the mean interatomic distance in the configuration of the activated molecule (which corresponds to a position on the peak of the centrifugal maximum for the representative point).

The equilibrium constant K_{eq} for reaction (40.7a) can be calculated by the usual formulae of statistical physics. In the given case, in order to stress its physical meaning, it is convenient to write K_{eq} as the volume in configuration space accessible to the stable molecule AB–C:

$$K_{eq} = \tfrac{4}{3}\pi(a^3 - b^3)\exp\left(\frac{\varepsilon_0}{kT}\right), \qquad (40.10)$$

where a and b are certain effective mean distances which limit the region of the stable complex AB–C in configuration space; for $a - b = \delta a \ll a$, the effective volume comes simply to $4\pi a^2 \delta a$.

The activated state of a stable triatomic (or polyatomic) system ABC corresponds to this scheme of dissociation, so that by $F^\star_{ABC}$ in (40.2) should be understood the partition function of a polyatomic activated complex:

$$F^\star_{ABC} = (F^\star_{A-BC})_{trans}(F^\star_{A-BC})_{vib}(F^\star_{A-BC})_{rot}, \qquad (40.11)$$

whose critical surface is defined by the condition $R_{A-BC} = R^\star$. Substituting (40.11) into (40.2) or (40.10) and (40.9) into (40.8) and identifying a with $R^\star$ we find

$$\kappa_{diss} = \pi(R^\star)^2\left(\frac{8kT}{\pi\mu_{A-BC}}\right)^{\frac{1}{2}}\exp\left(-\frac{D-\varepsilon_0}{kT}\right). \qquad (40.12)$$

It is clear that the effective energy of dissociation via the complex is less than the bond energy ε_0, which in principle enables the experimentally-observed negative temperature-dependence of κ_{rec} to be explained.

The reverse process to the dissociation (40.7) is described by the scheme

$$B + C \rightleftarrows BC \qquad (40.13a)$$

$$BC + A \to AB - C, \qquad (40.13b)$$

in which the first of the reactions is assumed to be an equilibrium reaction.

Rice [436], [439] investigated the mechanism (40.13) in detail and came to the conclusion that existing experimental data on the recombination of atoms at low temperatures (300–500 K) are satisfactorily explained by such a recombination mechanism. The nature of the forces forming the bond B–C is insufficiently understood at the present time; however, investigation of the dependence of κ_{rec} on the properties of the molecule C definitely indicates that the forces are long-range. The values of the bond energy ε_0, found from the temperature dependence, vary for different recombining pairs and inert molecules from 0 kcal mol^{-1} to 4 kcal mol^{-1} [409], [439], [455].

In calculating the equilibrium constant of the reactions (40.7a) or (40.13a), it should be borne in mind that the relatively-simple expression (40.10) is valid only when the bond energy of the complex is large in comparison with kT. For $\varepsilon_0 \approx kT$, the very concept of a complex becomes vague, so that, to calculate the equilibrium concentration of complexes, it is necessary to introduce additional assumptions about the region of phase space which corresponds to the complexes effectively participating in recombination. This arbitrariness, which can lead to very considerable differences in the values of ε_0, is connected with an incomplete understanding of the processes taking place in triple collisions.†

It should be emphasized that the contribution of the dissociation mechanism via the complex becomes smaller with an increase in temperature and becomes less important with a decrease in the polarization and an increase in the ionization potential of the molecule C. In this connection, dissociation via the formation of a complex cannot be enlisted to explain the negative temperature dependence of the pre-exponential factor of the rate constant for the decomposition of diatomic molecules at high temperatures. The theory of the decomposition of diatomic molecules, under these conditions ($kT \gg \varepsilon_0$), must be based on a detailed investigation of the mechanism of activating collisions, taking into account the possible disturbance of the equilibrium distribution of the vibrational states below the dissociation limit, caused by the decomposition. Though differences between the mechanisms of strongly-activating collisions and stepwise excitation are not very substantial in the decomposition of complex molecules, in the case of the decomposition of diatomic molecules they lead to basically different results.

The ideas of the equilibrium theory of dissociation and recombination can be extended without difficulty to the case of recombination of ions‡

$$A^+ + B^- + C \to AB + C. \tag{40.14}$$

The original theory of Thomson [48], [522] started from the assumption that the recombination rate may be calculated as the product of the current of ions A^+ and B^-, within a sphere of radius $R < R^*$, and the probability that a deactivating collision of A^+ (or B^-) with C takes place; in this case the relative motion of A^+B^- in a hyperbola goes over into motion in an ellipse, and the pair A^+B^- is considered bound.

In terms of the transition state theory, the following partition function of the activated complex corresponds to this model:

$$F^\star_{A^+B^-C} = F^\star_{A^+B^-}(F^*_{A^+C} + F^*_{B^-C}), \tag{40.15}$$

† Therefore a detailed investigation of the mechanism of a triple collision is necessary for the construction of an equilibrium theory under the condition $\varepsilon_0 \approx kT$ [343], [344].

‡ We will limit ourselves to the consideration of the triple recombination of ions which occurs at pressures less than 100–1000 mm of mercury. At high pressures, to a certain degree, the diffusion of the ions occurs, so that an increase in the pressure leads to a reduction in the recombination rate. An account of the theory of recombination in such conditions is given in the book by Massey and Burhop [48] and in the article by Natanson [51].

where F^*_{A+C} is the partition function of the pair A^+C, extending over the volume V^* within which the collisions $A^+ + C$ lead to a loss in the energy of A^+.

The activated complex $(A^+B^-)^*$ is given by the condition $R_{A+B^-} = R_T$, where R_T is determined (according to Thomson) by the relation

$$\frac{e^2}{R_T} = \tfrac{3}{2}kT. \tag{40.16}$$

A transmission coefficient χ, giving the probability of the deactivation of A^+ (or B^-) on going through the critical surface $R_{A+B^-} = R_T$, must be introduced into the general formula (40.2). χ can be represented approximately as the ratio of the effective collision cross-section σ of A^+ with C to the cross-section of the critical surface

$$\chi = \frac{\sigma}{\pi R_T^2}. \tag{40.17}$$

Substituting (40.15)–(40.17) into (40.6) we get

$$K_{\text{rec}} = \tfrac{4}{3}\pi R_T^3 \left(\frac{8kT}{\pi\mu_{AB}}\right)^{\frac{1}{2}} (\sigma_{AC} + \sigma_{BC}). \tag{40.18}$$

The dynamics of the pair collisions A^+–C and B^-–C should be considered explicitly to determine σ_{AC} and σ_{BC}. However, without detailed calculations, it is clear that the σs must be close to the gas-kinetic cross-sections, if the masses of the colliding partners are of the same order of magnitude [232].

Recently, Mahan and Person [342] considered the dynamics of the collision of A^+ (or B^-) with C explicitly and calculated the mean energy which the ion loses in collision with the neutral molecule. Moreover, they took it into account that deactivating collisions (i.e. collisions displacing a particle from a hyperbolic orbit into an elliptical orbit) can take place when $R > R_T$. The calculation shows that the probability of deactivation decreases very slowly on increase in R_{AB}, so that, even for $R_{AB} \approx 1500$ Å, it is very large. Under these conditions, it becomes necessary to take into account the possibility of repeated activation, as a result of which an ion trapped in an elliptical orbit will be displaced into an hyperbolic orbit again. The calculation of such collisions forms the substance of non-equilibrium theory (see § 44). Repeated dissociation can be approximately accounted for, within the limits of equilibrium theory, by introducing either some critical radius R_T (as in Thomson's theory) or some critical energy E^*, which determines the condition of capture by the relation $E^{AB}_{\text{kin}} - (e^2/R_{AB}) < E^*$.

41. The contribution of various degrees of freedom of a dissociating molecule to the decomposition rate constant

In the collision of diatomic molecules, excitation and deactivation of the rotational, vibrational, and electronic degrees of freedom are possible. If the cross-sections of all these processes are known, then the rate constants of the elementary stages $\kappa(l, n, j \to l', n', j')$ necessary to write down the kinetic equations for the population of the corresponding states are obtained by averaging the cross-sections over the velocities of the colliding molecules. Since, at present, the constants $\kappa(l, n, j \to l', n', j')$ cannot be calculated in the general case, we will discuss qualitatively the contribution of the various degrees of freedom to the activation of a diatomic molecule. Moreover, to be specific, we will have in mind dissociation, although all the arguments also carry over to the recombination of atoms. In addition, we will assume that the dissociation does not destroy the Maxwellian distribution of the velocities, so that the system under consideration can be characterized by the temperature T of the translational motion.

Vibrational transitions

The quantum-mechanical, and even the classical, problem of the excitation of vibrations in molecular collisions can be solved only for certain relatively simple models. As indicated in Chapter 2, all these models are classified basically by the magnitude of the dimensionless parameter ξ. For large values of this parameter ($\xi \gg 1$) the collisions hardly induce transitions and lead only to adiabatic displacement of the vibrational levels during the collision. The transition probability depends exponentially on ξ.

If the conditions are such that $\xi \ll 1$, the collisions are of a strongly-nonadiabatic nature. From the estimate, given in § 7, it follows that such a situation can arise either on very intense vibrational excitation of the dissociating molecule at high levels $(D - E \sim kT)$ or on vibrational excitation of a heavy diatomic gas at low levels, induced by collisions with molecules of a light gas ($\mu \ll M$). The condition for strong nonadiabaticity $\xi \ll 1$ enables the use of a model of instantaneous collisions, for example a model of rigid spheres, in considering the relaxation process and calculating the cross-sections of inelastic collisions.

Methods for calculating transition probabilities, based on the use of the specific conditions imposed on ξ, are given in Chapter 2. We will only note that, at present, there is apparently no satisfactory physical model which would enable the dependence of the vibrational excitation probability on the parameter ξ to be studied over its entire range of values. This is connected with the fact that, in most versions, the calculations to some extent draw upon a perturbation theory based on the use of a small parameter expressed in terms of ξ. Therefore, to calculate the dissociation rate, recourse must be made to considerable simplifications of the real situation, which nevertheless enable specific features of non-equilibrium decomposition to be explained.

Rotational transitions

The dissociated and bound states of a diatomic system can be represented by representative points on a plane whose coordinates are the two obvious integrals of motion of a pair of atoms: the total energy E and the angular momentum j. Instead of these variables, it is convenient to introduce the rotational energy of the non-vibrating molecule $E_{rot} = j^2/2Mr_e^2$ and the vibrational energy of the rotating molecule $E_{vib,rot} = E - E_{rot}$. For a non-rotating molecule, E_{vib} can be calculated, for example, by formula (7.25) for the model of a Morse oscillator. In Fig. 35, the phase plane of molecule AB

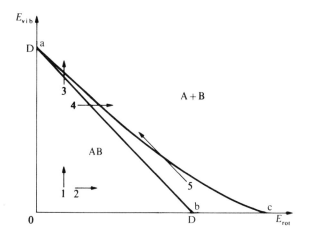

FIG. 35. Phase plane of the bound and dissociated states of a diatomic molecule.

is shown. The bound states of AB correspond to the part of the E_{rot}, E_{vib} plane within the triangle a0b, where the line ab is the line of constant energy $E = D$. The dissociated states correspond to the portion of the plane outside curve ac. The points inside the triangle abc correspond to metastable states with energies $E > D$. Within the limits of the classical approximation, the decomposition channel from these states is closed due to the existence of a rotational barrier. Its height and point of maximum r_m are determined from the condition of an extremum in the effective potential:

$$\frac{\partial}{\partial r}U_{eff}\bigg|_{r=r_m} = \frac{\partial}{\partial r}\left[U(r) + \frac{j^2}{2Mr^2}\right]\bigg|_{r=r_m} = 0. \qquad (41.1)$$

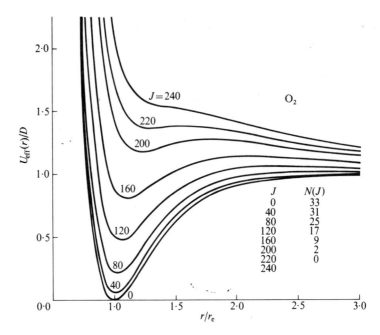

FIG. 36. Effective potential curves of the electronic ground state of an oxygen molecule (Bauer and Tsang [145]).

The above may be illustrated by the example of the O_2 molecule. In Fig. 36, the effective potential curves for this molecule, corresponding to various momenta j, are portrayed as functions of the dimensionless internuclear distance ρ. The number of vibrational levels in each of these potentials is also shown. It is not difficult to see that an increase in j leads to a reduction in the effective dissociation energy, which is determined by the difference between the maximum and minimum values of the effective potential energy. However, the decrease δE in the effective dissociation energy proves to be less than the rotational energy of the molecule E_{rot}, and therefore, on averaging over all values of E_{rot}, the decrease in the population of the rotational states, described by the Boltzmann factor $\exp(-\beta E_{\text{rot}})$, exceeds the exponentially-increasing contribution $\exp(\beta \delta E)$ to the decomposition rate constant of the excited rotational states. The total effect of such an averaging shows that the contribution of states with very large j to the dissociation rate is very small. This just reflects the fact that the probability of rupturing the bond A–B as a result of centrifugal extension is small.

Escape from the metastable region for free molecules is possible only as a result of tunnel transmission under the centrifugal barrier. As a rule, this effect can be neglected in gas kinetics and the states of a molecule in the region abc can be regarded as true stationary states.

The result of collisions between AB and C can be represented as a transition of the representative point from a fixed state (E_{rot}, E_{vib}) to another (E'_{rot}, E'_{vib}). The arrows 1 and 2 (see Fig. 35) correspond to vibrational and rotational transitions in the stable molecule AB. The arrow 3 corresponds to dissociation due to vibrational excitation, and arrow 4 to dissociation due to rotational excitation. It is also clear from this diagram that, in AB + C collisions, escape from the metastable region is possible, conserving the total energy of the molecule (arrow 5), if the rotationally-excited molecule loses an appreciable fraction of its internal angular momentum, which is carried away in the form of the angular momentum of the relative motion.

Electronic transitions in a dissociating molecule

The calculation of the probability of nonadiabatic transitions between the electronic terms of diatomic molecules AB, induced by collision with the molecule C, represents a more complicated problem than the calculation of the vibrational transition probability. An additional difficulty here is that even the qualitative dependence of the system of electronic terms of the interacting AB + C molecules on the coordinates describing the relative orientation of AB and C is unknown. In connection with this, it is necessary to take approximate, even qualitative, account of the effect of electronic and rotational transitions on the kinetics of the vibrational relaxation and dissociation. This can be done on the basis of the following physical considerations.

As is shown in § 44, the dissociation rate is determined by the transition rate between vibrational levels, the energies of which E^* are slightly less than the dissociation energy D.† Below this energy $(E_{vib} \leqslant E^*)$ the distribution of the vibrational levels of the decomposing molecule approaches a Boltzmann distribution. We will now consider an electronic term of an excited electronic state, which tends asymptotically to the same dissociation limit as a term of the ground state. If the energy E_l, which corresponds to the minimum of the excited term, is appreciably less than E^*, then it can be assumed that, during the dissociation process, equilibrium will be maintained between the lower vibrational levels of the excited electronic state and the excited vibrational levels of the electronic ground state by means of transitions induced by collisions. The rate of these transitions, which are analogous to transitions in the case of induced predissociation, is determined by many factors, including the overlapping of the vibrational wavefunctions, which are perturbed at the moment of collision by the effect of atom C. An exact calculation of such a transition probability requires knowledge of the intermolecular interaction between AB and C in the excited electronic

---

† At least, this assertion is correct for the dissociation of diatomic molecules in the atmosphere of an inert gas (see § 46).

state of AB, about which very little is known at present. However, if the distribution of the lower vibrational levels of the excited state and of the excited levels of the ground state is assumed to be the equilibrium distribution, the transition rate between them must be appreciably greater than the dissociation rate. The ratio of these rates contains an exponential factor of the form $\exp\beta(E^*-E_l)$, which increases rapidly as the difference E^*-E_l increases. Hence, it may be concluded that equilibrium will be established for terms possessing a deep minimum in the potential energy. Then the dissociation rate from this state can be calculated in the same way as from the ground state, the corresponding contribution entering additively into the decomposition rate constant, with a weight equal to the ratio of the degeneracy of the excited states to the degeneracy of the ground state. In addition, the inequality $\exp(-\beta E_l) \ll 1$ must be fulfilled; only, in this case, dissociation via the electronically-excited state represents an alternative

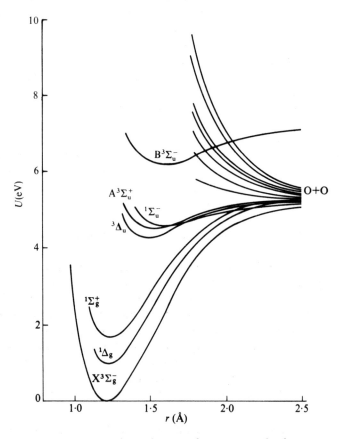

FIG. 37. Lower electronic states of an oxygen molecule.

decomposition channel, proceeding parallel to decomposition from the ground state. An estimate shows that, for example, the two lowest excited terms of oxygen $^1\Delta_g$ and $^1\Sigma_g^+$, at temperatures of up to several thousands of degrees, satisfy such a condition (Fig. 37). The statistical weights of these states equal 2 and 1 respectively. Since the statistical weight of the ground state $^3\Sigma_g^-$ equals 3, the contribution of these excited states leads to an increase, by a factor of approximately two, in the decomposition rate.

The establishment of equilibrium between the electronic terms during dissociation does not mean, of course, that it is established within a time of the order of the vibrational relaxation time. Apparently, in the case of oxygen, a Boltzmann distribution is established during vibrational relaxation in the lower vibrational levels of the ground state, from which slow transitions subsequently proceed to the vibrational levels of the states $^1\Delta_g$ and $^1\Sigma_g^+$.

For the electronic terms which lie comparatively close to the dissociation limit, an analogous estimate of their effect on the rate of non-equilibrium dissociation can be made, but with less certainty. On the one hand, the Boltzmann factor $\exp(E^* - E_l)$ here does not play so decisive a role. On the other hand, the probability of nonadiabatic transitions between such terms must be much greater than between the lower terms. This is due to the fact that the highly-excited terms and the ground-state terms usually arise from different electronic configurations, the nature of which determines the details of the intermolecular interaction of AB and C. If the interaction differs considerably in these two states, as is possible in the case of different electronic configurations, it is very probable that intersection of the electronic terms of AB will be induced by collisions with C. The presence of such an intersection greatly increases the probability of nonadiabatic transitions. In this connection, it might be expected that transitions between electronic states close to the dissociation limit would proceed at a rate of the order of the collision frequency between AB and C. If the excitation energy of these terms is less than the energy E^* characterizing the slow stage of decomposition, then the lower vibrational levels of excited states will be populated like the vibrational levels at the same energy in the electronic ground state. In this case, the electronically-excited states provide a further channel for parallel decomposition, which increases the rate of dissociation. If the excitation energy is greater than E^*, then equilibrium will not be established between the ground electronic states and the excited states, so that these latter states will not contribute to the total rate of dissociation. The terms $^3\Delta_u$, $^3\Sigma_u^+$, and $^1\Sigma_u^-$ of the oxygen molecule, the bond energies of which equal approximately 1 eV, are examples of such highly-excited states. These terms arise in the excited configuration $\sigma_g^2 \pi_u^3 \pi_g^3$; the two lower excited terms and the ground state term of oxygen arise from the configuration $\sigma_g^2 \pi_u^4 \pi_g^2$.

Thus the contribution of the electronically-excited states can be taken into account, very approximately, by multiplying the decomposition rate constant calculated for the ground state by the factor g_{el}, which is equal to the ratio of the total statistical weight of the excited electronic states with excitation energies less than E^* to the statistical weight of the ground state. For this case, the rate of a nonadiabatic transition between the excited states and the ground state must be greater than the rate of dissociation.

It should be borne in mind that the energy E^* depends on the temperature. This means that the factor g_{el} leads to an additional negative temperature dependence in the pre-exponential factor A, which must be particularly important at a temperature T, determined by the relation $E^*(T) \sim E_l$. The nature of the temperature dependence of g_{el} can only be elucidated in greater detail by a specific examination of the mechanism of nonadiabatic transitions.

We will note one more effect possibly important in the dissociation of diatomic molecules. If the molecule possesses weakly-bound excited electronic states, which converge as $r \to \infty$ to a common dissociation limit, then, under the effect of the intermolecular interaction, a bonding term may be transformed into an antibonding term, and the vibrational energy levels corresponding to these electronic states may be replaced by a continuous spectrum. In this case, dissociation from the excited electronic state does not necessarily precede vibrational excitation. Such a situation is particularly likely in the collision of a molecule with an atom possessing an unfilled electronic shell; the degeneracy of the electronic state of the atom is removed on interaction, and from each molecular term there arise several terms of the triatomic system $AB+C$, part of which will certainly be antibonding.

42. Variational theory of dissociation and recombination

The defect of the equilibrium theory of dissociation and recombination is that it does not take into account the perturbation of the Maxwell–Boltzmann distribution function of the reacting molecules caused by the reaction itself. Moreover, the arbitrary choice of critical surface results in great uncertainty in the values of the kinetic constants κ_{diss} and κ_{rec}. These defects are partially removed by the variational theory, which formulates the condition for the completion of the reaction in the form of the requirement that a varying critical surface be intersected by the representative point. Close to this surface, the distribution function is assumed to be the equilibrium distribution function, so that the total current across the surface can be calculated by the transition state method. It is not difficult to see that, since any critical surface may be intersected several times by the representative point, the condition that the critical surface is crossed is only necessary (but not sufficient) for the completion of a reaction.

The only requirement made of the critical surface is that there be no 'hole' in it through which a trajectory might go from the region of dissociated states into the region of bound states.

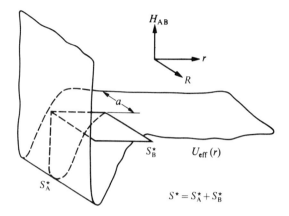

FIG. 38. Variational critical surface for the calculation of recombination and dissociation rates (Keck [298]).

An example of such a surface is given in Fig. 38, copied from the work of Keck [298]. The energy surface, shown in the system of coordinates r and R, corresponds to the effective potential $U_{\text{eff}}(r)$ of the interaction of the atoms A and B. Two plane parts $S_A^\star$ and $S_B^\star$ of the critical surface, which divide the dissociated and bound states in the space of r, R, H_{12} (where H_{12} is the total energy of the recombining atoms) intersect the energy surface. The bound states correspond to the portion of the space contained inside a cylinder directed along the R axis. The planes $S_A^\star$ and $S_B^\star$ are determined by the equations

$$R = a$$
$$H_{12} - B = 0, \tag{42.1}$$

where B is the height of the rotational centrifugal barrier.

The radius a serves as a variational parameter, determining the number of 'gas-kinetic' collisions of the pair $A+B$ with the third molecule C. The current across the plane $S_A^\star$ in the direction of the origin of coordinates is easily calculated, since it simply corresponds to the rate of decomposition of the AB molecule activated by the collisions with C, on condition that the total internal energy of the system exceeds the energy of dissociation D. The corresponding expression for κ_{diss} can be written down at once on the basis of the results of Chapter 5 (see also [31], [230]):†

$$\kappa_{\text{diss}}(S_A^\star) = \pi a^2 \sqrt{\left(\frac{8kT}{\pi\mu}\right)} \frac{(D/kT)^{s-1}}{(s-1)!} \exp\left(-\frac{D}{kT}\right), \tag{42.2}$$

† In a more exact investigation corrections must be made in this expression to the excluded volume, connected with the strong repulsion of A–C and B–C, and also to the effect of the centrifugal barrier not taken into account in the derivation of (42.2) [298].

where $s-1$ equals half the total number of terms in the Hamiltonian of the colliding partners quadratic in the momenta and coordinates.

In the given case, the contributions to the current across S_A^* come from the vibration of AB (two terms), the rotation of AB (two terms), and the relative rotation of AB and C (two terms), so that $s = 3$.[†] The expression (42.2), often called the formula of simple collision theory, is sometimes used to interpret experimental data on dissociation, with an empirical definition of the parameter s. It is not difficult to see, however, that such an interpretation is inconsistent. In fact, if the trajectory of the representative points which intersects the surface S_A^* in the direction of the origin of coordinates is followed, then for $R < a$ this trajectory must curve upwards, since it must go into the region of the dissociated atoms ($R \to \infty$, $r \to \infty$). On its way, the trajectory must intersect the plane $S_{B'}^*$, which is a continuation of the plane S_B^* for $R < a$. If the constant $\kappa_{\text{diss}}(S_{B'}^*)$ refers to the corresponding current, then $\kappa_{\text{diss}}(S_A^*)$ will be close to the actual rate of dissociation only on condition $\kappa_{\text{diss}}(S_A^*) \ll \kappa_{\text{diss}}(S_{B'}^*)$, since the slowest stage is determined in the subsequent process. In conditions typical of gas kinetics, this inequality is not satisfied (see below). This, in turn, implies that (42.2) for $a \approx R_0$ gives an overestimated upper limit for the dissociation rate constant.

As regards the corresponding recombination rate constant $\kappa_{\text{rec}}(S_A^*)$, it can be calculated in terms of the equilibrium constant K_{eq}, which, in the case under consideration (the classical approximation and taking account of a single electron state), can be written in the form

$$K_{\text{eq}} = \frac{v}{Z_0} \exp\left(-\frac{D}{kT}\right), \tag{42.3}$$

where v is the frequency of vibration of the AB molecules; Z_0 is the number of collisions of rigid spheres with a distance of closest approach r_e.

The dissociation mechanism considered above can be depicted in the following way on the phase plane of E_{rot}, E_{vib} (see Fig. 35): in collisions of AB and C the dissociation limit is reduced to a value equal to the sum of the energy of rotation of AB and the energy of the relative rotation of AB and C, and molecules whose representative points fall beyond the lower dissociation limit are regarded as dissociated. In connection with this mechanism, we will mention Light's theory [327], in which the magnitude of the displacement of the dissociation limit, at the moment of collision, is expressed in terms of the intermolecular potential. However, since this theory does not enable the whole process of vibrational excitation to be followed and a definite stage of decomposition to be isolated correctly, it cannot be applied immediately to the interpretation of the experimental data.

[†] The assumption is sometimes made that the relative rotation does not contribute to κ_{diss}, so that $s = 2$ [251].

In accordance with the assumptions of Light's theory, the contribution of the translational energy should be neglected, the contribution of rotation taken into account within the limits of the simplest theory of collisions, and the value connected with the displacement of the dissociation limit considered weakly dependent on the temperature. Thus Light's theory gives a temperature dependence of the form $T^{\frac{1}{2}}T^{-2}$ for the pre-exponential factor A. It should be noted, however, that such a dependence is obtained only when the displacement of the dissociation limit is determined by the statistically-averaged interaction of AB and C and does not depend on the relative velocity of AB and C. It can be shown that this is equivalent to assuming a large value for the Massey parameter ξ. This in turn implies that the rate of the vibrational excitation of the molecules must be very small (see §8), therefore the slow stage of the whole process will be determined by just the vibrational excitation and not by the transitions from the upper levels into the dissociated state, as assumed by Light. If non-quadratic terms in the potential energy of AB are important in the Hamiltonian of the system, the temperature dependence of the pre-exponential factor A will differ from that given above. Light and Arnstein [328] showed, in particular, that, on representing AB by a Morse oscillator, a weaker negative temperature-dependence is obtained, in comparison with the $T^{-\frac{3}{2}}$, which is correct for the model of a truncated harmonic oscillator (a harmonic potential well of finite depth D).

We will now consider the current across the surface $S_B^\star$. The intersection of this surface by the representative points is caused by two effects. Firstly, under the influence of the strong field of C, the radial motion of the atoms A and B in the unchanging effective potential $U_{\text{eff}}(j)$ is changed, which is the analogue to the vibrational excitation or deactivation of a diatomic molecule. Secondly, the interaction of AB with C induces a change in the angular momentum j of the pair AB. The change in j entails a change in U_{eff} and consequently a change in the position of the critical surface.

If it is assumed that $B = 0$ in (42.1), then the plane $S_B^\star$ will be stationary. The calculation of κ_{rec} for such a model was first carried out by Wigner [553], and the contribution of $\kappa_{\text{rec}}(S_A^\star)$ was, in general, neglected ($a = 0$). The law $\kappa_{\text{rec}} \sim T^{-\frac{1}{2}}$ was obtained for the temperature dependence. Calculating κ_{diss} in terms of the equilibrium constant, we find that the temperature dependence of the dissociation rate constant must be expressed by the formula

$$\kappa_{\text{diss}}(S_B^\star) \sim T^{-1} \exp\left(-\frac{D}{kT}\right). \tag{42.4}$$

This temperature dependence of the pre-exponential can be interpreted in the following way. The number dN of pairs A–B near the critical surface $H_{12} = E^\star$ is given by the equilibrium distribution function

$$\frac{dN}{dt} = \frac{1}{F_{\text{AB,vib}} F_{\text{AB,rot}}} \rho(E^\star) \exp\left(-\frac{E^\star}{kT}\right) dE. \tag{42.5}$$

The current through the stationary surface is hence obtained as dN/dt. Calculating dN/dt for the stationary critical surface, we find

$$\kappa_{\text{diss}}(S_B^*) = \left\langle \frac{dN}{dt} \right\rangle = \frac{\rho(E^*)}{F_{\text{AB,vib}} F_{\text{AB,rot}}} \exp\left(-\frac{E^*}{kT}\right) \left\langle \frac{dE^*}{dt} \right\rangle, \quad (42.6)$$

where the mean $\langle dE^*/dt \rangle$ is calculated for various realizations of the states of the pairs AB–C, and the rate of change of energy is determined by the work done by the external forces (in relation to the isolated pair AB), dependent on the intermolecular interaction:

$$\frac{dE^*}{dt} \approx \frac{\partial U_{\text{eff}}}{\partial r} \dot{r} \quad \text{for} \quad \frac{dE^*}{dt} > 0. \quad (42.7)$$

If the interaction of AB with C in the main leads to repulsion then, for an estimate of dE^*/dt, it may be assumed that

$$\left\langle \frac{dE^*}{dt} \right\rangle \sim \left\langle \left| \frac{\partial U_{\text{eff}}}{\partial r} \right| |\dot{r}| \right\rangle \sim kT\alpha \sqrt{\left(\frac{E^*}{M}\right)}. \quad (42.8)$$

Substituting (42.8) into (42.6) and assuming $E^* = D$, we find a dependence of the type (42.4) for high temperatures ($kT \gg \hbar\omega$). At low temperatures ($kT \ll \hbar\omega$), the pre-exponential factor in the expression for $\kappa_{\text{diss}}(S_B^*)$ turns out to be independent of the temperature.

Wigner's theory was made more precise by Keck [298] and Woznick [564], who chose the plane (42.1) as the critical surface. In this case, the temperature dependence obtained for $\kappa_{\text{rec}}(S_B^*)$ does not have a simple interpretation, since the relative contribution of the vibrational and rotational transitions to the current across S_B^* changes with a change in the temperature. The results of a calculation of the recombination of atomic oxygen on argon in the electronic ground state $^3\Sigma_g^-$ is given in Fig. 39. The O–O interaction was approximated by the Morse potential, the O–Ar interaction by the Lennard Jones potential at large distances and by the Mayson–Vanderslais potential at short distances. The curve κ_1 corresponds to the recombination rate constant, calculated within the framework of Wigner's model [553]. Curve κ_2 corresponds to Keck's model and the separate contributions of the vibrational and rotational transitions are shown. It is evident that taking the rotational potential barrier into account reduces the absolute magnitude of the recombination constant and slows down its fall with temperature. For example, in the range 500–10 000 K, κ_{rec} proves to be practically independent of the temperature. From the method of calculating $\langle dE^*/dt \rangle$, it is evident that a sufficiently-strong attraction between the recombining pairs and the atom C must lead to an increase in κ_{rec}, since, in the averaging, the preferred nature of configurations with negative energy is taken into account, the contribution of

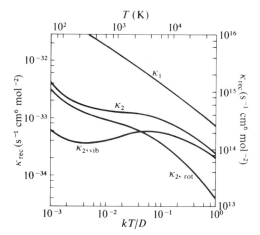

Fig. 39. Variational calculation of the recombination rate of oxygen (in the electronic ground state) on argon (Keck [298]).

which is determined by the Boltzmann factor $\exp[U(R)\beta]$. Detailed calculations show that this is actually so [565], the attraction, increasing the absolute magnitude of κ_{rec}, also increases its negative temperature dependence.

A more exact approximation to the true recombination rate constant, within the framework of the variational theory, can be obtained if the surface S_B^* is displaced downwards by the value ε in relation to the level $D+B$. Investigation of this approximation shows that $\kappa_{\text{diss}}(S_B^*)$, regarded as a function of ε, actually goes through a minimum for $\varepsilon_m \sim (0.2-0.4)kT$. For $\varepsilon < \varepsilon_m$ the current through S_B^* increases due to the increase in the statistical weight of the states of the O_2 molecules close to the dissociation limit. For $\varepsilon > \varepsilon_m$, it increases, in view of the exponential increase of the equilibrium distribution function.

The existence of a minimum in $\kappa_{\text{diss}}(S_B^*)$ clearly indicates that, in dissociation and recombination, there is a slowest stage, determining the rate of the whole process and localized somewhat below the dissociation limit. However, taking account of effects caused by the existence of the slow stage can be carried out consistently only within the limits of non-equilibrium theory. The variational theory, which does not take into account the possibility of a multiple intersection of the critical surface by the representative point in a single triple collision $A+B+C$, will also not take into account the possibility of the crossing of these surfaces in the reverse direction in subsequent collisions between the vibrationally highly-excited molecule AB and the atom C. But it is precisely this latter process (the so-called redissociation or reverse dissociation), which causes the disturbance of the equilibrium distribution of the vibrational states of AB close to the dissociation threshold, which leads to a reduction in the values of the non-equilibrium constants κ_{rec} or κ_{diss} in comparison with the equilibrium values. From what has been said, it thus follows that the constant κ_{rec}, calculated by Keck [298], should only be compared with the equilibrium rate constant $\kappa_{rec,eq}$ of the other theories.

Rice [439] made such a comparison and showed that Keck's theory gives a reasonable upper limit, which, though sensitive to the assumed form of the interatomic potential, can all the same be used for the calculation of $\kappa_{rec,eq}$. However, it is necessary, in addition, to take into account the possibility of recombination, with the formation of some electronically excited states of AB.

Improvement of the variational calculation can be made either by way of a more 'successful' choice of critical surface or by excluding from the calculation those trajectories which, repeatedly intersecting the critical surface, lead back into the region of configuration space which corresponds to the reactants. This can only be accomplished with the help of the numerical analysis of a large number of trajectories. Keck [299] carried out such a model calculation for the dissociation of the AB molecule, represented by a Morse potential, where the interaction of AC and BC was described by a positive exponentially-decreasing function. The classical equations of motion for the six coordinates and six momenta which describe the relative motion of the triatomic system were integrated, and the trajectories found were classified according to the number of their intersections with the surface $H_{12} = 0$. One of the results of a similar calculation is shown in Fig. 40. Each triple collision is characterized by two trajectories; (a) shows the distance between the recombining atoms r and the distance between C and the centre of mass of the recombining pair R. The upper diagram (a) is a calculation for the triple collision $Ar + H + H$. The time is plotted along the abscissa in units of the periods of the small vibrations of H_2, and along the ordinate are plotted the distances r and R in units of $1/\alpha'$ (where α' is the parameter in the Morse potential which, for hydrogen, equals 1.93×10^8 cm^{-1}). The line z is the coordinate of the most probable position of the rotational barrier at the temperature T, determined by the condition $kT/D = 0.01$. The solid lines represent the inelastic collisions of Ar with a vibrationally-excited H_2 molecule; the dash-dotted line refers to the triple collision $H + H + Ar$ without formation of a bound molecule of H_2; and the finely-dotted line represents triple collisions leading to recombination.

The lower diagram (b) shows the change in the relative energy of the recombining pair in a triple collision. Along the ordinate is the dimensionless ratio of the difference between the energy of the rotational barrier B and the energy E of the H_2 molecule to kT. The three curves correspond to the three pairs of trajectories in the upper diagram.

Statistical averaging over all trajectories whose initial states correspond to a bound state of $AB + C$ and whose final states correspond to a dissociated

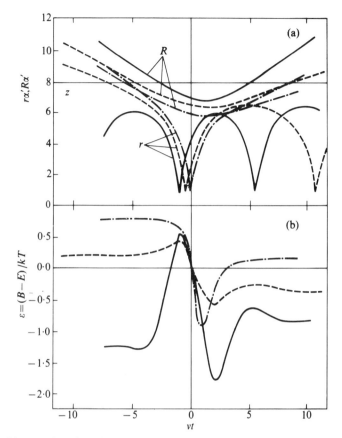

FIG. 40. Phase trajectories of triple collisions leading and not leading to recombination (Keck [299]).

state of $A+B+C$ enables the dissociation cross-section σ_{diss} to be obtained as a function of the state of AB, prior to the collision, averaged over the energies of the relative motion of AB and C. The dependence of σ_{diss} on the energy difference between the dissociation limit of the rotating molecule B and its vibrational energy E in the effective potential, found by Keck [299], has the form

$$\sigma_{\text{diss}} \sim (1+\varepsilon)^{-3.5}, \qquad \varepsilon = \frac{(B-E)}{kT}, \qquad (42.9)$$

and this dependence is not sensitive to the masses of the atoms.

At present, it is difficult to say to what extent this is a general result and how it would change if the interatomic interaction were approximated by other functions. The calculation of the cross-section σ_{diss} is a fundamental problem in the theory of dissociation and recombination; therefore the determination of σ_{diss}, even for the simple models within the limits of classical mechanics, is very important for the further development of the theory of elementary chemical processes. Determination of the disturbance of the equilibrium distribution on dissociation or recombination is a statistical problem which, however, cannot be completely solved without a preliminary determination of the cross-section of an elementary process. The specific form of the cross-section depends essentially on the detailed dynamics of the collision of AB and C.

43. The vibrational relaxation of diatomic molecules

A calculation of the rate of decomposition of diatomic molecules AB, with regard for the disturbance of the equilibrium distribution, must be based on the solution of kinetic equations of the type (23.37). There are two different reasons for the disturbance of the Boltzmann distribution. Firstly, departure from equilibrium can be caused by a sudden change in the external conditions, as takes place, for example, behind a shock front spreading in a gas. The violation of equilibrium in this case is not directly connected with the decomposition and arises even when there is a negligibly small reaction rate. Second, a perturbation of the equilibrium distribution is caused by the decomposition itself. Deferring discussion of this question to a following section, we will consider here vibrational relaxation processes in diatomic molecules, assuming an equilibrium distribution over the translational degrees of freedom which are characterized by the temperature T (a more general case is considered in [257]).

The quantity usually observed is the mean vibrational energy $\langle E_{\text{vib}} \rangle$, determined in terms of the non-equilibrium distribution function $x_n(t)$ by the expression

$$\langle E_{\text{vib}} \rangle = \sum_n E_n x_n(t). \tag{43.1}$$

It is impossible to obtain a closed equation for $\langle E_{\text{vib}} \rangle$, from the system of kinetic equations (23.37), in the general case. The system of equations is successfully reduced to one equation for $\langle E_{\text{vib}} \rangle$ only when definite restrictions imposed on the transition probability $P_{n,m}$ are fulfilled [193], [484], [487], [489], [490].

We will consider some model cases which will illustrate, in sufficient detail, the general situation, and for which a closed equation can be obtained for $\langle E_{\text{vib}} \rangle$.

Model of stepwise relaxation in a reservoir of monatomic gas

If the AB molecule is represented by a harmonic oscillator and the Landau–Teller approximation is taken for the transition probability $P_{n,m}$, then the system of kinetic equations for $x_n(t)$ takes the form (38.6). This model has been discussed in § 38 where, for $\langle E_{\text{vib}} \rangle$, the simple equation

$$\frac{d\langle E_{\text{vib}} \rangle}{dt} = -\frac{1}{\tau_{\text{vib}}}[\langle E_{\text{vib}} \rangle - E_{\text{vib.eq}}] \tag{43.2}$$

was obtained, in which the time of relaxation τ_{vib} was expressed in terms of the probability P_{10} of deactivation of the first vibrational level of the AB molecule on collision with the atom C as follows

$$\tau_{\text{vib}}^{-1} = Z_{\text{vib}} P_{10}\left[1 - \exp\left(-\frac{\hbar\omega}{kT}\right)\right]. \tag{43.3}$$

As regards the non-equilibrium distribution function $x_n(t)$, two types of initial distribution are of special interest. For an initial δ-function distribution, which possibly corresponds approximately to the formation of excited molecules in secondary reactions in flash photolysis [91], [97], the peak of the distribution function is gradually displaced to smaller n, and it is accompanied by a spreading of the bell-shaped distribution. For an initial Boltzmann distribution, x_n depends on the time only through the temperature T_{eff}, which decreases monotonically with time from the initial value T_0 to the temperature of the heat reservoir T [375]:

$$\theta_{\text{eff}} = \ln\left\{\frac{\exp(-t/\tau_{\text{vib}})[1 - \exp(\theta - \theta_0)] - \exp(\theta)[1 - \exp(-\theta)]}{\exp(-t/\tau_{\text{vib}})[1 - \exp(\theta - \theta_0)] - [1 - \exp(-\theta)]}\right\}, \tag{43.4}$$

where

$$\theta_{\text{eff}} = \frac{\hbar\omega}{kT_{\text{eff}}}, \quad \theta_0 = \frac{\hbar\omega}{kT_0}, \quad \theta = \frac{\hbar\omega}{kT}.$$

What has been said above refers, strictly speaking, to the relaxation of diatomic molecules with non-degenerate electronic states in an atmosphere of a monatomic inert gas. If the atom possesses an unfilled electronic shell or the molecule is in a degenerate state, then, as well as the vibrational relaxation, the electronic transitions between the low-lying electronic terms (for example, transitions between the fine-structure components) should be considered. If the latter processes are sufficiently fast so that, in the general relaxation picture, their effect on the vibrational relaxation can be neglected, then we again return to eqn (43.2) in which τ_{vib} must be calculated, taking possible nonadiabatic effects into account. Finally, if the atom C is the same as the atoms of the molecule undergoing relaxation, then an exchange reaction can contribute to the general relaxation process

$$A'A''(n_1) + A''' \to A' + A''A'''(n_2), \qquad (43.5)$$

the only result of which is a change in the vibrational quantum number n [145]. The change in n in this reaction depends on the details of the fundamental exchange process, and it is quite possible that the process (43.5) does not follow a stepwise mechanism but approximates to the mechanism of multiple-quantum relaxation (see below). It might be expected however, that, in order of magnitude, the vibrational relaxation time in the reaction (43.5) equals the inverse of the exchange rate. This mechanism was enlisted, in particular, to interpret the relaxation of oxygen molecules induced by collisions with oxygen atoms [303].

Model of stepwise relaxation in a diatomic gas reservoir

The vibrational relaxation of a mixture consisting of diatomic molecules AB and CD differs appreciably from the relaxation of AB in an atmosphere of a monatomic gas, since, together with the exchange of vibrational and translational energy, the exchange of the vibrational energy of the partners in collision makes an important contribution.

The system of kinetic equations for the distribution functions of the vibrational states $x_n(t)$ and $y_m(t)$ of the molecules AB and CD, respectively, reduces to a system of two equations for the mean energies $\langle E_{AB,vib} \rangle$ and $\langle E_{CD,vib} \rangle$ if the molecules are represented by harmonic oscillators and the interaction specifying the energy exchange is assumed proportional to the amplitude of the vibrations [92]. The relaxation equations, written in terms of the mean vibrational quantum numbers $\langle n_{AB} \rangle = \langle E_{AB,vib} \rangle / \hbar \omega_{AB}$ and $\langle m_{CD} \rangle = \langle E_{CD,vib} \rangle / \hbar \omega_{CD}$, have the form [100]:

$$\frac{d\langle n \rangle}{dt} = -\frac{1}{\tau_1}(\langle n \rangle - n_{eq}) + \frac{\zeta_1}{\tau_{12}}[\langle m \rangle \exp(\theta_2 - \theta_1)(\langle n \rangle + 1) - \langle n \rangle(\langle m \rangle + 1)],$$

$$\frac{d\langle m \rangle}{dt} = -\frac{1}{\tau_2}(\langle m \rangle - m_{eq}) + \frac{\zeta_1}{\tau_{21}}[\langle n \rangle(\langle m \rangle + 1) - \langle m \rangle \exp(\theta_2 - \theta_1)(\langle n \rangle + 1)].$$
$$(43.6)$$

Here τ_1 has the meaning of the time for the vibrational relaxation of AB, taking place on the assumption that the vibrational energy exchange between partners is forbidden. Collisions between the AB, as well as collisions between AB and CD, contribute to τ_1. Therefore τ_1 depends on the composition of the mixture, which is given by the relative concentrations

$$\zeta_1 = \frac{N_{AB}}{(N_{AB} + N_{CD})}, \qquad \zeta_2 = \frac{N_{CD}}{(N_{AB} + N_{CD})};$$

the meaning of τ_2 is analogous. The time τ_{12} is determined by the number of collisions Z_{12} of the molecules AB and CD between themselves and the probability of the deactivation of AB (AB($n = 1$) → AB($n = 0$)) with simultaneous activation of CD (CD($n = 0$) → CD($n = 1$)), where $\tau_{12} = \tau_{21}$. The parameters θ_1 and θ_2 equal $\hbar\omega_{AB}/kT$ and $\hbar\omega_{CD}/kT$ respectively.

The right-hand side of system (43.6) contains terms depending simultaneously on $\langle n \rangle$ and $\langle m \rangle$. They are determined by the vibrational-energy exchange processes between the molecules AB and CD. These equations, however, do not contain coefficients characterizing the redistribution of energy in collisions between two molecules of AB (or CD). This is completely understandable, as any redistribution of energy within the subsystem of AB molecules does not change the mean energy of this subsystem. As regards the non-equilibrium distribution function $x_n(t)$ (or $y_m(t)$), it, of course, depends essentially on the rate of elementary processes of the type

$$AB(n) + AB(n') \to AB(n \pm 1) + AB(n' \pm 1). \tag{43.7}$$

An analytic solution of the system (43.6) in the general form has not been obtained. However, a case of practical interest occurs when one of the times for vibrational energy exchange and translational energy exchange greatly exceeds the other; in this case, the connection between the two processes of relaxation, determined by the exchange of vibrational energy, emerges most clearly. Without dwelling on the details of the calculation, we will quote only the final results, based on the work of Osipov [87], [88], [90], [94], [95], [100] and Shuler [485].

Let $\tau_2 \gg \tau_1$, for example, so that the direct conversion of the vibrational energy of CD into the kinetic energy of the partners can be neglected. In this case, the only mechanism of vibrational relaxation of CD consists of the two-stage transfer of energy

$$CD(m) + AB(n) \to CD(m \pm 1) + AB(n \pm 1)$$

$$AB(n') + \begin{bmatrix} AB(n'') \\ CD(m'') \end{bmatrix} \to AB(n' \pm 1) + \begin{bmatrix} AB(n'') \\ CD(m'') \end{bmatrix}. \tag{43.8}$$

The kinetics of this process will differ as a function of the value of the ratio τ_{12}/τ_1.

For $\tau_{12}/\tau_1 \ll 1$, a quasi-equilibrium distribution of the vibrational states of AB and CD is established in the first, rapid stage.† The corresponding mean values $\langle n \rangle$ and $\langle m \rangle$ are found from the condition for the completion of

---

† Shuler and Weiss [489] noticed an interesting property appearing at this stage of relaxation: the non-equilibrium distribution function of the AB molecules does not depend on the distribution function of the CD molecules and is determined only by the mean energy of CD.

vibrational–vibrational exchange of energies

$$\langle m \rangle (\langle n \rangle + 1) \exp(\theta_2 - \theta_1) = \langle n \rangle (\langle m \rangle + 1). \tag{43.9}$$

The exchange between the vibrational energy of the whole system and the translational energy occurs in the second, slower stage. The relaxation equation for $\langle n \rangle$ takes the form

$$\frac{d\langle n \rangle}{dt} = -\frac{1}{\tau_a}(\langle n \rangle - n_{eq}),$$

$$\tau_a = \tau_1 \left(1 + \frac{N_{CD}}{N_{AB}} \cdot \frac{d\langle m \rangle}{d\langle n \rangle}\right), \tag{43.10}$$

where the derivative $d\langle m \rangle / d\langle n \rangle$ must be found from (43.9).

For $\tau_{12}/\tau_1 \gg 1$, the vibrational relaxation of the AB molecules is completed in the first stage, in accordance with the equation

$$\frac{d\langle n \rangle}{dt} = -\frac{1}{\tau_1}(\langle n \rangle - n_{eq}). \tag{43.11}$$

The vibrational relaxation of the CD molecules occurs at the second stage, and the distribution x_n is maintained close to the equilibrium distribution the whole time. The relaxation equation for $\langle m \rangle$ has the form

$$\frac{d\langle m \rangle}{dt} = -\frac{1}{\tau_b}(\langle m \rangle - m_{eq}),$$

$$\tau_b = \frac{\tau_{12}}{\zeta_1}[(n_{eq} + 1)\exp(\theta_2 - \theta_1) - n_{eq}]. \tag{43.12}$$

For conditions close to resonance, when $|\theta_2 - \theta_1| \ll 1$, the expressions for the relaxation times τ_a and τ_b take especially simple forms:

$$\tau_a = \frac{\tau_1}{\zeta_1}, \quad \tau_b = \frac{\tau_{12}}{\zeta_1}. \tag{43.13}$$

The first of these expressions implies, in particular, that the continuous rapid removal of vibrational energy from the system of AB molecules to the system of CD molecules slows down the vibrational relaxation of AB. The relative fraction of AB molecules can be expressed in terms of the thermal heat capacity C_1 of the subsystem of AB and the thermal heat capacity C_s of the total system by the expression $\zeta_1 = C_1/C_s$. Then from (43.13) we obtain

$$\tau_a = \frac{C_s}{C_1}\tau_1. \tag{43.14}$$

In this form, expression (43.14) can be used to calculate the vibrational relaxation time of a system with many degrees of freedom if the coupling between the system and the reservoir is via a vibration of type I (relaxation time of the initial system τ_1) which, in its turn, is sufficiently strongly coupled to the remaining vibrations. In particular, for polyatomic molecules, τ_a is equal to the vibrational relaxation time, τ_1 equals the time calculated on the assumption of the absence of coupling between the normal vibrations, and C_s is the heat capacity of the normal vibrations close in frequency to the normal vibration of I (heat capacity C_1).

We will mention one more often-encountered relaxation process, described by eqns (43.6). We will assume that, in the reservoir, which consists of diatomic AB molecules, a small fraction of the molecules is characterized by a distribution differing from the equilibrium distribution. We will attribute the value $\langle n \rangle$ to this fraction of the molecules. By $\langle m \rangle$ we will denote the mean number of vibrational quanta for the remaining molecules of the reservoir (such a situation arises, for example, in the optical excitation of vibrations or in conditions of flash photolysis). Assuming $\zeta_1 \ll 1$, $\zeta_2 \approx 1$, $\langle m \rangle = m_{eq} = n_{eq}$, and $\theta_1 = \theta_2$, we find the following relaxation equation for the fraction of molecules of interest:

$$\frac{d\langle n \rangle}{dt} = -\frac{1}{\tau}(\langle n \rangle - n_{eq}), \quad \frac{1}{\tau} = \frac{1}{\tau_1} + \frac{1}{\tau_{12}}. \tag{43.15}$$

Usually $\tau_1 \gg \tau_{12}$, so that the relaxation rate of the subsystem under consideration is determined by the rate of resonant exchange of vibrational energy with the remaining part of the reservoir. In addition, in each act of collision the molecule undergoing relaxation loses or gains one vibrational quantum. A comparison of the non-equilibrium distribution functions in the relaxation of diatomic molecules in monatomic and diatomic gases was made by Rich and Rehm [442].

The diffusion model

If the mean square energy transfer $\langle \Delta E^2 \rangle$ is small in comparison with $(kT)^2$, then the relaxation of an oscillator is described by the Fokker–Planck equation (36.12). If, in addition, it turns out that $\langle \Delta E^2 \rangle$ is proportional to the instantaneous value of the vibrational energy, then a closed kinetic equation of the form (43.2) can be obtained for $\langle E_{vib} \rangle$, in which it should be assumed that

$$\langle E_{vib} \rangle = kT, \quad \frac{1}{\tau_{vib}} = Z_0 \frac{\langle \Delta E^2 \rangle}{kT E_{vib}}, \tag{43.16}$$

where the first equality in (43.16) follows from the condition in the classical description of relaxation.

If the interaction of the oscillator with the reservoir is simulated by a potential of the type $xF(t)$ (see § 8), where the force $F(t)$ represents the perturbation caused by the reservoir, then the mean square energy transfer $\langle \Delta E^2 \rangle$ proves, in fact, to be proportional to E_{vib}.[†]

If it is now taken into account that the mean square energy transfer in this approximation is the same for quantum and classical oscillators, then the single interpolation formula proposed by Osipov and Generalov [101] can be written down for τ_{vib}

$$\frac{1}{\tau_{\text{vib}}} = Z_0 \sum_n P_{n0} \left[1 - \exp\left(-\frac{\hbar\omega}{kT}\right) \right]. \tag{43.17}$$

With such a definition of τ_{vib}, it is possible to make use of the relaxation equation (43.2), which now describes both single-quantum transitions for $\hbar\omega\beta \geqslant 1$ (in this case one term remains in (43.17)) and multiple-quantum transitions for $\hbar\omega\beta \ll 1$.

Model of multiple-quantum relaxation

It is of interest to study the case when the exchange of several vibrational quanta, in the act of collision, is possible. Since such a situation is most probable in a resonant interaction when τ_1 is known to exceed τ_{12}, the exchange of translational and vibrational energy can, in general, be neglected. Limiting discussion to a model of harmonic oscillators and assuming that their interaction, which determines the energy exchange, is proportional to the product of the amplitudes of the vibrations (see § 12), the multiple-quantum transition probabilities $P_{n,n',m,m'}$ can be found explicitly. Then for the energy $\langle E \rangle$ of a small group of AB molecules undergoing relaxation in a heat reservoir consisting of the same AB molecules, the following equation is obtained [71]:

$$\frac{d\langle n \rangle}{dt} = \frac{1}{\tau}(\langle n \rangle - n_{\text{eq}}), \qquad \frac{1}{\tau} = Z_0 \langle \sin^2 \theta \rangle, \tag{43.18}$$

where θ is defined by formula (12.10); the brackets $\langle \ldots \rangle$ denote an average over all the collision parameters (over the relative velocities and impact parameters).

The fact that there is a periodic function $\sin^2 \theta$ under the averaging sign in (43.18) is connected with the specific mechanism of vibrational energy exchange in the collision of the two molecules; as the partners in the bound system of the two oscillators approach, beats begin, as a result of which the

[†] From formula (8.14), it is evident that, for the forced harmonic oscillator the expression for $\langle \Delta E^2 \rangle$ contains the term ΔE_0^2 as well as $E \Delta E_0$. However, the former term should be neglected if we wish in general to use a diffusion equation to describe relaxation. In fact, relations (8.14) establish a link between $\langle \Delta E^2 \rangle$ and $\langle \Delta E \rangle$. On the other hand, from the Fokker–Planck equation, a relation connecting $\langle \Delta E^2 \rangle$ and $\langle \Delta E \rangle$ can be obtained. These two relations turn out to be different, which is ultimately due to representation of the reservoir by an external force $F(t)$. This defect of the representation in the Fokker–Planck equation is removed precisely because terms of order $\Delta E_0 / \Delta E$ are neglected in the expression for $\langle \Delta E^2 \rangle$.

energy periodically passes from one molecule to the other. Of course, after averaging, the periodic character of the factor $\langle \sin^2 \theta \rangle$ disappears, so that the time of relaxation depends monotonically on the magnitude of the interaction potential.

The model of two oscillators effectively bound at the time of collision of the partners is analogous to the Slater model in the theory of unimolecular reactions (see § 24). In the case under consideration, an element of statistics is introduced on averaging the dynamical factor $\sin^2 \theta$ over a random distribution of the lifetimes of the complex of colliding molecules.

In the limiting case, when $\theta \ll 1$, formula (43.18) goes over into (43.15), since the single-quantum transitions make the main contribution. In the other limiting case, when $\theta \gg 1$, on average, many beats occur during the lifetime of the complex. In these conditions $\langle \sin^2 \theta \rangle$ can be approximately replaced by $\frac{1}{2}$ in the averaging, which gives

$$\frac{d\langle n \rangle}{dt} = -\frac{Z_0}{2}(\langle n \rangle - n_{eq}). \tag{43.19}$$

This result is easy to understand: for a sufficiently-strong interaction between the partners, the energy is distributed, on average, with equal probability on the two oscillators. Since the distribution function of one of the oscillators is assumed to be the equilibrium distribution, the relaxation of the other takes place during a time of the order of the time between successive collisions.

For a strong interaction between the partners, it is of interest to consider a statistical model for the redistribution of energy analogous to Kassel's model in the theory of unimolecular reactions (see § 25). Disengaging ourselves from the dynamics of the exchange of the vibrational energy on collision, we will assume that the total vibrational energy of the partners $E_1 + E_2$ is redistributed at each collision equiprobably over all the accessible vibrational states of the partners, where the distribution function of one of them is the Boltzmann distribution. Under these assumptions, the transition probability $n \to n'$ for the other oscillator has the form [265]–[267]

$$P_{n,n'} = \begin{cases} (1-e^{-\theta}) e^{n\theta} \sum_{\kappa=n}^{\infty} \frac{\exp(-\kappa\theta)}{\kappa+1}, & n' \leq n \\ (1-e^{-\theta}) e^{n\theta} \sum_{\kappa=n'}^{\infty} \frac{\exp(-\kappa\theta)}{(\kappa+1)}, & n' \geq n. \end{cases} \tag{43.20}$$

The solutions of the relaxation equations with transition probabilities (43.20) were investigated by Hoare [265]–[267], and it was found that, for the mean vibrational energy, the relaxation equation (43.17) is correct.

From the above exposition, it is clear that, essentially, the different models of relaxation lead to one and the same relaxation equation for the mean energy. Therefore the solution of the problem in reverse—the recovery (even if approximate) of the individual transition probabilities from the experimental kinetical data—usually cannot be carried out unambiguously [190], [191].

Investigation of the non-equilibrium distribution function can give more definite information about the transition probabilities. In particular, for an initial δ-function distribution and a 'cold' reservoir ($\langle n(t = 0)\rangle \gg kT/\hbar\omega$), the distribution function $x_n(t)$ undergoes relaxation qualitatively differently for the above-mentioned mechanisms of single-step and multiple-step transitions. In the first case, the maximum of the distribution function is gradually displaced in the direction of small n ($n < n(t = 0)$), and in the second, the maximum appears straight away at the lowest levels, so that the population of the intermediate states remains small the whole time, in the relaxation process [267]. In this sense the situation is analogous to that considered earlier (see § 39) in connection with the calculation of the unimolecular reaction rate and the non-equilibrium distribution function; whereas the non-equilibrium distribution function essentially depends on the activation mechanism, the reaction rate is comparatively insensitive to it, if the theory operates with the effective number of deactivating collisions.

The relaxation of a distribution function for a model in which transitions take place between non-neighbouring levels, which, in a certain sense, is intermediate between the single-quantum transition model and the statistical model considered above, is discussed in the literature [193], [488].

44. Non-equilibrium theory of dissociation and recombination

Theoretical calculations of the non-equilibrium rate of dissociation of diatomic molecules, which pretend to comparison with experiment, refer, in the main, to the case of decomposition of molecules in an atmosphere of a monatomic inert gas. The various calculations differ in their assumptions about the details of the energy-exchange mechanism in the collision of an atom C with a vibrationally highly excited molecule AB. Since these assumptions lead to differences in the temperature dependence of the pre-exponential A, we will consider some variants of the theory.

Model of single-quantum stepwise excitation

First, a model which took account of the marked dependence of the vibrational transition probabilities on the quantum number n was considered in the literature [53], [59], [113], [115]. In these papers, the authors approximated the potential function of the non-rotating molecule by a Morse oscillator and calculated, in first-order perturbation theory, the transition probabilities between the vibrational levels in the collision of AB and C. From the calculations (and also from the results of Chapter 2) the following qualitative picture emerges of the distribution of the transition probabilities over the vibrational spectrum. In the lower region of the discrete spectrum, in practice, transitions are allowed only between neighbouring levels, and the transition probability increases rapidly with rise in vibrational quantum number. In the upper region of the spectrum, together with transitions to neighbouring levels, there occur transitions to non-neighbouring levels. Finally, close to the dissociation limit, the probability of transition into the continuous spectrum becomes appreciable. The energy E^*, which approximately separates the regions of single-quantum and multiple-quantum transitions can be determined from the Massey criterion

$$\frac{\omega(E^*)}{v\alpha} \approx 1. \tag{44.1}$$

For $E \ll E^*$ the probability of vibrational excitation is described by the Landau–Teller formula (see § 8). In this case, the mean energy transfer $\langle \Delta E^2 \rangle$ is very small and decreases exponentially with increase in the vibrational quanta. For $E > E^*$, when the condition $\omega/\alpha v \ll 1$ is satisfied, the vibrations of AB during the collision are unimportant and the mean energy transfer can be calculated within the bounds of the impulse approximation (see § 11).

To calculate the dissociation rate, it is very important to know in which region of the discrete spectrum the energy E^* lies. If E^* lies within a region of width kT close to the dissociation limit, then small variations in E^* within the limits of this region, will not greatly change the rate constant κ_{diss}. In fact although, for $E > E^*$, the rate of excitation, characterized by the mean energy transfer ΔE, increases rapidly due to the contribution of transitions between non-neighbouring levels, this does not change the constant κ_{diss} substantially, whose magnitude must be less than the vibrational excitation rate of any level.

The rate of equilibrium dissociation of the molecules from levels immediately adjacent to the dissociation limit can be written in the form $Z_0 \exp(-D/kT)$. On the other hand, the rate of multi-step vibrational excitation close to E^* can be approximated by an analogous expression $pZ_0 \exp(-E^*/kT)$, in which p denotes the probability of a single-quantum excitation divided by the number of quanta in an energy interval kT below the level E^*. An estimate of the maximum effect of the rapid, final stage of the decomposition can be made by decreasing the dissociation energy to E^*; then, owing to the relation $D - E^* \leqslant kT$, the corresponding change in κ_{diss} will be small.

In cases where E^* is comparatively far from the dissociation limit, it is more difficult to make an analogous estimate of the effect of the different stages on the rate of decomposition.

We will now consider the dissociation of AB molecules under the condition of the approximate equality of the masses M and μ. Returning to the Morse oscillator model, we will assume

$$\omega(E) = \alpha' \left[\frac{2(D-E)}{M} \right]^{\frac{1}{2}}, \quad v \sim \left(\frac{kT}{\mu} \right)^{\frac{1}{2}}, \quad \alpha' \sim \alpha, \qquad (44.2)$$

whence, using (44.1), we find

$$D - E^* \approx \frac{m}{\mu} kT. \qquad (44.3)$$

Thus, the simplest case of dissociation of the AB molecules in an atmosphere of an inert monatomic gas C, with approximately the same or greater mass than AB, corresponds approximately to a model in which only transitions between neighbouring vibrational levels are important, and the probability of dissociation from the final vibrational level is assumed to be much greater than the probability of vibrational transitions.

For a non-rotating molecule and a single electronic state, the system of kinetic equations describing stepwise vibrational excitation and dissociation has the form (38.1). The non-equilibrium rate constant κ_{diss}, corresponding to this system, is given by relation (38.10). Having used the fact that terms referring to vibrational levels near the dissociation limit, where $\hbar\omega(E^*)/kT \ll 1$, make the main contribution to the second sum in (38.10), the summation can be replaced by an integration. If it is now taken into account that the single-quantum transition probability in the region $E \approx E^*$ is close to unity and that the level density for the Morse oscillator is proportional to $(D-E)^{-\frac{1}{2}}$, then for κ_{diss} the following expression is obtained:

$$\begin{aligned}\kappa_{\text{diss}} &= \frac{1}{F_{\text{vib}}} Z_0 \frac{\hbar\omega^*}{kT} \exp\left(-\frac{D}{kT}\right) \\ &= Z_0 \frac{\hbar\omega}{(DkT)^{\frac{1}{2}}} \exp\left(-\frac{D}{kT}\right)(1 - e^{-\hbar\omega/kT}).\end{aligned} \qquad (44.4)$$

In many cases, it is convenient to discuss the temperature dependence of the pre-exponential A in (40.4) on the basis of a formal evaluation of the recombination rate constant from the relation:

$$\frac{\kappa_{\text{diss}}}{\kappa_{\text{rec}}} = K_{\text{eq}}, \qquad (44.5)$$

where K_{eq} is the equilibrium constant.

The calculation of K_{eq} can be carried out phenomenologically, without explicit consideration of the mechanism of energy exchange. The validity of the relation (44.5) for the actual connection between the non-equilibrium rate constants of dissociation and recombination will be considered below (see § 45). The temperature dependence of the pre-exponential factor in the equilibrium constant is determined by the ratio of the partition functions for the relative translational motion of the dissociated pair A and B to that for the vibrational–rotational motion of the bound AB molecule. The former is directly proportional to $T^{3/2}$ and the latter to $TF_{\text{vib}}(T)$. As $Z_0 \sim T^{1/2}$ on the basis of (44.4) and (44.5), we find

$$\kappa_{\text{rec}} \sim \frac{\hbar\omega^*}{kT}. \tag{44.6}$$

This relation takes the form $\kappa_{\text{rec}} \sim T^{-1/2}$ for the model of the non-rotating Morse oscillator. The rotation of the molecule and the presence of several electronic states can be taken into account approximately, in accordance with the discussion in § 41, by multiplying (44.4) or (44.6) by $g_{\text{el}}g_{\text{rot}}$. Thus the expression for the decomposition rate constant finally takes the form

$$\kappa_{\text{diss}} = Z_0 g_{\text{el}} g_{\text{rot}} \frac{\hbar\omega^*}{kT} \exp\left(-\frac{D}{kT}\right)\left[1 - \exp\left(-\frac{\hbar\omega}{kT}\right)\right]. \tag{44.7}$$

Numerical evaluation of the factor g_{rot} for the dissociation of Br_2, for example, on the assumption that, close to the dissociation limit, the Morse formula is still valid for the interatomic potential, shows that g_{rot} is practically independent of the temperature and lies approximately between 5 and 10 [53]. The factor g_{el} for Br_2 must be close to 5 and may depend on temperature, thus leading to a more negative dependence of the pre-exponential than that which follows from (44.4).

For the temperature dependence of κ_{rec}, eqn (44.6) gives the expression

$$\kappa_{\text{rec}} \sim g_{\text{el}} g_{\text{rot}} T^{-1/2}. \tag{44.8}$$

Instead of determining κ_{rec} from relation (44.6), a single-step activation scheme can be directly applied to calculate κ_{rec}.

Such a process was considered by Benson and Fueno [152]. In the evaluation of a sum, analogous to (38.10), the authors approximated the dependence of the vibrational energy of AB on the quantum number n of the vibrational level by a quadratic function (this type of dependence is valid for the Morse oscillator), and in calculating the effect of the rotation of the pair AB on the rate of recombination, the long-range polarization part of the potential, proportional to r^{-6}, was taken into account. The expression thus obtained for the recombination rate constant in one of the stable electronic states was

$$\kappa_{\text{rec}} = \lambda Z_0 \left(\frac{1}{g}\right)\left(\frac{1}{\Delta n^*}\right)^2 \frac{2}{3}\pi\langle r_m\rangle^3, \tag{44.9}$$

where Δ_n^* is the number of vibrational levels of AB lying in an energy range kT close to the dissociation limit; λ is the probability of single-quantum deactivation; $1/g$ is the probability of the electronic state under consideration of the molecule AB arising on the approach of the pair A + B (for pairs 2H, 2N, 2Br, and 2I, the values of $1/g$ for the electronic ground state equal $\frac{1}{4}$, $\frac{1}{16}$, $\frac{1}{16}$, and $\frac{1}{16}$ respectively); the factor $\frac{2}{3}\pi \langle r_m \rangle^3$ is equal to the volume of configuration space, accessible to the molecule AB close to the dissociation limit, and $\langle r_m \rangle$ is equal to the mean interatomic distance in the configuration of the molecule, which corresponds to the peak of the centrifugal potential barrier.

Benson and Fueno [153] and also Bunker [178] obtained $\langle r_m \rangle^3 \sim T^{-\frac{1}{2}}$ for the temperature dependence of $\langle r_m \rangle^3$. Moreover, they assumed that the increase in the number of collisions Z_0 with increasing temperature is largely compensated by a decrease in the effective collision cross-section of AB and C. In this case, the dissociation rate constant calculated in accordance with (44.5) must contain an additional factor $T^{-\frac{1}{2}}$, in comparison with expression (44.4).

As regards the probability factor λ, its calculation in an explicit form requires definite assumptions to be made with respect to the potentials of the interatomic interactions A–C and B–C. Some qualitative information about them can be obtained from data on the vibrational relaxation of molecules. This approach was employed in the papers of Bauer and Salkoff [144], who used the same potential to calculate both the transition probability between the lower vibrational levels of oxygen and the recombination probability. These authors did not consider multi-stage recombination, although they came to the conclusion that recombination proceeds mainly as a result of capturing the atoms in one of the highest vibrational levels of the bound state of O_2.

The diffusion model

The diffusion model enables multiple-quantum transitions to be taken into account; however, it imposes a restriction on the magnitude of the energy transfer near the dissociation limit:

$$\frac{\langle \Delta E^2 \rangle}{(kT)^2} \ll 1. \qquad (44.10)$$

Within the limits of the diffusion model, it is not difficult to establish the connection between the non-equilibrium and variational theories of dissociation. If, in each collision, the energy transfer is small in comparison with kT, then the general expression for the equilibrium current j_0, identified with the variational rate constant $\tilde{\kappa}_{\text{diss}}$, after proper normalization of the distribution function and crossing the energy level E,

$$j_0(E) = \tilde{\kappa}_{\text{diss}} = \frac{1}{F} \int_E^\infty dE'' \int_0^E Z_0 P(E' \to E'') \rho(E') \exp(-\beta E') \, dE' \qquad (44.11)$$

can be represented in the form [300]

$$j_0(E) = \tilde{\kappa}_{\text{diss}} = \frac{Z_0}{2F}\langle|\Delta E(E)|\rangle \exp(-\beta E)\rho(E). \tag{44.12}$$

Here, the averaged absolute magnitude of the energy transfer is determined by the relation

$$\langle|\Delta E(E)|\rangle = \int |E' - E| P(E \to E')\, dE'. \tag{44.13}$$

An expression for the rate constant, in accordance with variational theory, is obtained from (44.12), as was shown in § 42, by minimizing $\tilde{\kappa}_{\text{diss}}$ with respect to E. The occurrence of the slowest stage at $E = E_0^*$ in regard to the equilibrium current can also be used to calculate the rate constant of non-equilibrium dissociation. Turning to (35.16) it is not difficult to see that the integrand reaches a maximum at some energy $E = E^*$, when the denominator of the integrand, which has the meaning of a certain effective current, is a minimum. Taking the integrand outside at the point E^*, we rewrite (35.16) in the form

$$\kappa_{\text{diss}} = \frac{Z_0}{2F}\langle\Delta E^2(E^*)\rangle \frac{1}{\delta E} \exp(-\beta E^*)\rho(E^*), \tag{44.14}$$

where the mean square energy transfer is determined by the relation

$$Z_0\langle\Delta E^2(E)\rangle = \int (E' - E)^2 P(E \to E')\, dE', \tag{44.15}$$

and δE is the effective energy range over which the integration in (35.16) is extended. Since the rate of attenuation of the integrand in the expression in (35.16) is determined mainly by the exponential, it can usually be assumed that $\delta E \approx 1/\beta$.

In the comparison of (44.12) and (44.14), we will assume that approximately

$$\langle|\Delta E(E_0^*)|\rangle \approx \langle\Delta E^2(E^*)\rangle^{\frac{1}{2}}. \tag{44.16}$$

Then (44.12) means that the non-equilibrium rate constant will turn out to be approximately $\beta < \Delta E^2(E^*)\rangle^{\frac{1}{2}}$ times smaller than the variational rate constant.

The expression (44.14) enables the transition between the diffusion model and the model of single-quantum excitation to be traced. For this, δE in (44.14) should be identified with the magnitude of the quantum spacing of the vibrational spectrum close to the dissociation limit, assuming at the same time

$$\langle\Delta E^2(E^*)\rangle \approx (\delta E)^2. \tag{44.17}$$

Returning to the estimate (44.1), it is not difficult to see that multiple-quantum transitions in the region which makes the largest contribution to the integrand (38.13) (i.e. the region across which the transitions determine the rate of the total process) occur on condition $\mu \ll M$. This corresponds to the case of the dissociation of a heavy diatomic gas in an atmosphere of a light inert gas. A simple estimate of the factor $\langle \Delta E^2(E^*) \rangle \beta^2$, carried out for the model of a truncated harmonic oscillator within the limits of the impulse approximation, gives $\mu D/MkT$ (see §11). Substituting this value into (44.10), it is not difficult to see that the diffusion approximation for the model of a truncated harmonic oscillator is applicable only for very small ratios μ/M. For more realistic anharmonic models, condition (44.10) imposes less rigid restrictions. In fact, the mean value of the square of the momentum $\langle p_A^2 \rangle$, determining the value $\langle \Delta E^2 \rangle$, within the limits of the impulse approximation can be estimated in the following way:

$$\left[\frac{\langle \Delta E^2(E^*)\rangle}{(kT)^2}\right]_{\text{anh}} \sim \frac{\langle p^2(E^*)\rangle}{(kT)^2}\frac{\mu kT}{M^2} \sim \frac{\mu kT}{M}\omega(E^*)\oint p(r)\,dr \sim \frac{\mu}{M}\cdot\frac{\omega(E^*)}{kT}\cdot\frac{D}{\omega_0}$$

$$\sim \left[\frac{\langle \Delta E^2(E^*)\rangle}{(kT)^2}\right]_{\text{harm}} \frac{\omega(E^*)}{\omega_0}. \tag{44.18}$$

In this evaluation, the fact that the action $\oint p\,dr$ in order of magnitude is equal to the ratio D/ω_0 (ω_0 is the frequency of vibration close to the minimum), both for the harmonic and the anharmonic models, has been used.† Since $\omega(E^*)/\omega_0 \ll 1$, the restriction (44.10) for the anharmonic model is not so critical as for the harmonic model.

Although accounting for anharmonicity introduces an additional term $\omega(E^*)$ into the expression for $\langle \Delta E^2(E^*) \rangle$, which, in principle, may depend on temperature, the final temperature dependence of the pre-exponential in (44.14) turns out to be the same for both the harmonic and anharmonic models. This is connected with the fact that the level density $\rho(E^*)$ contains a factor $1/\omega(E^*)$ and thus the frequency $\omega(E^*)$ does not come into the expression for the rate of decomposition. Of course, this is only correct within the limits of the approximation under which formula (44.14) was obtained. Physically, such a situation is due to the mutual compensation of the increase in level density on approach to the dissociation limit and the decrease in the kinetic energy of the atoms in the molecule due to the anharmonicity.

Substituting (44.18) into (44.14), assuming $\rho(E^*) = 1/\hbar\omega(E^*)$, $F = kT/\hbar\omega_0$ and taking into account that, on condition $\mu \ll M$, the number of collisions Z_0 is determined only by the mass of the light atom C, we find, for the constant κ_{diss},

$$\kappa_{\text{diss}} \approx \pi R_0^2 \sqrt{\left(\frac{8kT}{\pi m_C}\right)} \cdot \frac{m_C}{m_A}\cdot\frac{D}{kT}\cdot\exp\left(-\frac{D}{kT}\right), \tag{44.19a}$$

† We recall that the action $\oint p\,dr$, expressed in units of $\hbar$, is equal to the number of quantum states of the system. For the Morse oscillator, for example, the number of vibrational states is twice the number of states in the potential well of a truncated oscillator.

with

$$\frac{m_C}{m_A} \cdot \frac{D}{kT} \cdot \frac{\omega(E^*)}{\omega_0} \ll 1. \qquad (44.19b)$$

In particular, for the Morse oscillator, condition (44.19b) takes the form

$$\frac{m_C\sqrt{D}}{m_A\sqrt{(kT)}} \ll 1. \qquad (44.19c)$$

In expression (44.19a) the contribution of the rotational and electronic degrees of freedom is not taken into account, and the effective radius of interaction R_0 is determined for a specific model of the molecule (a one-dimensional or three-dimensional model). The corresponding calculations are carried out in the literature [134]–[136], [300], [380].

The contribution of rotation to κ_{diss} is taken into account in a similar way for the diffusion model and the model of single-quantum transitions. Therefore the difference in the temperature dependence of κ_{diss} for these models may be considered on the basis of formulae (44.4) and (44.19a). It is not difficult to see that, in the region of high temperatures ($\beta\hbar\omega \ll 1$), the pre-exponential factors in (44.4) and (44.19a) differ by a factor proportional to $T^{\frac{1}{2}}$. This is a comparatively small difference if it is remembered that, in the one case, single-quantum transitions and, in the other, multiple-quantum transitions take place.

Model of multiple-quantum activation

Such a mechanism of energy transfer is possible, in principle, when a transition occurs between non-neighbouring vibrational levels at each collision and the mean energy $\langle \Delta E^2 \rangle^{\frac{1}{2}}$ is of the same order, or greater than, kT. In this case, it is necessary to study a system of kinetic equations of the type (23.4), in which $P(E, E')$ describes transitions between energetically-remote states E and E'. We will consider here the simplest model of Hoare [267], in which a statistical redistribution of the energy in a system of two colliding harmonic oscillators is proposed (see § 43). The decomposition rate constant, calculated as the inverse of the mean time for the first crossing of the boundary $E = E_0$, proves to equal

$$\kappa_{\text{diss}} = Z_0 \exp\left(-\frac{D}{kT}\right) \qquad (44.20)$$

for $D/kT \gg 1$. The equilibrium constant $\kappa_{\text{diss,eq}}$ corresponding to this model is determined by the relation

$$\kappa_{\text{diss,eq}} = Z_0 \sum_{n=N+1}^{\infty} \sum_{m=0}^{N} P_{mn} x_{m,\text{eq}}, \qquad (44.21)$$

where P_{mn} is defined by formula (43.20). A calculation of this sum, in the limit $D/kT \gg 1$, shows that, for sufficiently large values of the ratio D/kT, the rate constants (44.20) and (44.21) are equal

$$\kappa_{\text{diss}} = \kappa_{\text{diss,eq}}. \tag{44.22}$$

Such a relation is not satisfied for the diffusion or single-step models for any values of the ratio D/kT.

The assertions, sometimes met with, that, in the limit $D/kT \gg 1$, the equilibrium and non-equilibrium rate constants differ little (see, for example [201]) should be understood in application to the single-step model, only in the sense that, on increase in the ratio D/kT, dissociation processes differ from relaxation processes to a greater and greater extent, i.e. that the relation $\mu_0 \ll \mu_1, \mu_2, \ldots$ is satisfied with increasing accuracy. At the same time, the mean time to reach the dissociation limit $\bar{t}$ agrees to ever greater accuracy with the constant μ_0 which determines the rate of a simple exponential decay, and the latter, in its turn, converges to the value κ_{diss}, found on integration of the diffusion equation in the quasi-stationary approximation. As regards the value of κ_{diss}, for the diffusion mechanism it is always less than the equilibrium value $\kappa_{\text{diss,eq}}$.

To follow the change in κ_{diss} or $\bar{t}$, in passing from the single-quantum model to the multiple-quantum excitation model in the general form, is rather complicated, so that definite results can be obtained only for certain, simple models. We will not dwell on this problem, but refer readers to the literature [471], [488], [532].

An analytic expression for κ_{diss} can be obtained only when the transitions between neighbouring levels are the most probable. In particular, from the calculations of the electron–ion recombination rate carried out by Bates and Kingston [140], it follows that the expression for κ_{diss} retains the form (38.10); however, by $P_{n,n-1}$ should now be understood the effective probability of the particles leaving the level n, including transitions $n \to n'$, downwards ($n' < n$) and upwards ($n' > n$), with a weight proportional to the Boltzmann factor $\exp[-\beta(E_{n'} - E_n)]$.

We will finally mention a simple method of calculating the multiple-quantum excitation, proposed by Rice [440]. First of all, the rate constant of non-equilibrium decomposition is calculated for single-step excitation with an effective separation between levels equal to $\langle |\Delta E| \rangle$. The expression obtained is then summed over all parallel decomposition paths, which leads to its multiplication by the coefficient g_{vib}, equal to the number of vibrational levels in the energy interval $\langle |\Delta E| \rangle$ adjacent to the dissociation limit. For the rate constant of non-equilibrium dissociation, Rice obtained the expression

$$\kappa_{\text{diss}} = Z_0 g_{\text{el}} \frac{\langle |\Delta E| \rangle^{\frac{1}{2}} D^{\frac{1}{2}}}{\hbar \omega} \left(\frac{r_m}{r_e}\right)^2 \exp\left(-\frac{D}{kT}\right) \frac{e^{-a}(1-e^{-a})}{(1+e^{-a})F_{\text{vib}}} \tag{44.23}$$

where
$$a = \langle|\Delta E|\rangle/kT.$$

With the aim of comparing (44.23) with the preceding formulae, we note that, usually, $a \ll 1$, so that to sufficient accuracy we can write

$$\kappa_{\text{diss}} = \frac{1}{F_{\text{vib}}} Z_0 g_{\text{el}} \frac{\langle|\Delta E|\rangle^{\frac{1}{2}} D^{\frac{1}{2}}}{2\hbar\omega} \cdot \frac{\langle|\Delta E|\rangle}{kT} \left(\frac{r_m}{r_e}\right)^2 \exp\left(-\frac{D}{kT}\right). \quad (44.24)$$

The factor $(r_m/r_e)^2$ here has the meaning of the coefficient g_{rot}, occurring in (44.7). It is not difficult to see that Rice's approximation permits a simple interpretation within the limits of diffusion theory. For this, it is sufficient to rewrite (44.14) in the form

$$\kappa_{\text{diss}} \approx \frac{1}{F_{\text{vib}}} Z_0 \frac{\langle\Delta E^2\rangle^{\frac{1}{2}}}{\delta E} \cdot \frac{\langle\Delta E^2\rangle^{\frac{1}{2}}}{kT} \exp\left(-\frac{D}{kT}\right), \quad (44.25)$$

identifying the first factor with g_{vib}.

Recombination of ions

In the construction of a non-equilibrium theory of the recombination of ions in triple collisions, two additional difficulties arise. The first is connected with the dynamical part of the problem relating to the consideration of a binary collision, since, at this stage, an explicit expression for the interaction potential is necessary, which is usually known only very approximately. The second difficulty is with the statistical part, arising in connection with the need to take account of reverse dissociation. If the energy being transferred ΔE is of the order of the relative kinetic energy of the ion pair, then the system of kinetic equations can no longer be reduced to a diffusion equation, and the only simplification which can still be made is that transitions between states differing only in the values of the total energy E are considered.

One of the simplest cases of recombination, discussed by Bates and Moffett [141], corresponds to the recombination of ions in their own gas

$$A^+ + A^- + A \rightarrow A_2 + A. \quad (44.26)$$

The loss of energy of the ion colliding with the atom is due mainly to charge exchange, as a result of which the ion and atom 'change places'. The cross-section for resonant charge exchange depends very weakly on the energy (see § 17). This enables the transition probability $P(E \rightarrow E')$ of the ion pair, which comes into the integral equation determining the non-equilibrium distribution function $x(E)$, to be determined in an explicit form. Finally, the equation itself is solved numerically under the assumption of a stationary current with boundary conditions corresponding to an equilibrium population of the upper levels and an absence of population at some level ε, where the neutralization of the pair due to charge exchange is probable.

$$A^+ + A^- \to A + A. \tag{44.27}$$

The results of calculation show that the rate constant of the process (44.26), taking the reverse dissociation into account, is very close to the equilibrium constant $\kappa_{rec,eq}$ determined by the expression (40.8) if, in the latter, by σ_1 and σ_2 are understood the cross-sections of resonant charge exchange of the positive and negative ions per atom. The ratio $\kappa_{rec}/\kappa_{rec,eq}$ changes within the limits 1·02–0·97 for an increase in ε from $4kT$ to $12kT$, where the value 0·97 should be considered as an asymptotic value which does not change with further increase in ε.

The effect of reverse dissociation on the recombination rate is easier to elucidate if the dynamics of the ion pair is followed, taking account of the perturbing effect of the atom. This can be carried out, of course, only with the help of computers. The final result of such a calculation, reported in the work of Feibelman [224], is the determination of the transition probability of the ion pair from a hyperbolic orbit to an elliptical orbit, which would be stable with regard to reverse dissociation. The criterion for stability is the condition that, after a given number of collisions (N), the 'ion–neutral molecule' energy of the ion-pair bond should increase in comparison with the energy of the pair bond being formed after the first collision.† The probability P of such a process was calculated by the Monte Carlo method, the interaction of the ion–neutral molecule was represented by the interaction of rigid spheres and in the calculation of the trajectory of one ion the Coulomb field of the other was taken into account. It was shown that, already by $N = 3$, the probability goes out to the asymptote, the value of which depends strongly on the impact parameter b. For large b the probability P is very small, however, the rate of decrease of the function $P(b)$ for an increase in b is also small. This fact does not agree with the basic assumption of the Thomson recombination theory, namely, that a certain characteristic radius R^* can be introduced, beyond which the ion pair can be considered dissociated. A calculation for the model system gives recombination coefficients close to the values of Thomson's theory. However, this result is obtained only because of dissociating collisions. If no account is taken of reverse dissociation processes, then Thomson's theory, using a realistic estimate for R^*, gives recombination coefficients exceeding the experimental data by two orders of magnitude.

45. Connection between the rate constants of dissociation and recombination

Effect of relaxation on the reaction rate

Experimentally, the measured rate constants of decomposition and recombination usually refer to the initial chemical process, when the reverse

† The case for $N = 2$ was investigated in a paper by Brueckner [173], but this does not pay due regard to the correlation of the momenta of the ions [224].

reaction can still be neglected. In these conditions, a substantial disturbance of the Boltzmann distribution takes place over the energy levels near the dissociation limit of the diatomic molecules. Therefore the question arises whether the rate constants κ_{diss} and κ_{rec}, characterizing the kinetics of the initial stage, are connected by the relation

$$\frac{\kappa_{\text{diss}}}{\kappa_{\text{rec}}} = K_{\text{eq}}. \tag{45.1}$$

It is not difficult to see that, if the non-equilibrium constants κ_{diss} and κ_{rec} are calculated in the quasi-stationary approximation, then (45.1) is, in fact, satisfied. The two possible solutions to the system of kinetic equations differ only in the normalization of the population, which, in turn, is proportional to the partition function of the bound or dissociated state. Precisely this ratio of the partition functions enters into $\kappa_{\text{diss}}/\kappa_{\text{rec}}$. Hence follows the validity of (45.1). What has been said can easily be verified by the direct integration of the diffusion equation (36.12).

The non-equilibrium distribution function $x(E)$ depends on one arbitrary parameter, which can be chosen so that the boundary conditions corresponding to dissociation (absence of dissociated molecules; total absorption at the upper level) or recombination (absence of bound molecules at the lower level; total absorption at bottom of well) are satisfied. From physical considerations, it is clear that the quasi-stationary description is correct only when the decomposition or recombination is the only slow process not reaching equilibrium. In other words, the characteristic time of the chemical conversion must be much greater than the relaxation time over the vibrational and electronic states of the dissociating molecule or the states formed on recombination of the molecule. At present, there are a large number of papers [66], [102], [308], [417], [437], [440], [441], [454], [547], [548], [550] in which the limits of applicability of the relation (45.1) to specific examples are discussed. In addition, it is certainly necessary to take into account the method by which the rate constants of the direct and reverse processes are measured. The recombination rate, for example, can be measured either by the decrease in the atoms or by the increase in the concentration of molecules in the lowest vibrational levels of the electronic ground state. Generally speaking, these methods give different values for the rates, which, however, will become closer the more the reaction rate decreases in comparison with the relaxation rate. In practice, the process of recombination, measured by the decrease in the concentration of the atoms, will finish before stable molecules in the electronic ground state are formed. The recombination of CO and O in a CO flame is a clear example. The much-discussed cause of the 'latent' energy of the CO flame [11] is probably connected with the difficulty of the transition from the metastable $^3\Pi_g$ state of the CO_2 molecule to the ground state $^1\Sigma_g$. Experimental data indicating the effect of relaxation on the recombination rate are given in the paper by Christie [199]. In this work, the recombination of iodine in various inert gases was investigated. Since the experimental conditions are such that it is impossible to exclude the recombination of iodine atoms on collision with iodine molecules, the general kinetic scheme must be written in the form

$$I+I+C \to I_2+C$$
$$I+I+I_2 \to I_2+I_2. \qquad (45.2)$$

It follows from this scheme that the experimental recombination rate constant $\kappa_{rec,exp}$ must depend linearly on the ratio $[I_2]/[C]$ (at constant partial pressure of C). However, it was shown that, for small values of this ratio, the dependence of $\kappa_{rec,exp}$ is non-linear. This can be explained, for example, if, for small values of the ratio $[I_2]/[C]$, the slowest stage of recombination is some process unaccounted for in (45.2). Christie [199] assumes that the non-linear character of $\kappa_{rec,exp}$ is connected with the slowing down of the vibrational deactivation of the newly-formed molecule I_2^*, due to the reduction in the concentration of I_2^*. This means that the values of κ'_{rec} found from extrapolation (as $[I_2]/[C] \to 0$) of the linear region must be attributed to the recombination reaction $I+I$ in C, on condition that deactivation of I_2^* proceeds by collisions with I_2. On the other hand, extrapolation of the actual curves must give the constant κ''_{rec} for the recombination of $I+I$ and the deactivation of I_2^* by collisions with C. In the transitional region, it is not possible to separate these two processes, therefore relation (45.1) is inapplicable.

Estimates given in the work [66], show that the interpretation of the slow stage as a process of vibrational relaxation is incorrect for the recombination reactions taking place under typical conditions of flash photolysis. It should be assumed that, in the case of the recombination of iodine under consideration, the non-equilibrium process is connected with the slowing down of the transition between the excited and electronic ground states of I_2. The great difference between the effective relaxation of I_2^* in collisions with I and with atoms of the inert gas is explained thus: on formation of the complex $I_2^*-I_2$, the electronic terms of the combined system differ so much from the terms of two isolated I_2 molecules that the probability of nonadiabatic transitions between those terms which approach each other closely in the complex $I_2^*-I_2$ is very large. No marked change in the position of the electronic terms in the interaction of I_2^* with an atom of an inert gas is to be expected.

If the expression for the rate constant of vibrational relaxation and dissociation is formally extrapolated to high temperatures, then conditions are reached such that $1/\tau_{vib} \sim \kappa_{diss}$ or even $1/\tau_{vib} < \kappa_{diss}$ [569]. The latter relation contradicts the obvious physical requirement that the rate of the subsequent process (dissociation) may not exceed the rate of one of the stages (relaxation).

This contradiction is connected with the violation of the conditions under which the quasi-stationary description of the reaction is possible.

The slowing-down effect of the vibrational relaxation on the decomposition rate is discussed in a number of papers, under different assumptions with regard to the dependence of the dissociation probability on the quantum numbers of the vibrational level [170], [255], [277], [454], [526].

In the dissociation of diatomic molecules in an inert gas such a slowing-down of the effective rate arises as the temperature increases, when the constants κ_{diss} and $1/\tau_{vib}$, extrapolated into the region of high temperatures, approach one another. Dissociation begins to be overlapped by relaxation; the two processes cannot be separated so that the individual constants κ_{diss} and $1/\tau_{vib}$ lose their meaning. At the same time, the mean decomposition time $\bar{t}$ increases in comparison with the value which would be obtained if the overlap of the two processes were neglected. If $\bar{t}$ is identified with an effective rate constant κ_{eff} then, in comparison with κ_{diss}, extrapolated from the low-temperature region, this effective constant will contain an additional negative temperature-dependence.

In the dissociation of a pure diatomic gas, an additional (in comparison with that mentioned above) effect of relaxation on the dissociation process is possible. The rate constant of a diatomic gas depends not only on the temperature T of the translational motion but also on the nature of the non-equilibrium vibrational distribution function of the lower vibrational states, since the states of one of the partners in the collision may contribute to the decomposition of the other partner. As the simplest assumption, it may be taken that the vibrational distribution function is characterized by some temperature $T_{vib} \neq T$ and that κ_{diss} depends only on T_{vib}. Such a dependence is assumed, for example, within the framework of Hoare's model [267]. For such a model dissociation inevitably gives rise to a cooling of the vibrational heat reservoir, whose temperature is reduced as a result of vibrational-translational energy exchange. This process is described by the following equations, proposed by Kuznetsov [34]:

$$\frac{d\langle E_{vib}\rangle}{dt} = \frac{\langle E_{vib}\rangle - E_{vib,eq}}{\tau_{vib}} + (D - E_{vib})\frac{1}{\alpha}\frac{d\alpha}{dt},$$

$$\frac{d\alpha}{dt} = -\kappa_{rec}\alpha\delta + \kappa_{diss}(1-\alpha)^2\delta^2, \tag{45.3}$$

where $\langle E_{vib}\rangle$ and $E_{vib,eq}$ are the values of the vibrational energy at temperatures T_{vib} and T; α is the fraction of undissociated molecules; δ is the ratio of the density of the gas to its density under normal conditions.

Investigation of the system of equations (45.3) shows that the rate constant $\kappa_{diss}(T_{vib})$ begins to differ appreciably from $\kappa_{diss}(T)$ for $T > T_0$, where T_0 is determined by the equation

$$\kappa_{\text{diss}}(T_0)\left(\frac{D}{kT_0}\right)^2 \sim \frac{1}{\tau_{\text{vib}}(T_0)}. \qquad (45.4)$$

For $T > T_0$ an additional negative temperature-dependence appears in the expression for κ_{diss}; however, κ_{diss} still preserves the meaning of a true rate constant. Therefore, it should be expected that relation (45.1) will still be fulfilled in these conditions.

An investigation of the effect of relaxation on the rates of dissociation and recombination, carried out on the basis of other assumptions about the form of the elementary transition probabilities, is to be found in papers [41], [250], [515]. Sometimes, it turns out to be necessary to take into account the change in the temperature of the translational motion, which has been completely neglected in all foregoing considerations.

46. Thermal dissociation of oxygen

The thermal decomposition of oxygen in shock waves is one of the most fully-studied dissociation processes. The set of experimental data on the effectiveness of various molecules and, in particular, O_2, Ar, and O, in the activation of O_2, [13], [41], [187], [244], [567] may serve as a test of the theory concerning intermolecular energy transfer with participation of strongly vibrationally-excited molecules. Unfortunately, precisely this side of the problem is, at present the least developed. Therefore we will restrict ourselves to a discussion of just the simplest process,

$$O_2 + \text{Ar} \rightarrow O + O + \text{Ar}, \qquad (46.1)$$

for which the model of single-quantum excitation may serve as a satisfactory approximation [309]. It appears that the most exact measurement of the rate of process (46.1) was carried out by Wray [567]–[569], who studied the decomposition of O_2 in shock waves. In the temperature range 5000–10 000 K, κ_{diss} is described by the formula

$$\kappa_{\text{diss}} = 2 \cdot 5 \times 10^{16} T^{-\frac{1}{2}} \exp\left(-\frac{D}{kT}\right) \text{ cm}^3 \text{ mol}^{-1} \text{ s}^{-1}, \qquad (46.2)$$

$$D = 118 \text{ kcal mol}^{-1}.$$

The same experiments are described to the same exactness by other formulae

$$\kappa_{\text{diss}} = 2 \cdot 4 \times 10^{18} T^{-1} \exp\left(-\frac{D}{kT}\right), \qquad (46.3)$$

$$\kappa_{\text{diss}} = 10^{15} \exp\left(-\frac{D}{kT}\right)\left[1 - \exp\left(-\frac{\hbar\omega}{kT}\right)\right], \quad \frac{\hbar\omega}{k} = 2200 \text{ K}, \qquad (46.4)$$

the latter agreeing with the theoretical formula for the model of single-quantum stepwise excitation (see § 44). Expression (46.4) can be used to extrapolate κ_{diss} to low temperatures and for the calculation of κ_{rec} by relation (45.1). Returning to the equilibrium constant K_{eq}, given in the work of Wray [569], we obtain $\kappa_{\text{rec}} = 3.5 \times 10^{13} \text{ cm}^6 \text{ mol}^{-2} \text{ s}^{-1}$ for $T = 2000$ K, which should be compared with the experimental value of $2.5 \times 10^{13} \text{ cm}^6 \text{ mol}^{-2} \text{ s}^{-1}$ obtained by Wray.

For $T = 300$ K the calculation according to (46.4) and (45.1) gives $\kappa_{\text{rec}} = 1.4 \times 10^{14} \text{ cm}^6 \text{ mol}^{-2} \text{ s}^{-1}$, which agrees with the estimate of Morgan and Schiff [378], $\kappa_{\text{rec}} < 3.2 \times 10^{14} \text{ cm}^6 \text{ mol}^{-2} \text{ s}^{-1}$. We see that the theory which predicts the temperature dependence of κ_{diss}, enables a sufficient extrapolation to be carried out to calculate κ_{rec} (or κ_{diss}) satisfactorily from κ_{diss} (or κ_{rec}).

We will now consider the problem of the degree to which the model of single-step excitation agrees with the absolute value of A. At $T = 10\,000$ K, for the pre-exponential A in (46.2) we have

$$A = 24 \times 10^{13} \text{ cm}^3 \text{ mol}^{-1} \text{ s}^{-1}. \tag{46.5}$$

The theoretical expression for A has the form

$$A = \frac{1}{F_{\text{vib}}} Z_0 g_{\text{el}} g_{\text{rot}} \frac{\Delta E}{kT}. \tag{46.6}$$

The factor g_{rot} lies in the range 5–10. The factor g_{el} equals

$$g_{\text{el}} = \frac{g(^3\Sigma_g^-) + g(^1\Delta_g) + g(^1\Sigma_g^+) + g(^3\Delta_u) + g(^3\Sigma_u^+) + g(^1\Sigma_u^-)}{g(^3\Sigma_g^-)} = \frac{16}{3}. \tag{46.7}$$

The number of collisions Z_0 at $T = 10\,000$ K is close to $3 \times 10^{14} \text{ cm}^3 \text{ mol}^{-1} \text{ s}^{-1}$. The partition function F_{vib} at this temperature equals 4·5. Thus the ratio $\Delta E/kT$ equals 0·1–0·2, which agrees in order of magnitude with the estimate of this ratio in expression (44.1).

Thus the decomposition of oxygen in argon is apparently an example where the theory enables the rate constant to be estimated and its temperature dependence predicted satisfactorily.

If, instead of reaction (46.1), we consider the process

$$O_2 + O_2 \rightarrow O + O + O_2 \tag{46.8}$$

then first it is necessary to take into account the contribution of the additional degrees of freedom of O_2 (in comparison with Ar) to the decomposition rate constant.

On the basis of qualitative considerations, it might be thought that part of the internal energy of the undissociated molecule O_2 is transferred to the decomposing molecule, lowering the energy E^* corresponding to the limiting stage. If just the rotational energy is taken into account and its contribution is calculated on the basis of criterion (44.1), then the experimentally-observed increase in the dissociation rate in process (46.8), in comparison with the rate of the process (46.1), can be qualitatively explained [60]. The difference in the rates amounts to an order of magnitude, but does not agree completely with the experimental data; the ratio $\kappa_{diss}(O_2—O_2)/\kappa_{diss}(O_2—Ar)$ is reduced from 30–40 to 5–10 in the temperature range 3500–7000 K (see [114], § 18), whereas the estimate in paper [60] gives a value independent of temperature for this ratio. It is possible that this discrepancy is connected with disregarding the contribution of the vibrational degrees of freedom of the undissociated O_2 molecule.

Finally, in passing from Ar and O_2 to O, the situation becomes even more complicated. Firstly, the nonadiabatic effects and the effect of the displacement of the vibrational levels of O_2 beyond the dissociation limit can play a large role here. Second, to some extent an exchange reaction of the type (43.5) may contribute to the decomposition rate. In particular, if the rotational energy of the molecule O'O" is converted, with large probability, into the vibrational energy of the reformed O"O''' molecule, then this may lead to an increase in the observed decomposition rate of O_2 owing to its escape from the metastable region on the phase diagram shown in Fig. 35 (arrow 5).

8

BIMOLECULAR REACTIONS

47. Exchange as a bimolecular reaction

BY bimolecular reactions, we will understand chemical transformations accompanied by a change in the structure of the two molecules or ions participating in the reaction. The most general bimolecular reaction can be represented in the form of the following very simple scheme

$$X + Y \to Z + U, \qquad (47.1)$$

where X and Y are the molecules of the reactants; Z and U are the molecules of the reaction products.

In a particular case, it may turn out that only one kind of molecule arises in the reaction. Then process (47.1) describes the recombination of atoms, ions, or radicals. Recombination reactions are the reverse of dissociation reactions, which, in turn, are regarded as unimolecular reactions (see Chapter 4). Therefore the recombination rate constants can be calculated from the rate constants of decomposition in terms of the equilibrium constant, if, of course, the conditions specified in § 45 are satisfied.

In this chapter, reactions will be considered in which the molecules arising, Z and U, differ from the initial molecules, X and Y, so that the direct and reverse processes are bimolecular reactions.

In accordance with the definition given above, bimolecular reactions are not necessarily characterized by a second-order law. In this sense, an analogy can be developed with unimolecular reactions whose kinetic equations, also, do not necessarily correspond to first-order reactions. This is connected with the fact that the reaction rate is determined not only by the redistribution rate of the particles, but also by the energy exchange rate between the reacting molecules X and Y with the molecules M, found throughout the volume of the reacting gas.

In the theory of bimolecular reactions presented, it is expedient to preserve, as far as possible, the analogy between reactions of this type and unimolecular reactions. This is dictated not only by considerations of the convenience of a methodical approach but also by the physical similarity of the processes resulting from the interaction of the molecules X and Y in reaction (47.1) and the molecules AB and C in the unimolecular reaction. In this connection, it is convenient to introduce the idea of an *active* pair $X + Y$, by which we will understand a pair of molecules, X and Y, whose energy exceeds some value E_a and is sufficient for the reaction (47.1) to take place with appreciable probability. In addition, it is assumed that, during the entire time of the elementary process (47.1), the isolated pair $X+Y$ does not interact with the remaining molecules of the gas and that the development of the system $X+Y$ in time is completely determined by its Hamiltonian.

It is natural that, in the definition of the activation energy E_a given above, the limitations arising in connection with the neglect of tunnel corrections (see § 3) should be taken into account.

Considering the reaction (47.1) as the motion of a representative point in phase space, the reaction rate dn_X/dt can be written in the form of a current across a certain critical surface S^*. If, for this, a sphere is chosen enclosing the region of interaction of the molecules X and Y (or Z and U), then dn_X/dt is expressed in terms of the reaction cross-section $\sigma_{if}(v)$:

$$\frac{dn_X}{dt} = -(n_X n_Y) \sum_{if} f_{X-Y}(i) \int \sigma_{if}(v) v f_{X-Y,\text{trans}}(v) \frac{4\pi p^2 \, dp}{(2\pi\hbar)^3}$$
$$+ (n_U n_Z) \sum_{fi} f_{U-Z}(f) \int \sigma_{fi}(v) v f_{U-Z,\text{trans}}(v) \frac{4\pi p^2 \, dp}{(2\pi\hbar)^3}, \qquad (47.2)$$

where, for example, $f_{X-Y}(i)$ denotes the distribution function of the pair $X+Y$ for the internal states; $f_{X-Y,\text{trans}}(v)$ is the distribution function for the relative velocities; the summation over i means the average over all initial states of the colliding pair; summation over f takes account of all reaction channels differing in the set of quantum numbers f of the non-interacting molecules Z and U.

In relation (47.2), f denotes the non-equilibrium distribution function, which, generally speaking, depends on the concentrations of the molecules. If conditions are such that f_{X-Y} and f_{Z-U} do not depend on the instantaneous concentrations n_X, n_Y and n_Z, n_U, then (47.2) leads to the simple kinetic equation

$$\frac{dn_X}{dt} = -\kappa_1 n_X n_Y + \kappa_{-1} n_Z n_U, \qquad (47.3)$$

where κ_1 and κ_{-1} may be interpreted as rate constants of the exchange reaction in the forward and reverse directions of (47.1).

It is clear, from what has been said, that, in the calculation of the reaction rate of (47.1), the following main problems arise, namely 1 and 2.

1. An investigation of the dynamics of the binary collision of molecules, with the aim of calculating the reaction cross-section. One result of this study is an elucidation of the distribution of energy between the different degrees of freedom of the molecules under formation and the angular distribution of the molecules.

2. Determination of the non-equilibrium distribution function of the reacting molecules and clarification of the conditions under which the relation (47.2) leads to the equation of formal chemical kinetics (47.3).

To solve these two problems, knowledge of the adiabatic electronic terms of the system of interacting atoms is necessary. Shortage of information in this respect is particularly acute precisely for exchange reactions. In the case of unimolecular reactions, the basic parameter of the theory, the activation energy, can usually be estimated on the basis of the thermodynamic data, since the reaction path, as a rule, does not go through a potential barrier. For the dissociation of diatomic molecules, the situation is even better: here, the decomposition energy is known from independent spectroscopic measurements.

For bimolecular reactions, a typical situation is that the reaction path goes through a potential barrier, whose height E_0 enters as a parameter into the expression for the rate constant κ. A determination of E_0 from the experimental temperature dependence of $\kappa(T)$ runs into the difficulty that the variation of the rate constant with temperature is determined not only by the Arrhenius factor $\exp(-E_0/kT)$ but also by the pre-exponential factor. The dependence of the pre-exponential factor on temperature is determined by the dynamical details of the elementary process and the degree of violation of the Maxwell–Boltzmann distribution in the reaction process. Thus, we again return to the two points mentioned above and the necessity of constructing a general theory. The parameters in the theory can be determined by comparison with experiment only on condition that the general problem of calculating $\kappa(T)$, with regard to the dynamical and statistical aspects, is solved theoretically. Since, at present, the investigation of these problems has only just begun, obtaining independent information with regard to the basic features of the potential energy surfaces is extremely important for the theory. In this connection, the construction of the adiabatic terms of the simplest exchange reactions is discussed in some detail (see § 48), although this topic lies somewhat outside the general sphere of problems under consideration.

Important information about the dynamics of an elementary process can be obtained from the study of exchange reactions in crossed molecular beams or hot atom reactions. At present, these methods enable the angular distribution, the distribution of velocities, and the distribution of internal energies to be studied accurately. However, we will not consider these problems, but will limit ourselves to reactions taking place under thermal conditions. As regards the theory of reactions in molecular beams, the review of Herschbach can be recommended to the reader [259], [260].

48. The potential energy surfaces of bimolecular reactions

In this section, we will consider the qualitative features of the potential energy surfaces of the simplest exchange reaction in which three atoms take part,

$$A + BC \to AB + C, \qquad (48.1)$$

dwelling briefly on the theoretical and semi-empirical methods of calculation.

The London formula

The calculation of the potential energy surface for a system of three atoms is a many-electron problem in quantum mechanics. The most exact solution to such problems, at present, can be obtained by the variational method. We will assume that, in the valence shell of the atoms taking part in the reaction (48.1), there is one electron apiece and that a certain three-electron coordinate function $\psi(r_1, r_2, r_3)$ is chosen for a variational calculation of the electronic term. The Pauli principle requires that the function ψ possess definite symmetry properties with respect to a permutation of the spatial coordinates of the electrons.† On the basis of the general theory for the construction of spatial functions [28], [108], [123], [360] from an arbitrary function, functions corresponding to a given value of the total spin of the system of electrons (here, and in future, we neglect spin–orbit coupling) can be constructed. Since we are interested in the potential energy surface which correlates on an increase in the intermolecular separation with the terms of a free diatomic molecule and of an atom in the electronic ground state, the multiplicity of ψ is unambiguously determined by the value of the spin to which the total wavefunction of the two non-interacting systems corresponds. We will assume that the molecule BC is in the state $^1\Sigma$ and the atom A is in the state ^2S, so that the total spin $S = \frac{1}{2}$. In this case, there exist two linearly-independent functions $^2\psi_1$ and $^2\psi_2$, which are built up from the non-symmetric function $\bar{\psi}$ by applying the coordinate exchange operators of the electrons to it,

$$^2\psi_1 = N_1(1 - P_{13})(1 + P_{12})\bar{\psi},$$
$$^2\psi_2 = N_2(1 - P_{13})(1 + P_{23})\bar{\psi}, \qquad (48.2)$$

where N_1 and N_2 are normalizing factors.

The general doublet function $^2\psi$ is constructed in the form of a linear combination

$$^2\psi = A_1^2\psi_1 + A_2^2\psi_2, \qquad (48.3)$$

---

† Under interchange of spin and spatial coordinates, the total wavefunction, of course, changes sign.

where the coefficients A_i are found from the variational principle. (We will not consider here the possible variational parameters coming into $\bar{\psi}$.)

The standard solution of the variational problem for two states leads to the following equation for the adiabatic electronic terms:

$$\begin{vmatrix} H_{11} - U & H_{12} - S_{12}U \\ H_{21} - S_{21}U & H_{22} - U \end{vmatrix} = 0 \tag{48.4}$$

where

$$H_{ik} = \int {}^2\psi_i^* H_e {}^2\psi_k \, d\tau, \quad S_{12} = \int {}^2\psi_1^* \psi_2 \, d\tau.$$

The solution of (48.4) gives

$$U_{1,2} = \frac{H_{11} + H_{22} - 2H_{12}S_{12}}{2(1 - S_{12}^2)} \pm \frac{[(H_{11} - H_{22})^2 - 4H_{12}S_{12} + 4H_{12}^2(1 - S_{12}^2)]^{\frac{1}{2}}}{2(1 - S_{12}^2)}. \tag{48.5}$$

Here

$$H_{11} = N_1 2(2Q + 2\alpha - \beta - \gamma - 2\delta),$$
$$H_{12} = H_{21} = (N_1 N_2)^{\frac{1}{2}} 2(-Q - \alpha + 2\beta - \gamma + \delta),$$
$$H_{22} = N_2 2(2Q - \alpha - \beta + 2\gamma - 2\delta),$$
$$S_{12} = (N_1 N_2)^{\frac{1}{2}} 2(-q - a + 2b - c + d), \tag{48.6}$$

and the factors N_1 and N_2 are determined by the relations:

$$N_1 2(2q + 2a - b + c - 2d) = 1,$$
$$N_2 2(2q - a - b + 2c - 2d) = 1. \tag{48.7}$$

$Q, \ldots, \delta$ and $q, \ldots, d$ denote the integrals

$$Q = \int \bar{\psi}^* H_e \bar{\psi} \, d\tau \qquad q = \int \bar{\psi}^* \bar{\psi} \, d\tau$$

$$\alpha = \int \bar{\psi}^* H_e P_{12} \bar{\psi} \, d\tau \qquad a = \int \bar{\psi}^* P_{12} \bar{\psi} \, d\tau$$

$$\beta = \int \bar{\psi}^* H_e P_{13} \bar{\psi} \, d\tau \qquad b = \int \bar{\psi}^* P_{13} \bar{\psi} \, d\tau$$

$$\gamma = \int \bar{\psi}^* H_e P_{23} \bar{\psi} \, d\tau \qquad c = \int \bar{\psi}^* P_{23} \bar{\psi} \, d\tau$$

$$\delta = \int \bar{\psi}^* H_e P_{123} \bar{\psi} \, d\tau \qquad d = \int \bar{\psi}^* P_{123} \bar{\psi} \, d\tau. \tag{48.8}$$

We will consider the London approximation [126], [336], with the aim of investigating qualitatively the potential energy surface of the doublet state, as defined by formula (48.5). In this approximation, the electronic function of the system of three atoms is approximated by a product of three one-electron functions, referred to different centres, $\bar{\psi} = a(1)b(2)c(3)$. Moreover, it is assumed that the nonorthogonality of the atomic orbitals and triple exchanges may be neglected:

$$\delta = 0, \qquad a = b = c = d = 0. \tag{48.9}$$

In this case, the exchange of electrons is equivalent to an exchange of the orbitals, so that the parameters α, β, and γ take on the meaning of exchange integrals between pairs of atoms A–B, B–C, and A–C, and Q separates into a sum of Coulomb integrals of the three pairs. As for $^2\psi_1$ and $^2\psi_2$, they agree with the coordinate functions of the valence structures describing the chemical bond of A–B and B–C. For a system of three atoms the energy depends on three variables, determining the relative position of the atoms. Therefore, for a qualitative investigation of the properties of this function, it is convenient, on fixing the value of one of the variables, to represent U in the form of a surface in a three-dimensional space in a system of coordinates in which the energy lies along the z axis.

Expression (48.5), taking account of the assumptions (48.9), has the form

$$U_{1,2} = Q \pm \{\tfrac{1}{2}(\alpha-\beta)^2 + (\beta-\gamma)^2 + (\gamma-\alpha)^2]\}^{\tfrac{1}{2}}. \tag{48.10}$$

If the interaction of the structures is not taken into account, i.e. the energies of U_1 and U_2 are calculated as mean values of H_e with the functions $^2\psi_1$ and $^2\psi_2$, then we arrive at the expressions:

$$U_1 = Q + \alpha - \frac{\beta}{2} - \frac{\gamma}{2},$$

$$U_2 = Q + \gamma - \frac{\alpha}{2} - \frac{\gamma}{2}. \tag{48.11}$$

The same result can also be obtained from (48.10) if the radical is formally expanded in a series with the condition $\beta, \gamma \ll \alpha$ (preferable contribution of structure $^2\psi_1$) or $\alpha, \beta \ll \gamma$ (preferable contribution of structure $^2\psi_2$).

We will first consider the case of a linear arrangement of the three atoms, and will introduce as coordinates the internuclear distances R_{AB} and R_{BC}. We will fix the value R_{BC} close to the equilibrium position of the molecule BC and will reduce R_{AB}, drawing the atom A nearer to the molecule BC. At sufficiently large distances R_{AB}, the total energy can be calculated with the use of one structure's wavefunction $^2\psi_2$ and will be equal to H_{22}. For $R_{AB} \sim R_{BC}$, it is necessary to take into account the interaction of the configurations, i.e. to be based on solution (48.10). This means that the correct

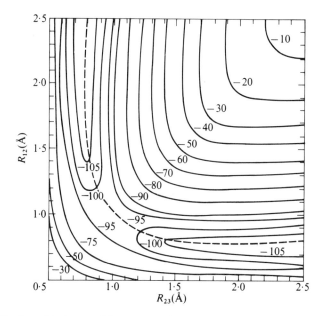

Fig. 41. Semi-empirical potential energy surface of the linear H_3 system (Cashion and Herschbach [194]): R_{12} and R_{23} are distances between protons; the energy is expressed in kcal mol^{-1}.

function ψ in this region is represented in the form of a linear combination of the functions of the two structures with approximately the same weight.

We will now increase R_{BC}, leaving R_{AB} constant. The coefficients A_1 and A_2 of the linear combination (48.3) will be reduced in such a way that asymptotically, as $R_{BC} \to \infty$, the function ψ will tend to $^2\psi_1$. This follows from the fact that, as the atom A goes off to infinity, the general state of the three particles can once again be described by the wavefunction of one structure. Thus it is evident that following the path $R_0 < R_{AB} < \infty$, $R_{BC} = R_0$, $R_0 < R_{BC} < R_\infty$, $R_{AB} = R_0$, on the plane of R_{AB}, R_{BC}, the wavefunction of one structure goes over continuously into the wavefunction of the other structure. This is accompanied by the particle exchange $A + BC \to AB + C$, where the multiplicity of the electronic state of the final molecule agrees with the multiplicity of the state of the initial molecule (both states are singlet $^1\Sigma$). A study of various paths of this type enables the potential energy surface to be constructed, which is usually represented by level curves (Fig. 41). The essential feature of this surface is that on it there is a saddle region from which a slope leads to the valley of the reactants and reaction products.

It is not difficult to explain the causes leading to such a structure for the potential energy surface. If the functions $^2\psi_1$ and $^2\psi_2$ are used independently in the calculation of $U_{1,2}$, then we will obtain two surfaces, each of which corresponds to the lowest electronic term either in the region of the reactants or in the region of the reaction products. These two surfaces intersect in some range of variation of the internuclear distances which corresponds to a sufficiently-close approach of the three atoms. If account is taken of the interaction between the states $^2\psi_1$ and $^2\psi_2$ (i.e. a wavefunction is sought for in the form of (48.3)) then, generally speaking, the intersection is replaced by a convergence of the adiabatic terms. On condition that the minimum splitting $\Delta U_{\min}$ is sufficiently large, in accordance with the Massey criterion, the upper term can be neglected and the motion of the atom considered over the lower surface. In its dependence on the nature of the terms in the zero-order approximation and the off-diagonal elements of the interaction, the region of closest approach may correspond to a potential barrier of the saddle type (as occurs in the $H_2 + H$ system) or may correspond to the start of a steeper slope or rise (as takes place for reactions with large thermal effects).

Returning to the investigation of an energy surface with a saddle, we will show that, in the space of the three coordinates R_{AB}, R_{BC}, R_{AC}, paths exist along which the profile of the potential energy surface has an essentially different form. To do this, we will consider the potential energy surface for a fixed internuclear distance $R_{BC} = R_0$. The position of atom A relative to the molecule BC is described by the coordinates R_{AC} and R_{AB}. By a variation of the two variables we can strive to make the expression under the square root in (48.10) tend to zero. The corresponding values of the variables are found from the equations

$$\alpha = \beta,$$
$$\beta = \gamma. \qquad (48.12)$$

The reduction of the square root to zero at one point means that the two surfaces U_1 and U_2 are in contact at that point. In the neighbourhood of a general point of contact, the surfaces have the form of double cones (see § 16). Any path in the R_{AC} and R_{AB} coordinates which passes through

$$R_{AC} = R_{AB} = R_C$$

takes the system from the lower potential energy surface to the upper, and the energy of the system along this path will be increasing all the time. Thus, we see that it is possible to go from one surface to the other by an adiabatic path. Such a potential energy surface for the system of three atoms of hydrogen is shown qualitatively in Fig. 42. Atoms B and C are fixed on the y axis on different sides of the x axis, and atom A moves in the xy plane. As the representative point on the y axis approaches the origin of coordinates, (atom A approaches the molecule BC), to begin with, the energy, which is determined by the effect of the activation barrier of the reaction $A + BC \rightarrow$

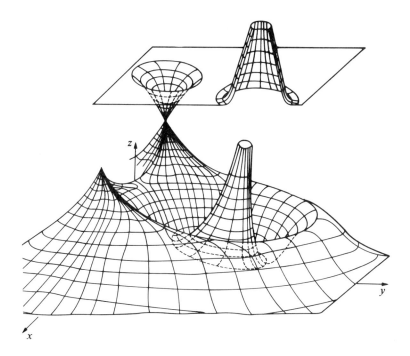

Fig. 42. Qualitative form of the potential energy surface of the non-linear H_3 system.

AB + C, increases. Then, after the saddle and a small drop, the energy once again increases rapidly, as a result of the Coulomb repulsion of the nuclei A and B. This increase is completely analogous to the increase in potential energy on the strong convergence of the nuclei in diatomic molecules. By a displacement from the y axis, the peak can be by-passed and the linear configuration C...A...B again reached when the atom A turns out to be 'embedded' in the BC molecule. If the initial fixed distance R_0 is sufficiently large, then a surface of the saddle type will again correspond to this configuration. However, in contrast to the previous example, the motion over the surface along the path of steepest ascent transfers the representative point to the other sheet of the adiabatic surface.

Since, for the given case, the number of dimensions of the manifold in which the contact of the surfaces takes place is two less than the number of dimensions of the manifold of the coordinates (intersection in a point for a manifold of two coordinates and intersection in a line for a manifold of three coordinates), it might be thought that the total contribution of processes describing the motion along a trajectory which goes over adiabatically from one sheet to another is negligibly small. Nevertheless it is not so: the dimension of the effective transition region, with regard for the nonadiabaticity, agrees with the dimension of the manifold of the coordinates, so that transitions of a similar type make a finite contribution to the general effect (see § 16).

The position of the point of intersection is easy to establish in the case of three identical atoms, since here, in addition, we can take advantage of symmetry considerations. It is not difficult to show [360] that, in the configuration of an equilateral triangle (symmetry group D_{3h}), the functions $^2\psi_1$ and $^2\psi_2$ form the basis of a doubly-degenerate irreducible representation E, so that, at the point $R_{AB} = R_{BC} = R_{AC}$, the equality $U_1 = U_2$ must be fulfilled. For the motion of the representative point over the surface U along the x axis, the symmetry of the framework C_{2v} is conserved the whole time, so that the adiabatic electronic terms can be classified in accordance with the irreducible representations of this group. In the passage of the point through the vertex of the cone, the symmetry of the function corresponding to the lower sheet of the potential energy surface changes: for $x > R_0/2$ it corresponds to the representation B_2 and for $x < R_0/2$ to the representation A_1. This change of symmetry of the function reflects the intersection of the terms and is connected with the change of sign of the function on passing round the vertex of the cone, as mentioned earlier (see § 14). Different projections of the hypersurface $U(R_{AB}, R_{BC}, R_{AC})$ onto the usual three-dimensional space, given in Figs. 41 and 42, enable the general qualitative behaviour of this function of three variables to be represented.

We will now consider the question of which electronic states of the triatomic system correspond to the upper sheet U_2. For example, let $R_{AB} \to \infty$ and $R_{BC} = R_0$. Then, as was shown earlier, the wavefunction which can be formed from the wavefunctions of atom A in the ground state and the molecule BC also in the ground state corresponds to the lower sheet. The wavefunction corresponding to the upper sheet must be orthogonal to the lower sheet's wavefunction. But such orthogonality at an infinite distance R_{AB} can be achieved only at the expense of the orthogonality of the molecular wavefunctions of CB. The excited triplet state of the CB molecule is the only state, in the chosen approximation, orthogonal to the singlet ground state. Thus it is clear that the reaction path going through the vertex of the cone ultimately changes two of the three atoms taking part in the reaction from the singlet state into the triplet. Moreover, such a transition can take place with or without a redistribution of the particles—this depends on the direction of the path near the vertex of the cone. As regards the electronic function, asymptotically, as one of the atoms goes off to infinity it will be expressed by the linear combination (48.3) with coefficients A_1 and A_2 not equal to zero.

In those cases where the redistribution of particles takes place at sufficiently large distances, the exchange interaction in the newly-formed molecule can be neglected, and the general formula (48.5), in a sense, simplified. Such a situation arises, for example, in the reactions of alkali metal atoms M with molecules of haloids X_2, when the interaction in the molecule, MX, which arises, can be described at large distances by a Coulomb potential corresponding to the ion-pair M^+X^-. In this case, it is expedient to look for the variational function $\bar{\psi}$ in the form of a linear combination of covalent and ionic functions:

$$\bar{\psi} = \psi_c + \lambda\psi_i = a(1)b(2)c(3) + \lambda d(1)b(2)c(3), \qquad (48.13)$$

where $d(1)$ is some one-electron function describing the localization of the external electron 1 on the molecule BC, in accordance with the assumed reaction mechanism

$$A + BC \rightarrow A^+ + (BC)^- \rightarrow A^+B^- + C \rightarrow AB + C. \qquad (48.14)$$

In the approximation (48.13), the potential energy surface U_1 can be represented in a form analogous to (48.5), where minimization with respect to the parameter λ must be carried out. This minimization, in turn, includes solution of a quadratic equation. If the off-diagonal element between the covalent and ionic functions is neglected, then we obtain approximately

$$U_1 = \min \begin{cases} \langle\psi_c|H_{el}|\psi_c\rangle \sim U_{BC}(R_{BC}) \\ \langle\psi_i|H_{el}|\psi_i\rangle \sim -e^2/R_{A-BC} + A_{BC} - I_A + U_{\overline{BC}}(R_{BC}), \end{cases} (48.15)$$

where U_{BC} and $U_{\overline{BC}}$ are the potentials of the neutral BC molecule and the corresponding negative ion; A_{BC} is the electron affinity of the BC molecule; I_A is the ionization potential of the atom A.

A characteristic feature of such a type of potential energy surface is that the slowly-decreasing Coulomb interaction in one of the valleys causes a very large channel width. Along the line of intersection determined by the condition $\langle\psi_c|H_{el}|\psi_c\rangle = \langle\psi_i|H_{el}|\psi_i\rangle$, the surface has a dislocation, from one side of which the steep descent begins (Fig. 43).

Method of Eyring and Polanyi

In the approximation under which London's formula (48.10) was derived it is possible to determine the Coulomb and exchange integrals, in terms of the known energies of the ground and excited states of the three pairs of atoms, for all internuclear distances. The terms of the ground states are retrieved from spectroscopic data. The excited terms are not bound and are therefore unknown, as a rule. Attempting to circumvent this difficulty, Eyring and Polanyi proposed [222] that, at all internuclear distances, the ratio ρ of the Coulomb integral to the exchange integral be a constant, equal to the value obtained for this ratio near the point of equilibrium for the

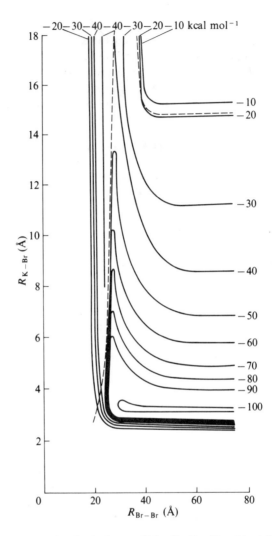

FIG. 43. Potential energy surface for the linear collision K + Br$_2$ (Herschbach [260]). The broken curve shows the dependence of the radius of intersection of the covalent and ionic terms on a change in the distance between the bromine atoms. (Energy is in kcal mol^{-1}.)

stable molecule. This assumption is not well founded physically; however, it is widely used to evaluate the height of potential barriers.

Method of diatomic complexes in molecules

This method originates in London's formula (48.10); however, all the parameters in (48.10) are determined on the basis of data referring to terms of the ground and excited states of each pair of atoms. At present the antibonding terms of only the simplest system—a molecule of hydrogen—are known with sufficient accuracy. Therefore in future, in considering this method, we will have in mind the potential energy surface for the reaction $H_2 + H \to H + H_2$. The potential energy surface for the linear configuration of the nuclei in coordinates R_{AB}, R_{BC} is symmetrical with respect to the diagonal $R_{AB} = R_{BC}$. Therefore, to determine the height of the barrier, it is sufficient to consider the cross-section of this surface along the straight line $R_{AB} = R_{BC}$. In this case, it should be assumed that $\alpha = \beta$ in (48.10), so that U_1 takes the form

$$U_1(R_{AB}) = \tfrac{3}{2} U(R_{AB}, {}^1\Sigma) + \tfrac{1}{2} U(R_{AB}, {}^3\Sigma) + U(2R_{AB}, {}^3\Sigma). \qquad (48.16)$$

The energy U_1 in (48.16) reaches a minimum for $R_{AB} = R_s$. This point corresponds to the saddle. To investigate the potential energy surface near the saddle, U_1 can be expanded in a series in powers of the displacement $R - R_s$:

$$U_1 = E_0 + \tfrac{1}{2} k_r (\Delta R_{AB}^2 + \Delta R_{BC}^2) + k_{rr} \Delta R_{AB} \Delta R_{BC} + \tfrac{1}{2} k_\gamma (\Delta \gamma)^2, \qquad (48.17)$$

where γ is the angle between the AB and BC bonds; k_i are force constants.

On the basis of the London formula it is not difficult to obtain an expression in which the force constants are connected with the potentials of the pair of atoms by the relations

$$k_r + k_{rr} = \tfrac{3}{4} U''(R_s, {}^1\Sigma) + \tfrac{1}{4} U''(R_s, {}^3\Sigma) + 2U''(2R_s, {}^3\Sigma), \qquad (48.18)$$

$$k_{rr} = U''(2R_s, {}^3\Sigma) + \frac{\tfrac{3}{4}[\alpha'(R_s)]^2}{[\alpha(2R_s) - \alpha(R_s)]},$$

$$\frac{k_\gamma}{R_s^2} = -\frac{U'(2R_s, {}^3\Sigma)}{2R_s},$$

where α is the exchange integral defined by the expression

$$\alpha = \tfrac{1}{2}[U({}^3\Sigma) - U({}^1\Sigma)].$$

Neglect of the overlap integrals in the derivation of this formula, in fact, does not introduce great error [216], since the errors brought in by the inaccuracy of the model are compensated for by an empirical choice of parameters.

Sato's method [457]

Instead of the adjustable parameter ρ used in the method of Eyring and Polanyi, in Sato's method another parameter κ is introduced, whose origin is due to the nonorthogonality of the atomic orbitals. In this method, the basic formula (48.5) can be written approximately in the form:

$$U_1 = \frac{1}{1+\kappa}\left\{Q_{AB}+Q_{AC}+Q_{BC}-\frac{1}{\sqrt{2}}[(\alpha-\beta)^2+(\alpha-\gamma)^2+(\beta-\gamma)^2]^{\frac{1}{2}}\right\}. \quad (48.19)$$

Sato [457] uses this formula to calculate the potential energy surface, on the assumption that κ is constant. To evaluate the Coulomb and exchange integrals from the experimentally-known ground state terms of the diatomic molecules AB, BC, and AC, the following is suggested. The energies of the bonding and antibonding states U_λ and U_λ^* of any pair of atoms are written in the form

$$U_\lambda = \frac{Q_\lambda(R_\lambda)+\alpha_\lambda(R_\lambda)}{1+\kappa}, \quad (48.20a)$$

$$U_\lambda^* = \frac{Q_\lambda(R_\lambda)-\alpha_\lambda(R_\lambda)}{1-\kappa}. \quad (48.20b)$$

The term U_λ is approximated by the Morse function

$$U_\lambda = D_\lambda\{\exp[-2\alpha_\lambda(R_\lambda-R_{\lambda e})]-2\exp[-\alpha_\lambda(R_\lambda-R_{\lambda e})]\} \quad (48.21a)$$

and the term U_λ^* by a function of the form

$$U_\lambda^* = \frac{D_\lambda}{2}\{\exp[-2\alpha_\lambda(R_\lambda-R_{\lambda e})]+2\exp[-\alpha_\lambda(R_\lambda-R_{\lambda e})]\}. \quad (48.21b)$$

Since all parameters of the excited term are determined on the basis of data relating to the ground term, the substitution of (48.21) into (48.20) enables the functions Q_λ and α_λ to be determined for given κ. Of course, the question still remains open, to what extent the approximation (48.21) corresponds to the actual shape of the term of the excited electronic state.

Although the one-parameter methods of Eyring–Polanyi (parameter ρ) and Sato (parameter κ) are very close in conception, they give a qualitatively different form of surface near the saddle. For the H_2+H system, the method of Eyring and Polanyi leads to a small depression at the peak of the potential barrier. In Sato's method, the depression is not obtained. As regards the configuration at the saddle point both methods give the same results: for three atoms of hydrogen, the saddle point corresponds to a linear configuration of the nuclei (symmetric in the first case and slightly asymmetric in the second).

By introducing a large number of empirical parameters in formula (48.10), a sufficiently-great flexibility can be achieved in the representation of the potential function by a formula of type (48.10). Unfortunately, the physical origin of these parameters becomes more difficult to trace, so that the expression ultimately gives an empirical description of the potential energy surface. The use of similar types of empirical potentials, proposed in a number of papers [405], [406], is important in the study of the special features of the dynamics of exchange reactions and, in particular, for the solution of the problem in reverse—the recovery of a surface from the kinetic data.

We will now consider, as an illustration of the accuracy of the different methods of calculating the adiabatic terms, the simplest systems, H_3^+ and H_3.

The system H_3^+

The adiabatic electronic term of the ground state of H_3^+ is calculated by the variational method in papers [200], [202], [203], [274], [289], [397]. Minimum energy corresponds to a symmetric configuration of the nuclei.

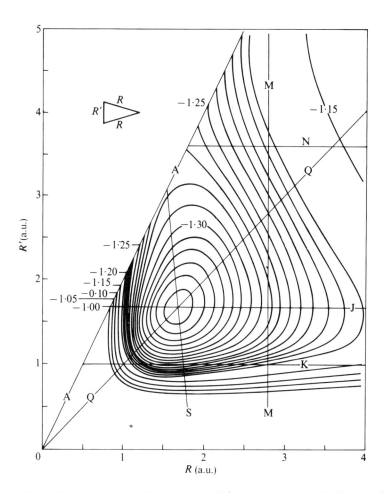

FIG. 44. Potential energy surface of the triangular H_3^+ system, calculated by the variational method (Conroy [202]). Distances and energy are in atomic units.

The frequencies of the small vibrations of the protons, which, in the case of the equilibrium configuration D_{3h}, correspond to symmetrical valence (a_{1g}) and degenerate deformation (e) vibrations, are characteristic of the surface near the minimum. The equilibrium position of the nuclei in the H_3^+ system coincides with the vertex of the cone in H_3 corresponding to the point of intersection (or contact) of the ground and excited terms of the system of three atoms of hydrogen. The term of the first excited state in H_3^+ does not have a minimum and, in the symmetric configuration of the nuclei, is doubly degenerate. In Fig. 44, taken from the paper by Conroy [202], is shown a diagram of the potential energy levels for the H_3^+ system in the configuration of an isosceles triangle. The line A corresponds to the linearly-symmetric configuration and the line Q corresponds to triangular symmetry. The remaining lines show the directions for which specific calculations have been carried out [202]. Some properties of the energy surface near the minimum of the symmetric system $H_3^+(D_{3h})$ are presented in Table 9.

H_3 system

The system of three hydrogen atoms has been the subject of numerous investigations over the last 30 years. It is of interest in connection with the elucidation of the mechanics of the simplest exchange reaction $H_2 + H \rightarrow H + H_2$. Discussion of a number of quantum-mechanical methods of calculating the potential energy surface can be found, for example, in the work of Bradley [168]. Only two results are important here: (a) the reaction path corresponding to minimum energy corresponds to the linear complex H ... H ... H; (b) the energy of the symmetric linear complex of H_3 goes through a simple minimum at the point $R_{AB} = R_{BC} = R_s$. In Fig. 41, the diagonal $R_{12} = R_{23}$ corresponds to the linear symmetric complex and the point of minimum to the point R_s. To investigate the form of the saddle, it is necessary to distort the symmetric configuration, shifting the nuclei by the value $\Delta R_{AB} = -\Delta R_{BC}$. It is precisely here that the different approximations give different qualitative results. The Eyring approximation, as was noted above, gives not just a simple parabolic form of saddle but a saddle with a small depression at the peak [223]. Subsequent calculations have not confirmed the existence of this local minimum [168], [204].

A diagram of the levels of the linear H_3 system, derived from the work of Conroy and Bruner [204], is presented in Fig. 45. In the same paper, a very exact calculation was carried out of the energy of the non-linear H_3 system and, in particular, the passage through the vertex of the double cone was followed (Fig. 46).

Since the question of the form of the saddle and an estimate of its height E_0 is very important in calculating the exchange reaction rate, the results of various calculations of the surface parameters near the symmetric linear configuration of atoms are presented in Table 9.

TABLE 9

Some properties of the potential energy surfaces of the H_3 and H_3^+ systems

	Linear H_3 system (term $^2\Sigma^+$)						Triangular H_3^+ system (term $^1A_1$)	
	Semiempirical methods			Theoretical calculations				
Parameters	Eyring's method [223]	Sato's method [457]	Method of complexes [194]	Shavitt [475]	Edmiston and Krauss [215]	Conroy and Bruner [204]	Parameters	Calculation of Christoffersen [200]
R_{12} (Å)	1·354	0·929	0·950	0·942	0·955	0·930	$R_{12} = R_{23} = R_{13}$ (Å)	0·88
R_{23} (Å)	0·753	0·929	0·950	0·942	0·955	0·930		
$R_{13} = R_{12} + R_{23}$								
$\omega_1(\sigma_g)$ (cm^{-1})	3626	2108	2144	1945	—	—	$\omega_1(a_1)$ (cm^{-1})	3354
$\omega_2(\pi_u)$ (cm^{-1})	665	877	823	952	—	—	$\omega_2(e)$ (cm^{-1})	2790
$\omega^*(\sigma_u)$ (cm^{-1})	630	1918	2464	1361	—	—		
E_0 (kcal mol^{-1})	—	—	12·3	—	14·3	7·74	$D = E(H_2) + E(H^+) - E(H_3^+)$ (kcal mol^{-1})	99·7
E_z^+ (kcal mol^{-1})	7·09	5·52	5·42	5·51	—	—		
E_a (kcal mol^{-1})	8·50	8·03	11·3	14·8	—	—		

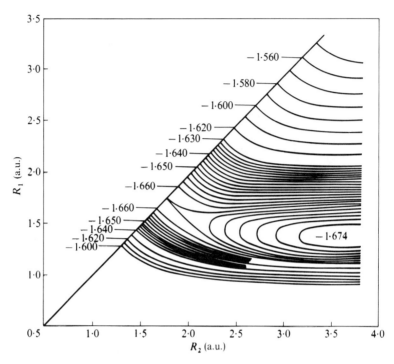

FIG. 45. Potential energy surface of the linear H_3 system, calculated by the variational method (Conroy and Bruner [204]). Distances and energy are in atomic units.

The parameter $\rho = 0.20$ in the Eyring method, selected from the condition that E_a agree with the experimental value of the activation energy, is taken equal to 8·50 kcal mol^{-1} [264]. The value of E_a, according to an analysis of the experimental data given in the paper [540], equals (8.0 ± 0.5) kcal mol^{-1}. The parameter κ, equal to 0·1475 in Sato's method [457], is chosen by the same considerations.

The values of R_s and E_0, found by the method of diatomic complexes in molecules, are based on the results of the work of Kolos and Wolniewicz [311], with reference to the $^1\Sigma_g^+$ and $^3\Sigma_u^+$ terms of H_2. The values of the frequencies are derived from the paper [194], in which an approximate expression for the energy of the triplet state is used and a small error has occurred in the calculation [216]. Since knowledge of the frequencies is necessary to determine E_z, the value of E_a is evaluated only approximately.

As regards the results of theoretical work, we note that the calculation of Shavitt [475] gives a surface with a simple saddle. The calculation of Edmiston and Krauss [215], carried out in the self-consistent field approximation with regard for the interaction of the configurations, assumes a fixed value of

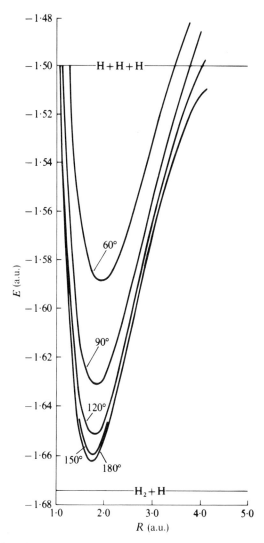

FIG. 46. Cross-section of the potential energy surface of the H_3 system, passing through the vertex of the double cone (Conroy and Bruner [204]). Configuration of an isosceles triangle with a fixed value of the angle between the vectors $\mathbf{R}_{12}$ and $\mathbf{R}_{23}$. The horizontal lines show the energy of the dissociated $H_2 + H$ and $H + H + H$ systems. The bottom curve, for an angle 180°, corresponds to the linear configuration of H_3.

$R_{HH} = 1.8$ a.u. Consequently, the value found for E_0 gives a lower bound to the height of the saddle. This lower bound can be reduced still further if the previously unaccounted-for contribution of the correlation energy is estimated by the semi-empirical method. Finally, the variational calculation of Conroy and Bruner [204] in which the electronic correlation is taken into account more completely, gives an appreciably lower value for E_0. In the symmetric configuration, at the saddle point the energy is equal to 7·74 kcal mol^{-1} at $R_s = 1\cdot74$ a.u.

We also note the work of Harris, Micha, and Pohl [252], in which it was found that the parabolic approximation to the barrier, in the reaction coordinate, gives a very poor approximation to the actual form of the saddle; if the force coefficient in the parabolic approximation is determined by requiring a satisfactory approximation of the barrier in a range equal to approximately 1 Å near the peak, then it turns out to be five times smaller than the value obtained for the curvature of the saddle at the maximum. This is very important in the correct evaluation of the contribution of tunnel corrections to the reaction rate.

For an approximate comparison of the values of E_a in Table 9 with the experimental values, we will quote the threshold value of the energy $E'_a = (7\cdot6 \pm 0\cdot5)$ kcal mol^{-1} for the reaction $D + H_2 \rightarrow HD + H$, found by Kuppermann and White [316] in an investigation of hot deuterium atom reactions. Exact comparison between this value and the theoretical values of E_0 (or E_a) is possible only after taking account of all kinetic processes contributing to the observed exchange reaction rate.

49. Equilibrium theory

The transition state method is the starting point for the equilibrium theory of bimolecular reactions, enabling the reaction rate constant to be expressed explicitly in terms of the characteristics of the potential energy surface and the critical surface.

Choice of the critical surface

The critical surfaces are chosen in a different way for reactions taking place with and without activation energies. For surfaces with a potential barrier, the reaction coordinate is defined as the line of steepest descent from the saddle point. We will assume that, in the initial system and in the activated complex, the rotation is separated from the vibrations and from motion along the reaction coordinate. Then the basic formula of the transition state method, as applied to reaction (47.1), can be represented in the following way:

$$\kappa = \frac{kT}{2\pi\hbar}\chi_g^\star \frac{F^\star_{XY,trans}F^\star_{XY,rot}F^\star_{XY,vib}\exp(-E_a/kT)}{(F_{X,trans}F_{X,rot}F_{X,vib})(F_{Y,trans}F_{Y,rot}F_{Y,vib})}. \quad (49.1)$$

Here, the rotational partition functions are calculated for rigid molecules X and Y and a rigid transition complex XY*, and the vibrational partition functions for a system of harmonic oscillators corresponding to the normal vibrations of the molecules X, Y and the complex XY*, in which the reaction coordinate is fixed and equal to q_r^*.

For the reaction of three atoms, the expression for κ takes different forms for the linear and triangular complexes. In the former case, there are three vibrational degrees of freedom and two rotational, in the latter case, two vibrational and three rotational. For the temperature dependence of κ we obtain

linear complex

$$\kappa \sim T^{\frac{1}{2}} \frac{[1 - \exp(-\hbar\omega_{BC}\beta)]}{\prod_{i=1}^{3}[1 - \exp(-\hbar\omega_i^*\beta)]} \exp\left(-\frac{E_a}{kT}\right)$$

$$= \begin{cases} T^{-\frac{1}{2}} \exp\left(-\frac{E_a}{kT}\right), & \hbar\omega_i\beta \gg 1, \quad (49.2a) \\ T^{\frac{3}{2}} \exp\left(-\frac{E_a}{kT}\right), & \hbar\omega_i\beta \ll 1, \quad (49.2b) \end{cases}$$

triangular complex

$$\kappa \sim \frac{[1 - \exp(-\hbar\omega_{BC}\beta)]}{\prod_{i=1}^{2}[1 - \exp(-\hbar\omega_i^*\beta)]} \exp\left(-\frac{E_a}{kT}\right)$$

$$= \begin{cases} \exp\left(-\frac{E_a}{kT}\right), & \hbar\omega_i\beta \gg 1, \quad (49.3a) \\ T \exp\left(-\frac{E_a}{kT}\right), & \hbar\omega_i\beta \ll 1. \quad (49.3b) \end{cases}$$

The transition between these two cases might be traced if the interaction of the vibrations with the rotation is taken into account in the calculation of F^*, but is not used in approximating the spectra from the 'rigid rotator–harmonic oscillator' model. The rate constant, written in the form of (49.3), is applicable, for example, only for a complex XY* in which the angle of deviation from the linear configuration appreciably exceeds the angular amplitude of the deformation vibrations.

In connection with the temperature dependence of the pre-exponential factors in (49.2) and (49.3) we note the following: in the low-temperature region, the dependence indicated is only fulfilled as long as the tunnel corrections can be neglected (see § 3). In the high-temperature region, there are restrictions connected with the requirement that the mean amplitudes of the vibrations be small in comparison with the effective width of the potential well $1/\alpha$. These restrictions can be expressed in the form of a requirement that the value of the total pre-exponential factor be appreciably smaller than the gas-kinetic number of collisions. If it is formally shown that the relation

$$\frac{kT}{2\pi h} \cdot \frac{F^{\star}_{XY}}{F_X F_Y} < Z_{0,XY} \tag{49.4}$$

is violated, then the factor $Z_{0,XY}$, which for the temperature dependence of the pre-exponential gives the expression $T^{\frac{1}{2}}$ instead of $T^{\frac{3}{2}}$ and T in (45.2) and (45.3), should be used as the pre-exponential.†

We will now consider the potential energy surface without an activation barrier. Such a situation is typical of ion–molecular reactions and is also possible in the reactions of certain atoms with molecules.

If the interaction between the partners X and Y, at sufficiently large distances, corresponds to an attraction, then in a number of cases it can be assumed that the reaction rate is determined by the rate of formation of a long-living complex XY*. In reality, of course, the complex XY* can decompose both in the direction $Z+U$ and in the direction $X+Y$; however, conditions are often fulfilled where the decomposition in the reverse direction can be neglected (see § 50). In such conditions, the reaction (47.1) can be described by the following model: the particles approach one another and, if the radial velocity is sufficient for summounting the centrifugal barrier, the formation of the complex XY* occurs, which then decomposes into the molecules $Z+U$. If the interaction potential is centrally symmetric, then, to this model, there corresponds a critical surface $S^{\star}$, a sphere of radius R_m corresponding to the coordinate of the centrifugal barrier.

A surface $S^{\star}$ of this type is determined not only by the potential but by the integrals of motion of the system under consideration, since R_m depends on the angular momentum of the relative motion of X and Y.

Returning to formula (31.27), it is not difficult to find $R_m(l)$ for the potential $U(R) \sim R^{-s}$. The subsequent averaging over l leads to expressions for the rate constant which, for the particular cases $s = 4$ (polarization interaction) and $s = 6$ (Van der Waals' interaction), have the form [432], [570]

$$\kappa = 2\pi e \left(\frac{\alpha}{\mu}\right)^{\frac{1}{2}}, \tag{49.5}$$

$$\kappa = 2^{\frac{2}{3}} 3^{\frac{1}{3}} \Gamma\left(\frac{2}{3}\right) \frac{\pi^{\frac{1}{2}}}{\sigma^{\star}} (kT)^{\frac{1}{6}} \left(\frac{\alpha_1 \alpha_2 I_1 I_2}{I_1 + I_2}\right)^{\frac{1}{3}} \mu^{-\frac{1}{2}}, \tag{49.6}$$

---

† This restriction enables the rotation to be separated from the vibrations and also enables the vibrations to be considered in the harmonic approximation. Numerous examples of reactions involving the exchange of a hydrogen atom, in which condition (49.4) is broken, are given in the paper by Mayer, Schieler, and Johnston [361].

where α is the polarizability of the neutral partner in the ion–molecular reaction characterized by the polarization potential $U(R) = -e^2\alpha/2R^4$; α_1 and α_2 are the polarizabilities of the neutral partners for which the long-range part of the potential is approximated by the expression

$$U(R) = \frac{3}{2}\frac{\alpha_1\alpha_2}{R^6}\left[\frac{I_1 I_2}{(I_1+I_2)}\right],$$

in which the ionization potentials I_1 and I_2 of the partners have been introduced.

It is essential that the partition functions of the internal motion of the partners do not come into (49.5) and (49.6). This is connected with the fact that, with the chosen critical surface S^*, the reaction (i.e. the intersection of S^*) takes place at such large intermolecular distances that the contributions of the internal motions to F and F^* are considered to be the same, so that they therefore drop out of the ratio F^*/F.

The neglect of the short-range part of the potential in the calculation of R_m is correct only on condition that R_m appreciably exceeds the gas-kinetic diameter R_0. This imposes a restriction on expressions (49.5) and (49.6) in the high-temperature region: for their validity, the relation

$$\kappa \ll Z_{0,XY} \tag{49.7}$$

must be satisfied, implying that only the long-range part of the potential contributes to the rate constant.

The exchange reaction $H + H_2 \to H_2 + H$

The following reactions are examples of one of the most-studied exchange processes

$$H + H_2 \to H_2 + H \quad (\kappa_1) \tag{49.8a}$$

$$H + D_2 \to HD + D \quad (\kappa_2) \tag{49.8b}$$

$$D + H_2 \to HD + H \quad (\kappa_3) \tag{49.8c}$$

whose rate constants have been measured over a fairly wide range of temperatures [443], [467]. The theoretical interpretation of the mechanisms of these reactions enables any nonadiabatic processes to be completely ignored, since all the excited states of the system H_3 lie appreciably higher than the ground state. The vertex of the double cone (see Fig. 42) corresponds to energies whose weight, in the Maxwell–Boltzmann distribution of the reactants, is negligibly small. Thus the rate constants of the reactions (49.8) are determined only by the parameters of the lower potential energy surface, and they differ only in the masses of the isotopes.

In Fig. 47, taken from the paper by Ridley, Schulz, and LeRoy [443], is shown the temperature dependence of the rate constants κ_1 (curve 1) and κ_3

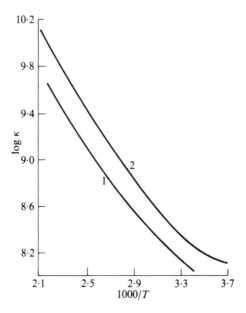

Fig. 47. Temperature dependence of the rate constants for the reactions $H + H_2 \rightarrow H_2 + H$ (curve 1) and $D + H_2 \rightarrow DH + H$ (curve 2) (Ridley, Schulz, and LeRoy [443]).

(curve 2). In the entire temperature-range studied (approximately 270–550 K) it can be assumed, with sufficient accuracy, that all vibrational partition functions are equal to unity, whereupon the basic formula (49.1) applied to (49.8) takes the form:

$$\kappa = \chi(T) g^\star \frac{I^\star \sigma_{BC}}{I_{BC} \sigma^\star} \left(\frac{m_A + m_{BC}}{m_A m_{BC}} \right)^{\frac{3}{2}} \times \hbar^2 \left(\frac{2\pi}{kT} \right)^{\frac{1}{2}} \exp\left(-\frac{E_a}{kT} \right). \tag{49.9}$$

A simple analysis shows that the deviation of the dependence of $\kappa(T)$ from the simple Arrhenius law, apparent in the appreciable curvature of the graphs in Fig. 47 at low temperatures, cannot be explained by the contribution of the pre-exponential factor $T^{-\frac{1}{2}}$ in (49.9). Therefore, staying within the limits of the transition state method, it is necessary to assume the existence of a sufficiently-strong tunnel effect. The approximation of a parabolic barrier gives formula (3.7) for the tunnelling transmission coefficient. Treating the experimental results represented in Fig. 47 by the method of least squares gives the following values for the frequency $\omega^\star$ [443]:

$$\omega_1^\star = 1178 \text{ cm}^{-1}, \tag{49.10a}$$

$$\omega_3^\star = 1136 \text{ cm}^{-1}. \tag{49.10b}$$

The ratio of the frequencies $\omega_3^*/\omega_1^* = 0.963$ is a little larger than the value calculated from the ratio of the reduced masses:

$$\left(\frac{\mu_1}{\mu_3}\right)^{\frac{1}{2}} = \left(\frac{m_{HH}m_H}{m_{HH}+m_H}\right)^{\frac{1}{2}}\left(\frac{m_{HD}m_H}{m_{HD}+m_H}\right)^{-\frac{1}{2}} = 0.943. \quad (49.11)$$

Such an estimate for a non-symmetric system (such as system (49.8c)) is correct only when the slope of the ascent on the saddle is somewhat steeper than the slope of the descent. The results of the theoretical work of Shavitt [475] permits such an approximation. After separating out the tunnel factor $\chi(T)$ which, for the reaction (49.8a), varies over the range 2·13 (444 K)– 12·51 (300 K), the constants κ_1 and κ_3 can be represented in the form

$$\kappa_1 = \chi_1(T)1\cdot 18 \times 10^{15} T^{-\frac{1}{2}} \exp\left(-\frac{9200}{RT}\right) \text{cm}^3 \text{ mol s}^{-1}, \quad (49.12a)$$

$$\kappa_3 = \chi_3(T)3\cdot 63 \times 10^{15} T^{-\frac{1}{2}} \exp\left(-\frac{9400}{RT}\right) \text{cm}^3 \text{ mol s}^{-1}. \quad (49.12b)$$

Formally, in the sense of their temperature dependence, these expressions agree with the formula of the transition state method (49.9); however, they differ from it in two respects. In the first place, the activation energy E_a for (49.8a) must be greater than the corresponding value for (49.8c). This follows since the energy of the zero-order vibration for the complex (DHH)* is less than for the complex (HHH)*, and the energy of the zero-order vibration of the reactants is the same in both cases. The values of E_a, entering (49.12a) and (49.12b), occur in inverse ratio. In the second place, the ratio of the pre-exponential numerical factors in (49.12a) and (49.12b), equal to 0·325, does not agree with the theoretical value 1·336, found on the basis of formula (48.9).†

These contradictions, which may be partly related to the incorrect allowance for tunnel corrections [443], probably indicate the fundamental inadequacies of the transition state method. One of these, the assumption of the existence in the complex of quantum energy levels with short lifetimes, is briefly considered in § 52. Another, the neglect of the possible violation of the equilibrium distribution caused by the reaction itself, is considered in § 53.

50. The statistical theory

Derivation of the general formula

The basic assumption of the statistical theory of bimolecular reactions is that the elementary process (47.1) takes place in two independent stages. In the first, the complex (or composite system) is formed, analogous in its

† The ratio g_1^*/g_3^*, equal to 2, enters into this value.

features to the active molecule (see Chapters 4 and 5). In the second stage, it decomposes by all possible channels, in accordance with the statistical theory of unimolecular reactions. It is evident that for the validity of the assumption about the independence of the two stages, the lifetime of the composite system must appreciably exceed times characteristic of intramolecular motion. An estimate of $\langle \tau \rangle$ can be obtained within the limits of the statistical theory of unimolecular reactions, in this way completely reducing all the assumptions of the statistical theory of bimolecular reactions to the case already considered in Chapter 5.

We will assume that the initial states of the molecules X and Y are characterized by a set of quantum numbers i. On formation of the complex, let the numbers i lose their meaning and characterize the active molecule by a new set of quantum numbers ε. Further, let the final products Z and U be characterized by quantum numbers f. Then the rate constant κ_{if} of an elementary process $i \to f$ is presented in the form

$$\kappa_{if} = \sum_{\varepsilon} \delta\kappa^c_{i\varepsilon} P_{\varepsilon f}, \qquad (50.1)$$

where $\delta\kappa^c_{i\varepsilon}$ is the rate of formation of the composite system (or, as we will say, the rate of capture of X and Y) in the state ε; $P_{\varepsilon f}$ is the probability of its decay by channel f; the summation in (50.1) goes over all states which can be formed from the state i.

The capture rate $\delta\kappa^c_{i\varepsilon}$ can be expressed in terms of the differential capture cross-section $d\sigma^c_{i\varepsilon}$ by the relation

$$\delta\kappa^c_{i\varepsilon} = v_i \, d\sigma^c_{i\varepsilon}. \qquad (50.2)$$

The probability of decomposition $P_{\varepsilon f}$ equals the normalized partial decay rate by channel f:

$$P_{\varepsilon f} = \frac{\kappa_{\varepsilon f}}{\sum_{f'} \kappa_{\varepsilon f'}}, \qquad (50.3)$$

where the sum in the denominator is taken over all channels in the direct and reverse directions. The decay rate $\kappa_{\varepsilon f}$ of the active molecule is calculated, in its turn, in accordance with the theory set out in Chapter 5. Thus the problem of the calculation of κ_{if} can be split into two stages, in accordance with (50.2) and (50.3).

Within the limits of classical mechanics, the differential capture cross-section is defined by the relation

$$d\sigma^c = 2\pi b \, db, \qquad (50.4)$$

where b is the impact parameter satisfying the condition $b < b_c$; b_c is the Maxwellian impact parameter at which capture occurs.

It is assumed, in such a formulation, of course, that the formation of the composite system does not involve surmounting the potential barrier, i.e. the first stage of the reaction proceeds without activation energy. It is convenient to express relation (50.4) in terms of the angular momentum l of the relative motion of the X and Y molecules

$$d\sigma^c = \frac{\pi}{p^2} 2l \, dl. \tag{50.5}$$

Although the capture condition is most simply formulated by the requirement $l < l_c$ in (50.5) it is expedient to introduce, in place of l, the total angular momentum of the system of colliding molecules $\mathbf{J}$, since it alone is an exact integral of motion. In future, for simplicity, we will have in mind a collision of an atom A with a molecule BC, so that the total angular momentum $\mathbf{J}$ is represented in the form of a vector sum

$$\mathbf{J} = \mathbf{l} + \mathbf{j}, \tag{50.6}$$

in which $\mathbf{j}$ denotes the internal angular momentum of BC.

The change from the variable $\mathbf{l}$ to the variable $\mathbf{J}$ (on condition that $\mathbf{j}$ be fixed) is most simply accomplished on the basis of a vector addition diagram (see Fig. 31), in the same way as was done in the change to the variable $\mathbf{J}$ in the derivation of the formulae of § 31. As the result of such a transformation, the differential capture cross-section $d\sigma^c$ is written in the form

$$d\sigma^c = \frac{\pi}{p^2} \phi(J, j, l_c) J \, dJ, \tag{50.7}$$

where the explicit form of the function ϕ may be easily found on the basis of Fig. 31, where it is clear that ϕ tends to zero for $J > l_c + j$. To determine l_c it is necessary to formulate a condition for capture. If the capture of A and BC takes place at sufficiently large distances, then the isotropic polarization interaction $U(R) = -Q(R_0/R)^n$, for which l_c depends only on the relative kinetic energy E_{trans} of the fragments and is determined by formula (31.27), makes the main contribution to the potential. In addition, of course, the restriction from below (31.29), connected with the neglect of the contribution of the short-range force, is imposed on the energy E_{trans}.

We will now make more specific the set of quantum numbers ε, having assumed that the only integrals of motion of the composite system are the total energy E and the total angular momentum $\mathbf{J}$ [73], [332], [398], [399], [561]. In addition to this, some assertions must be made concerning the electronic states between which a statistical redistribution of energy is possible. The nonadiabatic transition probabilities between these states are assumed to be sufficiently large, so that they do not limit the transition of the representative point from one potential surface to the other.

The total cross-section σ_1 of the reaction (48.1) can now be written in the form

$$\sigma_1 = \frac{\pi}{p^2} \int 2J \, dJ \, \phi(J, j, l_c) \frac{\kappa_1(E, J)}{\kappa_1(E, J) + \kappa_{-1}(E, J)}, \qquad (50.8)$$

where $\kappa_1(E, J)$ and $\kappa_{-1}(E, J)$ are the rates for the spontaneous decomposition of ABC* in the forward and reverse directions.

Calculating the reaction cross-section in a given channel (for example, in a state with given vibrational or rotational quantum numbers), $\kappa_1(E, J)$ in the numerator of (50.8) should be replaced by the corresponding partial rate constant. In the majority of cases, formula (50.8) permits a substantial simplification, since the relative angular momentum l, as a rule, appreciably exceeds the internal angular momentum of the molecule taking part in the reaction. In this case, it can be assumed that $J \sim l = p_{A-BC} b$ and (50.8) takes the form

$$\sigma_1 = \frac{\pi}{p^2_{A-BC}} \int 2J \, dJ \frac{\kappa_1(E, J)}{\kappa_1(E, J) + \kappa_{-1}(E, J)}$$

$$= 2\pi \int_0^{b_c(v)} b \, db \frac{\kappa_1(E, J)}{\kappa_1(E, J) + \kappa_{-1}(E, J)},$$

$$J \approx \mu v b, \qquad (50.9)$$

where μ and v are the reduced mass and the relative velocity of the reactants respectively.

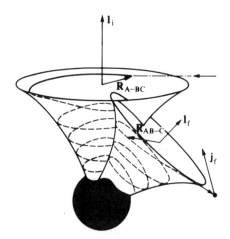

FIG. 48. Potential energy surface for a reaction taking place via the formation of an intermediate complex.

Schematically, a reaction of type (48.1) can be represented as the motion of the representative point over the potential energy surfaces depicted in Fig. 48. The relative motion of A and BC in the initial stage of capture occurs in the plane perpendicular to the relative angular momentum $\mathbf{l}_i$. If the energy axis is directed along the vector $\mathbf{l}_i$, then capture is represented by a trajectory described by the vector $\mathbf{R}_{A-BC}$, descending spirally into a 'funnel' corresponding to the potential U_{A-BC}. Then the representative point reaches the regions of strong interaction, where the redistribution of energy is described by statistical laws (black region). Finally, the representative point begins to ascend the second 'funnel', which corresponds to the potential U_{AB-C} and whose axis is directed along the vector of the relative motion of the reaction products $\mathbf{l}_f$. The direction of the vector $\mathbf{l}_f$ is connected with the direction of the internal angular momenta $\mathbf{j}$ of the fragments and the vector $\mathbf{l}_i$ by the conservation law $\mathbf{l}_i + \mathbf{j}_i = \mathbf{l}_f + \mathbf{j}_f$.

If, in the calculation of σ_1, the dependence of the decomposition rate of ABC* on the total angular momentum J is neglected, then (50.9) is further simplified

$$\sigma_1 = \frac{\kappa_1(E)}{\kappa_1(E) + \kappa_{-1}(E)} 2\pi \int_0^{b_c(v)} b\, db = \sigma_1^c(v) \frac{\kappa_1(E)}{\kappa_1(E) + \kappa_{-1}(E)}, \qquad (50.10)$$

where the first factor on the right-hand side of the equality is the total capture cross-section; the second is the probability of decomposition in the forward direction.

If it is taken into account that the $\kappa_i(E, J)$ are proportional to the number of states in the channel i, and this number is an increasing function of the energy ε_i referred to the zero-energy level in this channel, then it is clear that, of the two constants κ_1 and κ_{-1}, the larger will be that which corresponds to the exothermic reaction direction.

Let the reaction (48.1) be exothermic. If $\varepsilon_1 \gg \varepsilon_{-1}$ we expect $\kappa_1 \gg \kappa_{-1}$, so that from (50.9) we find

$$\sigma_1 = \sigma_1^c(v_i). \qquad (50.11)$$

Thus for an exothermic reaction the reaction cross-section coincides (under the restriction $\varepsilon_1 \gg \varepsilon_{-1}$) with the capture cross-section.

In the case of an endothermic reaction with energy just above the threshold, we have $\varepsilon_1 \ll \varepsilon_{-1}$, $\kappa_1 \ll \kappa_{-1}$, so that (50.9) gives

$$\sigma_1(\varepsilon_1) = \frac{\pi}{p_{A-BC}} \int 2J\, dJ \frac{\kappa_1(E, J)}{\kappa_{-1}(E, J)}. \qquad (50.12)$$

This equation will be used to derive the threshold energy dependence of $\sigma_1(\varepsilon_1)$. At a sufficiently low energy ε_1 of AB and C above the threshold the vibrations of BC cannot be excited. Therefore κ_1 is proportional to the number of rotational states of BC compatible with the requirement $J =$ constant, $E_{\text{rot}}^{\text{BC}} \leq \varepsilon_1$. Referring to Fig. 31 we see that two limiting cases can be distinguished according to two types of dashed curves dividing open and closed channels. The intersections of these curves with the j- and l-axes are $j_{fm}(\varepsilon_1)$ and $l_{fm}(\varepsilon_1)$. The two cases mentioned correspond either to $j_{fm} \gg l_{fm}$ or to $l_{fm} \gg j_{fm}$.

As typical values of J in the integrand of (50.12) are

$$J \approx J_{\max} \sim \max\{j_{fm}, l_{fm}\},$$

a rough estimate will give

$$\kappa_1 \sim \min\{j_{fm}^2(\varepsilon_1), l_{fm}^2(\varepsilon_1)\},$$

which correspond to the cross-hatched part 3 in Fig. 31(b) (if $j_{fm} \gg l_{fm}$) or to part 4 (if $l_{fm} \gg j_{fm}$). On the other hand, κ_{-1} is proportional to the area of part 1

$$\kappa_{-1} \sim Jj_{fm}.$$

Substituting these estimates into (50.12) we find

$$\sigma_1(\varepsilon_1) \sim \int_0^{J_{\max}} 2J \, dJ \frac{\kappa_1(\varepsilon_1)}{\kappa_{-1}(J)} \sim j_{fm}(\varepsilon_1) l_{fm}(\varepsilon_1) \min\{j_{fm}(\varepsilon_1), l_{fm}(\varepsilon_1)\}. \tag{50.13}$$

Now, j_{fm} is proportional to $\varepsilon_1^{\frac{1}{2}}$, and for a power-law interaction in the exit channel, l_{fm} is proportional to $\varepsilon_1^{\frac{1}{2} - 1/n}$ (see eqn (31.27)). Thus, depending on values of the moments of inertia of AB and ABC*, we have

$$\sigma_1(\varepsilon_1) \sim \varepsilon_1^{\frac{3}{2} - 1/n}$$

for energies ε_1 such that $j_{fm}(\varepsilon_1) < l_{fm}(\varepsilon_1)$;

$$\sigma_1(\varepsilon_1) \sim \varepsilon_1^{\frac{3}{2} - 2/n} \tag{50.14}$$

for energies such that $l_{fm}(\varepsilon_1) < j_{fm}(\varepsilon_1)$,

For ion–molecule reactions $n = 4$, and (50.14) gives $q_1 \sim \varepsilon_1^{\frac{3}{4}}$ or $\sim \varepsilon_1$, depending on the relation between j_{fm} and l_{fm}. One would expect that the $\varepsilon^{\frac{3}{4}}$ law would go over into the linear law with increasing energy, still under condition $\varepsilon_1 < \hbar\omega_{\text{AB}}$. This threshold dependence of the cross-section is connected essentially with the law of conservation of angular momentum. That it is so is evident from the following consideration. In the first stage of the reaction taking place in the endothermic direction, a composite system ABC* is formed, whose angular momentum is distributed in a certain range $0 < J < l_{fc}$. On decomposition in the direction of A + BC, the maximum angular momentum j of the molecule BC is not large and is determined by the excess of energy over threshold $\varepsilon = E - \Delta E$. On the other hand, the magnitude of l_{ic}, determining the limit of open channels with the direction A + BC, is also small for small ε. This means that the vector sum of $\mathbf{j}_i$ and $\mathbf{l}_i$

The model reaction $A_2^+ + A_2 \to A_3^+ + A$

As an illustration of the statistical theory of bimolecular reactions we will discuss an ion–molecular reaction of the type indicated above. The process

$$H_2^+(E_{trans}, n) + H_2 \to H_3^+ + H + \Delta E, \qquad \Delta E = 1\cdot 1 \text{ eV}, \qquad (50.15)$$

in which the H_2^+ ions possess a given vibrational excitation (quantum number n) and the relative kinetic energy E_{trans}, may serve as a specific example of such a reaction. The problem consists of calculating the cross-section of this reaction as a function of n and E_{trans}, i.e. $\sigma = \sigma(E_{trans}, n)$.

We will turn first to the calculation of the total reaction cross-section for the vibrational ground state of H_2^+. An estimate carried out by Firsov [122] on the basis of a simplified variant of the statistical theory (without regard for the conservation of total angular momentum) shows that, at an energy $E_{trans} \ll 0\cdot 1$ eV, the rate of decomposition of the complex H_4^+ in the forward direction (κ_1) substantially exceeds the rate of decomposition in the reverse direction (κ_{-1}). Then the reaction cross-section $\sigma(E_{trans}, 0)$, must agree with the capture cross-section $\sigma^c(E_{trans})$, which, in turn, is defined by the general formula (50.11) with the polarization potential $U(R) = -\alpha e^2/2R^4$ (where α is the polarizability of the neutral particle) [242]:

$$c = \frac{2\pi e}{v}\sqrt{\left(\frac{\alpha}{\mu}\right)}, \qquad v = \left(\frac{2E_{trans}}{\mu}\right)^{\frac{1}{2}}, \qquad (50.16)$$

where μ is the reduced mass of the partners.

An experimental investigation of the reaction (50.15) shows that, at small energies, the cross-section for the emergence of the H_3^+ ions obeys formula (50.16) [253], [433], [473]. At relative kinetic energies of the H_2^+ ions exceeding approximately $1\cdot 2$ eV, the cross-sections for the emergence of H_3^+ decrease with an increase in the energy more rapidly than the cross-section for the formation of the intermediate complex. This can be explained either by the dissociation of H_3^+ into H_2 and H^+ (in these conditions the emergence of H^+ ions is, in fact, observed) [433] or by the decomposition of the intermediate complex in the reverse direction.

We will now go over to study the dependence of σ on the vibrational quantum number n of the H_2^+ ion. In calculations using the general formulae of § 31, it should be taken into account that the initial angular momentum of the $H_2^+(n)$ ion is sufficiently small if the H_2^+ ion is obtained on ionization of the H_2 molecules with low energies. Therefore, in the calculation of σ, formula (50.9) can be employed, and for $n \neq 0$, the contribution of E_{trans} to the total energy of the composite system H_4^+ can be neglected, in general, if $n\hbar\omega^+ \gg E_{\text{trans}}$ (ω^+ is the frequency of vibrations of H_2^+). It is convenient to represent the cross-section σ_1 in the form $\sigma^c(E_{\text{trans}})P(E_{\text{trans}}, n)$, where $P(E_{\text{trans}}, n)$ should be considered as a mean reaction probability by the given channel. A calculation of this factor was carried out by Nikitin [77] and also by Tunitskiĭ and Kuprianov [119]. It is difficult to compare these calculations immediately, since the latter is based on a simplified variant of the theory proposed by Firsov [122], and it does not take the identical decay channels into account.

A theoretical investigation of different cases of addition of the angular momenta of the molecules, arising in the decomposition of the composite system H_4^+, leads to the following estimate for the relative effectiveness of decomposition in the forward direction at energy $E_{\text{trans}} = 0.04 \text{ eV}$ [77]:

$$P(n=0):P(n=1):P(n=2):P(n=3):P(n=4) = 1:0.9:0.75:0.62:0.52.$$
(50.17)

Qualitatively, this is in agreement with the first terms of the analogous sequence describing the experimental data [539]:

$$1:0.5:0.4:0.5:0.$$
(50.18)

The difference between theory and experiment and, in particular, the fact that the statistical theory does not explain the very strong reduction in the reaction probability for $n \geq 4$ is probably connected with the violation of the basic assumption of the theory concerning the formation of a long-living complex at sufficiently high energies and with the disregarding of the effect of the anisotropic interaction on the determination of the boundary between the open and closed channels.† Moreover, it must be taken into account that the interpretation of the experiment giving the sequence (50.18) is carried out on the assumption that the Frank–Condon principle holds good for the ionization of H_2. However, it is possible that, near threshold, this principle is not satisfied [171].

It is evident from examples (50.17) and (50.18) that, as the energy of the composite system increases, the charge exchange channels may successfully compete with the exothermic reaction channels, accompanied, of course, by a redistribution of the vibrational energy. The fact that the charge exchange cross-section decreases sufficiently slowly with increase in the relative energy

† Such an anisotropy appears most clearly in the reactions of an ion with a dipolar molecule [377], [456]. In this case, the field of the ion inhibits the rotation of the molecule and, in the limiting case, orientates it completely in the direction of the line of collision, so that the classification of the channels into open and closed must be carried out with regard for the interaction of the angular momenta **l** and **j**.

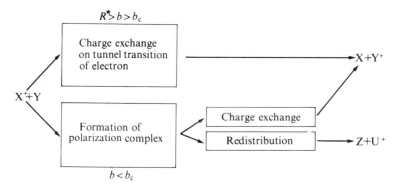

FIG. 49. Competition of charge exchange and redistribution of atoms in ion–molecular process, taking place via an intermediate complex.

of the partners also contributes to this competition. The scheme of this form of competition is shown in Fig. 49 for the quasi-resonant charge-exchange channel. At the first branch, either direct charge exchange or formation of a polarization complex is possible. The relative contribution of these processes is proportional to their corresponding cross-sections, which are determined by the range of variation of the impact parameters ($b_c(v)$ and $R^*(v)$ denote the critical impact parameters for capture and charge exchange respectively). The relative effectiveness of the processes at the second branch is determined by the nature of the redistribution of energy in the complex, and depends decisively on the sign and magnitude of the energy effect of the process with redistribution.

This type of experimental competition can be observed most easily for a process whose charge–exchange channel can be distinguished from the initial channel. Markin and Talrose [45] studied the reaction

$$H_2^+ + O_2 \to (H_2O_2)^* \begin{cases} HO_2^+ + H & (a) \\ O_2^+ + H_2 & (b) \end{cases} \quad (50.19)$$

which satisfies this condition. It was found that the ratio of the cross-sections σ_a/σ_b decreases with an increase in energy; for a relative energy of the ions of about 10 eV the cross-section of reaction (a) becomes very small. This immediately suggests that, with an increase in the energy, the lifetime of the composite system is reduced, so that ultimately, the very concept of the intermediate complex loses its meaning [29], [116].

Limits of applicability of the mechanism of the composite system

The rate constants of any process can be calculated comparatively simply within the limits of the statistical theory of reactions. However, it is necessary to note the following limitation on the model of an intermediate complex (a composite system).

Firstly, for the most abundant ion–molecular reactions, at least one of the partners in the collision turns out to be in a degenerate electronic state. This means that, in the composite system, there is known to be a series of electronic terms, the splitting between which decreases as the intermolecular distance increases. In this connection, the question arises as to the region in the configuration space of the nuclei in which the nonadiabatic transitions are located. If these transitions are localized at small internuclear distances (i.e. 'within' the composite system), then the splitting between terms is large, generally speaking, and effective exchange between electronic and vibrational energies is possible only in regions of intersection of the multi-dimensional potential energy surfaces.

Evidence for the existence of such regions and their position on the energy scale requires the construction of correlation diagrams for the electronic terms of a polyatomic system, i.e. ultimately requires a sufficiently-detailed investigation of the adiabatic terms. But it is precisely this stage, which the statistical theory tries to by-pass, which is not incorporated into the theory in an explicit form.

If the nonadiabatic transitions are localized at large internuclear distances (i.e. on the 'boundary' or outside the composite system), where the splitting between the adiabatic terms is already small, then the extent to which the final result of such transitions may be described in terms of the statistical redistribution over the states remains unclear. In calculating the function $P(E_{trans}, n)$ for the reaction (50.15), it was assumed, for example, that the charge–exchange channel contains an additional factor $g_1 = 2$, which takes into account the existence of the two electronic states for the system $H_2^+ + H_2$ [77]. Of course, such a method of taking several electronic states into account cannot be considered firmly based. To the difficulties indicated above is added yet another, namely that, in the composite system, those electronic states which correspond in the various decay channels to the electronically-excited states of the fragments may, in principle, be excited. An estimate of the transition probabilities into these states is even more complicated, although even for such processes attempts have been made at a description in terms of statistical theory [399].

Secondly, the mean lifetime $\langle \tau \rangle$ of the composite system may prove to be insufficiently long for the achievement of the equilibrium distribution. It should be noted here that the model of the intermediate complex can apply not just to processes taking place without activation energy, at the first stage. The presence of an activation energy only complicates the formation of the complex, so that expression (50.4) for the differential capture cross-section remains correct also for processes taking place accompanied by a surmounting of the barrier, provided the capture probability, $P_c(b) < 1$, is introduced into (50.4). At the same time, the activation barrier inhibits the decomposition of the complex, increasing $\langle \tau \rangle$. Therefore, for such processes, the application of statistical theory is justified in the calculation both of reaction rates and distribution functions of the energies of the fragments [305], [335]. The recombination of polyatomic radicals or the pre-association of complex molecules may serve as the simplest specific examples. In the latter case, the small decomposition probability is due to the weak coupling of electronic terms of different symmetry, which makes the description of the intermediate complex in terms of the partial level widths especially convenient [74], [386].

One of the difficulties in the estimation of $\langle \tau \rangle$ is that, usually, the bond energy E_0 of the composite system is unknown. For the ion–molecular reactions, a reasonable estimate of E_0 leads to the value 0·5–3 eV at which $\langle \tau \rangle$ is still sufficiently large for the validity of the basic assumptions of the theory. In particular, the question of the calculation of E_0 for the complex H_4^+ is discussed in the paper of Light and Shuler [329]. For reactions of neutral particles, E_0 is probably substantially smaller than the value indicated above, therefore the question of the formation of the composite system can only be resolved experimentally. In particular, a study of the angular distribution of the reaction products in molecular beams yields qualitative information about the lifetime of the complex, since, here, there is a natural time scale—the period of rotation of the complex. The strongly-anisotropic distribution which is observed in reactions of alkali metals M with molecules of X_2 or XR (where X is a halide atom and R is a CH_3 or C_2H_5 radical) indicates that these processes cannot be described as the result of the formation and decomposition of a composite system; the time for the reorganization of the whole system is less than the period of rotation, and amounts to a quantity of the order 5×10^{-13} s [258], [259]. However, since this time is, nevertheless, greater than the vibration period, the statistical theory may describe some limiting model of the energy redistribution, useful in a qualitative interpretation. The model calculations of Pechukas, Light, and Rankin [399] are convincing, having shown that, in a calculation of the energy redistribution for the model of a radially-vibrating rigid sphere, the statistical approximation gives results agreeing qualitatively with the exact solution of the model problems of Shuler and Zwanzig (see § 11) [486].

For the reaction

$$K + HBr \rightarrow H + KBr + \Delta E_0 \tag{50.20}$$

the statistical theory [399] predicts a distribution function for the rotational–vibrational energies of the KBr molecules agreeing qualitatively with the experimental data [258] and with the model dynamical calculations [166]; a large part of the energy ΔE being released in the reaction appears in the form of the internal energy of KBr, and the maximum of the theoretical distribution corresponds to an energy $E_{KBr,vib,rot} \approx 0.25$ eV at $E_{trans} \approx 0.08$ eV and $\Delta E = 0.30$ eV (the experimental value of $E_{\text{KBr,vib,rot}} \approx 0.21$ eV [245]). At the same time, the statistical theory always leads to a monotonic energy distribution function over the vibrational states, indicating the preferential population of the lower levels. This deduction is often not in agreement with the experimental results for the reactions of neutral particles, which indicates the inapplicability of statistical theory to these reactions.

A necessary condition for an exchange reaction to take place via the formation of an intermediate complex is that the probability of its decay by a given channel depends only on the total energy E and not on the individual contributions of the kinetic and internal energies of the colliding partners. For endothermic reaction channels, this implies that the preliminary internal excitation of the molecules must lower the effective reaction threshold in relation to the translational energy. Such a situation was actually observed for the reaction

$$CH_4^+ + CH_4 \rightarrow CH_5^+ + CH_3 \qquad (50.21a)$$
$$\hookrightarrow CH_3 + H_2, \qquad (50.21b)$$

the first of which is exothermic, and the second, endothermic [238]. The change in the energy of the electrons ionizing CH_4, from 13.5 eV to 80 eV, leads to a lowering of the threshold of process (50.21b) from 1 eV to 0.4 eV. Of course, as to the quantitative agreement of theory with experiment it is impossible to speak here, since the appreciable contribution of the kinetic energy to the internal energy of the complex $(CH_4-CH_4)^+$, clearly points to its sufficiently long lifetime. As for the total reaction cross-section of (50.21), including both channels, it is described sufficiently well by the polarization capture model, both as to absolute value and velocity dependence [117], [238].

In contrast to (50.21) the simplest reactions,

$$H_2^+ + He \rightarrow HHe^+ + H \qquad (50.22)$$

$$H_2 + He^+ \diagup \begin{matrix} HHe^+ + H & (50.23a) \\ H + H^+ + He, & (50.23b) \end{matrix}$$

it seems, take place without the formation of an intermediate complex [231], [306]. For the first of them (an endothermic reaction), the kinetic energy does not contribute to the surmounting of the barrier, but for the second (an exothermic reaction) the total cross-section σ proves to be significantly smaller than the polarization capture cross-section σ_c (at a thermal energy σ amounts to about 6 per cent of σ_c). These features are probably explained by the fact that the reaction rate is limited by the non-adiabatic transitions, whose probability for the case under consideration may be very small, due to the large distance between the electronic terms of the system $(HeH_2)^+$ correlated with terms of the initial and final molecules.

51. Theory of direct reactions

Experimental study of reactions in molecular beams shows that many exchange reactions, to which, in particular, are attributed the reactions of alkali metal atoms with halide and hydrogen halide molecules, take place without the formation of a long-living intermediate complex, i.e. take place by a direct mechanism. In future, by a direct mechanism we will understand a mechanism of interaction of the partners such that the time of their stay in a volume of molecular dimensions is of the order of several periods of vibration of the nuclei or less. The theoretical problem of the calculation of the cross-section of similar processes or of the determination of the distribution function of the states of the reaction products is very complicated. Even for a system of three atoms, practical results can be obtained only with the aid of computers. Therefore very simple models of direct reactions, enabling experiments to be interpreted qualitatively, acquire particular value.

Some qualitative results can be obtained if consideration is confined to the motion of the representative point close to the minimum of the potential energy of the system. For example, for the reaction (48.1), with a linear configuration for the activated state, it is possible to limit consideration, to some extent, to the two coordinates R_{AB} and R_{BC} which describe the relative motion of the atoms along a fixed line. If, in these coordinates, the kinetic energy of the system were written in the form of a sum of squares of the velocities $\dot{R}_{AB}$ and $\dot{R}_{BC}$ multiplied by some effective mass, then the redistribution of the energy in the reaction might be interpreted as the motion of a heavy point over a surface whose level curves are analogous to those represented in Fig. 41. However, in the coordinates R_{AB} and R_{BC}, the kinetic energy T has the form

$$T = \frac{1}{2(m_A+m_B+m_C)}[m_A(m_B+m_C)\dot{R}_{AB}^2 + 2m_A m_C \dot{R}_{AB}\dot{R}_{BC} + m_C(m_B+m_A)\dot{R}_{BC}^2]. \tag{51.1}$$

It is expedient to go over to new coordinates, ξ_1 and ξ_2 or ζ_1 and ζ_2, by means of a linear transformation of the coordinates—by a change of scale and a rotation of the axes—having demanded that, in the new system, the kinetic energy be expressed by the following relation:

$$T = \frac{\mu}{2}(\dot{\xi}_1^2 + \dot{\xi}_2^2) = \frac{\mu}{2}(\dot{\zeta}_1^2 + \dot{\zeta}_2^2) \tag{51.2}$$

and the ξ_1 axis should coincide with the R_{AB} axis and the ζ_2 axis with the R_{BC} axis. If ξ_1, ξ_2 and ζ_1, ζ_2 are considered as a Cartesian system of coordinates, then R_{AB}, R_{BC} must be treated as an obliquely-angled system, with an angle β between the axes. The transformation taking (51.1) into (51.2) was accomplished in a paper by Smith [496].

The angle of rotation β and the effective mass μ are given by the relations

$$\tan^2 \beta = \frac{m_B(m_A + m_B + m_C)}{m_A m_C}, \tag{51.3a}$$

$$\mu = \left(\frac{m_A m_B m_C}{m_A + m_B + m_C}\right)^{\frac{1}{2}}. \tag{51.3b}$$

Although representing the reaction by the motion of a heavy material point is sufficient graphically, the obliquely-angled system does not give any real advantages in specific calculations, since the nature of the potential in the region of the abrupt turn in the trajectory remains unknown. Only when the motion of the point is close to motion in a channel with vertical walls and to a sharp turn with a break may the redistribution of the energy be found from simple kinematic considerations [453]. Apparently, such a situation sometimes occurs in reactions taking place by the so-called rebound mechanism. By this term is meant reactions in which the resulting diatomic molecule is scattered mainly backwards (in the centre of mass system) in relation to the direction of flight of atom A. The cross-section of these reactions, as a rule, does not exceed 10 Å^2 [259].

The fact that, for the rebound mechanism, small impact parameters play a role and that the scattering takes place mainly backwards makes it possible to assume that the exchange takes place on close approach of the atoms, and the rigid sphere model may serve as a satisfactory approximation for a qualitative explanation of the distribution of the energies in an exothermic reaction. Simple treatments, based on a study of the kinematics of the collisions of hard spheres and the elastic-spectator model† lead to the following expressions for the relative kinetic energy E'_{trans} and the internal energy $E'_{\text{vib,rot}}$ of the reaction products [259]

$$E'_{\text{trans}} = E_{\text{trans}} \cos^2 \beta + E_{\text{vib,rot}} \sin^2 \beta,$$
$$E'_{\text{vib,rot}} = E_{\text{trans}} \sin^2 \beta + E_{\text{vib,rot}} \cos^2 \beta + \Delta D, \tag{51.4}$$

where E_{trans} and $E_{\text{vib,rot}}$ are the relative kinetic energy and the internal energy of the reactants; ΔD is the difference between the binding energies of the molecules BC and AB.

Relations (51.4), which are strictly satisfied within the limits of the linear collision in a real three-dimensional collision, determine energies corresponding to the peaks of the corresponding distribution functions. If it is assumed that the energy ΔD is released chiefly on the approach of A to BC, then this term in (51.4) should be dropped, replacing E_{trans} approximately by ΔD. In

---

† This term denotes the situation when the participation of the atom C in the exchange is limited to the role of a spectator elastically scattered by the newly-formed molecule AB.

this case, the fraction of the reaction energy appearing in the form of the vibrations of AB equals $\sin^2 \beta$.

A calculation of the non-linear collisions, carried out within the limits of the same model by Bunker [179], shows that configurations A–B–C, for which $\beta = \pi/2$ and $E'_{\text{vib,rot}} = \Delta D$, always exist. Of course, the question of what is the real contribution of the configurations to the thermal reaction rate remains open, since the dependence of the structure of the surface on the angle ABC is unknown in the general form. Analogous relations are correct for the redistribution of the relative angular momentum l and internal angular momentum j of the molecules [259].

A model of rigid spheres is applicable, strictly speaking, on condition that, during the effective time of collision of the two atoms, their displacement due to the intramolecular potential can be neglected. This, of course, is a very rigid restriction, and it is the worse satisfied the smaller the relative energy of the partners. Specific calculations for this model are carried out in papers [140], [330], on the assumption of the applicability of classical mechanics to the description of rearrangement reactions. A generalization of the theory to the quantum case was made by Ivanov and Sayasov [26]. These authors formulated an assumption which enabled a problem with three (or more) particles to be reduced to problems with two particles. Apart from the assumption of the smallness of the collision time, which takes the form of the condition

$$\tau\omega_{\text{AB}}, \tau\omega_{\text{BC}} \ll 1, \tag{51.5}$$

it should be assumed that the amplitude of the pair scattering F_{ik} is appreciably smaller than the distances between the atoms and molecules

$$F_{ik} \ll R_{ik}. \tag{51.6}$$

Since F_{ik} is usually of the order of an effective atomic radius in molecules, this condition also limits the region of applicability of the model sufficiently rigidly. And, finally, regarding the potential V of the interaction of the three particles, it is assumed to be of an additive nature:

$$V = V_{\text{AB}}(R_{\text{AB}}) + V_{\text{BC}}(R_{\text{BC}}) + V_{\text{AC}}(R_{\text{AC}}). \tag{51.7}$$

As for the pair potentials, they remain completely arbitrary; only one restriction is placed on them, connected with condition (51.5). It should be noted that the approximation of an additive potential (51.7) generally does not allow the potential barrier and the saddle region and also the transition region from one surface to the other to be described. Therefore this approximation can be applied only to reactions in which neither the activation energy nor the adiabatic effects play an important part.

Condition (51.5) limits from below the relative energy of the partners in collision to the value of 5–10 eV. At such energies, the exchange reactions

$$A + BC \to AB + C \qquad (51.8)$$

begin to be accompanied by dissociation reactions

$$A + BC \to A + B + C. \qquad (51.9)$$

With regard to the assumptions (51.5)–(51.7) the theory [26] enables cross-sections of the exchange reaction σ_{exch} and dissociation σ_{diss} to be calculated, taking account of the quantum character of the motion of the nuclei in the reactants and reaction products. We will limit ourselves here to a short account of the classical variant of the theory, describing in rough outline the dependence of the cross-section on the energy in the region whose lower bound lies at 5–10 eV.

First, we note that, as the energy increases, the interaction mechanism changes, so that a transition via a configuration corresponding to a saddle point becomes even less preferable. If, for example, the relative energy of the partners is appreciably less than the difference in activation energies for the linear and triangular configurations, then nearly-collinear collisions make the main contribution to the reaction cross-section. An exchange process, in this case, can be described in the following way: A approaches BC, the chemical bond AB is formed, and then C is repelled from the molecule AB under formation. Precisely this situation is considered in the transition state method. If the relative energy is appreciably higher than the activation energy, so that the configuration ABC of the system does not play a significant role, the other exchange mechanism is preferable: A is incident onto atom C of the BC molecule, knocks out C, and, having slowed down, is found to be bound to atom B.† The formulae quoted below refer to precisely this case.

Assuming, for simplicity, that the target molecule is at rest until the collision, we will consider that the stable molecule AB is formed when the relative energy E_{AB} of the atoms A and B after knocking out C is less than the bond energy D_{AB}.

The formula for the differential cross-section of an energy loss ΔE by the atom A, incident with energy E onto the stationary atom C,‡ serves as the starting point for the derivation of an expression for σ_{exch}. For the rigid sphere model [36], we have

$$d\sigma = \pi a^2 \frac{d\Delta E}{\Delta E_{max}}, \qquad \Delta E_{max} = E \frac{4\mu_{AC}}{m_A + m_C}. \qquad (51.10)$$

---

† The so-called direct knock-out reaction model.

‡ The more general case of collisions in which the motion of the target is taken into account was considered by Porter [411].

The energy E'_{AB} of atom A after scattering on C, expressed in the centre of mass of the system A–B, is given by

$$E'_{AB} = (E - \Delta E)\frac{\mu_{AB}}{m_A}. \qquad (51.11)$$

Expressing $d\Delta E$ in (51.10) in terms of dE'_{AB} and identifying the differential retardation cross-section with the differential capture cross-section, we find

$$d\sigma_{exch} = d\sigma = \pi a^2 \frac{m_A}{\mu_{AB}} \cdot \frac{dE'_{AB}}{\Delta E_{max}}. \qquad (51.12)$$

The total exchange cross-section is hence obtained by integration over dE_{AB}

$$\sigma_{exch} = \int_{E_{AB}^{min}}^{E_{AB}^{max}} \frac{d\sigma}{dE'_{AB}} dE_{AB} = \pi a^2 \frac{m_A}{\mu_{AB}} \cdot \frac{1}{\Delta E_{max}} \left[\min\left\{\begin{array}{c}\frac{\mu_{AB}}{m_A}E \\ D_{AB}\end{array}\right\} - \frac{\mu_{AB}}{m_A}(E - \Delta E_{max})\right]. \qquad (51.13)$$

Hence it follows that, for $E \leqslant (m_A/\mu_{AB})D_{AB}$, the exchange cross-section simply equals πa^2. This can be understood: the relative energy of A–B, from the very beginning, is such that each knocking-out of C enables a stable molecule to be formed. For $E > (m_A/\mu_{AB})D_{AB}$, the exchange cross-section decreases due to the opening of a dissociation channel:

$$\sigma_{exch} = \frac{\pi a^2}{4}\left[\frac{(m_A + m_C)^2(m_A + m_B)}{m_A m_B m_C} \cdot \frac{D_{AB}}{E} - \frac{(m_A - m_C)^2}{m_A m_C}\right]. \qquad (51.14)$$

For that value of the energy E which makes the square bracket on the right tend to zero, the exchange channel is completely closed, and each collision of A + BC leads to the dissociation of BC.

It is evident that, for a small difference in the mass of the atoms A and B, the limiting energy E^* greatly exceeds D_{AB}. For $m_A = m_C$ the exchange cross-section does not tend to zero, in general, but decreases like $1/E$, since the atoms A and C simply exchange momenta.

Although the ideas behind the impulse approximation given above can hardly be applied to calculate rate constants in thermal reactions or even in reactions with energies ~ 10 eV, nevertheless, they enable important limiting cases to be considered, when the solution of the many-body problem can be expressed in terms of the characteristics of pair scattering. Moreover, a consistent theory must take into account repeated scattering [26], the abrupt change in the potential [26], the initial vibrational and rotational excitation of the target molecule, and also the dependence of the effective dissociation energy on the angular momentum of the newly-forming molecule. The theory [26], taking into consideration the vibrational excitation of the molecules (the quantum distribution over momenta in the vibrational ground state or the quasi-classical distribution in a strongly-excited vibrational state), gives the qualitative features of the ion–molecular reaction of an argon ion with molecules of hydrogen and deuterium [240]. However, the interpretation of this reaction as a direct knock-out process cannot be considered definitive.

We will now discuss another type of reaction, taking place by the so-called stripping mechanism. This terminology refers to reactions in which the incident atom A mainly interacts with the one atom of the molecule which directly takes part in the exchange.† The interaction of A with C in the reaction (51.8) can be neglected, for example, in the case of fast collisions, when the time of interaction of A with B is less than the time necessary for a substantial exchange of momentum from B to C. It is possible that processes of a similar type, together with direct knock-out processes, play a part at energies higher than 5–10 eV. For usual thermal reactions ($T < 10\,000$ K) the ratio of the collision time to the intramolecular vibration time is such that the interaction of A with BC is out of the question. Nevertheless there exists a definite type of reaction, even at low temperatures, whose formal characteristics satisfy the stripping conditions. The cross-section of these reactions is usually very large ($\sigma \geq 100$ Å^2), and the diatomic molecule which arises is scattered mainly forwards with respect to the direction of motion of the atom A. Since large cross-sections indicate that exchange takes place at large impact parameters ($b \geq 6$–8 Å), a model with a short-range potential (for example, the rigid sphere model) cannot be enlisted in the interpretation of these reactions.

The reactions

$$M + X_2 \rightarrow MX + X \qquad (51.15)$$

serve as examples of processes of this type, where M is an alkali metal atom and X is an atom of a halogen.

The kinematics of such reactions is approximately described by a 'spectator–stripping' model, according to which atom M strips an X atom from the X_2 molecule and the momentum of the other X atom is practically unchanged. An interpretation of the dynamics of these reactions, proposed by Herschbach

† The difference between the direct knock-out and stripping models is explained most simply by comparing the limiting energy E^* of the incident atom at which the exchange channel is closed. In the first case

$$E_A^* = \left(\frac{m_A + m_C}{m_A - m_C}\right)^2 \left(\frac{m_A + m_B}{m_B}\right) D_{AB},$$

in the second case

$$E_A^* = \left(\frac{m_A + m_B}{m_B}\right) D_{AB}.$$

[259], [260], proceeds from a 'harpooning' mechanism, namely, in the first stage of the process dissociative charge exchange takes place and, in the second, the formation of the bound ion-pair M^+X^-

$$M+X_2 \to M^+ +X_2^- \to M^+ +X^- +X \to (M^+X^-)+X. \qquad (51.16)$$

The first stage—charge exchange—is the only nonelastic process taking place at large distances, so that elastic processes make the main contribution to scattering without redistribution of particles. The fact that all nonelastic channels without redistribution of particles can be excluded from consideration makes it comparatively simple to describe the reaction and the elastic scattering in terms of the so-called optical model. The essence of the optical model is that in place of a real potential $W(R)$, leading to elastic scattering, a complex potential $\mathfrak{W}(R)$ is introduced, whose imaginary part (in a quantum-mechanical treatment of the scattering process) causes the partial absorption of the scattered particles. This absorption, showing up as a weakening of the beam of incident particles, should, of course, be attributed to the reaction. Therefore, if the elastic scattering of the particles can be interpreted in terms of the optical model, it gives direct information about the dependence of the reaction probability on the impact parameter. The optical model describing the reaction as an absorption of incident particles naturally does not give any indication as to the nature of the distribution of energy between the degrees of freedom of the reaction products. In this sense, it gives less information about reactions than does the detailed dynamical model (see § 53). However, it possesses the advantage of allowing a simple interpretation to be made of elastic scattering, in terms of the interaction potential and the interaction causing charge exchange.

An energy diagram illustrating the dissociative charge-exchange process is given in Fig. 50. On the right are represented the terms of the X_2 molecule

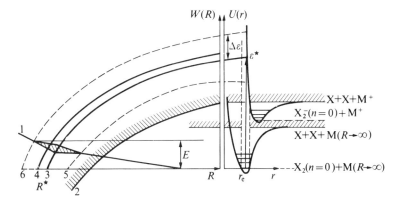

FIG. 50. Electronic terms, corresponding to dissociative charge exchange.

and X_2^- ion (internuclear distance r) at an infinite intermolecular distance R ($R \to \infty$). On the left, the dependence on R of the terms is represented. Term 1 corresponds to the effective interaction of X_2 and M in the initial state for a fixed value of the angle γ between the vector $\mathbf{R}$ and the molecular axis $\mathbf{r}$ of the X_2 molecule. At large distances, of interest in the problem under consideration, this interaction $W_c(R)$ is characterized by a weak attraction. However, with the centrifugal energy taken into account, the effective potential $W_{c,\mathrm{eff}} = W_c(R) + l^2/2\mu R^2$ corresponds to a repulsion. The set of effective ionic terms corresponds to the Coulomb attraction. To each value of the energy ε, measured from the dissociation level of the system X_2 ($n = 0$) + M, there corresponds a term $W_i(R, \varepsilon)$, and $W_{i,\mathrm{eff}} = W_i(R, \varepsilon) + l^2/2\mu R^2$. In Fig. 50 are shown the following terms: 2 denotes the boundary of the continuous spectrum ($\varepsilon = \varepsilon_0$); 3 is a term corresponding to a vertical transition ($\varepsilon = \varepsilon^*$); 4 is a term corresponding to an arbitrary value of the energy ε; 5 and 6 are the boundaries of the Franck–Condon region ($\varepsilon = \varepsilon^* - \Delta\varepsilon$ and $\varepsilon = \varepsilon^* + \Delta\varepsilon$).

The interaction between the covalent and ionic states can be described by different means. First, one can speak of nonadiabatic transitions between the covalent terms and the continuous system of ionic terms. Secondly, by analogy with the arguments given in § 16, one can speak of the broadening of the covalent term immersed in the continuum of ionic terms and of the transformation of the stationary state into a quasi-stationary state. The second interpretation is represented in Fig. 50, where the covalent term 1 is shown broadened within the Franck–Condon region.

We will assume that the decay of each vibrational state of the X_2 molecule in the region $R \sim R^*$ can be considered independently and that the relative motion of X_2 and M can be described within the limits of classical mechanics. It is precisely the first of these assumptions which makes it possible to simplify the general problem significantly, reducing it, in fact, to a two-body problem. The possibility of an independent description of the decay only arises when the interaction of X_2 and M does not induce vibrational transitions between the levels of X_2, during the time of going through the Franck–Condon region (the condition for the adiabaticity of the collisions), and when it is sufficiently weak in absolute value for overlapping of the broadened vibrational levels of the molecule X_2, in the region $R \sim R^*$ not to occur. These conditions take the form

$$\omega_n \tau^* \gg 1, \tag{51.17a}$$

$$\frac{\overline{\Gamma_n}}{\hbar \omega_n} \ll 1, \tag{51.17b}$$

where τ^* is the characteristic time for transmission through the Franck–Condon region; $\overline{\Gamma_n}$ is the mean width of the vibrational level in the region $R \sim R^*$.

On fulfilment of condition (51.17), the decay probability of each level can be calculated by the formula of the adiabatic quasi-stationary approximation:

$$P_n = \exp\left[-\frac{1}{\hbar}\int_{-\infty}^{\infty} \Gamma_n(\mathbf{R})\,dt\right]. \tag{51.18}$$

Expression (51.18) gives an attenuation of the beam of elastically-scattered particles moving along the trajectory $\mathbf{R} = \mathbf{R}(t)$. A rough estimate of the exponent of the exponential in (51.18) gives

$$\frac{1}{\hbar}\int_{-\infty}^{\infty} \Gamma_n(\mathbf{R})\,dt \approx \frac{\overline{\Gamma_n}\tau^*}{\hbar} = \frac{\overline{\Gamma_n}}{\hbar\omega_n}\omega_n\tau^*. \tag{51.19}$$

We will discuss this relation in connection with conditions (51.17). Firstly, it is clear that conditions (51.17) do not restrict the value of the integral (51.19); in the expression on the right, the first factor is small but the second is large. This means that, in the approximation of noninteracting vibrational states, a case of appreciable absorption is possible, i.e. with a comparatively-large reaction probability. Clearly, for sufficiently-large impact parameters relations (51.17) will always be satisfied. However, the total contribution of the collisions for which, under conditions (51.17), the condition for a comparatively-strong absorption $\overline{\Gamma_n}\tau^*/\hbar \geq 1$ is also satisfied depends on the features of the interaction determining the charge exchange. If only the first of the two conditions (51.17) is infringed, then this implies that the vibrational excitation channel can compete with the dissociation channel. If only the second is infringed, then the vibrational excitation probability is small but, on calculating the transition probability, it is necessary to take into account the mutual effect of the different vibrational states of the covalent electronic term. These complications, however, are not crucial in the sense that an approximate solution of the problem can still be obtained as a result of considering the interaction of two particles.† If neither condition in (51.17) is satisfied, then the problem of calculating the cross-section, both in elastic scattering and scattering accompanied by reactions, leads to a general three-body problem, and only one electronic term arising on adiabatic mixing of the covalent and ionic states can be considered.

The chief characteristics of the potential energy surface of the system for which covalent and ionic states exist were discussed in § 48. Here we will note the basic features of the representative point's motion over such a type of surface. The existance of a sharp dislocation on the surface, corresponding to a transition from a covalent to an ionic state of the partners, is very clearly illustrated by the idea of 'electronic harpooning' [259], [260], [339]. Atom A

---

† An analogy to such an approximation, in the theory of vibrational transitions, is the model of a forced oscillator (see § 8), often used in calculations of the vibrational relaxation time.

approaches molecule BC and, at a sufficiently large distance, when the relative motion of the neutral pair A and BC can still be considered as free, an electron of atom A goes over to the BC molecule. After the electronic 'harpoon' has secured the bond of the pair A^+ and BC^- owing to the Coulomb attraction, it turns out to be a possible rearrangement of the system, leading to the formation of the pair A^+B^- and C as a result. The rearrangement process itself can take place in different ways. In particular, the electronic term of the BC^- system can correspond to an unstable state. Should the dissociation of the ion take place during a time small in comparison with the time during which the distance between A^+ and BC^- changes appreciably then, formally, the angular distribution of the particles and the energy distribution might be calculated on the basis of the laws of conservation of angular momentum and energy. However, in contrast to the actual stripping reaction, these relations would have to take into account, in addition, the changes in energy and relative momentum of the pair BC, arising in the nonadiabatic transition of the BC molecule from the neutral to the ionic state.

These ideas were used to interpret the experimental studies of reactions between atoms of K, Rb, Cs and molecules of Br_2 and I_2 [220]. If the dissociation of the ion cannot be considered instantaneous, then the general problem cannot be divided into a set of processes between two particles and, for its solution, it is necessary to consider the interaction of three particles explicitly. In the following paragraph, we will mention some results of calculation of the dynamics of a system, for which a realization of the stripping mechanism might be assumed.

52. Dynamics of exchange reactions

Correlation of the dynamical characteristics with the structure of the potential energy surface

The chief problem which is raised in a study of the dynamics of an elementary exchange process, i.e. in the study of the detailed motion of particles in the region of their intensive interaction, is to determine a correlation between the nature of the potential energy surface and the parameters which characterize the state of the partners up to, and after, the collision. The solution of this problem enables us to regenerate, to some extent, a portion of the potential energy surface, according to the experimental data on scattering accompanied by exchange. As regards the solution of the direct problem— a non-empirical calculation of the reaction cross-section or the rate constant —it is, at present, practically impossible because of the absence of reliable results on the interaction potentials of polyatomic systems. Therefore we will stop to consider work in which the motion of the system is studied by numerical methods, assuming the potential energy surface as given. As regards the qualitative characteristics of different types of surface, here, the following classification, proposed by Polanyi [405]–[407], for exothermic reactions is convenient: a surface is called attractive if the energy of the reaction begins to be released on motion of the point in the valley of the reactants and is called repulsive if the energy is released in the valley of the reaction products. In other words, in the first case, descent begins before the region of the trajectory's turning on the R_{AB}, R_{BC} plane and, in the second, after. Moreover, an intermediate case is possible, a surface of so-called mixed type, where the release of energy occurs in the region of the trajectory's turn.

On the basis of simple qualitative considerations connected with representing the reaction by the motion of the representative point, it might be expected that, for an attractive surface, the heat of reaction would be mainly represented in the form of vibrational energy of the newly-forming molecule [219], [404], [405]. In an attempt to determine a quantitative correlation between E'_{vib}, E'_{rot}, and the quantities A, M, R, expressing the fraction of the heat of reaction ΔD being liberated on various portions of the surface, it is necessary first to define A, M, R unambiguously as quantities physically characterizing the general structure of the surface sufficiently clearly. If such a definition is completely successful, some connection between E'_{vib}, E'_{rot}, ΔD, and A, M, R might then be expected to appear. The determination of such a type of correlation is one of the problems in the study of the dynamics of exchange reactions. Referring the reader to the work of Kuntz, Nemeth, Polanyi, Rosner, and Young [315] with regard to the choice of quantities A, M, R, we will only mention three possibilities: (a) the fractions A, M, R are defined with reference to motion along a right-angled trajectory on the R_{AB}, R_{BC} plane formed by the lines $R_{AB} = r_e^{AB}$ and $R_{BC} = r_e^{BC}$; (b) the fractions A, M, R are defined with reference to motion along the line of minimum energy, asymptotically converging to the line mentioned above; (c) the fractions A, M, R are defined relative to the motion along the actual trajectory which corresponds to the minimum energy sufficient to surmount the barrier. In accordance with the last choice, the triad A, M, R is defined not only by the nature of the surface but also by the ratio of the masses of the particles, since, in the definition of the fraction, an element of dynamics is included.

The numerical study of various models enables the approximately-linear dependence between the fractions of vibrational energy ($E'_{vib}/\Delta D$) and the sum of the coefficients $A + M$ to be ascertained. This implies that the part of the heat of reaction which is set free on motion of the representative point in the valley of the reactants, and at the time of 'rounding the bend', goes over, in preference, into the vibrational energy of the products. Since a massive point rounds the bend more smoothly than a light point, it might be expected that, for the same potential energy surface, the final molecule LH, arising in the reaction L + HH, will be less vibrationally excited than an HH molecule, arising in the reaction H + HH (L and H here denote light and heavy atoms, for example, hydrogen and bromine). This was, in fact, found and was interpreted in terms of the above-mentioned correlation between E'_{vib} and $A + M$ as a result of the relation $M_{\text{L}+\text{HH}} < M_{\text{H}+\text{HH}}$ (the so-called light atom anomaly [315]). On the other hand, this correlation enables one to predict that, for a sequence of potential energy surfaces the height of whose potential barriers are gradually increased (a monotonic decrease in A), E'_{vib} must gradually diminish.

At present, there are a number of papers in which the dynamics of the simplest exchange reactions $A + BC \to AB + C$ are investigated. Unfortunately, it is still impossible to formulate sufficiently clearly even the qualitative connection between the chemical characteristics of the atoms A, B, and C and the reaction rate constant. Therefore, we will restrict ourselves to just the mention of pioneering works on the treatment of linear reactions (two degrees of freedom) [142], [222], [362], [536], [538], planar reactions (four degrees of freedom) [165], [166], [197], [315], [392], [405], [406] and reactions in space (six degrees of freedom) [182], [293], [295], [364], [365], [366], [421], [422], [537].

As a general comment, we will mention that the investigation of the one- and two-dimensional models is of interest in so far as the possibility of interpreting one or other of the features of the reaction, within the limits of the simplified representation, is investigated, in comparison with the more realistic three-dimensional models. Such a comparison is carried out only within the limits of a classical calculation. The solution of the corresponding quantum-mechanical problems presents far greater difficulties; therefore, here, one is restricted to very simple models, for which it is often not easy to establish limits of applicability. In such a situation, it is of interest to carry out classical calculations, imitating the quantum problem to the maximum extent. To this end, the canonical 'action-angle' variables should be used as coordinates, permitting a quantum interpretation in the quasi-classical limit. For the simplest exchange reaction of atom A with a diatomic molecule BC there are six coordinates in all (after separating-out the straight line motion of the centre of gravity of the whole system) characterizing the relative motion of the atoms (for example, the coordinates R, r, γ) and the rotation in space of the system as a whole (for example, the three Eulerian angles of the moving system, determining the orientation of the moving system relative to a stationary system). Thus, for the determination of the trajectory in phase space, twelve first-order equations are obtained, connecting the first derivatives of the generalized coordinates and momenta with their instantaneous values. Use of the laws of conservation of the total energy E and of the total angular momentum vector $\mathbf{J}$ (conservation of the three components J_x, J_y, J_z or of the two canonical action variables J, J_z and one angle variable α_z, conjugate to the action J_z, in the fixed system of coordinates) enables the number of simultaneous equations to be reduced. In particular, on going over into the rotating system of coordinates (see § 7) the Hamiltonian of the three particles can be represented as a function of nine variables: the coordinates r, R, γ, ψ and the momenta p_r, P_R, P_γ, P_ψ, and J. However, the rotating system of coordinates is not convenient for the calculation of the differential reaction cross-section whose angular dependence is referred to a system of coordinates whose axes are fixed in space. Therefore, for numerical calculations, it is more advantageous to leave 12 equations and to use the conservation laws to check the calculations.

In the determination of the phase trajectories of a system of three particles, it is convenient to introduce the generalized momenta and coordinates characterizing the system A + BC at fixed distance r: $l, l_z, j, j_z, \alpha, \alpha_z, \alpha', \alpha'_z, P_R, R$ (see § 7). Going over from the internuclear distance r and conjugate momentum p_r to the canonical variable gives the action variable n and the angular coordinate ϕ.

At an infinite separation between the molecules, the momenta l, l_z, j, j_z, and n do not depend on the time; the meaning of initial quantum numbers of the system can be attributed to them and, in accordance with this, the index i is attached to them. If the free molecule BC is represented by an harmonic oscillator and rigid rotator, then such an identification is developed on the basis of the following relations:

$$E_{BC,rot} = \frac{\hbar^2 j_i(j_i+1)}{2M_{BC}r_e^2}, \quad (52.1a)$$

$$E_{BC,vib} = \hbar\omega_0(n_i+\tfrac{1}{2}), \quad (52.1b)$$

$$(\mu_{A-BC}vb)^2 = \hbar^2 l_i(l_i+1), \quad (52.1c)$$

where $E_{BC,rot}$, $E_{BC,vib}$ are the initial rotational and vibrational energies of the molecule BC; v is the relative velocity of A and BC at infinity ($v = P_R/\mu$); b is is the impact parameter.

From (52.1a), it is clear that the closest approach to the quantum-mechanical treatment is obtained when the dimensionless actions j, n, and l are ascribed integral values. The angular coordinates corresponding to the generalized momenta j and n depend linearly on the time;

$$\alpha' = \omega_{rot}(j_i)t + \alpha'_0, \quad (52.2a)$$

$$\phi' = \omega_0 t + \phi_0 \quad (52.2b)$$

describing the rotation and vibration of the free molecule BC. As for the angles α'_z, α, and α_z, for $R \to \infty$, they do not depend on the time.

Integration of the equations of motion begins with initial conditions, corresponding to fixed values of l_i, n_i, j_i, l_{zi}, j_{zi}, α'_0, α'_z, α, α_z, P_R, ϕ_0, and a sufficiently large value of R, at which the interaction between A and BC can be considered negligibly small. The integration is completed when the variables $n(t)$, $j(t)$, $l(t)$, $l_z(t)$, $j_z(t)$ cease to depend on the time and grow to the asymptotic values $n'_f, j'_f, l'_f, l'_{zf}, j'_{zf}$ and the angles α' and ϕ' become linear functions of the time with initial phases α'_f and ϕ'_f. Such a comparatively simple integration procedure is, strictly speaking, applicable only when redistribution of the particles does not take place, so that, for example, $n(t)$ retains the meaning of an effective quantum number of the BC molecule the whole time. Under exchange, a final value $n'(t)$ describes a vibrational state of the newly-formed AB molecule; therefore, in finding the phase trajectory, the integration must be accompanied by a transformation of the coordinates and momenta. Nevertheless, the final result can always be expressed in the form of a set of numbers $n'_f, \ldots, \phi'_f$ which we will attribute, in future, to the reaction products of AB+C. The set of number $l'_f, l'_{zf}, j'_f, j'_{zf}$, and n'_f can be interpreted approximately as a set of the final quantum numbers l_f, l_{zf}, j_f, j_{zf}, and n_f. Since the latter are integral numbers, the dashed values $l'_f, \ldots, n'_f$ are regarded as coinciding with $l_f, \ldots, n_f$, when the $l'_f, \ldots, n'_f$ fall into a unit interval near $l_f, \ldots, n_f$. Thus the rule for the final ascription is:

$$l'_f = l_f \quad \text{for } l_f - \delta l \leqslant l'_f \leqslant l_f + 1 - \delta l$$
$$\cdots\cdots\cdots\cdots\cdots\cdots\cdots\cdots\cdots\cdots\cdots\cdots\cdots\cdots$$
$$n'_f = n_f \quad \text{for } n_f - \delta n \leqslant n'_f \leqslant n_f + 1 - \delta n. \qquad (52.3)$$

Here, $\delta l, \ldots, \delta n$ are any numbers satisfying the condition $|\delta l| \leqslant \frac{1}{2}$. The uncertainty in their choice indicates the uncertainty in the quantum interpretation of the result of a classical calculation. In quasi-classical conditions, when $l'_f, l_f \gg 1, \ldots, n'_f, n_f \gg 1$, this uncertainty is small.

Since the sets of quantum numbers **i** and **f** include less parameters than the number of initial conditions, there are an infinite number of trajectories connecting the initial and final states. However, not all trajectories with the same values of the generalized momenta $l_i, \ldots, n_i$, but different initial phases $\alpha_0 \ldots \phi_0$ lead to that region of the phase space of the three atoms which is characterized by the set of final generalized momenta $l_f, \ldots, n_f$. The total number of those phase trajectories which are defined initially by the values of the numbers $l_i, \ldots, n_i$ and finally fall into the phase volume determined by the given set $l_f, \ldots, n_f$ can be considered proportional to the probability of the exchange reaction for the channel $(l_i, \ldots, n_i) \to (l_f, \ldots, n_f)$. If this number is normalized to the total number of trajectories emerging from the region $l_i, \ldots, n_i$, then the ratio obtained will simply give the reaction probability $P_L(\mathbf{i}, \mathbf{f})$, with reference to the trajectory L of the three-dimensional space in which the convergence of the reactants takes place (see § 6). Finding and summing different trajectories implies averaging over the angular coordinates $\alpha_0, \ldots, \phi_0$, which are considered to be uniformly distributed over the allowed range 0–2π.

Thus, for the exchange process under consideration, the assignment of a trajectory L and the set of 'quantum numbers' **i** and **f** of the initial and final states implies the following ascription:

$$L \leftarrow v, l, l_z$$
$$i \leftarrow n_i, j_i, j_{zi}$$
$$f \leftarrow n_f, j_f, j_{zf}. \qquad (52.4)$$

The averaging over the components of the momenta simplifies this ascription as follows:

$$L \leftarrow v, l \quad \text{(or } v, b)$$
$$i \leftarrow n_i, j_i$$
$$f \leftarrow n_f, j_f. \qquad (52.5)$$

As a result, $P_L(\mathbf{i}, \mathbf{f})$ reduces to the probability $P_L(i, f)$, which can be employed in the calculation of reaction cross-sections $\sigma_{if}(v)$ by the formulae of § 7. If the cross-sections $\sigma_{if}(v)$ are known, then the partial reaction cross-sections $\sigma(n_i, j_i, v)$ corresponding to the assignment of only the initial states n_i, j_i, and v and the total cross-section $\sigma(v)$ are given by the relations

$$\sigma(n_i, j_i, v) = \sum_f \sigma_{if}(v),$$

$$\sigma(v) = \sum \sigma(n_i, j_i, v) f_{BC}(n_i, j_i), \qquad (52.6)$$

where $f_{BC}(n_i, j_i)$ is the initial distribution function over the internal states of the original molecule BC.

The distribution function over the internal states of the resulting molecule is determined from the relation:

$$f_{AB}(n_f, j_f, v) = \frac{\sum_i \sigma_{if}(v) f_{BC}(n_i, j_i)}{\sigma(v)}. \qquad (52.7)$$

An averaging over the relative velocities of the molecules A and BC, carried out with the help of the function $f_{A-BC}(v)$, enables $\sigma(v)$ to be transformed into a rate constant or mean cross-section $\langle \sigma \rangle$:

$$\kappa = \bar{v}\langle \sigma \rangle = \int_0^\infty \sigma(v) v f_{A-BC}(v)\, dv. \qquad (52.8)$$

An analogous averaging of the function f_{AB} from (52.7) yields the mean distribution function over the internal states:

$$\langle f_{AB}(n_f, j_f) \rangle = \frac{\int f_{AB}(n_f, j_f, v) \sigma(v) v f_{A-BC}(v)\, dv}{\int \sigma(v) v f_{A-BC}(v)\, dv}. \qquad (52.9)$$

Reaction with a potential barrier. The model reaction $H + H_2 \to H_2 + H$

The cross-section of this reaction was calculated by Karplus, Porter, and Sharma [294], [295], within the limits of the above-mentioned variant of the classical approximation for $n = 0$; $j = 0, 1, 2, 3, 4, 5$, and $6 \leqslant E_{\text{trans}} \leqslant 22$ kcal mol^{-1}. The potential energy surface used in the calculation was characterized by a simple peak without a depression at the summit of the barrier. The absolute magnitude of E_0, equal to 9·13 kcal mol^{-1} for the surface chosen, is not of special interest, since the purpose of the discussion is an investigation of the applicability of the one-dimensional model and an evaluation of the accuracy of the transition state method. Various energy characteristics of the reaction

$$H + H_2(n, f) \to H_2 + H \tag{52.10}$$

are presented in Fig. 51. The calculation of the reaction probability for a one-dimensional model (collinear collision) with the initial condition $E_{\text{vib}} = 0$ corresponds to a step function with an effective threshold E'_1 somewhat exceeding the height of the potential barrier E_0. This excess is due to the fact that the trajectory of the representative point does not coincide with

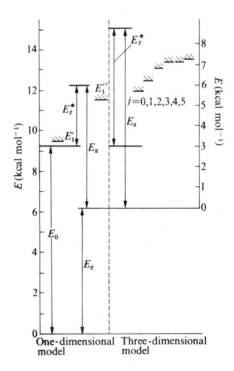

FIG. 51. Energy diagram of reaction $H + H_2 \to H_2 + H$.

the line of the saddle, as a result of the effective influence of the centrifugal forces acting in the region of the trajectory's turn. If the initial vibrational energy is chosen equal to $E_{\text{vib}} = E_z = \frac{1}{2}\hbar\omega$, which, in a quantum-mechanical treatment, corresponds to the zero-point energy of vibrations, i.e. $n = 0$, then the minimum of the translational energy $E_{\text{trans}}^{\min}$ which is necessary for the classical surmounting of the barrier equals $E_0 - E_z$. In the quantum-mechanical treatment of the transition state method $E_{\text{trans}}^{\min}$ is identified with the activation energy E_a, renormalized by the contribution of the energies of the zero-point vibrations of the reactants and of the activated complex, i.e. $E_a = E_0 - E_z + E_z^\star$. Such a definition of the threshold is not completely satisfactory, since the energy $E_z^\star$ corresponds to some quasi-stationary state possessing, in reality, a whole spectrum of energies. It is quite possible that, as a result of nonadiabatic effects (the dynamical coupling of the reaction coordinate with the deformation vibrations of the activated complex), there exists an appreciable probability of the particle's transmission under the adiabatic barrier E_a. This effect is not taken into account in the transition state method, which possibly partly explains the discrepancy between theory and experiment discussed in § 49.

The classical calculation of the reaction probability for the linear model, with the initial condition $E_{\text{vib}} = E_z$ (i.e. $n = 0$), corresponds to a step function with an effective threshold E_1'' lying higher than $E_0 - E_z$ but lower than $E_0 - E_z + E_z^\star$.

The first condition means that, at the time of transition of the representative point through the barrier, the vibration normal to the reaction coordinate q_r changes almost adiabatically, so that the action variable $J = E_{\text{vib}}(H \ldots H_2)/\hbar\omega(H \ldots H_2)$ changes but little. From the condition for the constancy of J, we find $E_{\text{vib}}^\star = E_{\text{vib}} \hbar\omega^\star / \hbar\omega$, i.e. $E_{\text{trans}}^{\min} > E_0 - E_z$.

The second condition indicates that the vibrational energy, to some extent, contributes to the surmounting of the potential barrier, as a result of the direct conversion of vibrations into translational energy along the reaction coordinate (a nonadiabatic effect). This contribution should be distinguished from that taken into account in the classical and quantum-mechanical formulations of the transition state method, which is due simply to the *widening* of the reaction channel on excitation of the vibrations (an adiabatic effect). It is not clear whether, in general, a quantum-mechanical interpretation of such a type of nonadiabatic classical effect is possible in the case when the excitation of the highest vibrational levels is energetically impossible. If such an interpretation is possible, then it must take account of the quasi-stationary character of the level E_a.

We will now go over to consider the three-dimensional model. The energy levels $E(j)$, $j = 0, 1, 2, 3, 4, 5$, shown in Fig. 51, correspond to the thresholds of the cross-sections $\sigma(E_{\text{trans}}, n = 0, j)$. The cross-sections increase monotonically from threshold, converging to an asymptotic value $\sim 1.5 \text{ Å}^2$ for

$E_{\text{trans}} \approx 22$ kcal mol^{-1}. Qualitatively, such a dependence is obtained even for the simplest model, within the bounds of which it is assumed that, in surmounting the barrier, only the component of the velocity of the relative motion along the line of impact is effective.

For the three-dimensional model, the dependence of the reaction cross-section on the relative energy is obtained most simply by making use of the general relation (6.18), since the rate constant is easily calculated by the transition state method. The partition functions of the internal degrees of freedom in the reactants and the activated complex should be considered identical so that, in general, they drop out of the ratio $F^\star_{\text{ABC}}/F_A F_{\text{BC}}$. The remaining pre-exponential factor is equal to the average current directed into a sphere of radius $R^\star$. The mean current is given by the relation (5.7), where m_r should be identified with the reduced mass of the partners μ. In this way, we obtain the following relation:

$$4\pi(R^\star)^2 (2\pi\mu\beta)^{-\frac{1}{2}} \exp(-\beta E_0) = \left(\frac{\pi\mu\beta}{8}\right)^{-\frac{1}{2}} \int_0^\infty \sigma(E) \exp(-\beta E) \beta^2 E \, dE. \quad (52.11)$$

Considering it as an equation in $\sigma(E)$, and applying the inverse Laplace transform to both parts of the equality (52.11), we find

$$\sigma(E) = \begin{cases} \pi(R^\star)^2 \left(1 - \dfrac{E_0}{E}\right) & \text{for } E > E_0 \\ 0 & \text{for } E < E_0. \end{cases} \quad (52.12)$$

The values of σ found in the paper [295] reach the asymptote faster than follows from (52.12). Qualitatively, this can be understood within the framework of ideas behind the transition state method. On an increase in the relative velocity, not only the component along the line of impact (the normal component) increases but also the tangential component. This leads to an increase in the amplitude of the deformation vibrations of the transition complex and consequently to a widening of the channel. The widening of the channel is reflected in the appearance of an additional factor T in the formula of the classical variant of the transition state method (49.2b) in comparison with the formula of the collision theory which, for a cross-section of type (52.12), gives the dependence

$$\kappa(T) \sim T^{\frac{1}{2}} \exp\left(-\frac{E_0}{kT}\right). \quad (52.13)$$

As regards the comparatively-weak dependence of the threshold of the cross-sections on the initial angular momentum of H_2, physically, it is explained by the difficulty in achieving the relative orientation of the partners necessary for the accomplishment of exchange.

The results obtained by Karplus, Porter, and Sharma [295] enable the accuracy of the transition state method, or at least its classical variant, to be tested directly. The direct averaging of the calculated cross-section over the rotational states of H_2 and over the relative velocities enables the rate constant κ to be represented over the temperature range $300\,K < T < 1000\,K$, in the form of either of the formulae:

$$\kappa(T) = 4\cdot3 \times 10^{13} \exp\left(-\frac{7\cdot435}{RT}\right) \text{cm}^3 \text{ mol}^{-1} \text{s}^{-1}, \tag{52.14a}$$

$$\kappa(T) = 7\cdot37 \times 10^{9} T^{1\cdot176} \exp\left(-\frac{6\cdot234}{RT}\right) \text{cm}^3 \text{ mol}^{-1} \text{s}^{-1}. \tag{52.14b}$$

These expressions should be compared with formula (49.1), in which the classical-rotational and quantum-vibrational partition functions have been substituted. Neglecting the contribution of the vibrational partition functions for the symmetrical vibrations of $(H_3)^*$ and H_2, explicitly separating out the contribution of the deformation vibrations, and completely neglecting the tunnel effects (disregarded in the quasi-classical approximation under consideration), we will present $\kappa(T)$ in the form:

$$\kappa(T) = g^\star \left(\frac{3m_H}{2}\right)^{\frac{1}{2}} \frac{I^\star}{I} \frac{\sigma_{HH}}{\sigma^\star} \left(\frac{2\pi}{kT}\right)^{\frac{1}{2}} \hbar^2 \left[2 \sinh\left(\frac{\hbar\omega_2}{2kT}\right)\right]^{-2}$$

$$\times \exp\left[-\frac{E_0 - E_z + (E_z)^\star}{kT}\right]. \tag{52.15}$$

Here, $g^\star = 2$ if $\sigma_{HH} = \sigma^\star = 2$, or $g^\star = 1$ if $\sigma_{HH} = 2$, $\sigma^\star = 1$ (see §4). The energy of activation, which comes into the exponential, agrees with the E_a of the one-dimensional model $E_a = 9\cdot13 - 6\cdot2 + 3\cdot12 = 6\cdot05$ kcal mol^{-1}. Rewritten in a form analogous to (52.14), the expressions take the form

$$\kappa(T) = 7\cdot413 \times 10^{13} \exp\left(-\frac{8\cdot812}{RT}\right), \tag{52.16a}$$

$$\kappa(T) = 2\cdot163 \times 10^{11} T^{0\cdot797} \exp\left(-\frac{7\cdot998}{RT}\right). \tag{52.16b}$$

Comparison of (52.16) with the corresponding expressions (52.14) shows that the approximation of the transition state method differs significantly from the result of an exact calculation. To explain the true reasons for this difference, a careful investigation of the various forms of error introduced by the basic assumptions of the transition state method is necessary.

Reactions with a potential well

Investigation of the trajectories of the representative points for the reaction considered above, and also for a number of other processes taking place

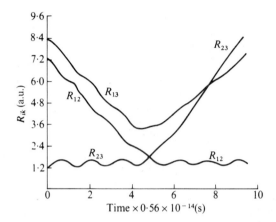

FIG. 52. Typical phase trajectory $R_{ik}(t)$ of the reaction $H + H_2 \to H_2 + H$, illustrating the direct mechanism (Karplus, Porter, and Sharma [294]). R_{23} is the oscillating curve up to exchange; R_{12} is the oscillating curve after exchange. Initial conditions: $j = 0, n = 0, v = 1.32 \times 10^6$ cm s^{-1}.

with an energy of activation, shows that, qualitatively, these processes correspond to a direct mechanism when the rearrangement time proves to be of the order of the frequency of vibrations of the nuclei (Fig. 52). In a qualitative form, this mechanism can be considered as a limiting case of the impulsive interaction, although the actual dynamics of the direct mechanism essentially contains a simultaneous interaction of three particles.

If there is a small depression at the peak of the potential barrier, or the barrier is generally replaced by a potential well, then the formation of an intermediate complex whose lifetime appreciably exceeds the period of one vibration is possible on the close approach of the three atoms. In this case, the transition to the statistical model of exchange reactions would be brought about. A similar type of process was considered by Raff and Karplus [421] in the example of the reaction

$$K + CH_3I \to KI + CH_3. \tag{52.17}$$

In a dynamical analysis of the exchange, the group CH_3 was replaced by a point mass and the potential energy surface was formed by combining the ionic and covalent potentials (see § 48). The level lines of this surface for the linear configuration K–I–CH$_3$ are shown in Fig. 53. The potential well on the surface can be interpreted qualitatively as a term of the ion $(CH_3I)^-$, arising after the electronic harpooning and strongly distorted by the influence of the Coulomb field of K^+. As was noted in § 51, reactions accompanied by charge exchange can be interpreted, in a number of cases, within the limits of the stripping model. In fact, however, for the surface shown in Figs. 43 and 53, the real situation is very far from the dynamics of

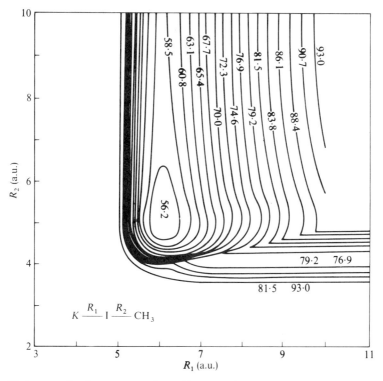

FIG. 53. An example of an energy surface with a potential well near the saddle point. System K + CH$_3$I (Raff and Karplus [421]).

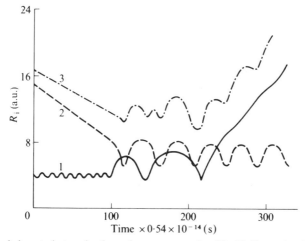

FIG. 54. Typical phase trajectory for the surface, represented on Fig. 53, illustrating the formation of an intermediate complex. Curve 1 is $R_{\text{I-CH}_3}(t)$; curve 2 is $R_{\text{K-I}}(t)$; curve 3 is $R_{\text{K-CH}_3}(t)$ (Raff and Karplus [491]).

stripping. As is evident from Fig. 54, the exchange time τ by far exceeds the period of vibration of the reactant molecule, moreover, to an approximate extent, it is possible to speak of the existence of the complex $(K-I-CH_3)^*$. Of course, the formation of such an intermediate complex is essentially reflected in the distribution of the energy over the degrees of freedom of the reaction products and in their angular distribution. For a more detailed study of these problems we should turn to the original paper [421]. Here, we will mention that computer calculations for similar types of process enable a lower limit to be established for the time τ at which the statistical description of the exchange reaction becomes adequate.

53. Disturbance of the equilibrium distribution in bimolecular reactions

Bimolecular reactions, like unimolecular reactions, can, to some extent, cause a disturbance of the equilibrium distribution. In contrast to the latter, however, this violation does not inevitably appear as a change in the order of the reaction. This is connected with the fact that both redistribution of atoms and relaxation can take place in collisions of one and the same molecules. In order that the disturbance of the equilibrium distribution should cause a change in the order of the reaction, as regards the concentration of the reacting molecules, it is necessary to bring about those conditions under which relaxation and exchange would take place due to the collisions of different molecules. We will mention an example of such a process below, but first will pause to consider the simplest bimolecular reaction

$$X + X \to Z + U. \qquad (53.1)$$

We will consider the initial stage of the reaction (53.1), neglecting the effect of the reverse process and assuming that the reaction cross-section σ only depends on the relative kinetic energy of the partners E_{trans}. At present, it is difficult to make definite assertions as to the reactions for which this, in fact, takes place. For example, for the reaction discussed, $H + H_2$, this approximation is satisfactory, since the contribution of the excited vibrational states of H_2 can be neglected for $T < 4000$–6000 K, and the rotation makes almost no contribution to the surmounting of the activation barrier.

The kinetic equation (or Boltzmann equation) for a single-particle distribution function f in an isothermic homogeneous system is derived on the basis of the balance of particles in an element of phase volume $d\Gamma$.† Since, in the region $d\Gamma$, the change in the number of particles takes place only due

---

† The basis of this derivation is contained in the papers of Bogolyubov [2], [3].

to collisions, it is possible to write

$$\frac{\partial}{\partial t} f(1) \, d\Gamma_1 = -T_a + T_b - R, \tag{53.2}$$

where the index 1 signifies that the distribution function depends on the phase coordinates of the molecule 1. The terms $-T_a$ and $-R$ give the reduction in the number of molecules in $d\Gamma_1$ as a result of elastic collisions and reactions, and the term T_b gives the increase in the number of molecules as a result of elastic collisions. The reduction in the number of molecules in $d\Gamma_1$ as a result of collisions (elastic and inelastic) takes place due to the escape of other molecules out of the element $d\Gamma_2$. Therefore for $T_a + R$ we can write

$$T_a + R = \int_{\Gamma_2} f(1) f(2) \, d\Gamma_1 \left(\frac{d\Gamma_2}{dt} \right), \tag{53.3}$$

and the differentiation $d\Gamma_2/dt$ must be referred to the relative distance R_{12}. Following the discussion of §6, we will represent $d\Gamma_2/dt$ in the form

$$\frac{d\Gamma_2}{dt} = 2\pi b \, db v_{12} \, d^3 v_2 = d\sigma_t v_{12} \, d^3 v_2, \tag{53.4}$$

where $d\sigma_t$ is the total differential scattering cross-section. (Here, and in future, we will drop unimportant factors $((2\pi\hbar)^3)$ in the denominator.)

We will now rewrite (53.3) in the form

$$T_a + R = f(1) \, d\Gamma_1 \int f(2) \, d\sigma_t(1, 2) v_{12} \, d^3 v_2, \tag{53.5}$$

where it is noted that the differential cross-section depends on the velocities of molecules 1 and 2.

We will denote the velocities of molecules 1 and 2 after the elastic collision by $\mathbf{v}'_1$ and $\mathbf{v}'_2$. Then T_b can be represented in the form

$$T_b = \int f'(1) f'(2) \frac{d\Gamma'_1 \, d\Gamma'_2}{dt}, \tag{53.6}$$

where the integration is taken over the phase space of Γ'_1 and Γ'_2 subject to a well-defined connection between $\mathbf{v}'_1, \mathbf{v}'_2$ and $\mathbf{v}_1, \mathbf{v}_2$, which follows from the dynamics of the pair collision. The contribution due to the inelastic collisions must be eliminated from the current in the integrand in (53.6). This can be done by introducing, for each trajectory characterized by parameters b and v_{12}, a factor $[1 - P_r(b, v_{12})]$, in which P_r denotes the reaction probability. Introducing the differential reaction cross-section

$$d\sigma_r = 2\pi b P_r(b, v_{12}) \, db \tag{53.7}$$

and using Liouville's theorem $d\Gamma_1 d\Gamma_2 = d\Gamma'_1 d\Gamma'_2$, we will rewrite T_b in the form

$$T_b = \int f'(1)f'(2) \, d\sigma_t(1,2) v'_{12} \, d^3 v_2 - \int f'(1)f'(2) \, d\sigma_r(1,2) v'_{12} \, d^3 v_2, \quad (53.8)$$

where, in the integrands, it is necessary to take account of the connection between $\mathbf{v}'_1, \mathbf{v}'_2$ and $\mathbf{v}_1, \mathbf{v}_2$. Thus the kinetic equation (53.2) can be written in the form

$$\frac{\partial f(1)}{\partial t} = \int [f'(1)f'(2) - f(1)f(2)] \, d\sigma_t(1,2) v_{12} \, d^3 v_2$$

$$- \int f'(1)f'(2) \, d\sigma_r(1,2) v_{12} \, d^3 v_2, \quad (53.9)$$

where the differential cross-sections for the total, elastic, and inelastic (in the given case, leading to a reaction) scattering are connected by the relation

$$d\sigma_t = d\sigma_e + d\sigma_r. \quad (53.10)$$

This kinetic equation differs from eqn (23.4) in that it is non-linear in the function sought for f. Since this complicates the solution substantially, the first step in the simplification consists of an approximate reduction of (53.9) to a linear equation. With this aim, we will assume that f can be represented as a linear combination of a Maxwellian distribution function $f_0(v)$ and a small correction $f_0(v)\phi(v)$:

$$f(v) = Af_0(v)[1 + \phi(v)], \qquad \phi(v) \ll 1. \quad (53.11)$$

In the general case, the time dependence of $f(v)$ is contained in both the coefficient A and in the unknown function ϕ, where the type of relaxation process under study determines which of these makes the chief contribution.

For simplicity, we will first consider processes in which, in general, there are no chemical reactions, i.e. $\sigma_r = 0$. In this case, the change in the distribution function takes place owing to the translational relaxation and, for $t \to \infty$, $f \to f_0$. Therefore A in (53.11) should be identified with the time-independent mean density of the molecules n_0, and $\phi(v)$ must be considered time dependent. Substituting (53.11) into (53.9) and neglecting terms of the second order in ϕ, we obtain the equation [401]:

$$f_0(\mathbf{v}) \frac{\partial \phi(\mathbf{v}, t)}{\partial t} = \int f_0(\mathbf{v}) f_0(\mathbf{u}) [\phi(\mathbf{v}', t) + \phi(\mathbf{u}', t) - \phi(\mathbf{v}, t) - \phi(\mathbf{u}, t)]$$

$$\times d\sigma_e(\mathbf{u} - \mathbf{v}) |\mathbf{u} - \mathbf{v}| \, d^3 u. \quad (53.12)$$

Just as in the case of the differential equation of diffusion theory, this integral equation has a solution written down in the form of a sum of terms exponentially decreasing with time. Only the most slowly-decreasing term, which also determines the decay of the perturbation for large values of the time, is of interest. Therefore we will assume

$$\phi(\mathbf{v}, t) = \phi_1(\mathbf{v}) \exp(-\mu_1 t). \tag{53.13}$$

Substituting (53.13) into (53.12), we obtain an eigenvalue integral equation for the determination of μ_1 and ϕ_1:

$$\mu f_0(\mathbf{v})\phi_1(\mathbf{v}) = \int f_0(\mathbf{v}) f_0(\mathbf{u}) [\phi_1(\mathbf{v}') + \phi_1(\mathbf{u}') - \phi_1(\mathbf{v}) - \phi_1(\mathbf{u})] \, d\sigma_e |\mathbf{v} - \mathbf{u}| \, d^2\mathbf{u} \tag{53.14}$$

whose approximate solution can be found by a variational method. The corresponding variational principle is written in the form

$$\mu_1 = \min \frac{\int f_0(\mathbf{v}) f_0(\mathbf{u}) y(\mathbf{v}) [y(\mathbf{v}') + y(\mathbf{u}') - y(\mathbf{v}) - y(\mathbf{u})] \, d\sigma |\mathbf{v} - \mathbf{u}| \, d^3\mathbf{u} \, d^3\mathbf{v}}{\int y^2(\mathbf{v}) f_0(\mathbf{v}) \, d^3\mathbf{v}}, \tag{53.15}$$

where $y(\mathbf{v})$ is the trial function approximating $\phi(\mathbf{v})$.

The lowest eigenvalue μ_0 of eqn (53.14) equals zero. For the isotropic case, there are two eigenfunctions ϕ_0' and ϕ_0'' corresponding to this eigenvalue,

$$\phi_0' = a_0', \tag{53.16a}$$

$$\phi_0'' = a_0''(\tfrac{3}{2} - \xi^2), \qquad \xi^2 = \frac{\mu v^2}{2kT}. \tag{53.16b}$$

On condition $\phi_0', \phi_0'' \ll 1$, both these functions can be incorporated into the exponential Maxwellian factor $\exp(-\mu v^2/2kT)$; the constant terms only change the normalization and the term $a_2 \mu v^2/2kT$ in (53.16b) leads to a small change of temperature. If the normalization of the equilibrium function f_0 is chosen correctly beforehand and the temperature is defined in terms of the average kinetic energy by the usual relation

$$\tfrac{3}{2}kT = \int \frac{\mu v^2}{2} f_0(\mathbf{v}) \, d^3\mathbf{v}, \tag{53.17}$$

then the correction determined by the functions ϕ_0' and ϕ_0'' should be discarded, having assumed $a_0' = a_0'' = 0$. To exclude the contribution of these effects to the zero eigenvalue μ_1, the trial function in (53.15) must be chosen orthogonal to the functions ϕ_0' and ϕ_0'', i.e. must, in addition, satisfy the relations

$$\int y(\mathbf{v}) f_0(\mathbf{v}) \phi_0'(\mathbf{v}) \, d^3\mathbf{v} = \int y(\mathbf{v}) f_0(\mathbf{v}) \phi_0''(\mathbf{v}) \, d^3\mathbf{v} = 0. \tag{53.18}$$

The simplest trial function, written in the form of a polynomial in the parameter ξ^2 and satisfying conditions (53.18), has the form

$$y = a_1(\tfrac{15}{8} - \tfrac{5}{2}\xi^2 + \tfrac{1}{2}\xi^4), \qquad (53.19)$$

where the coefficient a_1 remains arbitrary and is determined by the initial degree of perturbation in the distribution function. Substitution of (53.19) into (53.15) enables (53.15) to be calculated immediately, without finding the minimum, since (53.11) does not contain a single variational parameter. An application of the variational method to the model of rigid spheres gives [401]

$$\mu_1 = \frac{8}{15} n \left(\frac{8kT}{\pi\mu}\right)^{\frac{1}{2}} \pi d^2, \qquad (53.20)$$

which is smaller than the gas-kinetic number of collisions by approximately a factor of two.

We will now assume that the reaction cross-section σ_r differs from zero, and that the reaction threshold E_0 is sufficiently large. In this case, there are characteristic times τ_1 and τ_2 in the system, the first of which characterizes the rate of the re-establishment of the Maxwellian distribution due to elastic collisions, and the second characterizes the chemical reaction rate. On condition $\tau_1 \ll \tau_2$ (which is satisfied if E_0 is sufficiently large) and $\tau_1 \ll t \approx \tau_2$, it might be expected that the non-equilibrium distribution function f would exhibit quasi-stationary behaviour, i.e. would have the form (53.11), where the time dependence is contained in the factor A only. This situation is analogous to the discussion earlier on, in connection with the calculation of the non-equilibrium distribution function in the diffusion theory of reactions. As was shown above, for the proper determination of the number of particles and temperature for $\phi(\mathbf{v})$, the equalities (53.18), in which $y(\mathbf{v})$ is replaced by $\phi(\mathbf{v})$, can be assumed fulfilled. To an accuracy up to terms in ϕ^2, the coefficient A must then be identified with the instantaneous number of particles $n(t)$. Substituting the function f, thus chosen, into eqn (53.9) and neglecting the squared terms in ϕ, we obtain the integral equation for $\phi(\mathbf{v})$. Its chief difference from (53.12) will be that, on the left, in place of $f_0(\partial\phi/\partial t)$, will be the term

$$\frac{\partial f}{\partial t} = \frac{dn}{dt}(f_0 + f_0\phi) \approx \frac{dn}{dt} f_0. \qquad (53.21)$$

The last approximate equality is correct because the change in the number of particles dn/dt is due to the reaction and is consequently considered of the same order of magnitude as ϕ. It is convenient to represent the equation for ϕ, obtained with regard for this approximation, in a symmetric form, multiplying both parts by an arbitrary function $Q(\mathbf{v})$ and integrating over $d^3\mathbf{v}$.

Straightforward calculations give [413]

$$\frac{dn}{dt}\int Q(1)f_0(1)\,d^3v_1 = \frac{1}{4}\int [Q(1')+Q(2')-Q(1)-Q(2)]$$
$$\times f_0(1)f_0(2)[\phi(1')+\phi(2')-\phi(1)-\phi(2)]\,d\sigma_e|\mathbf{v}_1-\mathbf{v}_2|\,d^3v_1\,d^3v_2$$
$$-\int Q(1)f_0(1)f_0(2)\,d\sigma_r|\mathbf{v}_1-\mathbf{v}_2|\,d^3v_1\,d^3v_2. \tag{53.22}$$

If the unknown function $\phi(\mathbf{v})$ is now represented in the form of an expansion in a complete set of functions $S_m(\mathbf{v})$, orthogonal to $f_0(\mathbf{v})$,

$$\phi(\mathbf{v}) = \sum_m a_m S_m(\mathbf{v}), \tag{53.23}$$

and each of the functions of this set is taken as $Q(\mathbf{v})$, then (53.22) leads to an infinite system of linear nonhomogeneous equations for the expansion coefficients a_m of the function $\phi(\mathbf{v})$ in terms of $S_m(\mathbf{v})$. In fact, the solution of this system can be found if it is restricted to a finite number of terms, and the crudest approximation consists of just a single remaining term from the series in (53.23). In calculations of the corrections carried out up to the present, the function in (53.19) was used as the single term.

As to the models of elastic and inelastic interactions, the total collision cross-sections were chosen in accordance with the rigid sphere model

$$\sigma_e = \pi d^2, \tag{53.24}$$

and, for the reaction cross-section, various energy-dependencies were assumed:

$$\sigma_r = \pi d^2 \left[1 - \exp\left(-\frac{E}{E_0}\right)\right] \quad [414] \tag{53.25a}$$

$$\sigma_r = \begin{cases} 0, & E < E_0 \\ \pi d^2, & E > E_0 \end{cases} \quad [415] \tag{53.25b}$$

$$\sigma_r = \begin{cases} 0 & E < E_0 \\ \pi d^2(1-E_0/E), & E > E_0 \end{cases} \quad [412]. \tag{53.25c}$$

Of these, only the last is, to some extent, physically based (the component of velocity along the line of impact is considered effective in surmounting the barrier).

After ϕ is found, the non-equilibrium rate constant κ of a second-order reaction can be calculated:

$$\kappa = -\frac{1}{n^2}\frac{dn}{dt},$$
$$\kappa = \int f_0(1)f_0(2)[1+\phi(1)+\phi(2)]\,d\sigma_r|\mathbf{v}_1-\mathbf{v}_2|\,d^3v_1\,d^3v_2. \tag{53.26}$$

The calculation of the integral (53.26) gives, for the cross-sections (53.24) and (53.25c),

$$\kappa = \pi d^3 \left(\frac{8kT}{\pi\mu}\right)^{\frac{1}{2}} \exp\left(-\frac{E_0}{kT}\right)(1-\eta), \tag{53.27}$$

$$\eta = \frac{1}{32}\left(x^4 - 2x^3 + \frac{x^2}{2} + \frac{x}{2} - \frac{1}{16}\right)\exp(-x), \qquad x = \frac{E_0}{kT}, \tag{53.28}$$

Here η is the non-equilibrium correction to the equilibrium rate constant, where, by agreement, in the derivation it is assumed that $\eta \ll 1$. This condition makes it possible to estimate roughly the effect of the disturbance of the equilibrium distribution on the reaction rate. At $x = 5$, for example, the reaction rate is 8 per cent smaller than its equilibrium value; this correction, of course, is smaller than the errors introduced by the model itself.

The estimate quoted for the disturbance of the equilibrium distribution cannot be considered conclusive for two reasons. First, the linearization of eqn (53.9) restricts the range of applicability of the theory to small perturbations and the method of solution of the already-linear equation does not enable one to elucidate the problem of when one may speak of a settled quasi-stationary approximation. Secondly, a too-rough approximation is taken for the required function ϕ. The restriction of the series (53.23) to the first few terms, containing low powers of the parameter ξ, is justified in calculating transport coefficients (see, for example, Chapman and Cowling's book [125] since, there, the finding of the distribution function close to the maximum proves to be important. In chemical reactions, however, it is the molecules found in the 'tail' of the Maxwellian distribution which make the main contribution to the rate, therefore, high powers of ξ, which may give the distortion of the equilibrium function for $E/kT \gg 1$ with greater accuracy, should be taken into account.

These difficulties fall away, to a considerable extent, if the linearization of the Boltzmann equation is connected not with the small perturbation of the distribution function f_X but with the possibility of neglecting the collisions of the X molecules among themselves. Such a situation arises in considering reactions of the type

$$X + Y \rightarrow Z + U, \tag{53.29}$$

in which the X component is just a small impurity in the gas Y. The greatest perturbation of the distribution function f_X is obtained when the masses of the molecules X and Y are strongly differentiated. If $m_X \ll m_Y$ then the Boltzmann equation reduces to the differential equation of diffusion theory, which enables the solution to the problem to be obtained comparatively easily. In a number of cases, it is possible to obtain a sufficiently-good approximation on solution of the integral equation (53.9), which serves as a base for the derivation of the diffusion equation [310]. Thus, for example, it turns out that the reaction

$$H + Br_2 \rightarrow HBr + Br, \qquad (53.30)$$

taking place in a Xe atmosphere, causes substantial disturbance of the Maxwellian distribution of the H atoms if the relative concentration of the Br_2 molecules exceeds 20 per cent [310].

One more effect of the disturbance of the equilibrium distribution in a reaction of type (53.29), on condition $m_X \ll m_Y$, should be mentioned [101]. This effect is made manifest in that the distribution functions of each of the components approximately preserving the form of a Maxwellian distribution correspond to different temperatures. The intensive loss of energy by the light gas in the reaction (53.29) and the comparatively-slow process of restoration of energy in the elastic collisions of the X and Y molecules mean that the temperature of the translational degrees of freedom of the light gas proves to be appreciably lower than that of the heavy gas. As an example having, however, no immediate relation to the exchange reaction but clearly illustrating the effect mentioned above, the appreciable difference (up to 300 per cent) in temperature of the electrons and atoms behind a shock front, arising as a result of the loss of energy by the electrons in ionizing collisions, can be indicated.

The methods of calculation of the non-equilibrium distribution function mentioned above are limited to the case of a small admixture of reacting molecules in the gas Y. Following the review [103], we will indicate a more general method applicable to mixtures with an arbitrary concentration and proposed by Stupochenko [111], [112]. Stupochenko's method is based on the use of special (discontinuous) properties of the transition probabilities, calculated for the model of rigid elastic spheres, and enables the problem of solving an integral gas-kinetic equation to be reduced to the solution of a certain differential equation. The differential equation thus obtained differs, in essence, from the Fokker–Planck equation. It is more convenient to illustrate the results obtained by Stupochenko by the example of the problem of the distribution of kinetic energy of the reaction products (the fast particles Z) arising in the gas U. It is considered that the concentration of the gas Z is small and that the masses of the gases Z and U are equal ($m_Z = m_U$). This problem permits an exact solution (one of the few exact solutions of the gas-kinetic Boltzmann equation), and the results describe the behaviour of the distribution function in arbitrary systems with fast particle sources. The solution of this problem enabled an exact expression to be found for the distribution function, correct for all ranges of the kinetic energy and for any value of the initial energy of the reaction products. The following deduction can be made from this solution. A source of particles with initial kinetic energy E^*_{trans} leads to a disturbance (finite) of the Maxwellian distribution function, which is mainly concentrated in the energy region $E_{\text{trans}} < E^*_{\text{trans}}$. In the energy region $E_{\text{trans}} > E^*_{\text{trans}}$, the perturbation function dies away exponentially (with an unimportant factor). The method proposed by Stupochenko is applicable to the solution of the problem with both positive and negative sources.

Returning to the methods for taking account of the disturbance of equilibrium, set out above and represented by expressions (53.27) and (53.28), we note that they, in fact, use the reaction cross-section σ_r as the small parameter in the perturbation theory. However, in principle, another formulation of the problem is possible when the ratio of the reaction rate to the mean number of collisions is a small parameter [103]. It is clear that such a situation is possible when this parameter is still small (for example, the number of particles in the range of velocities covered by an appreciable perturbation is small in comparison with the total number of particles) but σ_r still cannot be considered as a small perturbation. It turns out that this method of calculation gives a greater disturbance of the equilibrium distribution than the simple method of perturbation theory in σ_r. In particular, for the model reaction with the cross-section (53.25c), the reaction rate at $x = 5$ proves to be 11 per cent less than the equilibrium rate [103] and not the 8 per cent as follows from (53.28).

The calculation of non-equilibrium effects is even more complicated if the internal degrees of freedom of the reacting partners is taken into consideration in the kinetic equation. The problems arising here, on which we will not dwell, are discussed in a number of papers [419], [497], [499]. Such a situation occurs, in particular, in exchange reactions, where the vibrational transitions in the molecule–reactants, induced by collisions with molecules of the inert gas, form the slow stage. As an example, reactions of the type

$$HX + DY \rightarrow DX + HY, \qquad (53.31)$$

studied by Bauer and coworkers, can be indicated [146], [147]. In contrast to the simple exchange reactions considered above, here, two bonds are formed and disappear in a single elementary process. It might be expected that the rate of rupture of the bonds in the molecules HX and DY depends strongly on the degree of their vibrational excitation, i.e., in other words, that the vibrational excitation of the partners makes the main contribution to the surmounting of the activation barrier. Since the rate of vibrational relaxation is appreciably smaller than the translational relaxation, it might be thought that the reaction rate (53.29) would be limited by just the first process. If now, in addition, such conditions are created that the collision of HX with DY, directly responsible for the redistribution, does not contribute substantially to the vibrational-relaxation rate, then the overall kinetics of process (53.29) will be determined both by the concentrations of the reactants and also by the concentrations of the molecules of the heat reservoir. A situation arises analogous to that in unimolecular reactions in the low- and average-pressure regions.

The fact of the change in the order of the reaction, itself, is very important, since it opens up new possibilities for investigating non-equilibrium bimolecular reactions in different pressure regions. For the simplest process (53.1), the disturbance of the equilibrium distribution only appears in the value of $\kappa(T)$, in no way affecting the kinetics of the exchange. In studying the reaction (53.31) in an argon atmosphere, a discrepancy from a second-order law was, in fact, discovered, made manifest, in particular, in that the reaction rate proved to be dependent on the partial pressure of the argon. In principle, this result can be interpreted within the limits of the following multi-stage scheme

$$HX + Ar \to HX^* + Ar \quad \text{(vibrational excitation)} \qquad (53.32a)$$

$$DY + Ar \to DY^* + Ar \quad \text{(vibrational excitation)} \qquad (53.32b)$$

$$HX^* + DY^* \to HY + DX \quad \text{(exchange).} \qquad (53.32c)$$

Under definite assumptions, the constant κ in (53.31) is expressed in terms of the vibrational relaxation time of HX and DY, and the exchange probability $P(n_1, n_2)$ is expressed as a function of the numbers of vibrational levels of the partners HX* and DY*. Comparison of a similar calculation with experiment enables the degree of vibrational excitation n_i^* of the particles to be estimated, at which exchange already proceeds with the gas-kinetic cross-section. Such an estimate, conducted for the reaction

$$H_2 + D_2 \to HD + HD, \qquad (53.33)$$

gives $n^*(D_2) = 4$ [147].

Unfortunately, it is difficult to say to what extent this result is exact. It is evident, however, that the disturbance of the equilibrium distribution of the internal states of the reacting molecules plays a very important role in the determination of both the absolute value of the rate and the order of the reaction. For a correct derivation of the kinetic equations and the determination of their parameters, the general problem of relaxation and exchange should be solved. At present, it is difficult even to formulate the conditions under which both these processes may be considered independent.

In such a situation, the theoretical investigation of the dynamics of exchange reactions plays an exceptionally-important role, since it enables the relative effectiveness of the excitation energy of different degrees of freedom in the surmounting of the activation barrier to be predicted. Going over from the simplest exchange reaction, with three atoms taking part, to the more complex will make it possible to distinguish the most active degrees of freedom in the elementary processes and will thus allow reasonable assumptions to be introduced, simplifying the general problem of the calculation of the reaction rate in non-equilibrium conditions.

APPENDIX

Problems of elementary processes in thermal gas reactions

THE ultimate problem in investigating elementary processes is the obtaining of a set of kinetic constants, which would enable the formation of an equation describing the change in time and space of the concentrations of the particles of interest. The number of elementary processes specifically investigated is very large; however, so far, when we turn to a real system, it is often possible to predict the chemical effects only in the most qualitative form. Several problems are discussed, in this appendix, which stand in the way of a quantitative theory of chemical reactions, and the latest developments in the theory during the last 2–3 years are briefly considered, also. The appendix is provided with a supplementary list of references. Moreover, a list of recently-published books and reviews is added [1]–[9], which relate to the theory of elementary processes.

The problems and resources of the theory in its present state for the most part lead not so much to the *a priori* calculation of effective cross-sections and rate constants of various elementary processes as to the derivation of formulae, enabling the kinetic parameters of real systems to be calculated on the basis of experimental data obtained in model conditions.

The basic kinetic parameter which is necessary for the incorporation of some elementary reaction or other into the general scheme of a complex process is the rate constant κ, connected with the reaction cross-section σ and the distribution function of the reacting partners in the phase space f by the relation

$$\kappa = \int \sigma v f \, d\Gamma.$$

To return to the experiments, they, in practice, reduce to two types, according to the elementary processes studied. In one of them, the cross-section and its energy dependence is measured, in the other, the rate constant is measured. In accordance with this, the problem in the theory of elementary processes in gases is either the calculation of the cross-sections σ or the calculation of the rate constants κ. A general scheme of such a calculation, illustrating its basic steps, is presented in Fig. 55.

The advantage of experiments of the first type is a stricter identification of the products and, in principle, the possibility of carrying the results over to any conditions for which the distribution function f is known, in the general case, a non-equilibrium distribution function. Unfortunately, the real value of results obtained in this way for gaseous thermal reactions is very limited. This is connected with the fact that many of them proceed by way of stepwise vibrational or electronic excitations. Thus, in attempting to apply the indicated method for calculating the constants from the experimental data on cross-sections, consistently, the cross-sections of a great number of elementary processes would have to be determined experimentally, and an even greater number of combinations of them taken into account in the real system.

Therefore, for a long time, the main line of development of the theory of elementary thermal processes has been the direct calculation of the rate constant by the transition

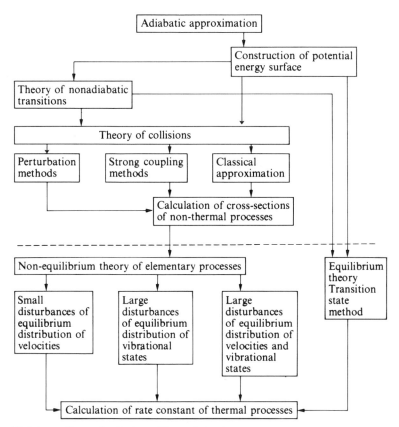

FIG. 55. Scheme illustrating basic problems of the theory of elementary gas phase processes.

state method, which seemingly represents the best 'roundabout way', enabling the microscopic characteristics of the reacting partners to be linked with the macroscopic rate constant.

The recent development of the transition state method, has consisted chiefly of a more detailed investigation of tunnel corrections, an elucidation of the accuracy of the adiabatic separation of variables, and an investigation of the connection between this method and the approach formulated in collision theory [10]–[23], [163]–[169]. As in the case of the direct route (see Fig. 55), in the calculation of κ in terms of σ the adiabatic approximation usually serves as the starting point, since, under conditions typical of gas reactions (300 K $< T <$ 20 000 K), the velocities of the molecules are sufficiently small and the Massey parameter is large for the overwhelming majority of pairs of electronic states. In such an approach, the transitions between the electronic states of colliding partners are studied in the theory of nonadiabatic transitions.

In many cases, however, a 'roundabout way' is completely unsuitable, and then recourse is had to the methods of collision theory.

The basic methods of collision theory, referred to in the scheme shown in Fig. 55, are presented in the book in connection with the calculation of the vibrational transition probability (perturbation theory) and with the probability of a change of electronic state (strong binding) in collisions and also in connection with a study of the dynamics of exchange reactions (the use of the classical approximation in the many-body problem). A calculation of the cross-sections of nonthermal processes completes the first dynamical stage of the theory of elementary processes.

The second, statistical stage considers a large group of elementary acts (collisions) and, in principle, consists of the calculation of the non-equilibrium distribution function of the reacting particles and the use of this function to calculate the macroscopic kinetic constants. The resources of the theory, at this stage, essentially depend on the relation between the relaxation times τ_r and the characteristic times τ_t of some transformation or other of the particles. If τ_r and τ_t are of the same order, then the resources of the theory are very small; however, on condition $\tau_r \ll \tau_t$ considerable progress is possible. It should be emphasized that the meaning of τ_r and τ_t in the inequality $\tau_r \ll \tau_t$ depends on the type of transformation under discussion. If, for example, it is a chemical transformation of the dissociation or exchange type, in which electronically-excited particles do not take part, then τ_r is the vibrational relaxation time of the reacting particles.

The condition $\tau_r \ll \tau_t$ enables the analysis of processes in gases to be substantially simplified. In this case, processes leading to the redistribution of energy over the degrees of freedom of the reacting molecules can be separated in time from chemical processes. Moreover, only under this condition is it possible to describe dissociation, recombination, and exchange by the usual chemical kinetic equations, in which only the concentrations of the reactants enter. For times t appreciably exceeding the time between the collisions of the molecules $1/Z_0$, it is possible to consider the gas as characterized by a definite value of the kinetic temperature T_{trans} and the corresponding Boltzmann distribution functions. Then the chemical processes and also vibrational and electronic relaxation are described by a system of kinetic equations, depending on T_{trans}, for the population of the electronic–vibrational states of the various components. For times t exceeding the vibrational and electronic relaxation times, this system can be further simplified and reduced to a system of equations for the concentrations. As a result of this transformation, also, the meaning of the kinetic coefficients, which are known as the rate constants of elementary chemical reactions, is clarified. We pause to consider here the separate types of processes, referring the reader to the relevant paragraph in the book where the corresponding details are to be found.

Vibrational relaxation of diatomic molecules over the levels of the non-degenerate electronic ground state

For this process, as regards the electronic states, complete adiabaticity is characteristic (see § 7). A calculation of the vibrational transition probability is carried out by the transition state method (see § 3), and the transmission coefficient is calculated by perturbation theory (see § 9). The vibrational relaxation time τ_{vib} is found, within the limits of the diffusion model (see §§ 35, 38). The parameters of the potential entering into the expression for τ_{vib} can be determined approximately from independent experiments.

In Fig. 56 is shown a scheme of the evolution of the Landau–Teller model, first proposed for the calculation of the vibrational relaxation time of diatomic molecules.

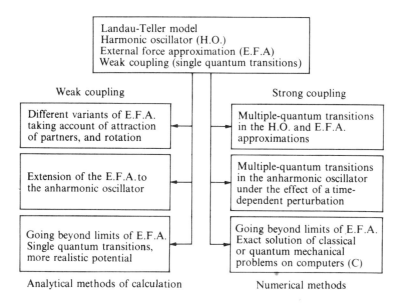

FIG. 56. Scheme illustrating methods of calculation of the probability of vibrational and translational energy exchange.

Nowadays, two trends in the theory of inelastic molecular collisions have appeared: the use of perturbation theory (weak coupling) for more-or-less realistic models of a collision (for example, taking account of the angular dependence of the potential) [24]–[44], [170]–[173] and an escape beyond the limits of the perturbation theory (strong coupling) or the rather simple models [45]–[56], [174]–[187]. Progress in the first direction is possible with the help of analytical methods whereas, for the second, the use of numerical methods is characteristic. Common to both trends is the tendency to relinquish the approximation of an external force (see §8), within whose limits the reaction of the molecule undergoing relaxation on the heat reservoir is disregarded.

The problem of the vibrational excitation of molecules in the strong-coupling approximation is formulated in the form of a system of coupled equations (quantum-mechanical or classical), whose solution is possible only with the use of computers. The final outcome of the calculation is the determination of a set of transition probabilities $n \to m$ or, in the simplest case, the mean square energy transfer $\langle \Delta E^2 \rangle$, appearing as a 'coefficient of diffusion' over the energy states of the molecule. In addition to this important problem, there remains the elucidation of the question, to what degree the rejection of the approximation of an external force changes the value of $\langle \Delta E^2 \rangle$.

The results of theoretical calculations of the vibrational transition probabilities were used in a number of papers [188]–[194] to derive relaxation equations describing the kinetics of the energy exchange between the translational and vibrational degrees of freedom of both colliding partners, represented by harmonic or anharmonic oscillators. In particular, the relaxation theory satisfactorily describes the interesting appearance of a vibrational pumping on exchange of energy between two components of a gas [195].

Vibrational relaxation of polyatomic molecules on the levels of a nondegenerate electronic ground state

In comparison with previous cases, an important complication arises here, connected with the contribution of intramolecular redistribution of energy to the process of intermolecular relaxation. The transition between the ground state and first few vibrational states can still be described by the theory developed for diatomic molecules. In describing transitions between higher states, regard for the simultaneous change of the quantum states of different normal vibrations is necessary. The detailed theory of such a type of relaxation for triatomic molecules has recently begun to be developed. This is connected with the substantial progress made in the experimental study of the separate relaxation stages, which proves to be possible owing to the use of laser sources of excitation. An investigation of the relaxation of CO_2 [57]–[68], [196]–[201] is the best illustration in this respect. In particular, an indication as to the definite contribution of the long-range forces to the process of energy exchange (compare § 3) was first obtained in the investigation of quasi-resonant energy exchange between CO_2 and N_2, and the negative temperature-dependence of the quasi-resonant transition probability was discovered [65], [200], [201]. For more complex molecules and higher states, the situation in the theory is so much more complicated that a very crude representation of the mechanism of intramolecular redistribution of energy has to be introduced. The simplest suggestion is that different normal vibrations of the molecules can be broken down into groups, each of which is characterized by a Boltzmann distribution with some effective temperature. The relaxation of the mean energy of each of the groups takes place on collision and is determined, as for the harmonic oscillator model, by the deactivation probability of the first vibrational level of some normal vibration or other [202]–[205]. The next step must be based on some kinetic model of the energy redistribution. In this connection we will mention the attempt to describe this process by introducing an effective 'friction' [69]–[71], which implies passing over to statistical theory, to a certain extent. Even the mechanism of intermolecular energy redistribution, by its very nature, recalls a statistical process. In this respect, however, the experimental data are too scanty for the construction of a realistic model.

Vibrational relaxation of diatomic models on levels of a degenerate electronic state

For these processes, the transmission coefficient of the transition state method is calculated by perturbation theory, taking account of the nonadiabatic coupling between the electronic states arising on the removal of the degeneracy of the electronic term of the molecule due to the interaction (see § 20). As regards the vibrational relaxation time, it is calculated by diffusion theory. It appears that similar types of process, where the electronic–vibrational interaction is important, are considerably more widespread than was earlier supposed. It is possible, for example, that the vibrational relaxation of nitrogen in an oxygen atmosphere, which displays an anomalous rate [72], takes place by just such a nonadiabatic mechanism.

The transfer and conversion of energy in atomic collisions

The theory of nonadiabatic transitions (Fig. 57) lies at the basis of the theory of inelastic atomic collisions. Within the framework of the semiclassical approximation, when the relative motion of the partners in collision is considered classically, the

APPENDIX

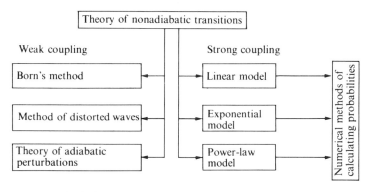

Fig. 57. Scheme illustrating methods of calculating nonadiabatic transition probabilities.

mathematical part of the problem of calculating the probability of a change in the electronic state is formulated in the form of a system of first-order temporal equations. Very often, the usual methods of perturbation theory, the Born approximation, or the distorted wave approximation prove to be unsuitable. The theory of adiabatic perturbations, whose small parameter is not the absolute value of the interaction but its characteristic rate of change, finds a greater range of applicability in comparison with the Born method or the method of distorted waves. However, the most important results, from the point of view of applications in the theory of nonadiabatic transitions, were obtained on the investigation of certain model situations enabling an exact solution to be found for the transition probability as a function of the model's parameters. The type of model is determined by the dependence of the matrix elements of the Hamiltonian on the nuclear coordinates. The characteristics of the interatomic interaction lead, in a natural way, to the linear (see §§ 15, 16) and non-linear (see § 17) models. These models often serve as a starting point for more complicated numerical calculations and for the interpretation of the detailed physical mechanism of a process.

It is convenient to divide all processes involving conversion of electronic energy into translational energy and processes of electronic energy transfer, in the collisions of atoms, into two groups.

To the first are referred reactions in which the change in the electronic state of the partners takes place at interatomic distances R, large in comparison with the atomic radius a. Recently, the theory of such reactions has been intensively developed. This is due to the fact that, at sufficiently large interatomic distances, the interaction can be calculated sufficiently accurately, so that one of the obstacles to the development of the theory of inelastic collisions is removed. Knowledge of the approximate wavefunction of the outer electrons of the colliding atoms is often required for these calculations. Resonant charge exchange, charge exchange and the transfer of excitation with small resonance defect, and transitions between the fine-structure components of the colliding atoms belong to this type of process, whose cross-section can be calculated comparatively accurately. Reactions in which the change in the electronic state takes place on close approach of the partners ($R \approx a$), charge exchange and transfer of excitation with large resonance defect, quenching of fluorescence of atoms, etc. belong to the second group of processes. Since the interaction of atoms in these cases is insufficiently well known, undetermined parameters have to be introduced into the theory. In principle, these parameters might be determined by the experimental data, and the result of a theoretical calculation might be used for a temperature extrapolation of the rate constant. Unfortunately, at present it is impossible to indicate even a single process for which such a programme might be carried out.

Conversion of the electronic energy of an atom into the vibrational energy of a molecule

For this type of process, the nonadiabatic transitions causing electronic deactivation take place in the regions of convergence of the potential energy surfaces (see § 14). For a system of three atoms and more complicated systems, information about the electronic terms is very limited, and the detailed form of terms in the region of nonadiabaticity is usually not known. However, even if this part of the problem is assumed solved in some approximation or other, then great difficulties arise in the calculation of the transition probabilities. One of these is the necessity for taking account of the representative point's multiple transits of the regions of nonadiabaticity, in the formation of the 'excited atom–molecule' complex (see § 19). It appears that it is impossible to calculate the distribution function of the energy being released over the vibrational, rotational, and translational degrees of freedom of the colliding particles, without taking this into account. As for the calculation of the total cross-section, here, at best, only the order of magnitude can be predicted, and even then only when we have some information available about the interaction of the partners in the electronically-excited states.

The process $Na^* + N_2$, considered in § 19, is very instructive as an illustration of the general situation in the collision of an atom with a diatomic molecule. In one of the first attempts at interpreting the mechanism of this reaction, it was proposed that the configuration of the interacting atoms making the chief contribution to the total rate constant is close to the configuration of an isosceles triangle. For this configuration, intersection of the electronic terms occurs; however, an estimate of the position of the point of intersection, carried out with regard to just the covalent states of the partners ($Na(^2P)$, $Na(^2S)$, $N_2(^1\Sigma_g^+)$), shows that the attainment of the point of intersection requires expenditure of energy, i.e. it is linked with the surmounting of the activation barrier. Enlisting the ionic states ($Na^+(^1S)$, N_2^-) in addition leads to the appearance of new possibilities: 'ionic' potential energy surfaces intersect 'covalent' surfaces, facilitating, in this way, the transition of the system from the initial ($Na^* + N_2$) state to the final, ($Na + N_2$) covalent, state. Calculations carried out for a linear collision [73], [206], [207] enable the value of the total reaction cross-section to be understood qualitatively. Recently, calculations carried out for the terms of the system $Li + H_2$, with regard for the ionic states of Li^+, H_2^-, show [74] that the minimum in the potential energy corresponds to the triangular configuration, where the attainment of the point of intersection of the excited state (which appears as an 'adiabatic mixture' of the covalent state $Li^* + H_2$ and the ionic state $Li^+ + H_2^-$) and the ground state does not require the surmounting of the barrier. It is therefore very probable that, for the analogous reaction $Na^* + N_2 \to Na + N_2$, the system of adiabatic terms corresponds qualitatively to the one given in Fig. 15, with the difference, however, that the interaction between the excited covalent state and the ionic state lowers the energy and displaces the vertex of the cone below the energy level of the initial state $Na^* + N_2$.

Further investigation of similar processes in a plasma [208], molecular beams [209], and shock waves [210] will, without doubt, further the development of the theory.

Thermal decomposition of diatomic molecules

The thermal decomposition of diatomic molecules (decomposition proceeding with conservation of the Maxwellian distribution of velocities) is the simplest process involving redistribution of the heavy particles. In comparison with processes without redistribution, the difficulties of the theoretical calculation increase sharply in the process under consideration, and the experimental data begin to play a greater and greater role in the theory, enabling the adequacy of some particular assumption to be estimated.

Equilibrium theory (see §40), which can be included in the general scheme for the calculation (Fig. 58) of various kinetic characteristics, proves to be completely unsuitable. Making it more precise with the aid of the variational approximation (see §42) does not bring about substantial improvement. However, the variational approximation can give useful information about the values of the mean energy transfer, which is the analogue of an 'energy' coefficient of diffusion within the limits of the diffusion theory of reactions (see Chapter 6). Recently, there has been a revival of interest in diffusion theory, in connection with the recombination of ions, which represents a very favourable objective from the point of view of the calculation of the value of $\langle \Delta E^2 \rangle$ on the basis of pair interactions [75]–[77] and in connection with the dissociation of a heavy diatomic gas in a light gas atmosphere [78]–[80]. Finally, the direct analysis of the trajectories of three particles, with the aim of calculating dissociation and recombination rates, begins

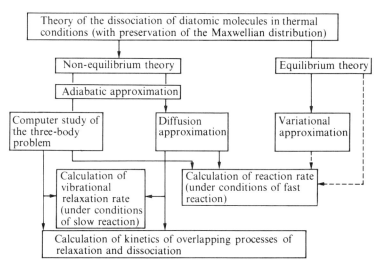

FIG. 58. Scheme illustrating the state of the theory of the thermal decomposition of diatomic molecules.

to be an ever-increasing occurrence. In this way, the following step, from the investigation of non-equilibrium non-overlapping relaxation and dissociation processes (relaxation under the condition of slow decomposition and decomposition under the condition of fast relaxation) to the investigation of the kinetics of overlapping processes [81]–[85], [211] is successfully made.

As regards the experimental results, at present, the dissociation rates of the molecules H_2, O_2, N_2, NO and the halide molecules have been measured. However, the accuracy of the measurements, especially the temperature dependence of the rate constants, leaves much to be desired. In a number of cases, for example in the dissociation of chlorine in an argon mixture, different authors give the expressions $T^{-\frac{3}{2}}$, T^{-2}, and even $T^{-\frac{5}{2}}$ for the temperature dependence of the pre-exponential of the rate constant, results which noticeably differ from the prediction of the simplest variant of the theory T^{-1} (see §44). It is difficult to point to the reason for the discrepancy: it is possible that, in these experiments, the condition $\tau_r \ll \tau_t$ (see §45) was not satisfied, or that the simplest variant of the theory takes incorrect account of the contributions of the various degrees of freedom to the decomposition rate. Also, the possibility is not excluded that, in the thermal dissociation of chlorine, a two-stage process of the type

$$Cl_2 + Cl_2 \rightarrow Cl_3 + Cl$$

$$Cl_3 \rightarrow Cl_2 + Cl$$

plays a role.

The formation of intermediate complexes sometimes proves to be essential in recombination processes [212], [213].

Unimolecular reactions

The present-day state of the theory of unimolecular reactions is illustrated in Fig. 59.

In regard to the redistribution of vibrational energy of active molecules, alternative mechanisms exist both for intramolecular and intermolecular processes. Dynamical models laying claim to a detailed description of the process of intramolecular redistribution are based either on the harmonic approximation of Slater (see §24) or on the numerical analysis of phase trajectories for molecules with a small number of atoms and a more-or-less realistic anharmonic potential [214]–[218]. Statistical models generally bypass the question of the kinetics of the intramolecular redistribution of energy, and the recent development of models is directed towards making the characteristics of the energy spectrum of polyatomic molecules more precise [219]–[224].

The numerous attempts to construct a model of an intermediate type, which includes some 'degree of statistics' and the rate of redistribution of the energy between normal vibrations as a parameter (the so-called quasi-harmonic model), should be mentioned [86]–[94], [225]. This model has still not found application to experiment although, in principle, it is more general than Slater's or Kassel's model.

So far the activation mechanism of polyatomic molecules (the intermolecular redistribution of energy) remains to a considerable extent obscure. In the creation of a theory of unimolecular reactions, the mechanism of strong collisions (see §23) has always been the generally-adopted mechanism. The mechanism of stepwise excitation, the limiting case of which is described by diffusion theory, began to be enlisted only

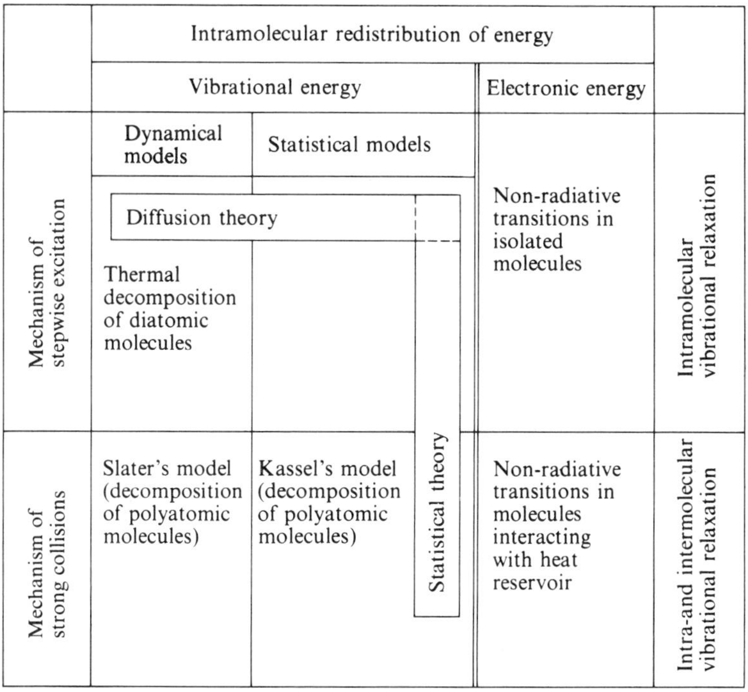

	Intramolecular redistribution of energy			
	Vibrational energy		Electronic energy	
Mechanism of stepwise excitation	Dynamical models	Statistical models	Non-radiative transitions in isolated molecules	Intramolecular vibrational relaxation
	Diffusion theory			
	Thermal decomposition of diatomic molecules			
Mechanism of strong collisions	Slater's model (decomposition of polyatomic molecules)	Kassel's model (decomposition of polyatomic molecules)	Statistical theory	Non-radiative transitions in molecules interacting with heat reservoir	Intra-and intermolecular vibrational relaxation

FIG. 59. Scheme illustrating the present state of the theory of unimolecular processes.

after it was found impossible to describe some reactions within the limits of the traditional approach.

In the light of recent analysis of the experimental data [95], it seems very important that the mechanism of thermal decomposition of triatomic molecules corresponds to a region of 'intersection' of the statistical and diffusion theories (see Fig. 59).

As regards the redistribution of the electronic energy of a polyatomic molecule and the conversion of the electronic energy into vibrational, the development of the theory in this region is characterized by joint consideration of the nonadiabatic transitions and intramolecular relaxation. This theory will be the basis for a description of photochemical reactions. The calculation of molecular spectra exhibiting the features of 'beats' connected with the temporal migration of excitation in the molecule [226] should be mentioned as one of the interesting applications of the theory.

Bimolecular reactions

Bimolecular exchange reactions are the most widespread of all known types of chemical reactions. Despite this, the theory of these reactions has only just begun to be developed.

It is true that, about 35 years ago, the question of the dependence of the exchange-reaction rate constants on temperature was considered in the general form, on the basis of transition state theory (see Chapter 1). Later, calculations of the magnitude of the pre-exponential of the rate constant for specific reactions (see, for example, [227]) were carried out on the basis of this method. For some reactions, the theory leads to satisfactory agreement with experiment; for other reactions there is no agreement.

Undoubtedly, the reason for this is the inadequacy of the transition state method. To begin with, the postulate, lying at the basis of the method, that the rate constant of an elementary reaction remains the same during its course, both in the absence of equilibrium and in its presence, cannot be considered as always satisfied. Likewise, another often-assumed postulate of the transition state method—the postulate that the motion of the nuclei comply with the laws of classical mechanics—cannot be considered as sufficiently strict.

The deficiencies of the transition state method can only be overcome by way of a clear discussion of the dynamical stages of the process leading to the calculation of the reaction cross-section. In certain cases, the calculation can be simplified substantially, at this stage, by introducing the statistical approximation (see § 50). This approximation turns out to be sufficiently reasonable, especially when its usual variant (see §§ 28–30) is supplemented by taking account of the conservation of the angular momentum of the intermediate complex [228]–[231] (see § 31). The statistical approximation is only applicable, however, to those reactions which take place via the formation of a long-lived complex. Ion–molecular reactions at low energies or reactions between neutral molecules may serve as examples [96]–[105].

The theory of direct reactions applied to processes in which the intermediate complex has a very short lifetime arose as an alternative to the statistical theory. It proved possible to detail the mechanisms of these reactions: direct knock-out reactions, rebound reactions, stripping reactions. Simple kinematical considerations, which can be expressed on the basis of some mechanism or other, enable the distribution of energy in these reactions [106]–[113], [232]–[238] to be predicted in rough qualitative form.

The statistical model and stripping model provide limiting cases which occur at sufficiently-low or sufficiently-high energies of the partners. In this connection, it was proposed [114] to characterize qualitatively any chemical process by the two quantities E_{compl}^{max} and E_{strip}^{min}. The first of these represents an upper bound for the kinetic energy, below which reaction takes place via the formation of an intermediate complex. The second, a lower bound for the kinetic energy, above which the reaction already begins to be described by the stripping mechanism. Of course, the values of E_{compl}^{max} and E_{strip}^{min}, which are not defined, strictly speaking, can only be used for a very rough classification of reactions. It is however important that these values correlate with some qualitative characteristics of the interaction of the reacting partners, and, in particular, with the character of the interaction between the atoms B and C in the reaction $A + BC \to AB + C$, long after the formation of the AB molecule (the interaction on that part of the potential energy surface which is located after the turn of the representative point into the valley of the reaction products). On an increase in the relative energy of the partners A and BC, the stripping mechanism sets in earlier than the weaker interaction between B and C, independent of whether it is attractive or repulsive. In such a case, when the atoms B and C are repelled, it appears the complex cannot be formed, in general, and for this case values of E_{compl}^{max} do not exist, in general. If the atoms B and C are attracted, then E_{compl}^{max} increases with the increase in this attraction.

A more specific classification of exchange reactions can be made only on the basis of an analysis of the dynamics of collisions. Here, the three chief directions along which the theory of processes accompanied by the redistribution of the heavy particles develops should be mentioned:

1. Classical calculations with more-or-less realistic potential energy surfaces. These calculations are organized to preserve, to the greatest extent, the analogy with the wave picture of the process (a solution of the problem in 'action-angle' variables) [114]–[124], [239]–[241]. The outcome of the calculation is complete information about the distribution functions of the states of the products and of the angular distribution of the particles.

2. Construction of effective potentials enabling elastic scattering in the presence of a reaction (so-called optical models) to be described. Within the framework of this approach, it is possible to introduce a parametric description of a number of reactions and, in this way, construct a semi-empirical basis for the classification of the processes [126]–[136], [242], [243].

3. Quantum-mechanical calculations, based either on appreciably-simplified models or on the use of certain small parameters which arise due to the specific properties of a reaction (for example, the small ratio of the masses of the interacting atoms, etc.) [137]–[150], [244]–[256].

Furthermore, the use of computers to formulate 'calculated experiments' should be pointed out, whose purpose is not to compare experiment with theory but to estimate the validity of some theoretical representation or other, on the basis of a comparison between the exact, computed solution to a problem and the approximate solution, obtained on the basis of some reasonable physical assumptions.

If the reaction cross-section is known, then a statistical stage of the calculation of the rate constant arises, which must take into account the effect of the disturbance of the equilibrium distribution caused by the reaction itself on its mean rate. Recent work on this problem [151]–[154] is devoted mainly to the investigation of cases where this disturbance is small, although definite methods for the investigation of a strong disturbance have been planned [153]–[156], [257]–[260]. The experimental discovery of the inversion population of the vibrational–rotational states of molecules arising in exothermic reactions and the creation of chemical lasers [157]–[162], [261]–[264] clearly show that the rate of the reverse endothermic processes may be very susceptible to changes in the distribution function of the internal states of the initial partners in a reaction [265]–[269]. Hence, in particular, the problem arises of the selective initiation or acceleration of endothermic reactions by laser beams [270].

The review of recent work presented clearly shows the very rapid development of the theory of elementary processes in the gaseous phase.

It is to be expected that, within the next few years, important steps will be made towards a deeper understanding of the mechanisms of processes connected with the redistribution of the energy and particles in molecular collisions.

REFERENCES

1. BATES, D. R. *Atomic and molecular processes.* Pure and Applied Physics Ser. Vol. 13. Academic Press, New York (1962).
2. BOGOLYUBOV, N. N. *Zh. éksp. teor. Fiz.* **16**, 691 (1946).
3. BOGOLYUBOV, N. N. *Problems of dynamical theory in statistical physics.* Moscow–Leningrad (1946).
4. BORN, M. and HUANG, KUN. *Dynamical theory of crystal lattices.* Clarendon Press, Oxford (1954).
5. BȲKHOVSKIĬ, V. K., NIKITIN, E. E., and OVCHINNIKOVA, M. YA. *Zh. éksp. teor. Fiz.* **47**, 750 (1964).
6. BȲKHOVSKIĬ, V. K. and NIKITIN, E. E. *Optika Spektroskp.* **16**, 111 (1964).
7. BȲKHOVSKIĬ, V. K. and NIKITIN, E. E. *Optika Spektroskp.* **17**, 615 (1964).
8. VAN DER POL, B. and BREMMER, H. *Operational calculus.* Cambridge University Press (1964).
9. VAĬNSHTEĬN, L., PRESNYAKOV, L., and SOBEL'MAN, I. *Zh. éksp. teor. Fiz.* **43**, 518 (1962).
10. WILSON, E. B., DECIUS, J. C., and CROSS, P. C. *Molecular vibrations.* McGraw-Hill, New York (1955).
11. GAYDON, A. *The spectroscopy of flames.* Chapman and Hall, London (1957).
12. GENERALOV, N. A. and LOSEV, S. A. *Dokl. Akad. Nauk SSSR* **148**, 552 (1963).
13. GENERALOV, N. A. and LOSEV, S. A. *J. quant. Spectrosc. radiat. Transfer* **6**, 101 (1966).
14. HERZBERG, G. Spectra of diatomic molecules, Vol. I (2nd edn.), *Molecular spectra and molecular structure.* Van Nostrand–Reinhold, New York (1950).
15. HERZBERG, G. *Vibrational and rotational spectra of polyatomic molecules.* Foreign Literature Press (1949).
16. GLASSTONE, S., LAIDLER, K., and EYRING, H. *Theory of rate processes.* McGraw-Hill (1941).
17. GOL'DANSKIĬ, V. I. *Dokl. Akad. Nauk SSSR* **124**, 1037 (1959).
18. GOL'DANSKIĬ, V. I. *Dokl. Akad. Nauk SSSR* **124**, 1261 (1959).
19. GOL'DMAN, I. I. and KRIVCHINKOV, V. K. *Problems in quantum mechanics.* State Technical Publishers (1957). Infosearch (1964).
20. DASHEVSKAYA, E. I. and NIKITIN, E. E. *Optika Spektroskp.* **22**, 866 (1967).
21. DEMKOV, YU. N. *Zh. éksp. teor. Fiz.* **45**, 195 (1963).
22. DEMKOV, YU. N. *Dokl. Akad. Nauk SSSR* **166**, 1076 (1966).
23. DRUKAREV, G. F. *Zh. éksp. teor. Fiz.* **45**, 1484 (1963).
24. IVANOV, G. K. and SAYASOV, YU. S. *Zh. éksp. teor. Fiz.* **45**, 1456 (1963).
25. IVANOV, G. K. and SAYASOV, YU. S. *Dokl. Akad. Nauk SSSR* **154**, 1314 (1964).
26. IVANOV, G. K. and SAYASOV, YU. S. Letters to the Editor, *Zh. éksp. teor. Fiz.* **3**, 40 (1966).

27. IZRAÏLEV, F. M. and CHIRIKOV, B. V. *Dokl. Akad. Nauk SSSR* **166**, 57 (1966).
28. KAPLAN, I. G. *Symmetry of many-electron systems*. Foreign Literature Press 'Nauka' (1969).
29. KARACHEVTSEV, G. V., MARKIN, M. I., and TAL'ROZE, V. L. *Kinet. Katal.* **5**, 377 (1964).
30. KASSEL, L. C. *Kinetics of homogeneous gas reactions*. United Scientific and Technical Publishers (1937). Am. chem. Soc. Monograph Ser. 57, New York (1932).
31. KONDRAT'EV, V. N. *Kinetics of chemical gas reactions*. Foreign Literature Press, Akademii Nauk (1958).
32. KONDRAT'EV, V. N. and NIKITIN, E. E. *Usp. Khim.* **36**, 2007 (1967).
33. KORST, N. N. and NIKITIN, E. E. *Teor. éksp. Khim.* **1**, 11 (1965).
34. KUZNETSOV, N. M. *Dokl. Akad. Nauk SSSR* **164**, 1097 (1965).
35. KUKSENKO, B. V., LEONAS, V. B., LOSEV, S. A., and OSIPOV, A. I. *Dokl. Akad. Nauk SSSR* **167**, 1280 (1966).
36. LANDAU, L. D. and LIFSHITZ, E. M. *Mechanics* (2nd edn.). Pergamon Press, Oxford (1969).
37. LANDAU, L. D. and LIFSHITZ, E. M. *Quantum mechanics* (2nd edn.). Pergamon Press, Oxford (1965).
38. LANDAU, L. D. and LIFSHITZ, E. M., *Statistical physics*. Pergamon Press, Oxford (1969).
39. LEONAS, V. B. *Zh. prikl. Mekh. tekhn. Fiz.* **6**, 124 (1963).
40. LIFSHITZ, E. M. *Zh. éksp. teor. Fiz.* **17**, 1017 (1947).
41. LOSEV, S. A. and GENERALOV, N. A. *Dokl. Akad. Nauk SSSR* **141**, 1072 (1961).
42. LOSEV, S. A. and OSIPOV, A. I. *Usp. fiz. Nauk.* **74**, 394 (1961).
43. MAZHUGA, V. V. and SOKOLOV, N. D. *Kinet. Katal.* **6**, 961 (1965).
44. MAZHUGA, V. V. *Dokl. Akad. Nauk SSSR* **167**, 1012 (1966).
45. MARKIN, M. I. and TAL'ROSE, V. L. *Symposium: 'The elementary processes of chemistry at high energies'*. 'Nauka' (1965).
46. MAYANTS, L. S. *Dokl. Akad. Nauk SSSR* **151**, 624 (1963).
47. MAYANTS, L. S. *Dokl. Akad. Nauk SSSR* **164**, 852 (1965).
48. MASSEY, H., BURHOP, E., and GILBODY, H. *Electronic and ionic impact phenomena* (2nd edn.). Clarendon Press, Oxford (1969).
49. MNATSAKANYAN, A. KH. *Optika Spektroskp.* **18**, 161 (1965).
50. MOTT, N. and MASSEY, H. *Theory of atomic collisions*. Oxford University Press (1965).
51. NATANSON, G. L. *Zh. tekhn. Fiz.* **29**, 1486 (1959).
52. NIKITIN, E. E. *Dokl. Akad. Nauk SSSR* **116**, 584 (1957).
53. NIKITIN, E. E. *Dokl. Akad. Nauk SSSR* **119**, 526 (1958).
54. NIKITIN, E. E. *Zh. fiz. Khim.* **33**, 572 (1959).
55. NIKITIN, E. E. *Zh. fiz. Khim.* **33**, 1893 (1959).
56. NIKITIN, E. E. *Optika Spektroskp.* **6**, 141 (1959).
57. NIKITIN, E. E. *Dokl. Akad. Nauk SSSR* **129**, 157 (1959).
58. NIKITIN, E. E. *Dokl. Akad. Nauk SSSR* **124**, 1085 (1959).
59. NIKITIN, E. E. and SOKOLOV, N. D. *Izv. Akad. Nauk SSSR Ser. Fiz.* **24**, 996 (1960).
60. NIKITIN, E. E. *Dokl. Akad. Nauk SSSR* **132**, 395 (1960).

61. NIKITIN, E. E. *Dokl. Akad. Nauk SSSR* **135**, 1442 (1960).
62. NIKITIN, E. E. *Optika Spektroskp.* **9**, 16 (1960).
63. NIKITIN, E. E. *Dokl. Akad. Nauk SSSR* **136**, 1376 (1961).
64. NIKITIN, E. E. *Optika Spektroskp.* **11**, 452 (1961).
65. NIKITIN, E. E. *Fiz. Prob. Spektroskp.* **1**, 184 (1962).
66. NIKITIN, E. E. *Kinet. Katal.* **3**, 830 (1962).
67. NIKITIN, E. E. *Optika Spektroskp.* **13**, 761 (1962).
68. NIKITIN, E. E. *Kinet. Katal.* **3**, 322 (1962).
69. NIKITIN, E. E. *Izv. Akad. Nauk SSSR, Ser. Fiz.* **27**, 996 (1963).
70. NIKITIN, E. E. *Dokl. Akad. Nauk SSSR* **152**, 1395 (1963).
71. NIKITIN, E. E. *Dokl. Akad. Nauk SSSR* **148**, 1298 (1963).
72. NIKITIN, E. E. Doctorate Dissertation, *Inst. khim. Fiz., Akad. Nauk SSSR* (1965).
73. NIKITIN, E. E. *Kinet. Katal.* **6**, 17 (1965).
74. NIKITIN, E. E. *Kinet. Katal.* **6**, 377 (1965).
75. NIKITIN, E. E. *Teor. éksp. Khim.* **1**, 135 (1965).
76. NIKITIN, E. E. *Teor. éksp. Khim.* **1**, 144 (1965).
77. NIKITIN, E. E. *Teor. éksp. Khim.* **1**, 428 (1965).
78. NIKITIN, E. E. *Optika Spektroskp.* **19**, 161 (1965).
79. NIKITIN, E. E. *Teor. éksp. Khim.* **2**, 19 (1966).
80. NIKITIN, E. E. *Optika Spektroskp.* **22**, 689 (1967).
81. OVCHINNIKOVA, M. YA. *Optika Spektroskp.* **17**, 822 (1964).
82. OVCHINNIKOVA, M. YA. *Zh. éksp. teor. Fiz.* **49**, 275 (1965).
83. OVCHINNIKOVA, M. YA. *Dokl. Akad. Nauk SSSR* **161**, 641 (1965).
84. OVCHINNIKOVA, M. YA. Master's Thesis, *Inst. khim. Fiz. Akad. Nauk SSSR* (1966).
85. OSIPOV, A. I. *Nauch. Dokl. Vÿssh. Shk., Ser. Fiz.* **4**, 149 (1958).
86. OSIPOV, A. I. *Vest. mosk. gos. Univ. Ser. Mat., Mekh. Astron. Fiz. Khim.* **4**, 97 (1958).
87. OSIPOV, A. I. *Dokl. Akad. Nauk SSSR* **130**, 523 (1960).
88. OSIPOV, A. I. *Vest. mos. gos. Univ., Ser.* III **4**, 96 (1960).
89. OSIPOV, A. I. and STUPOCHENKO, E. V. *Izv. Akad. Nauk SSSR Ser. Fiz.* **8**, 992 (1960).
90. OSIPOV, A. I. *Dokl. Akad. Nauk SSSR* **139**, 351 (1961).
91. OSIPOV, A. I. *Zh. fiz. Khim.* **35**, 1524 (1961).
92. OSIPOV, A. I. *Dokl. Akad. Nauk SSSR* **137**, 833 (1961).
93. OSIPOV, A. I. *Dokl. Akad. Nauk SSSR* **143**, 1392 (1962).
94. OSIPOV, A. I. *Zh. fiz. Khim.* **36**, 1798 (1962).
95. OSIPOV, A. I. *Vest. mosk. gos. Univ. Ser.* III, **2**, 41 (1962).
96. OSIPOV, A. I. *Zh. fiz. Khim.* **37**, 2757 (1963).
97. OSIPOV, A. I. *Kinet. Katal.* **4**, 487 (1963).
98. OSIPOV, A. I. and STUPOCHENKO, E. V. *Usp. fiz. Nauk* **79**, 81 (1963).
99. OSIPOV, A. I. *Izv. Akad. Nauk. SSSR, Ser. Fiz.* **28**, 134 (1964).
100. OSIPOV, A. I. *Zhurnal prikladnoĭ mekhaniki i tekhnicheskoĭ fiziki* (1964).
101. OSIPOV, A. I. and GENERALOV, N. A. *Fiz. gor. vzr.* **2**, 83 (1966).
102. OSIPOV, A. I. *Teor. éksp. Khim.* **2**, 649 (1966).
103. OSIPOV, A. I. *Fiz. gor. vzr.* **4**, 42 (1966).
104. OSIPOV, V. I. *Zh. éksp. teor. Fiz.* **49**, 1157 (1965).

105. Pitaevskiĭ, L. P. *Zh. éksp. teor. Fiz.* **42**, 1326 (1962).
106. Pokrovskiĭ, V. L. and Khalatnikov, I. M. *Zh. éksp. teor. Fiz.* **40**, 1713 (1961).
107. Prigozhin, I. *Non-equilibrium statistical mechanics.* Wiley, New York (1963).
108. Slater, J. *Electronic structure of molecules.* McGraw-Hill, New York (1963).
109. Smirnov, B. M. *Zh. éksp. teor. Fiz.* **46**, 1017 (1964).
110. Smirnov, B. M. *Dokl. Akad. Nauk SSSR* **155**, 1294 (1964).
111. Stupochenko, E. V. *Dokl. Akad. Nauk SSSR* **67**, 635 (1949).
112. Stupochenko, E. V. *Zh. éksp. teor. Fiz.* **19**, 493 (1949).
113. Stupochenko, E. V. and Osipov, A. I. *Zh. fiz. Khim.* **32**, 1673 (1958).
114. Stupochenko, E. V., Losev, S. A., and Osipov, A. I. *Relaxation of shock waves.* Springer-Verlag, Berlin (1967).
115. Stupochenko, E. V. and Osipov, A. I. *Teor. éksp. Khim.* **2**, 76 (1967).
116. Tal'roze, V. L. *Izv. Akad. Nauk SSSR, Ser. Fiz.* **24**, 1001 (1960).
117. Tal'roze, V. L. and Frankevich, K. L. *Zh. fiz. Khim.* **34**, 2709 (1960).
118. Semenov, N. N. *Some problems in chemical kinetics and reactivity.* Princeton University Press (1958).
119. Tunitskiĭ, N. N. and Kupriyanov, S. E. *Mezhdunarodní simposium po svoistvam nizkotemperaturnoí plazmí pri XX Kongresse Yu. P.A.K.* (1965).
120. Filimonov, V. N. *Usp. fiz. Nauk* **69**, 565 (1959).
121. Firsov, O. B. *Zh. éksp. teor. Fiz.* **32**, 1001 (1951).
122. Firsov, O. B. *Zh. éksp. teor. Fiz.* **42**, 1307 (1962).
123. Fok, V. A. *Zh. éksp. teor. Fiz.* **10**, 961 (1940).
124. Chandrasekhar, S. *Stochastic problems in physics and astronomy* (1947).
125. Chapman, S. and Cowling, T. *Mathematical theory of non-uniform gases* (3rd edn.). Cambridge University Press (1970).
126. Eyring, H., Walter, D., and Kimboll, D. *Quantum chemistry.* Wiley, New York (1944).
127. Allen, R. T. and Feuer, P. J. *J. chem. Phys.* **4**, 2810 (1964).
128. Alterman, E. B. and Wilson, D. J. *J. chem. Phys.* **42**, 1957 (1965).
129. Andres, R. P. and Boudart, M. *J. chem. Phys.* **42**, 2057 (1965).
130. Baetzold, R. C. and Wilson, D. J. *J. phys. chem.* **68**, 3141 (1964).
131. Baetzold, R. C. and Wilson, D. J. *J. chem. Phys.* **43**, 4299 (1965).
132. Bak, T. A. and Andersen, K. Technical Note No. 4, University of Copenhagen, 1960.
133. Bak, T. A. and Plesner, I. W. *Acta chem. scand.* **14**, 1310 (1960).
134. Bak, T. A. and Lebowitz, J. L. *Discuss. Faraday Soc.* **33**, 189 (1962).
135. Bak, T. A. Contribution to *The theory of chemical kinetics.* Benjamin, New York (1963).
136. Bak, T. A. and Lebowitz, J. L. *Phys. Rev.* **131**, 1138 (1963).
137. Bartlett, M. S. and Moyal, J. E. *Proc. Camb. phil. Soc.* **45**, 545 (1949).
138. Basco, N., Collear, A. B., and Norrish, R. G. W. *Proc. R. Soc.* **A 260**, 429 (1961).
139. Bates, D. R. *Proc. R. Soc.* **A 257**, 22 (1960).
140. Bates, D. R. and Kingston, A. E. *Proc. phys. Soc.* **83**, 43 (1964).
141. Bates, D. R. and Moffett, R. J. *Proc. R. Soc.* **A 291**, 1 (1966).
142. Bauer, E. and Wu, T. Y. *J. chem. Phys.* **21**, 726 (1953).

143. BAUER, H., KNESER, H., and SITTIG, E. *J. chem. Phys.* **30**, 1119 (1959).
144. BAUER, E. and SALKOFF, M. *J. chem. Phys.* **33**, 1202 (1960).
145. BAUER, S. H. and TSANG, S. C. *Phys. Flqids* **6**, 182 (1963).
146. BAUER, S. H. and RESLER, E. L. *Science* **164**, 1045 (1964).
147. BAUER, S. H. and OSSA, E. *J. chem. Phys.* **45**, 434 (1966).
148. BAZLEY, N. B., MONTROLL, E. W., RUBIN, R. J., and SHULER, K. E. *J. chem. Phys.* **28**, 700 (1958).
149. BEAHN, T. J., CONDELL, W. J., and MANDELBERG, H. I. *Phys. Rev.* **141**, 83 (1966).
150. BELL, R. P. *Trans. Faraday Soc.* **55**, 1 (1959).
151. BEN–REUVEN, A. *Phys. Rev.* **145**, 7 (1966).
152. BENSON, S. W. and FUENO, T. *J. Chem. Phys.* **36**, 1597 (1962).
153. BENSON, S. W., Berend, G. C., and Wu, J. C. *J. Chem. Phys.* **38**, 25 (1963).
154. BENSON, S. W. and BEREND, G. C. *J. chem. Phys.* **39**, 2777 (1963).
155. BENSON, S. and BEREND, G. *J. chem. Phys.* **40**, 1289 (1964).
156. BENSON, S. W. and BEREND, G. C. *J. chem. Phys.* **44**, 470 (1966).
157. BENSON, S. W. and BEREND, G. C. *J. chem. Phys.* **44**, 4247 (1966).
158. BERGER, R. T. and HERZFELD, C. M. *J. chem. Phys.* **39**, 1122 (1963).
159. BIGELEISEN, J. and GOEPPERT-MAYER, M. *J. chem. Phys.* **15**, 261 (1947).
160. BIGELEISEN, J. and WOLFSBERG, M. *J. chem. Phys.* **22**, 1264 (1954).
161. BIGELEISEN, J., and WOLFSBERG, M. *Adv. chem. Phys.* **1** 15 (1958).
162. BIGELEISEN, J. *Pure appl. Chem.* **8**, 217 (1964).
163. BISHOP, D. M. and LAIDLER, K. J. *J. chem. Phys.* **42**, 1688 (1965).
164. BLACKMAN, V. *J. Fluid Mech.* **1**, 61 (1956).
165. BLAIS, N. C. and BUNKER, D. L. *J. chem. Phys.* **37**, 2713 (1962).
166. BLAIS, N. C. and BUNKER, D. L. *J. chem. Phys.* **39**, 315 (1963).
167. BRABBS, T. A., BELLES, F. E., and ZLATARICH, S. A. *J. chem. Phys.* **38**, 1939 (1963).
168. BRADLEY, J. N. *Trans. Faraday Soc.* **60**, 1353 (1964).
169. BRAUNER, J. W. and WILSON, D. J. *J. phys. Chem.* **67**, 1134 (1963).
170. BRAY, K. N. C. and PRATT, N. H. Conditions for significant gasdynamically induced vibration–recombination coupling. Paper presented at the discussion on energy transfer at the Eleventh (International) Symposium on Combustion, Berkeley (1966).
171. BRIGLIA, D. D. and RAPP, D. *Phys. Rev. Lett.* **14**, 960 (1964).
172. BRINKMAN, H. C. *Physica* **22**, 29, 149 (1956).
173. BRUECKNER, K. A. *J. chem. Phys.* **40**, 439 (1964).
174. BUFF, F. P. and WILSON, D. J. *J. chem. Phys.* **32**, 684 (1960).
175. BUFF, F. P. and WILSON, D. J. *J. Am. chem. Soc.* **84**, 4063 (1962).
176. BUFF, F. P. and WILSON, D. J. *J. chem. Phys.* **45**, 1444 (1966).
177. BUNKER, D. L. and DAVIDSON, N. *J. Am. Chem. Soc.* **80**, 5085 (1958).
178. BUNKER, D. L. *J. chem. Phys.* **32**, 1001 (1960).
179. BUNKER, D. L. *Nature* **194**, 1277 (1962).
180. BUNKER, D. L. and BLAIS, N. C. *J. chem. Phys.* **37**, 393 (1962).
181. BUNKER, D. L. *J. chem. Phys.* **40**, 1946 (1964).
182. BUNKER, D. L. and BLAIS, N. C. *J. chem. Phys.* **41** 2377 (1964).
183. CALLAWAY, J. and BAUER, E. *Phys. Rev.* **140**, A 1072 (1965).
184. CALLEAR, A. B. and NORRISH, R. G. W. *Proc. R. Soc.* A **266**, 299 (1962).

185. CALLEAR, A. B. and WILLIAMS, G. J. *Trans. Faraday Soc.* **60**, 2158 (1964).
186. CALLEAR, A. B. *Appl. Optics* Suppl. No. 2, 145 (1965).
187. CAMAC, M. and VAUGHAN, A. *J. chem. Phys.* **34**, 460 (1961).
188. CARERI, G. *J. chem. Phys.* **21**, 749 (1953).
189. CARERI, G. *Adv. chem. Phys.* **1**, 119 (1958).
190. CARRINGTON, T. *J. chem. Phys.* **35** 807 (1961).
191. CARRINGTON, T. *Discuss. Faraday Soc.* **33**, 44 (1962).
192. CARRINGTON, T. *J. chem. Phys.* **41**, 2012 (1964).
193. CARRINGTON, T. *J. chem. Phys.* **43**, 473 (1965).
194. CASHION, J. K. and HERSCHBACH, D. R. *J. chem. Phys.* **40**, 2358 (1964).
195. CHAPMAN, G. D., CZAIKOWSKI, M. RAE, A. G. A., and KRAUSE, L. *Abstracts of the IVth International Conference on the Physics of Electronic and Atomic Collisions*, Quebec (1965).
196. CHARTERS, P. E. and POLANYI, J. C. *Discuss. Faraday Soc.* **33**, 107 (1962).
197. CHILD, M. S. *Proc. R. Soc.* A **292**, 272 (1966).
198. CHOW, C. C. and GREENE, E. F. *J. chem. Phys.* **43**, 324 (1965).
199. CHRISTIE, M. *J. Am. chem. Soc.* **84**, 4066 (1962).
200. CHRISTOFFERSEN, R. E. *J. chem. Phys.* **41**, 960 (1964).
201. CLARKE, J. F. and McCHESNEY, M. *The Dynamics of real gases*. London (1964).
202. CONROY, H. *J. chem. Phys.* **40**, 603 (1964).
203. CONROY, H. *J. chem. Phys.* **40**, 3121 (1964).
204. CONROY, H. and BRUNER, B. L. *J. chem. Phys.* **47**, 921 (1967).
205. COTTRELL, T. L. and REAM, N. *Trans. Faraday Soc.* **51** 159 (1955).
206. COTTRELL, T. L. and McCOUBREY, J. C. *Molecular energy transfer in gases*. London, 1961.
207. COTTRELL, T. L. and MATHESON, A. J. *Proc. chem. Soc., March.* 114 (1961).
208. COTTRELL, T. L. and MATHESON, A. J. *Proc. chem. Soc., January.* 17 (1962).
209. COTTRELL, T. L. and MATHESON, A. J. *Trans. Faraday Soc.* **58**, 2336 (1962).
210. COTTRELL, T. L., DOBBIE, R. C., McLAIN, J., and READ, A. W. *Trans. Faraday Soc.* **60**, 241 (1964).
211. COWAN, G. C. and HORNIG, D. F. *J. chem. Phys.* **18**, 1008 (1950).
212. DICKENS, P. G., LINNETT, J. W. and SOVERS, O. *Discuss. Faraday Soc.* **33**, 52 (1962).
213. DORER, F. H., RABINOVITCH, B. S., and PLACZEK, D. W. *J. chem. Phys.* **41**, 3995 (1964).
214. DORER, F. H. and RABINOVITCH, B. S. *J. phys. Chem.* **69**, 1973 (1965).
215. EDMISTON, C. and KRAUSS, M. *J. chem. Phys.* **42**, 1119 (1965).
216. ELLISON, F. O. *J. chem. Phys.* **41**, 2198 (1964).
217. ERDÉLYI, A. *et al. Higher transcendental functions*, Vol. 1. McGraw-Hill, New York (1953).
218. ERDÉLYI, A. *et al. Higher transcendental functions*, Vol. 2. McGraw-Hill, New York (1953).
219. EVANS, M. G. and POLANYI, M. *Trans. Faraday Soc.* **35** 178 (1939).
220. EVETT, A. A. and MARGENAU, K. *Phys. Rev.* **90**, 1021 (1953).
221. EU, B. and ROSS, J. *J. chem. Phys.* **44**, 2467 (1966).
222. EYRING, H. and POLANYI, M. *Z. phys. Chem.* B **12**, 279 (1931).

223. EYRING, H. *J. chem. Phys.* **3**, 107 (1935).
224. FEIBELMAN, P. J. *J. chem. phys.* **42**, 2462 (1965).
225. FERMI, E., PASTA, J., and ULAM, S. *Los-Alamos Sci. Lab., Rept. LA*—1940 (1955).
226. FISHBURNE, E. S. and EDSE, R. *J. chem. Phys.* **41**, 1297 (1964).
227. FIUTAK, J. and FRACKOWIAK, M. *Acta phys. Pol.* **26**, 353 (1964).
228. FORD, J. *J. math. Phys.* **2**, 387 (1961).
229. FORD, J. and WATERS, J. *J. math. Phys.* **4**, 1293 (1963).
230. FOWLER, R. H. and GUGGENHEIM, E. A. *Statistical thermodynamics.* Cambridge University Press (1939).
231. FRIEDMAN, L. and MORAN, T. F. *J. chem. Phys.* **42**, 2624 (1965).
232. FUENO, T., EYRING, H., and REE, T. *Can. J. chem.* **38**, 1693 (1960).
233. GATZ, C. R. *J. chem. Phys.* **44**, 1861 (1966).
234. GATZKE, J. *Z. phys. Chem.* **223**, 321 (1963).
235. GAYDON, A. G. and HURLE, I. R. *Eighth Symposium (International) on Combustion,* Pasadena, 1959, 309 (1960).
236. GAYDON, A. G. and HURLE, I. R. *Proc. R. Soc.* A **262**, 38 (1961).
237. GAYDON, A. G., KIMBELL, G. H., and PALMER, H. G. *Proc. R. Soc.* A **279**, 313 (1964).
238. GIARDINI-GUIDONI, A. and FRIEDMAN, L. *J. chem. Phys.* **45**, 937 (1966).
239. GIDDINGS, J. C. *J. chem. Phys.* **22**, 538 (1954).
240. GIESE, C. F. and MAIER, W. B. *J. chem. Phys.* **39**, 739 (1963).
241. GILL, E. K. and LAIDLER, K. *J. Can. J. Chem.* **39**, 1371 (1958).
242. GIOUMOUSIS, G. and STEVENSON, D. P. *J. chem. Phys.* **29**, 294 (1958).
243. GOLIKE, R. C. and SCHLAG, E. W. *J. chem. Phys.* **38**, 1886 (1963).
244. GREEN, E. F. and TOENNIS, J. P. *Chemical reactions in shock waves.* Arnold, London (1964).
245. GROSSER, A. E., BLYTHE, A. R., and BERNSTEIN, R. B. *J. chem. Phys.* **42**, 1268 (1965).
246. GURNEE, E. F. and MAGEE, J. L. *J. chem. Phys.* **26**, 1237 (1957).
247. HAARHOF, P. C. *Molec. Phys.* **6**, 337 (1963).
248. HAARHOF, P. C. *Molec. Phys.* **7**, 101 (1963–64).
249. HALL, G. G. and LEVINE, R. D. *J. chem. Phys.* **44**, 1567 (1966).
250. HAMMERLING, P., TEARE, J. D., and KIVEL, B. *Phys. Fluids* **2**, 422 (1959).
251. HANSEN, C. F. *A.I.A.A. Jl.* **3**, 61 (1965).
252. HARRIS, F. E., MICHA, D. A., and POHL, H. A. *Arkiv. Fys.* **30**, 259 (1965).
253. HARRISON, A. G., IVKO, A., and SHANNON, T. W. *Can. J. Chem.* **44**, 1351 (1966).
254. HARTER, R. J., ALTERMAN, E. B., and WILSON, D. J. *J. chem. Phys.* **40**, 2137 (1964).
255. HEIMS, S. P. *J. chem. Phys.* **38**, 603 (1963).
256. HENDERSON, M. C. *J. Acoust. Soc. Am.* **34**, 349 (1962).
257. HERMAN, R. and RUBIN, R. *J. Phys. Fluids* **2**, 547 (1959).
258. HERSCHBACH, D. R. *Discuss. Faraday Soc.* **33**, 149 (1962).
259. HERSCHBACH, D. R. *Appl. Optics.* Suppl No. 2, 128 (1965).
260. HERSCHBACH, D. R. *Adv. Chem. Phys.* **10**, 319 (1966).
261. HERZBERG, G. and LONGUET-HIGGINS, H. C. *Discuss. Faraday Soc.* **35**, 77 (1963).
262. HERZFELD, K. F. and LITOVITZ, T. A. *Absorption and dispersion of ultrasonic waves.* Academic Press, New York—London (1959).
263. HERZFELD, K. F. *Discuss. Faraday Soc.* **33**, 22 (1962).

264. HIRSCHFELDER, J. O., EYRING, H., and TOPLEY, B. *J. chem. Phys.* **4**, 170 (1936).
265. HOARE, M. *Nature* **192**, 63 (1961).
266. HOARE, M. *Molec. Phys.* **4**, 465 (1961).
267. HOARE, M. *J. chem. Phys.* **41**, 2356 (1964).
268. HOFACKER, G. L. *J. chem. Phys.* **43**, 208 (1965).
269. HOLMES, R., SMITH, F. A., and TEMPEST, W. *Proc. Phys. Soc.* **83**, 769 (1964).
270. HOLMES, R., JONES, G. R., and LAWRENCE, R. *Trans. Faraday Soc.* **62**, 1 (1966).
271. HORIE, T. and KURATA, K. *J. chem. Phys.* **31**, 783 (1959).
272. HORIE, T. and KASUGA, T. *J. phys. Soc., Japan* **19**, 1194 (1964).
273. HORIE, T. and KASUGA, T. *J. chem. Phys.* **40**, 1683 (1964).
274. HOYLAND, J. R. *J. chem. Phys.* **41**, 1370 (1964).
275. HUBBARD, S. *Rev. mod. Phys.* **33**, 249 (1961).
276. HUDSON, G. H., MCCOUBREY, J. C., and UBBELOHDE, A. R. *Proc. R. Soc.* A **264**, 289 (1961).
277. HUETZ-AUBERT, M., DANG, N. M., and HUETZ, J. C. *hebd. Séanc. Acad. Sci., Paris* **258**, 6077 (1964).
278. HUNG, CH. N. and WILSON, D. J. *J. phys. Chem.* **66**, 342 (1962).
279. HUNG, CH. N. and WILSON, D. J. *J. chem. Phys.* **38**, 828 (1963).
280. HUNT, G. R., MCCOY, E. F., and ROSS, I. G. *Aust. J. Chem.* **15**, 591 (1962).
281. HURLE, I. R. *J. chem. Phys.* **41**, 3911 (1964).
282. HURLE, I. R. and RUSSO, A. L. *J. chem. Phys.* **43**, 4434 (1965).
283. JACKSON, E. A. *J. math. Phys.* **4**, 551, 668 (1963).
284. JACKSON, J. M. and MOTT, N. F. *Proc. R. Soc.* A **137**, 703 (1932).
285. JENKINS, D. R. *Chem. Commun.* No. 6, 171 (1966).
286. JOHNSTON, H. S. *J. chem. Phys.* **19**, 663 (1951).
287. JOHNSTON, H. S. and RAPP, D. J. *J. Am. chem. Soc.* **83**, 1 (1961).
288. JOHNSTON, H. S. and HEICKLEN, J. *J. phys. Chem.* **66**, 532 (1962).
289. JOSHI, B. D. *J. chem. Phys.* **44**, 3627 (1966).
290. KARL, G. and POLANYI, J. C. *J. chem. Phys.* **38**, 271 (1963).
291. KARL, G., KRUUS, P., and POLANYI, J. C. *J. chem. Phys.* **46**, 224 (1967).
292. KARL, G., KRUUS, P., POLANYI, J. C., and Smith, I. W. M. *J. chem. Phys.* **46**, 244 (1967).
293. KARPLUS, M. and RAFF, L. M. *J. chem. Phys.* **41**, 1267 (1964).
294. KARPLUS, M., PORTER, R. N., and SHARMA, R. D. *J. chem. Phys.* **40**, 2033 (1964).
295. KARPLUS, M., PORTER, R. N., and SHARMA, R. D. *J. chem. Phys.* **43**, 3259 (1965).
296. KASHA, M. *Discuss. Faraday Soc.* **9**, 14 (1950).
297. KASSEL, L. S. *Proc. natn. Acad. Sci., U.S.A.* **16**, 358 (1930).
298. KECK, J. C. *J. chem. Phys.* **32**, 1035 (1960).
299. KECK, J. C. *Discuss. Faraday Soc.* No. 33, 173 (1962).
300. KECK, J. and CARRIER, G. *J. chem. Phys.* **43**, 2284 (1965).
301. KELLEY, J. D. and WOLFSBERG, M. *J. chem. Phys.* **44**, 324 (1966).
302. KERNER, E. *Can. J. Phys.* **36**, 317 (1958).
303. KIEFER, J. H. and LUTZ, R. W. The effect of oxygen atoms on the vibrational relaxation of oxygen. Paper presented at the *Discussion on Energy Transfer of the Eleventh International Symposium on Combustion, Berkeley* (1966).
304. KIM, S. K. *J. chem. Phys.* **28**, 1057 (1958).
305. KLOTS, C. E. *J. chem. Phys.* **41**, 117 (1964).
306. KOCH, VON H. and FRIEDMAN, L. *J. chem. Phys.* **38**, 1115 (1963).

307. KOHLMAIER, G. H. and RABINOVITCH, B. S. *J. chem. Phys.* **38**, 1692 (1963); *J. chem. Phys.* **39**, 490 (1963).
308. KOLKER, H. J. *J. chem. Phys.* **44**, 582 (1966).
309. KONDRATIEV, V. N. and NIKITIN, E. E. *J. chem. Phys.* **45**, 1087 (1966).
310. KOSTIN, M. P. *J. chem. Phys.* **43**, 2679 (1965).
311. KOLOS, W. and WOLNIEWICZ, L. *J. chem. Phys.* **43**, 2429 (1965).
312. KRAMERS, H. A. *Physica* **7**, 284 (1940).
313. KRAUSE, L. *Appl. Optics* **5**, 1375 (1966).
314. KRAUSS, M. and MIES, F. H. *J. chem. Phys.* **42**, 2703 (1965).
315. KUNTZ, P. J., NEMETH, E. M., POLANYI, J. C., ROSNER, S. D., and YOUNG, C. E. *J. chem. Phys.* **44**, 1168 (1966)
316. KUPPERMAN, A. and WHITE, J. M. *J. chem. Phys.* **44**, 4352 (1966).
317. LAIDLER, K. J. *J. chem. Phys.* **10**, 34, 43 (1942).
318. LAIDLER, K. J. *Chemical kinetics of excited states.* Oxford University Press (1955).
319. LANDAU, L. *Phys. Z.* **2**, 46 (1932).
320. LANDAU, L. and Teller, E. *Phys. Z.* **10**, 34 (1936).
321. LANDAU, L. *Phys. Z.* **10**, 67 (1936).
322. LEVINE, R. D. *J. chem. Phys.* **44**, 2029 (1966).
323. LEVINE, R. D. *J. chem. Phys.* **44**, 2035 (1966).
324. LEVINE, R. D. *J. chem. Phys.* **44**, 3597 (1966).
325. LEVITT, B. P. *Trans. Faraday Soc.* **59**, 59 (1963).
326. LIFSCHITZ, C. and WOLFSBERG, M. *J. chem. Phys.* **41**, 1879 (1964).
327. LIGHT, J. C. *J. chem. Phys.* **36**, 1016 (1962).
328. LIGHT, J. C. and ARNSTEIN, R. *J. chem. Phys.* **37**, 2240 (1962).
329. LIGHT, J. C. and SHULER, K. E. *J. chem. Phys.* **38**, 1880 (1963).
330. LIGHT, J. C. and HORROCKS, J. *Proc. phys. Soc.* **84**, 527 (1964).
331. LIGHT, J. C. *J. chem. Phys.* **40** 3221 (1964).
332. LIGHT, J. C. and Lin, J. *J. chem. Phys.* **43**, 3209 (1965).
333. LIGHT, J. C., ROSS, J. and SHULER, K. E. Rate coefficients, reaction cross-sections and microscopic reversibility, *Kinetic processes in gases and plasmas* (1969).
334. LIN, S. H. and EYRING, H. *J. chem. Phys.* **39**, 1577 (1963); **43**, 2153 (1965).
335. LIN, J. and LIGHT, J. *J. chem. Phys.* **45**, 2545 (1966).
336. LONDON, F. *Z. Elektrochem.* **35**, 552 (1929).
337. LONGUET-HIGGINS, H. C. In *Advances in spectroscopy*, vol. 2, p. 429. New York (1961).
338. LUKASIK, S. J. and YOUNG, J. E. *J. chem. Phys.* **27**, 1149 (1957).
339. MAGEE, J. L. *J. chem. Phys.* **8**, 687 (1940).
340. MAGEE, J. L., SHAND, W., and EYRING, H. J. *J. Am. chem. Soc.* **63**, 677 (1941).
341. MAHAN, B. H. and SOLO, R. B. *J. chem. Phys.* **37**, 2669 (1962).
342. MAHAN, B. H. and PERSON, J. C. *J. chem. Phys.* **40**, 2851 (1964).
343. MANN, A. In *Atomic collision processes*, p. 988. North-Holland, Amsterdam (1964).
344. MANN, A. *Ann. Phys.* **16**, 381 (1965).
345. MARCUS, R. A. and RICE, O. K. *J. phys. collord. Chem.* **55**, 894 (1951).
346. MARCUS, R. A. *J. chem. Phys.* **20**, 359, 364 (1952).
347. MARCUS, R. A. *J. chem. Phys.* **41**, 603 (1964).
348. MARCUS, R. A. *J. chem. Phys.* **41**, 610 (1964).
349. MARCUS, R. A. *J. chem. Phys.* **41**, 2614 (1964).

350. MARCUS, R. A. *J. chem. Phys.* **41**, 2624 (1964).
351. MARCUS, R. A. *J. chem. Phys.* **43**, 1598 (1965).
352. MARCUS, R. A. *J. chem. Phys.* **43**, 2658 (1965).
353. MARCUS, R. A. *J. chem. Phys.* **45**, 2138 (1966).
354. MARGENAU, H. *Phys. Rev.* **66**, 303 (1944).
355. MARRIOTT, R. *Proc. phys. Soc.* **83**, 159 (1964).
356. MARRIOTT, R. *Proc. phys. Soc.* **84**, 877 (1964).
357. MARRIOTT, R. *Proc. phys. Soc.* **86**, 1041 (1965).
358. MARRIOTT, R. *Proc. phys. Soc.* **88**, 617 (1966).
359. MATLAND, C. G. *Phys. Rev.* **92**, 637 (1953).
360. MATSEN, F. A. *J. phys. Chem.* **68**, 3282 (1964).
361. MAYER, S. W., SCHIELER, L., and JOHNSTON, H. S. *J. chem. Phys.* **45**, 385 (1966).
362. MAZUR, J. and RUBIN, R. J. *J. chem. Phys.* **31**, 1395 (1959).
363. MCCARROLL, B. and EHRLICH, G. *J. chem. Phys.* **38**, 523 (1963).
364. MICHA, D. A. *Ark. fys.* **30**, 411 (1965).
365. MICHA, D. A. *Ark. fys.* **30**, 437 (1965).
366. MICHA, D. A. *Ark. fys.* **30**, 425 (1965).
367. MICHEL, K. W., OLSCHEWSKI, H. A., RICHTERING, H., and WAGNER, H. G. *Z. phys. Chem.* **39**, 129 (1963).
368. MIES, F. H. and SHULER, K. E. *J. chem. Phys.* **37**, 177 (1962).
369. MIES, F. H. *J. chem. Phys.* **40**, 523 (1964).
370. MIES, F. H. *J. chem. Phys.* **41**, 903 (1964).
371. MIES, F. H. *J. chem. Phys.* **42**, 2709 (1965).
372. MILLIKAN, R. C. and WHITE, D. R. *J. chem. Phys.* **39**, 98 (1963).
373. MILLIKAN, R. C. and WHITE, D. R. *J. chem. Phys.* **39**, 3209 (1963).
374. MONTROLL, E. W. and SHULER, K. E. *J. chem. Phys.* **26**, 452 (1957).
375. MONTROLL, E. W. and SHULER, K. E. In *Advances in chemical physics*, Vol. I, p. 361 (1958).
376. MOORE, C. B. *J. chem. Phys.* **43**, 2979 (1965).
377. MORAN, T. F. and HAMILL, W. H. *J. chem. Phys.* **39**, 1413 (1963).
378. MORGAN, J. E. and SCHIFF, H. I. *J. chem. Phys.* **38**, 1495 (1963).
379. MORI, Y. *Bull chem. Soc., Japan* **35**, 1584 (1962).
380. NIELSEN, S. E. and BAK, T. A. *J. chem. Phys.* **41**, 665 (1964).
381. NIKITIN, E. E. and SOKOLOV, N. D. *J. chem. Phys.* **31**, 1387 (1959).
382. NIKITIN, E. E. *Eighth symposium on combustion, Pasadena, 1959* (1960).
383. NIKITIN, E. E. *Discuss. Faraday Soc.* **33**, 14 (1962).
384. NIKITIN, E. E. *Molec. Phys.* **7**, 389 (1963–64).
385. NIKITIN, E. E. *Molec. Phys.* **8**, 65 (1964).
386. NIKITIN, E. E. *Molec. Phys.* **8**, 473 (1964).
387. NIKITIN, E. E. *J. quant. Spectrosc. Radiat. Trans.* **5**, 436 (1965).
388. NIKITIN, E. E. *J. chem. Phys.* **43**, 744 (1965).
389. NIKITIN, E. E. *Combust. Flame* **10**, 381 (1966).
390. NIKITIN, E. E. Theory of non-adiabatic transitions. Recent development on the Landau–Zener model. In *Elementary processes and chemical reactions*. Berlin—Heidelberg—New York (1968).
391. NORRISH, R. G. W. and SMITH, W. *Proc. phys. Soc.* A **176**, 295 (1949).
392. NYELAND, C. and BAK, T. A. *Trans. Faraday Soc.* **61**, 1293 (1965).

393. OLSCHEWSKI, H. A., TROE, J., and WAGNER, H. GG. *Ber. Bunsengesell. phys. Chem.* **70**, 450 (1966).
394. PARKER, J. G. *Phys. Fluids* **2**, 449 (1959).
395. PARKER, J. G. *J. chem. Phys.* **34**, 1763 (1961).
396. PARKER, J. G. *J. chem. Phys.* **41**, 1600 (1964).
397. PEARSON, A. G., POSHUSTA, R. D. and BROWNE, J. C. *J. chem. Phys.* **44**, 1815 (1966).
398. PECHUKAS, PH. and LIGHT, J. C. *J. chem. Phys.* **42**, 3281 (1965).
399. PECHUKAS, PH., LIGHT, J. C., and RANKIN, CH. *J. chem. Phys.* **44**, 794 (1966).
400. PELZER, H. and WIGNER, E. *Z. phys. Chem.* B **15**, 445 (1932).
401. PHILLIPS, V. J. *Proc. phys. Soc.* **73**, 800 (1959).
402. PLACZEK, D. W., RABINOVITCH, B. S., WHITTEN, G. Z., and TSCHUIKOW-ROUX, E. *J. chem. Phys.* **43**, 4071; **44**, 3646 (1966).
403. PLACZEK, D. W., RABINOVITCH, B. S., and DORER, F. H. *J. chem. Phys.* **44**, 279 (1966).
404. POLANYI, J. C. *J. chem. Phys.* **31**, 1338 (1959).
405. POLANYI, J. C. and ROSNER, S. D. *J. chem. Phys.* **38**, 1028 (1963).
406. POLANYI, J. C. *J. quant. Spectrosc. Radiat. Trans.* **3**, 471 (1963).
407. POLANYI, J. C. *Appl. Optics*, Suppl. No. 2, 109 (1965).
408. PORTER, G. B. and CONNELLY, B. T. *J. chem. Phys.* **33**, 81 (1960).
409. PORTER, G. B. and SMITH, J. A. *Proc. R. Soc.* A **261**, 28 (1961).
410. PORTER, R. N. and KARPLUS, M. *J. chem. Phys.* **40**, 1105 (1964).
411. PORTER, R. N. *J. chem. Phys.* **45**, 2284 (1966).
412. PRESENT, R. D. *Proc. natn. Acad. Sci., U.S.A.* **41**, 415 (1955).
413. PRESENT, R. D. *J. chem. Phys.* **31**, 747 (1959).
414. PRIGOGINE, I. and XHROUET, E. *Physica* **15**, 913 (1949).
415. PRIGOGINE, I. and MAHIEU, M. *Physica* **16**, 51 (1950).
416. PRIGOGINE, I. and BAK, T. *J. chem. Phys.* **31**, 1368 (1959).
417. PRITCHARD, H. O. *J. phys. Chem.* **65**, 504 (1961).
418. PRITCHARD, H. O. *J. phys. Chem.* **66**, 2111 (1962).
419. PYUN, CHONG WHA and ROSS, J. *J. chem. Phys.* **40**, 2572 (1964).
420. RABINOVITCH, B. S. and SETSER, D. W. In *Advances in photochemistry*, Vol. 3, p. 1 (1964); RABINOVITCH, B. S. and FLOWERS, M. C., *Quart. Rev.* **18**, 122 (1964).
421. RAFF, L. M. and KARPLUS, M. *J. chem. Phys.* **44**, 1212 (1966).
422. RAFF, L. M. *J. chem. Phys.* **44**, 1202 (1966); **50**, 2276 (1969).
423. RAPP. D. *J. chem. Phys.* **32**, 735 (1960).
424. RAPP. D. and FRANCIS, W. E. *J. chem. Phys.* **37**, 2631 (1962).
425. RAPP, D. and SHARP, T. E. *J. chem. Phys.* **38**, 2641 (1963).
426. RAPP, D. and ENGLANDER-GOLDEN, P. *J. chem. Phys.* **40**, 573 (1964).
427. RAPP, D. *J. chem. Phys.* **40**, 2813 (1964).
428. RAPP, D. In *Atomic collision processes*. North-Holland, Amsterdam (1964).
429. RAPP, D. and ENGLANDER-GOLDEN, P. *J. chem. Phys.* **40**, 3120 (1964).
430. RAPP, D. *J. chem. Phys.* **43**, 316 (1965).
431. RAPP, D. and SHARP, T. E. Large transition probabilities. Paper presented at the *Discussion on Energy Transfer of the Eleventh (International) Symposium on Combustion, Berkeley* (1966).

432. REE, T. S., REE, T., EYRING, H., and FUENO, T. *J. chem. Phys.* **36**, 281 (1962).
433. REUBEN, B. G. and FRIEDMAN, L. *J. chem. Phys.* **37**, 1636 (1962).
434. RICE, O. K. *J. chem. Phys.* **9**, 258 (1941).
435. RICE, O. K. *J. chem. Phys.* **21**, 750 (1953).
436. RICE, O. K. *M. Chem.* **90**, 330 (1959).
437. RICE, O. K. *J. phys. Chem.* **64**, 1851 (1960).
438. RICE, O. K. *J. phys. Chem.* **65**, 1588 (1961).
439. RICE, O. K. In *Transfert d'énergie dans les gas.* New York—Bruxelles (1962).
440. RICE, O. K. *J. phys. Chem.* **67**, 6 (1963).
441. RICE, O. K. *J. phys. Chem.* **67**, 1733 (1963).
442. RICH, J. W. and REHM, R. G. Population distribution during vibrational relaxation of diatomic gases. Paper presented at the *Discussion on Energy Transfer of the Eleventh (International) Symposium on Combustion, Berkeley* (1966).
443. RIDLEY, B. A., SCHULZ, W. R., and LE ROY, D. J. *J. chem. Phys.* **44**, 3344 (1966).
444. ROBBEN, P. *J. chem. Phys.* **31**, 420 (1959).
445. ROBERTS, CH. S. *Phys. Rev.* **131**, 203 (1963).
446. ROBINSON, G. W. and FROSCH, R. P. *J. chem. Phys.* **38**, 1187 (1963).
447. ROSEN, N. and ZENER, C. *Phys. Rev.* **40**, 502 (1932).
448. ROSENSTOCK, H. M., WALLENSTEIN, M. B., WAHRHAFTIG, A. L., and EYRING, H. *Proc. natn. Acad. Sci., U.S.A.* **38**, 667 (1952).
449. ROSENSTOCK, H. M. *J. chem. Phys.* **34**, 2182 (1961).
450. ROSENSTOCK, H. M. and KRAUSS, M. In *Advances in mass spectrometry,* Vol. 2, p. 251. Oxford—London—New York—Paris (1963).
451. ROSS, J. and MAZUR, P. *J. chem. Phys.* **35**, 19 (1961).
452. RUBIN, R. J. and SHULER, K. E. *J. chem. Phys.* **25**, 59 (1958).
453. RUBIN, R. J. *J. chem. Phys.* **40**, 1069 (1964).
454. RUSH, D. G. and PRITCHARD, H. O. Vibrational disequilibrium in chemical reactions. Paper presented at the *Eleventh (International) Symposium on Combustion, Berkeley* (1966).
455. RUSSELL, K. E. and SIMONS, J. *Proc. R. Soc.* A **217**, 271 (1953).
456. RYAN, K. R. and FUTRELL, J. H. *J. chem. Phys.* **43**, 3009 (1965).
457. SATO, S. *J. chem. Phys.* **23**, 592, 2465 (1955).
458. SCHEEL, N. and GRIFFING, V. *J. chem. Phys.* **36**, 1953 (1962).
459. SCHLAG, E. W., RABINOVITCH, B. S., and SCHNEIDER, F. W. *J. chem. Phys.* **32**, 1599 (1960).
460. SCHLAG, E. W. and SANDSMARK, R. A. *J. chem. Phys.* **37**, 168 (1962).
461. SCHLAG, E. W. *J. chem. Phys.* **38**, 2480 (1963).
462. SCHLAG, E. W., SANDSMARK, R. A., and VALENCE, W. G. *J. chem. Phys.* **40**, 1461 (1964).
463. SCHLAG, E. W., SANDSMARK, R. A., and VALANCE, W. G. *J. phys. Chem.* **69**, 1431 (1965).
464. SCHLAG, E. W. and HALBER, G. L. *J. chem. Phys.* **42**, 584 (1965).
465. SCHNEIDER, F. W. and RABINOVITCH, B. S. *J. Am. chem. Soc.* **84**, 4215 (1962).
466. SCHNEIDERMAN, S. B. and MICHELS, H. H. *J. chem. Phys.* **42**, 3706 (1965).
467. SCHULZ, W. R. and LE ROY, D. J. *J. chem. Phys.* **42**, 3869 (1965).
468. SCHWARTZ, R. N. and SLAWSKY, Z. J., Herzfeld, K. F. *J. chem. Phys.* **20**, 1591 (1952).

469. SCHWARTZ, R. N. and HERZFELD, R. F. *J. chem. Phys.* **22**, 767 (1954).
470. SERAUSKAS, R. V. and SCHLAG, E. W. *J. chem. Phys.* **42**, 3009 (1965).
471. SERAUSKAS, R. V. and SCHLAG, E. W. *J. chem. Phys.* **43**, 898 (1965).
472. SETSER, D. W. and RABINOVITCH, B. S. *J. chem. Phys.* **40**, 2427 (1964).
473. SHANNON, T. W. and HARRISON, A. G. *J. chem. Phys.* **43**, 4206 (1965).
474. SHARP, T. E. and RAPP, D. *J. chem. Phys.* **43**, 1233 (1965).
475. SHAVITT, I. *J. chem. Phys.* **31**, 1359 (1959).
476. SHERR, M. D. and FINE, J. *J. chem. Phys.* **36**, 1269 (1962).
477. SHIN, H. *J. chem. Phys.* **39**, 2934 (1963).
478. SHIN, H. *J. chem. Phys.* **41**, 2864 (1964).
479. SHIN, H. *Can. J. Chem.* **42**, 2351 (1964).
480. SHIN, H. *J. phys. Chem.* **69**, 1424 (1965).
481. SHIN, H. *J. chem. Phys.* **42**, 59 (1965).
482. SHULER, K. E. *Seventh Symposium (International) on Combustion*, p. 87 (1958).
483. SHULER, K. E. *J. chem. Phys.* **31**, 1375 (1959).
484. SHULER, K. E. *Phys. Fluids* **2**, 442 (1959).
485. SHULER, K. E. *J. chem. Phys.* **32**, 1692 (1960).
486. SHULER, K. E. and ZWANZIG, R. *J. chem. Phys.* **33**, 1778 (1960).
487. SHULER, K. E., WEISS, G. H. and ANDERSEN, K. *J. math. Phys.* **3**, 550 (1962).
488. SHULER, K. E. and WEISS, G. H. *J. chem. Phys.* **38**, 505 (1963).
489. SHULER, K. E. and WEISS, G. H. *J. chem. Phys.* **45**, 1105 (1966).
490. SHULER, K. E. and WEISS, G. H. *J. chem. Phys.* **45**, 1110 (1966).
491. SLATER, N. B. *J. chem. Phys.* **24**, 1256 (1956).
492. SLATER, N. B. *Theory of unimolecular reactions.* New York (1959).
493. SLATER, N. B. *J. phys. Chem.* **64**, 476 (1960).
494. SLATER, N. B. *J. chem. Phys.* **35**, 445 (1961).
495. SLATER, N. B. In *The transition state.* Special Publ. of the Chem. Soc. No. 16 (1962).
496. SMITH, F. T. *J. chem. Phys.* **31**, 1352 (1959).
497. SNIDER, R. F. *J. chem. Phys.* **32**, 1051 (1960).
498. SNIDER, N. S. *J. chem. Phys.* **42**, 548 (1965).
499. SNIDER, N. S. and ROSS, J. *J. chem. Phys.* **44**, 1087 (1966).
500. STARR, W. L. and SHAW, TH. M. *J. chem. Phys.* **44**, 4181 (1966).
501. STEARN, A. E. and EYRING, H. *J. chem. Phys.* **3**, 778 (1936).
502. STEEL, C. *J. chem. Phys.* **31**, 899 (1959).
503. STEEL, C. and LAIDLER, K. J. *J. chem. Phys.* **34**, 1827 (1961).
504. STRETTON, J. L. *Trans. faraday Soc.* **61**, 1053 (1965).
505. STUECKELBERG, E. C. C. *Helv. phys. Acta* **5**, 369 (1932).
506. TAKAYANAGI, K. *Prog. theor. Phys.* **8**, 497 (1952).
507. TAKAYANAGI, K. *J. phys. Soc., Japan* **14**, 75 (1959).
508. TAKAYANAGI, K. and MIYAMOTO, Y. *Sci. Rep. Saitama Univ.* A **3**, 101 (1959).
509. TAKAYANAGI, K. and KANEKO, S. *Sci. Rep. Saitama Univ.* A **3**, 167 (1960); A **4**, 15 (1961).
510. TAKAYANAGI, K. *Sci. Rep. Saitama Univ.* A **4**, 49 (1962).
511. TAKAYANAGI, K. Suppl. to the *Prog. theor. Phys.* **25**, 1 (1963).
512. TAKAYANAGI, K. *Sci. Rep. Saitama Univ.* **5**, 14 (1965).

REFERENCES

513. TAKAYANAGI, K. In *Advances in atomic and molecular physics* (ed. Bates and Esterman), Vol. 1, p. 149. Academic Press, New York (1965).
514. TANNENWALD, L. M. *Proc. phys. Soc.* **87**, 109 (1966).
515. TAYLOR, B. L., CAMAC, M. and FEINBERG, R. M. Measurements of vibrational-vibrational coupling in gas mixtures. Paper presented at the *Discussion on energy transfer of the Eleventh (International) Symposium on Combustion, Berkeley* (1966).
516. TELLER, E. *J. phys. Chem.* **41**,109 (1937).
517. THIELE, E. and WILSON, D. J. *J. chem. Phys.* **35**, 1256 (1961).
518. THIELE, E. *J. chem. Phys.* **36**, 1466 (1962).
519. THIELE, E. *J. chem. Phys.* **38**, 1959 (1963).
520. THIELE, E. and WILSON, D. J. *J. phys. Chem.* **64**, 473 (1960).
521. THIELE, E. *J. chem. Phys.* **39**, 3258 (1963).
522. THOMSON, J. J. *Phil. mag.* **47**, 337 (1924).
523. THORSON, W. R. *J. chem. Phys.* **34**, 1744 (1961).
524. TODA, M. *J. phys. Soc., Japan* **13**, 1266 (1958).
525. TODA, M. *J. phys. Soc., Japan* **14**, 722 (1959).
526. TREANOR, CH. E. and MARRONE, PAUL V. *Phys. Fluids* **5**, 1022 (1962).
527. TREANOR, CH. E. *J. chem. Phys.* **43**, 532 (1965).
528. TREANOR, CH. E. *J. chem. Phys.* **44**, 2220 (1966).
529. TREDGOLD, R. H. *Proc. phys. Soc.* A **68**, 920 (1955).
530. TROE, J. and WAGNER, H. GG. Unimolecular thermal dissociation of small molecules. Paper presented at the 28th propulsion and energetics panel meeting on *Recent advances in aerothermochemistry*, Oslo, May 1966 (1966).
531. TSUCHIYA, S. and KURATANI, K. *Combust. Flame* **8**, 299 (1964).
532. VALANCE, W. G. and SCHLAG, E. W. *J. chem. Phys.* **45**, 216 (1966).
533. VANDERSLICE, J. T. and WEISSMAN, S. *J. chem. Phys.* **37**, 2247 (1962).
534. VESTAL, M. L. and ROSENSTOCK, H. M. *J. chem. Phys.* **35**, 2008 (1961).
535. VESTAL, M. L., WAHRHAFTIG, A. L., and JOHNSTON, W. H. *J. chem. Phys.* **37**, 1276 (1962).
536. WALL, F. T., HILLER, L. A, and MAZUR, J. *J. chem. Phys.* **29**, 255 (1958).
537. WALL, F. T., HILLER, L. A., and MAZUR, J. *J. chem. Phys.* **35**, 1284 (1961).
538. WALL, F. T. and PORTER, R. *J. chem. Phys.* **39**, 3112 (1963).
539. WEINGARTSHOFER, A. and CLARKE, E. M. *Phys. Rev. Lett.* **12**, 591 (1964).
540. WESTON, R. E. *J. chem. Phys.* **31**, 892 (1959).
541. WHITE, D. R. and MILLIKAN, R. C. *J. chem. Phys.* **39**, 2107 (1963).
542. WHITE, D. R. and MILLIKAN, R. C. *J. chem. Phys.* **39**, 1807 (1963).
543. WHITTEN, G. Z. and RABINOVITCH, B. S. *J. chem. Phys.* **38**, 2466 (1963).
544. WHITTEN, G. Z. and RABINOVITCH, B. S. *J. chem. Phys.* **41**, 1883 (1964).
545. WIDOM, B. *J. chem. Phys.* **28**, 918 (1958); **30**, 238 (1959).
546. WIDOM, B. *J. chem. Phys.* **31**, 1027 (1959).
547. WIDOM, B. *J. chem. Phys.* **31**, 1387 (1959).
548. WIDOM, B. *J. chem. Phys.* **34**, 2050 (1961).
549. WIDOM, B. *Discuss. Faraday Soc.* **33**, 37 (1962).
550. WIDOM, B. *Adv. Chem. Phys.* **5**, 535 (1963).
551. WIEDER, G. M. and MARCUS, R. A. *J. chem. Phys.* **37**, 1835 (1962).
552. WIGNER, E. *Z. phys. Chem.* B **19**, 203 (1932).

553. WIGNER, E. *J. chem. Phys.* **5**, 720 (1937).
554. WIGNER, E. *Trans. Faraday Soc.* **33**, 29 (1937).
555. WILDE, K. J. *J. chem. Phys.* **41**, 448 (1964).
556. WILSON, D. J. *J. phys. Chem.* **64**, 323 (1960).
557. WILSON, D. J., NOBLE, B., and LEE, B. *J. chem. Phys.* **34**, 1392 (1961).
558. WILSON, D. J. *J. chem. Phys.* **38**, 1098 (1963).
559. WILSON, D. J. and THIELE, E. *J. chem. Phys.* **40**, 3425 (1964).
560. WITTEMAN, W. J. *Philips Res. Rep.* **18**, Suppl. No. 2, 94 (1963).
561. WOLF, F. A. *J. chem. Phys.* **44**, 1619 (1966).
562. WOLFSBERG, M. *J. chem. Phys.* **36**, 1072 (1962).
563. WOOD, J. L. *J. chem. Phys.* **33**, 1587 (1960).
564. WOZNICK, B. J. and KECK, J. C. *J. chem. Phys.* **40**, 1166 (1964).
565. WOZNICK, B. J. *J. chem. Phys.* **42**, 1151 (1965).
566. WRAY, K. L. *J. chem. Phys.* **36**, 2597 (1962).
567. WRAY, K. L. *J. chem. Phys.* **38**, 1518 (1963).
568. WRAY, K. L. and FREEMAN, TH. S. *J. chem. Phys.* **40**, 2785 (1964).
569. WRAY, K. L. *Tenth Symposium (International) on Combustion*, U.S.A., p. 523 (1965).
570. YANG, K. and REE, T. *J. chem. Phys.* **35**, 588 (1964).
571. ZENER, C. *Proc. R. Soc.* A **137**, 696 (1932).
572. Zener, C. *Proc. Camb. Phil. Soc.* **29**, 136 (1932).

REFERENCES IN APPENDIX

1. JOHNSTON, H. S. *Gas phase reaction rate theory*. New York (1966).
2. STEVENSON, B. *Collisional activation in gases*. Pergamon Press, New York (1967).
3. *Fast reactions and primary processes in chemical kinetics*. Stockholm (1967).
4. Molecular dynamics of the chemical reaction in gases, *Discuss. Faraday Soc.* **44**, 1967.
5. *Chemische Elementarprozesse*. Springer–Verlag, Berlin (1968).
6. LEVINE, R. D. *Quantum mechanics of molecular rate processes*, Clarendon Press, Oxford (1968).
7. Chemische Elemantarprozesse, *Ber. Bunsengesell phys. Chem.* **72**, (1968).
8. *Kinetic processes in gases and plasmas*. New York–London (1969).
9. *Issledovaniya s molekulyarnyi puchkami*. 'Mir', U.S.S.R. (1969).
10. CHILD, M. S. *Molec. Phys.* **12**, 401 (1967).
11. CONNOR, J. N. L. *Molec. Phys.* **15**, 37 (1968).
12. COULSON, C. A. and LEVINE, R. D. *J. chem. Phys.* **47**, 1235 (1967).
13. EU, B. C. and ROSS, J. *J. chem. Phys.* **46**, 411 (1967).
14. HOCHSTIM, A. R. and SHULER, K. E. *J. chem. Phys.* **47**, 1894 (1967).
15. MARCUS, R. A. *J. chem. Phys.* **45**, 2630 (1966).
16. MARCUS, R. A. *Discuss. Faraday Soc.* **44**, 7 (1967).
17. MARCUS, R. A. *J. chem. Phys.* **46**, 959 (1967).
18. MARCUS, R. A. *J. chem. Phys.* **49**, 2610 (1968).
19. MORTENSON, E. M. *J. chem. Phys.* **48**, 4029 (1968).

20. Shavitt, I. *J. chem. Phys.* **49**, 4048 (1968).
21. Stern, M. and Wolfsberg, M. *J. chem. Phys.* **45**, 4105 (1966).
22. Tschuikow-Roux, E. and MacDonald, D. D. *J. chem. Phys.* **49**, 5342 (1968).
23. Yao, S. J. and Zwolinski, B. J. *J. phys. Chem.*, **72**, 373 (1968).
24. Nikitin, E. E. *Teor. éksp. Khim.* **3**, 185 (1967).
25. Kapralova, G. A., Nikitin, E. E., and Chaikin, A. M. *Kinet. Katal.* **10**, 974 (1969).
26. Berne, B. J., Jortner, J., and Gordon, R. J. *J. chem. Phys.* **47**, 1600 (1967).
27. Calvert, J. B. and Amme, R. C. *J. chem. Phys.* **45**, 4710 (1966).
28. Kapralova, G. A., Nikitin, E. E., and Chaikin, A. M. *Chem. Phys. Lett.* **2**, 581 (1968).
29. Mahan, B. H. *J. chem. Phys.* **46**, 98 (1967).
30. Shin, H. K. *J. chem. Phys.* **46**, 3688 (1967).
31. Shin, H. K. *J. chem. Phys.* **47**, 3302 (1967).
32. Shin, H. K. *Chem. phys. Lett.* **2**, 83 (1968).
33. Shin, H. K. *J. Am. chem. Soc.* **90**, 3025 (1968).
34. Shin, H. K. *J. chem. Soc.* **48**, 3644 (1968).
35. Shin, H. K. *J. Am. chem. Soc.* **90**, 3029 (1968).
36. Shin, H. K. *Int. J. quant. Chem.* **2**, 265 (1968).
37. Shin, H. K. *J. chem. Phys.* **49**, 3964 (1968).
38. Shin, H. K. *Chem. Phys. Lett.* **2**, 83 (1968).
39. Steinfeld, J. I. *J. chim. Phys.* **64**, 17 (1967).
40. Steinfeld, J. I. *J. chem. Phys.* **46**, 4550 (1967).
41. Steinfeld, J. I. *J. chem. Phys.* **46**, 4550 (1967).
42. Thompson, S. L. *J. chem. Phys.* **49**, 3400 (1968).
43. Zelechow, A., Rapp, D., and Sharp, T. E. *J. chem. Phys.* **49**, 286 (1968).
44. Taylor, A. I. and Guanturco, F. A. *Chem. Phys. Lett.* **4**, 376 (1969).
45. Kuksenko, B. V. and Losev, S. A. *Dokl. Akad. Nauk SSSR* **178**, 1289 (1968).
46. Kuksenko, B. V. and Losev, S. A. *Dokl. Akad. Nauk SSSR* **185**, 69 (1969).
47. Chan, S. K., Light, J. C., and Lin, J. *J. chem. Phys.* **49**, 86 (1968).
48. Diestler, D. J. and McKoy, V. *J. chem. Phys.* **48**, 2941 (1968).
49. Dubrow, R. and Wilson, D. J. *J. chem. Phys.* **50**, 1553 (1968).
50. Kelley, J. D. and Wolfsberg, M. *J. chem. Phys.* **45**, 3881 (1966).
51. Lick, W. *J. chem. Phys.* **47**, 2438 (1967).
52. Rapp, D. and Kassal, T. *J. chem. Phys.* **48**, 5287 (1968).
53. Riley, M. E. and Kuppermann, A. *Chem. Phys. Lett.* **1**, 537 (1968).
54. Secrest, D. and Johnson, B. R. *J. chem. Phys.* **45**, 4556 (1966).
55. Thiele, E. and Weare, J. *J. chem. Phys.* **48**, 2324 (1968).
56. Weare, J. and Thiele, E. *J. chem. Phys.* **48**, 513 (1968).
57. Kalenov, Yu. A. and Yushchenkova, N. I. *Dokl. Akad. Nauk SSSR* **189**, 1041 (1969).
58. Gordietz, B. F., Sobolov, N. N., Sokovikov, V. V., and Shelepin, L. A. *Phys. Lett.* **A 25**, 173 (1967).
59. Grimaldi, J. J., Enderes, P. F., and Wilson, D. J. *J. chem. Phys.* **50**, 1627 (1969).
60. Gross, R. W. F., *J. chem. Phys.* **50**, 1889 (1969).
61. Herzfeld, K. F. *J. chem. Phys.* **47**, 743 (1967).
62. Rhodes, C. K., Kelly, M. J., and Javan, A. *J. chem. Phys.* **48**, 5730 (1968).

63. SHARMA, R. D. *J. chem. Phys.* **49**, 5195 (1968).
64. SHARMA, R. D. *J. chem. Phys.* **50**, 919 (1969).
65. SHARMA, R. D. and BRAU, C. A. *J. chem. Phys.* **50**, 924 (1969).
66. SHIN, H. K. *Chem. Phys. Lett.* **2**, 629 (1968).
67. TAYLOR, R. L. and BITTERMAN, S. *J. chem. Phys.* **50**, 1720 (1969).
68. YARDLEY, J. T. and MOORE, C. B. *J. chem. Phys.* **46**, 4491 (1967).
69. NIKITIN, E. E. *Teor. éksp. Khim.* **3**, 185 (1967).
70. NIKITIN, E. E. and KOHLMAIER, G. H. *Ber. Bunsengesll phys. Chem.* **72**, 1021 (1968).
71. SERAUSKAS, R. V. and SCHLAG, E. W. *J. chem. Phys.* **45**, 3706 (1966).
72. BRESHEARS, W. D. and BIRD, P. F. *J. chem. Phys.* **48**, 4272 (1968).
73. BJERRE, A. and NIKITIN, E. E. *Chem. Phys. Lett.* **1**, 179 (1967).
74. KRAUSS, M. *J. Res. natn. Bur. Stand.* **72A**, 553 (1968).
75. BATES, D. R. and JUNDI, Z. *J. phys. Soc.* **B1**, 1145 (1968).
76. BATES, D. R. and FLANNERY, M. R. *Proc. R. Soc.* **A302**, 367 (1968).
77. LANDON, S. A. and KECK, J. C. *J. chem. Phys.* **48**, 374 (1968).
78. STUPOCHENKO, E. V. and SAFARYAN, M. N. *Teor. éksp. Khim.* **2**, 783 (1966).
79. SAFARYAN, M. N., STUPOCHENKO, E. V., and PRUCHKINA, N. M. *Teor. éksp. Khim.* **5**, 173 (1969).
80. VALANCE, W. G. and SCHLAG, E. W. *J. chem. Phys.* **47**, 3292 (1967).
81. GENERALOV, N. A., KUKSENKO, B. V., LOSEV, S. A. and OSIPOV, A. I. *Teor. éksp. Khim.* **4**, 311 (1968).
82. KUKSENKO, B. V. and LOSEV, S. A. *Teor. éksp. Khim.* **5**, 475 (1969).
83. BRAU, C. A., KECK, J. C., and CARRIER, G. F. *Phys. Fluids* **9**, 1885 (1966).
84. SNIDER, N. S. *J. chem. Phys.* **45**, 3299 (1966).
85. BRAU, C. A. *J. chem. Phys.* **47**, 1153 (1967).
86. MAYANTS, L. S. *Izv. Akad. Nauk SSSR, Ser. Khim.* **5**, 1026 (1969).
87. HOARE, M. R. *Discuss Faraday Soc.* **44**, 30 (1967).
88. SLATER, N. B. *Molec. Phys.* **12**, 107 (1967).
89. ŠOLC, M. *Molec. Phys.* **11**, 579 (1966).
90. ŠOLC, M. *Molec. Phys.* **12**, 101 (1967).
91. ŠOLC, M. *Z. phys. Chem. (DDR)* **234**, 185 (1967).
92. ŠOLC, M. *Chem. Phys. Lett.* **1**, 160 (1967).
93. ŠOLC, M. *Z. phys. Chem. (DDR)*, **236**, 213 (1967).
94. ŠOLC, M. *Collect. Czechosl. chem. Commun.* **34**, 177 (1969).
95. TROE, J. and WAGNER, H. GG. *Ber. Bunsengesell phys. Chem.* **71**, 937 (1967).
96. MORDVINOV, YU. P. and SMIRNOV, B. M. *Teplofiz. vỹs. Temp.* **5**, 219 (1967).
97. DUGAN, J. V. and MAGEE, J. L. *J. chem. Phys.* **47**, 3103 (1967).
98. DUGAN, J. V., RICE, J. H., and MAGEE, J. L. *Chem. Phys. Lett.* **2**, 219 (1968).
99. DUGAN, J. V. JR., RICE, J., and MAGEE, J. L. *Chem. Phys. Lett.* **3**, 323 (1969).
100. DUNBAR, R. C. *J. chem. Phys.* **47**, 5445 (1967).
101. KAUFMAN J. J. and KOSKI, W. S. *J. chem. Phys.* **50**, 1942 (1969).
102. LIGHT, J. C. *Discuss Faraday Soc.* **44**, 14 (1967).
103. MEISELS, G. G. *J. chem. Phys.* **72**, 3746 (1968).
104. MILLER, W. B., SAFRON, S. A., and HERSCHBACH, D. R. *Discuss. Faraday Soc.* **44**, 108 (1967).

105. WOLF, F. A. and TURNER, B. R. *J. chem. Phys.* **48**, 4226 (1968).
106. BUNKER, D. L. and CHANG, T. S. *J. chem. Phys.* **73**, 943 (1969).
107. DING, A., HENGLEIN, A., and LACMANN, K. *Z. Naturf.* **23a** 779 (1968).
108. DING, A., HENGLEIN, A., and LACMANN, K. *Z. Naturf.* **23a**, 780 (1968).
109. GENTRY, W. R., GISLASON, E. A., LEE, Y. T., MAHAN, B. H., and TSAO, C. *Discuss Faraday Soc.* **44**, 137 (1967).
110. HERMAN, Z., KERSTETTER, J., ROSE, T., and WOLFGANG, R. *Discuss Faraday Soc.* **44**, 123 (1967).
111. MAHAN, B., *Acc. chem. Res.* **1**, 217 (1968).
112. MATUS, L., OPAUCZKY, I., HYATT, D., MASSON, A., BIRKINSHAW, K., and HENCHMAN, M. J. *Discuss Faraday Soc.* **44**, 146 (1967).
113. SUPLINSKAS, R. J. *J. chem. Phys.* **49**, 5046 (1967).
114. POLANYI, J. C. *Discuss Faraday Soc.* **44**, 293 (1967).
115. ANLAUF, K. G., KUNTZ, P. J., MAYLOTTE, D. H., PACEY, P. D., and POLANYI, J. C. *Discuss Faraday Soc.* **44**, 183 (1967).
116. KUNTZ, P. J., NEMETH, E. M., and POLANYI, J. C. *J. chem. Phys.* **50**, 4607 (1969).
117. BLAIS, N. C. *J. chem. Phys.* **49**, 9 (1968).
118. KUNTZ, P. J., MOK, M. H., and POLANYI, J. C. *J. chem. Phys.* **50**, 4623 (1969).
119. GODFREY, M. and KARPLUS, M. *J. chem. Phys.* **49** 3602 (1968).
120. CSIZMADIA, I. G., POLANYI, J. C., and ROACH, A. C. *Can. J. Chem.* **47**, 4097 (1969).
121. KARPLUS, M. and GODFREY, M. *J. Am. chem. Soc.* **88**, 5332 (1968).
122. MARCUS, R. A. *J. chem. Phys.* **45**, 4500 (1966).
123. MOROKUMA, K., PEDERSEN, L., and KARPLUS, M. *J. Am. chem. Soc.* **89**, 5064 (1967).
124. SMITH, I. W. M. *Discuss Faraday Soc.* **44**, 194 (1967).
125. NIKITIN, E. E. *Teor. éksp. Khim.* **4**, 593 (1968).
126. NIKITIN, E. E. *Teor. éksp. Khim.* **4**, 751 (1968).
127. BERNSTEIN, R. B. and LEVINE, R. D. *J. chem. Phys.* **49**, 3872 (1968).
128. CHILD, M. S. *Molec. Phys.* **16**, 313 (1969).
129. EU, B. C. and ROSS, J. *J. chem. Phys.* **51**, 159 (1969).
130. EU, B. C. and ROSS, J. *Discuss Faraday Soc.* **44**, 39 (1967).
131. MARRIOTT, R. and MICHA, D. A. *Phys. Rev.* **180**, 120 (1969).
132. MICHA, D. A. *J. chem. Phys.* **50**, 722 (1969).
133. NIKITIN, E. E. *Chem. Phys. Lett.* **1**, 266 (1967).
134. NYELAND, C. and ROSS, J. *J. chem. Phys.* **49**, 843 (1968).
135. ROSENFELD, J. L. and ROSS, J. *J. chem. Phys.* **44**, 188 (1966).
136. SUN, H. Y. and ROSS, J. *J. chem. Phys.* **46**, 3306 (1967).
137. BAZ', A. I. and GOL'DANSKII, V. I. *Dokl. Akad. Nauk SSSR* **182**, 364 (1968).
138. BAZ', A. I. and GOL'DANSKII, V. I. *Dokl. Akad. Nauk SSSR* **186**, 1110 (1969).
139. BLUM, L. *J. phys. Chem.* **70**, 2758 (1966).
140. BLUM, L. *Molec. Phys.* **11**, 63 (1966).
141. BLUM, L. *Nuovo cim.* **B 45**, 32 (1966).
142. BLUM, L. *Molec. Phys.* **11**, 105 (1966).
143. CHILD, M. S. *Discuss Faraday Soc.* **44**, 68 (1967).
144. DIESTLER, D. J. and MCKOY, V. *J. chem. Phys.* **48**, 2951 (1968).

145. Marcus, R. A. *Discuss Faraday Soc.* **44**, 7 (1967).
146. Suplinskas, R. J. and Ross, J. *J. chem. Phys.* **47**, 321 (1967).
147. Tang, K. T., Kleinman, B., and Karplus, M. *J. chem. Phys.* **50**, 1119 (1969).
148. Karplus, M. and Tang, K. T. *Discuss Faraday Soc.* **44**, 80 (1967).
149. Marcus, R. A. *J. chem. Phys.* **45**, 4493 (1966).
150. Pirkle, J. C. and McGee, H. A. *J. chem. Phys.* **49**, 3532 (1968); *J. chem. Phys.* **49**, 4504 (1968).
151. Present, R. D. *J. chem. Phys.* **48**, 4875 (1968).
152. Present, R. D. and Morris, B. M. *J. chem. Phys.* **50**, 151 (1969).
153. Pyun Chong Wha. *J. chem. Phys.* **48**, 1306 (1968).
154. Pyun Chong Wha. *J. chem. Phys.* **50**, 2782 (1969).
155. *Primenenie vychislitel'noĭ matematiki v khimicheskoĭ i fizicheskoĭ kinetike, Sbornik stateĭ.* 'Nauka', SSSR (1969).
156. *Chislennye metody v teorii razrezhennykh gazov. Sbornik stateĭ, Vychislitel'nyĭ Tsentr.* Akad. Nauk SSSR (1969).
157. Tal'roze, V. L. *Kinet. Katal.* **5**, 11 (1964).
158. Batovskiĭ, O. M., Vasil'ev, G. K., Makorov, E. F. and Tal'roze, V. L. Letters in *Zh. éksp. teor. Fiz.* **9**, 341 (1969).
159. Basov, N. G., Kulakov, L. V., Markin, E. P., Nikitin, A. I., and Oraevskiĭ, A. N. Letters in *Zh. éksp. teor. Fiz.* **9**, 613 (1969).
160. Anlauf, K. G., Moylotte, D. H., Pacey, P. D., and Polanyi, J. C. *Chem. Phys. Lett.* **A 24**, 208 (1967).
161. Kasper, J. V. V., Parker, J. H., and Pimentel, G. C. *J. chem. Phys.* **43**, 1827 (1965).
162. Kompa, K. L., Parker, J. H., and Pimentel, G. C. *J. chem. Phys.* **49**, 4257 (1968).
163. Hirschfelder, J. O. *Int. J. quant. Chem.* **3**, 17 (1969).
164. Kostin, M. D. *J. chem. Phys.* **51**, 2455 (1969).
165. LeRoy, R. L. *J. phys. Chem.* **73**, 4338 (1969).
166. Marcus, R. A. *J. chem. Phys.* **49**, 2617 (1968).
167. Melton, R. A. and Gordon, R. G. *J. chem. Phys.* **51**, 5449 (1969).
168. Morokuma, K., Eu, B. C., and Karplus, M. *J. chem. Phys.* **51**, 5193 (1969).
169. Snider, N. S. *J. chem. Phys.* **51**, 4075 (1969).
170. Atterneyer, M. and Marcus, R. A. *J. chem. Phys.* **52**, 393 (1970).
171. Child, M. S. *Proc. R. Soc.* **A 315**, 1521 (1970).
172. Gianturco, F. A. and Marriott, R. *J. Phys.* **B 2**, 1332 (1969).
173. Nyeland, C. and Hunding, A. *Chem. Phys. Lett.* **5**, 143 (1970).
174. Kuksenko, B. V. and Losev, S. A. *Teor. éksp. Khim.* **5**, 468 (1969).
175. Breig, E. L. *J. chem. Phys.* **51**, 4539 (1969).
176. Berend, G. C. and Benson S. W. *J. chem. Phys.* **51**, 1480 (1969).
177. Cheung, A. S. and Wilson, D. J. *J. chem. Phys.* **51**, 3448 (1969).
178. Cheung, A. S. and Wilson, D. J. *J. chem. Phys.* **51**, 4733 (1969).
179. Curry, B. P. *J. chem. Phys.* **51**, 4951 (1969).
180. Diestler, D. J. *J. chem. Phys.* **52**, 2280 (1970).
181. Shin, H. K. *Chem. Phys. Lett.* **3**, 560 (1970).
182. Levine, R. D. *Chem. Phys. Lett.* **2**, 76 (1968).
183. Locker, D. J. and Endres, P. F. *J. Chem. Phys.* **51**, 5482 (1969).

184. RAPP, D. and KASSAL, T. *Chem. Rev.* **69**, 61 (1969).
185. RAZNER, R. *J. chem. Phys.* **51**, 5602 (1969).
186. SECREST, D. *J. chem. Phys.* **51**, 421 (1969).
187. SHIN, H. K. *Chem. Phys. Lett.* **3**, 137 (1970).
188. GORDIETS, B. F., OSIPOV, A. I., and SHELEPIN, L. A. *Kinetika kolebatel'nogo obmena v angarmonicheskom priblizhenii*. Preprint No. 17, Opticheskaya Lab. F.I. Akad. Nauk SSSR (1970).
189. DASHEVSKAYA, E. I. *Dokl. Akad. Nauk SSSR* **187**, 1025 (1969).
190. BRAY, K. N. G. *J. Phys.* **B 1**, 705 (1968).
191. FISHER, E. R. and KUMMLER, R. H. *J. chem. Phys.* **49**, 1075 (1968).
192. FISHER, E. R. and KUMMLER, R. H. *J. chem. Phys.* **49**, 1085 (1968).
193. HSU, C. T. and MAILLIE, F. H. *J. chem. Phys.* **52**, 1767 (1970).
194. TREANOR, C. E., RICH, J. W., and REHM, R. G. *J. chem. Phys.* **48**, 1798 (1968).
195. TEARE, J. D., TAYLOR, R. L., and ROSENBERG, C. W. *Nature* **225**, 5229 (1970).
196. HELLER, D. F. and MOOR, C. B. *Chem. Phys.* **52**, 1005 (1970).
197. REES, T. and BHANGU, J. K. *J. Fluid Mech.* **39**, 601 (1969).
198. ROSSER, W. A., WOOD, A. D., and GERRY, E. T. *J. chem. Phys.* **50**, 4996 (1969).
199. TRIPODI, R. *J. chem. Phys.* **52**, 3298 (1970).
200. ROACH, J. F. and SMITH, W. R. *J. chem. Phys.* **50**, 4114 (1969).
201. SHARMA, R. D. *Phys. Rev.* **177**, 102 (1969).
202. GORDIETS, B. F., SOBOLEV, N. N., and SHELEPIN, L. A. *Zh. éksp. teor. Fiz.* **53**, 1822 (1967).
203. BIRYUKOV, A. S., GORDIETS, B. F., and SHELEPIN, L. A. *Zh. éksp. teor. Fiz.* **55**, 1456 (1968).
204. BIRYUKOV, A. S., GORDIETS, B. F., and SHELEPIN, L. A. *Zh. éksp. teor. Fiz.* **57**, 585 (1969).
205. SATO, Y. and TSUCHIYA, S. *Chem. Phys. Lett.* **5**, 293 (1970).
206. B'ERRE, A. *Teor. éksp. Khim.* **4**, 383 (1968).
207. BAUER, E., FISHER, E. R., and GILMORE, F. R. *J. chem. Phys.* **51**, 4173 (1969).
208. MNATSAKANYAN, A. KH. *Teplofiz. vȳs. Temp.* **7**, 385 (1965).
209. ANDERSON, R. W., AQUILANTI, V., and HERSCHBACH, D. R. *Chem. Phys. Lett.* **4**, 5 (1969).
210. TSUCHIYA, S. and SUZUKI, I. *J. chem. Phys.* **51**, 5725 (1969).
211. DENISYUK, S. A., LEBEDEV, S. N., MALAMA, YU. G., and POLAK, L. S. *Khim. vȳs. Energiĭ* **2**, 304 (1968).
212. STEPUKHOVICH, A. D. and UMANSKIĬ, V. M. *Usp. Khim.* **38**, 1356 (1969).
213. BURNS, G., LEROY, R. J., MORRISS, D. J., and BLAKE, J. A. *Proc. R. Soc.* **A 316**, 81 (1970).
214. BUNKER, D. L. and PATTENGILL, M. *J. chem. Phys.* **48**, 772 (1968).
215. FREY, R. B. and THIELE, E. *J. chem. Phys.* **48**, 3240 (1968).
216. MIES, F. H. and KRAUSS, M. *J. chem. Phys.* **45**, 4455 (1966).
217. MIES, F. H. *J. chem. Phys.* **51**, 798 (1969).
218. MIES, F. H. *J. chem. Phys.* **51**, 787 (1969).
219. FORST, W., PRASIL, Z., and ST. LAURENT, P. *J. chem. Phys.* **46**, 3736 (1967).
220. FORST, W. *J. chem. Phys.* **48**, 3665 (1968).
221. FORST, W. *Chem. Phys. Lett.* **1**, 687 (1968).

222. FORST, W. and PRASIL, Z. *J. chem. Phys.* **51**, 3006 (1969).
223. TOU, J. C. *J. phys. Chem.* **71**, 2721 (1967).
224. TARDY, D. C., RABINOVITCH, B. S., and WHITTEN, G. Z. *J. chem. Phys.* **48**, 1427 (1968).
225. TIMASHEV, S. F., KAMINSKII, V. A. and VRUNOV, P. A. *Teor. éksp. Khim.* **5**, 38 (1969).
226. JORTNER, J. and BERRY, R. S. *J. chem. Phys.* **48**, 2757 (1968).
227. STEPUKHOVICH, A. D. *Kinet. i mekhanizm termicheskogo krekinga alkanov.* (1965).
228. MILLER, W. H. *J. chem. Phys.* **52**, 543 (1970).
229. PEERS, A. *Int. J. Mass. Spectrom. in Ion Phys.* **3**, 99 (1969).
230. TRUHLAR, D. G. and KUPPERMANN, A. *J. phys. Comm.* **73**, 1722 (1969).
231. TRUHLAR, D. G. *J. chem. Phys.* **51**, 4617 (1969).
232. BRODSKIĬ, A. M., LEVICH, V. G., and TOLMACHEV, V. V. *Dokl. Akad. Nauk SSSR* **182**, 1036 (1968).
233. BRODSKIĬ, A. M., LEVICH, V. G. ,and TOLMACHEV, V. V. *Dokl. Akad. Nauk SSSR* **183**, 852 (1968).
234. BERNSTEIN, R. B. *Int. J. quant. Chem.* **3**, 41 (1969).
235. GEORGE, T. F. and SUPLINSKAS, R. J. *J. chem. Phys.* **51**, 3666 (1969).
236. KUNTZ, P. J. *Chem. Phys. Lett.* **4**, 129 (1969).
237. SHIN, H. K. *Chem. Phys. Lett.* **5**, 232 (1970).
238. WOLFGANG, R. *Acc. chem. Res.* **8**, 248 (1969).
239. ANDERSON, J. B. *J. chem. Phys.* **52**, 3849 (1970).
240. OTTINGER, CH. *J. chem. Phys.* **51**, 1170 (1969).
241. RUSSELL, D. and LIGHT, J. C. *J. chem. Phys.* **51**, 1720 (1969).
242. EU, B. C. *J. chem. Phys.* **52**, 3021 (1970).
243. LEVINE, R. D. and BERNSTEIN, R. B. *Israel. J. Chem.* **7**, 315 (1969).
244. BAZ, A. I. and GOLDANSKII, V. I. *J. chem. Phys.* **51**, 5401 (1969).
245. CHILD, M. S. *Molec. Phys.* **16**, 313 (1969).
246. DION, D. R., MILLEUR, M. B., and HIRSCHFELDER, J. O. *J. chem. Phys.* **52**, 3179 (1970).
247. MCCULLOUGH, E. A. and WYATT, R. E. *J. chem. Phys.* **51**, 1253 (1969).
248. KLEINMAN, B. and TANG, K. T. *J. chem. Phys.* **51**, 4587 (1969).
249. KUPPERMANN, A. *Israel J. Chem.* **7**, 303 (1969).
250. DIESTLER, D. J. *J. Chem. Phys.* **50**, 4746 (1969).
251. TRUHLAR, D. G. and KUPPERMANN, A. *J. chem. Phys.* **52**, 3841 (1970).
252. O'MALLEY, T. F. *J. chem. Phys.* **52**, 3269 (1970).
253. RANKIN, C. S. and LIGHT, J. C. *J. chem. Phys.* **51**, 1701 (1969).
254. ROBINSON, P. D. *J. chem. Phys.* **52**, 3175 (1970).
255. WILSON, D. J. *J. chem. Phys.* **51**, 5008 (1969).
256. WYATT, R. E. *J. chem. Phys.* **51**, 3489 (1969).
257. LEBEDEV, S. N., MALAMA, YU. G., and OSIPOV, A. I. *Teor. éksp. Khim.* **6**, 206 (1970).
258. LIFSHITZ, A. *Israel J. Chem.* **7**, 329 (1969).
259. MORRIS, B. M. and PRESENT, R. D. *J. chem. Phys.* **51**, 4862 (1969).
260. SHIZGAL, B. and KARPLUS, M. *J. chem. Phys.* **52**, 4262 (1970).
261. PARKER, J. H. and PIMENTEL, G. C. *J. chem. Phys.* **51**, 91 (1969).

262. *Tezisy dokl. na mezhdunarodnom simpoziume po khimcheskim lazeram.* Moscow (1969).
263. TAL'ROZE, V. L., VASIL'EV, G. K., and BATOVSKIĬ, O. M. *Kinet. Katal.* **11**, 277 (1970).
264. DZHIDOEV, M. S. *et al. Usp. fiz. Nauk* **100**, 641 (1970).
265. RADȲLEVSKAYA, M. A. *Vest. Leningrad. Gos. Univ. Ser. Fiz. Khim.* No. 19, **86**, (1969).
266. ANLAUF, K. G., MAYLOTTE, D. H., POLANYI, J. C., and BERNSTEIN, R. B. *J. chem. Phys.* **51**, 5716 (1969).
267. POLANYI, J. C. and TARDY, D. C. *J. chem. Phys.* **51**, 5717 (1969).
268. POLANYI, J. C. and WONG, W. H. *J. chem. Phys.* **51**, 1439 (1969).
269. MOK, M. H. and POLANYI, J. C. *J. chem. Phys.* **51**, 1451 (1969).
270. ARTAMONOVA, N. D., *et al. Zh. éksp. teor. Fiz.* **58**, 2195 (1970).

INDEX

Action-angle variables 14, 35, 190, 290
Activated complex 7, 13, 18, 180
Active molecule 179, 322
  anharmonic model of 253
  harmonic model of 244
  Slater model of 196
Active pair 367
Activation, mechanism of stepwise 315
Adiabatic approximation 1, 8
  basis 2
  collisions 55
  criterion 6
  electronic functions 2
  electronic-nuclear functions 4
  electronic terms 2
  potentials of M–X system 150
  terms of two atom system 133
Anharmonicity, effect on intramolecular energy exchange rate 221
  level density 255
  reaction rate 217
  spontaneous decomposition rate 221
  intermolecular 70
  molecular 82
  of vibrations 70, 82
Arrhenius dependence 202, 205
  factor 115
Atom and infinite chain of identical atoms 219

Balance equations 192
Bimolecular reactions 179, 367
  disturbance of equilibrium distribution in 424
  rate of 12, 35
Boltzmann equation (kinetic equation) 424
Bond, rupture of, in N_2O 227

Charge exchange, non-resonant 139
  probability 140, 359
  resonant 134, 137
Coefficient of friction 285
  large 299
  small 301
Coefficient of viscosity 288
Collision time 283
  for exponential potential 74
  for Lennard-Jones potential 74

Collisions, effective number of deactivating 185
Complex, intermediate 324, 391, 444
  linear 387
  triangular 387
Composite system (intermediate complex) 324, 391, 444
Configuration space 8
Coordinate exchange operators 370
Coriolis interaction 113
Correlation diagrams 158, 175
Coupled harmonic oscillators 93
Critical surface 8, 386
  multiple transits of 13
  projection of 187
  variation of 334, 339
Curvilinear coordinates 19

Damping of wave packet 206
Decomposition, thermal 179
Degrees of freedom 10
  active 215, 238
  adiabatic 238, 240, 254
  contribution to decomposition rate constant 328
  in Slater and Kassel models 212
  effective number of vibrational 213, 217–221
  inactive 215, 238, 240
  rotational 254, 256
  translational 25
Density matrix (statistical matrix) 23, 189, 306
Detailed balance, principle of 34, 39, 57, 184, 192, 285, 311
Diatomic complex 378, 384
Diffusion coefficient 285
  equation 285
  model 347, 354
  theory 283, 304
  through energy levels 303
  states 294
Direct knock-out reaction model 406
Direct reactions, theory of 403
Dissociation of diatomic molecules 322
  rate constant 322
  thermal, of oxygen 364
Distorted waves, method of 76

Distribution function 7, 11
   equilibrium 8
   Maxwell–Boltzmann 7

Eckart potential 22
Effective cross-section 38
Effective mass of representative point 14
Elastic spectator model 404
Electron and photon impact 269
Electronic excitation 41, 149
   harpooning 409, 411
Electronic-nuclear functions 4
Electronic transitions, non-adiabatic 331
Energy, activation 179
   threshold 200
Energy exchange, in collisions of diatomic molecules 92
   polyatomic molecules 96
   translational 41
   vibrational 41
Equilibrium theory 322, 386
Exchange reaction 389, 412
   resonant 92
   with potential barrier 418
Eyring and Polanyi, method of 377, 380

Fine structure components 9
   transitions between 140, 149, 157
Flash photolysis 362
Fluorescence, quantum yield 181, 320
Fokker–Planck, diffusion equation 285
   constant coefficient of friction in 289
   transmission across potential barrier 292
Forced harmonic oscillator 52, 86
Frequency factor 227

Generalized coordinate 10
   momentum 10
Group representations 376

Hamiltonian 1
Harmonic oscillator model of active molecule 244
   forced 52, 86
   generalization to three dimensions 58
   with external force (forced) 52, 86
Harpooning, electronic, mechanism of 409, 411
Heat reservoir (heat bath) 56, 185, 216, 283
Hoare's model 357, 363
H_3 system 382
H_3^+ system 381
Hypersurface 18, 103, 186

Impact equilibrium theory of dissociation 324
   parameter 37, 59, 112

Impulse approximation 86, 356, 407
Inertia tensor of AB–C system 44
Integrals of motion 20, 238
Intermediate complex 324, 391, 444
Intermolecular anharmonicity 70
Intramolecular anharmonicity 70
Intermultiplet mixing 140
Ion–molecular reaction 397
Irreducible representations 150, 376
Isomerization of substituted cyclopropane molecules 272
   thermal 179
Isotope effect 28, 271
   direct 32
   inverse 32
   inverse secondary 273
   normal primary 31, 272
   normal secondary 33, 272

Kassel, model of 200, 208
   quantum generalization 211
Kinetic equation 181, 188, 424
Knock-out reactions 406
Kramer, diffusion theory of 281
Kuznetsov equations 363

Landau, statistical theory of 241
Landau–Teller equations 311
   formula 57
   refinements to 57, 73, 75
Landau–Zener formula 107, 110, 115–117, 126, 129, 169
   extension to multi-term system 118–121
   extension to system with 2 degrees of freedom 121–124
Lennard-Jones potential 73, 75
Level width 120
Light atom anomaly 414
Light's theory 336
Linear model 107
   generalization 117
Liouville's equation 22
   theorem 182, 426
London's formula 370, 378

Massey parameter 6, 41, 46–49, 102, 107, 124
   adiabatic collisions, in 55
   rotational excitation of diatomic molecules, in 49
   single quantum vibrational transitions, in 48
Maxwell–Boltzmann distribution function 7, 39, 65
Mixing parameter 108
Modified wave number, method of 59
Morse oscillator 47, 84, 218, 352
   potential 46, 75

INDEX

Moving system of coordinates 43
Multiple collisions 86
  quantum activation 357
    relaxation 348
    transitions 86, 354

Non-adiabatic coupling operator 5, 101, 104, 112, 124, 151, 167
  reactions 227
  transitions 99
Non-adiabaticity, electronic excitation and deactivation in M–X system 149
  fast transit through region of 131
Non-adiabaticity in atomic collisions 132, 148
  slow transit through region of 129
Non-equilibrium distribution function 181, 188, 315
  theory of dissociation and recombination 350
Non-linear terms, model of 124
  triatomic molecule 317
Non-resonant charge exchange 139
  electronic deactivation reactions 147
  processes with large energy transfer 147
  processes with small energy transfer 139
  transitions between fine structure components 140, 157

Optical model 409
Over-barrier reflection 16, 20, 115

Pair potentials 51
Parabolic barrier approximation 21
Parker's model 75
Partial reaction rate 35
Partition function (sum of states) 12
  rotational 27
  rotational–nuclear 26
  translational 25
  vibrational 26
Phase space 6
Phase volume, element of 11, 35, 43, 45
  for active molecule by combinatorial method 244
  for active molecule by Laplace transform method 248
Polarization complex 137, 238, 264
  method of 238, 264
Potential, doubly symmetric well 203
  types of 200
Potential energy surfaces 1, 6
  attractive 413
  bimolecular reactions, in 370
  covalent 440
  exchange reactions, in 412
  H_3, H_3^+ systems, of 383

Potential energy surfaces—*cont.*
  ionic 440
  mixed type 413
  repulsive 413
Pressure regions, high 274
  intermediate 280
  low 276

Quantum partition function 23, 32
  in dissociation type reactions 25
  in exchange type reactions 25
Quasi-classical approximation 15
Quasi-equilibrium (statistical) theory 237
Quasi-stationary approximation 293
Quenching of resonance fluorescence of excited mercury atoms 161
  potassium by nitrogen 157
  sodium by argon 147

Radially vibrating diatomic molecule model 58
  sphere model 66, 97
Random interaction with heat reservoir 305
Random walk over discrete energy levels 310
Rate constant 7, 11
  bimolecular reaction, of 12, 39
  pressure dependence of 181
  unimolecular reaction, of 12
Reaction channel 265
  coordinate 11, 17, 36, 386
  criterion, classical 8
    quantum 14, 18
  first-order 274
  second-order 274, 276
Recombination of atomic oxygen on argon 338
  atoms 322
  iodine 362
  ions 259
Redlich–Teller formula 31
Reduced mass 30, 194
Reflection rate 27
Relaxation, effect on reaction rate 360
  intermolecular 182
  of classical oscillator 308
  of oscillator interacting with heat reservoir 299, 310
  vibrational 436
Representative point 6
Resonant processes 134
  charge exchange in 134, 137
  exchange of vibrational energy in 92
  excitation exchange in S–P transitions 137
Rice's theory 323
Rigid sphere model 287, 405, 428
Rotational energy 329
  energy level density 260
  excitation 49

Rotational energy—*cont.*
 transitions   329
Rough molecule model   67

Sato, method of   379
Schrödinger equation   1
Semi-classical approximation   4
Simple collision theory   336
Single quantum stepwise excitation   350
Slater, model of   193, 201, 210
Spectator stripping model   408
Spin matrices (Pauli)   306
Spin orbit coupling (interaction)   1, 2
 strong   174
 weak   167
Statistical factor   27
Statistical matrix (density matrix)   23, 189, 306
Statistical theory (quasi-equilibrium theory)   237
 general formula of   391
 Landau's   241
 reaction rate in   241
 with angular momentum conservation   258
 without angular momentum conservation   267
Statistical weight (degree of degeneracy)   24, 26
Stepwise excitation (activation)   181, 274, 315
Stepwise relaxation in diatomic gas reservoir   344
 in monatomic gas reservoir   343
 multiple-quantum   348
 single-quantum   350
Stripping mechanism   408, 444
Strong collision assumption (mechanism)   186, 274, 316
Strongly activating collisions   181
Sudden perturbations, method of   129, 143
Symmetry of irreducible representations of point groups
 C_{2v} and C_s 151
 D_{3h} and C_{2v} 376

Symmetry number   23, 26

Thermal decomposition   179
 of nitrous oxide   281
Thermal dissociation of oxygen   364
Thermal isomerization   179
Thomson's theory   326, 360
Trajectory of nuclear motion   4
 of representative point   9
Transition probabilities, harmonic oscillator   54
Transition state method   1, 7, 15
 quantum variant   18
Transmission coefficient   13, 20
 for one-dimensional potential barrier   20
Truncated quantum harmonic oscillator   216, 304, 315, 337, 356
Tunnel corrections   16
 effects   115, 179, 232, 390
 transmission   17, 18, 203
 in double potential well   305
Two channel decomposition   319
Two-state approximation   102, 107

Uncertainty principle   18, 23, 188, 192
Unimolecular decomposition of triatomic molecule   223
 reaction rate   12, 179

Variational theory   10, 334
Vibrational excitation   42
Vibrational relaxation   41
 of diatomic molecule   342
 of nitric oxide   175
 of oxygen–oxygen collisions   79
 of oxygen and nitrogen   78
Vibrational transitions in diatomic molecules   58, 166
 molecular collisions   328

Wigner, distribution function   23, 201